The Keys to Happiness

Also by Laura Engelstein

Moscow, 1905

THE KEYS
TO HAPPINESS

*Sex and the Search for Modernity
in Fin-de-Siècle Russia*

Laura Engelstein

Cornell University Press

Ithaca and London

PUBLICATION OF THIS BOOK WAS ASSISTED BY A GRANT FROM
THE PUBLICATIONS PROGRAM OF THE NATIONAL ENDOWMENT
FOR THE HUMANITIES, AN INDEPENDENT FEDERAL AGENCY.

First published 1992 by Cornell University Press.
First printing, Cornell Paperbacks, 1994.

International Standard Book Number 0-8014-2664-2 (cloth)
International Standard Book Number 0-8014-9958-5 (paper)
Library of Congress Catalog Card Number 92-52751
Printed in the United States of America
*Librarians: Library of Congress cataloging information appears
on the last page of the book.*

♾ The paper in this book meets the minimum requirements of the
American National Standard for Information Sciences—Permanence
of Paper for Printed Library Materials, ANSI Z39.48-1984.

Contents

Contents

Illustrations

vii

Acknowledgments

I thank the institutions whose financial support made the research and writing of this book possible: the Mary Ingraham Bunting Institute of Radcliffe College, the American Council of Learned Societies, the Princeton University Committee on Research in the Humanities and Social Sciences, and the Kennan Institute for Advanced Russian Studies. The work was also supported in part by a grant from the International Research and Exchanges Board (IREX), with funds provided by the National Endowment for the Humanities and the United States Information Agency. None of these organizations is responsible for the views expressed in the book. In addition, the Philip and Beulah Rollins Preceptorship at Princeton University supported generous leave time, and the Institute for Advanced Study at Princeton offered me the luxury of a congenial environment near my home.

This book is based almost entirely on published sources, and I have had the pleasure of working in a number of remarkable libraries: the Helsinki University Slavonic Library, the Harvard University Law Library, the Library of Congress (particularly the legal collection), the National Library of Medicine, the New York Public Library, the V. I. Lenin State Library in Moscow, and in St. Petersburg the M. E. Saltykov-Shchedrin State Public Library and the Library of Theater Arts. I would like to express my particular appreciation to the Helsinki staff, who make that library an ideal place to work and to learn about Russian history. Special thanks also to Ann Toohey at the National Library of Medicine and to the staff at the Saltykov-Shchedrin (especially the director of the newspaper collection, Dr. Nikolai Nikolaev), who graciously made exceptions on my behalf. The staff of the Saltykov-Shchedrin Department of Graphics kindly allowed me to take photographs from their holdings.

At the end of the project, in May 1990, I was also able to make some use of archival resources, under the auspices of the scholarly exchange conducted by IREX and the Soviet Academy of Sciences. I am happy to say that I received the fullest cooperation from archival personnel on this trip. In Moscow I worked at the Central State Military-Historical Archive of the USSR and in St. Petersburg at the Central State Historical Archive of the USSR, where the staff was particularly helpful. My special thanks to Svetlana Gimein for her intelligent and energetic assistance and to Larissa Zakharova for putting me in touch with her.

Among the individual scholars, friends, and colleagues who contributed to this project in many ways, I will start by thanking the Department of History at Princeton University for providing a lively and encouraging atmosphere for intellectual exchange. Richard Wortman distinguished himself by his early enthusiasm for the subject of syphilis and by his gifts as a close colleague. My special gratitude to Natalie Davis for persuading me to try Princeton and to Lawrence Stone for welcoming me once I was there. I am also grateful to the historians who read and commented on sections of the draft and discussed the project's ideas: Gerald Geison, Carla Hesse, Isabel V. Hull, Thomas Laqueur, David L. Ransel, Joan W. Scott, Daniel Todes, William G. Wagner, Judith R. Walkowitz, and Reginald E. Zelnik. Richard Stites and Richard Wortman were exemplary "anonymous" readers, both encouraging and critical. Leopold Haimson has always encouraged my work, even when it puzzled him, and he made it possible for me to bring my unconventional theme to the attention of colleagues in St. Petersburg. In Moscow Neia Zorkaia's enthusiasm for my ideas gave me the courage to speak more openly with Soviet historians. For arranging for me to meet this extraordinary person, I thank the equally generous and intellectually independent Georgii Knabe.

As I have wandered through cities, libraries, and odd scholarly byways, personal and intellectual ties have often converged. My life in Helsinki would have been poorer and lonelier indeed were it not for the friendship and hospitality of Mauri and Anna-Liisa Vihko. Esther Fischer-Homberger in Bern was at first an intellectual inspiration and has since become a dear friend. William I. Bennett was the first person with any expertise (an M.D.) to endorse my views on syphilis, and he counts as a patron saint and psychological mainstay of this project. John G. Ackerman, friend, editor, and historian, also gets credit for hovering in the wings. Conversations with Itsie Hull about the history of sexuality and the peculiarities of central and eastern Europe helped get me started. The support of feminist colleagues has been crucial to this enterprise, but in particular the example and friendship of Judy Walkowitz have encouraged me through many difficult times. Above all, I owe an enormous personal and professional debt to Reggie Zelnik, who has remained a sympathetic and critical guide over the years, often saving me from myself as I pursued the ins and outs of a willful career. Carla Hesse, more than anyone else, lived through the final emergence of this book, which she encouraged with the wisdom of a comrade-historian and the unflagging concern of a friend.

This book incorporates, in modified form, the following previously published material:

"Morality and the Wooden Spoon: Russian Physicians View Syphilis, Social Class, and Sexual Behavior, 1890–1905." *Representations* 14 (1986). Reprinted in *The Making of the Modern Body: Sexuality and Society in the Nineteenth Century*, ed. Catherine Gallagher and Thomas Laqueur. Berkeley: University of California Press, 1987.

"Gender and the Juridical Subject: Prostitution and Rape in 19th-Century Russian Criminal Codes." *Journal of Modern History* 60:3 (1988): 458–95.

"Abortion and the Civic Order: The Legal and Medical Debates, 1911–1914." In *Russia's Women: Accommodation, Resistance, Transformation*, ed. Barbara Engel, Barbara Clements, and Christine Worobec. Berkeley: University of California Press, 1991.

LAURA ENGELSTEIN

Princeton, New Jersey

Note on Transliteration
and Spelling

In transliterating Russian titles, quotations, and names, I have used the Library of Congress system—except in the case of well-known persons (such as Leo Tolstoy) whose names are familiar in other spellings—and have changed the old-style orthography to conform to modern usage.

Abbreviations

ch.	*chast'* (part)
d., dd.	*delo, dela* (file, files)
ed. khr.	*edinitsa khraneniia* (storage unit)
f.	*fond* (collection)
l., ll.	*list, listy* (folio, folios)
op.	*opis'* (inventory)
TsGIA	Tsentral'nyi gosudarstvennyi istoricheskii arkhiv (Central State Historical Archive)
TsGVIA	Tsentral'nyi gosudarstvennyi voenno-istoricheskii arkhiv (Central State Military-Historical Archive)

The Keys to Happiness

Introduction

Sex was a political subject in late imperial Russia. When Alexander Herzen wrote the novel *Who Is to Blame?* (1846) about the injustices of Russian society under serfdom and Nikolai Chernyshevskii wrote *What Is to Be Done?* (1863) about how to pursue the fight for justice after the system had been abolished, both centered their narratives on the domestic relationship of women and men. Both examined the conflict between personal autonomy and conventional social relations; both used the hierarchy of sexual power and subordination to represent structures of domination and submission in the larger social world. When Leo Tolstoy indicted the corrupt values of a modernizing society under a coercive old regime, he invoked the specter of gender confusion and sexual debauch: *Anna Karenina* (1877), *The Kreutzer Sonata* (1889), and *Resurrection* (1899) embody the problem of moral decline in the shape of privileged women who smoked and indulged their sensual passions, of common women victimized by male lust and driven to prostitution, and of men destroyed by their own desires. When the eccentric Christian philosopher Vasilii Rozanov attacked the spiritual rigidity of the Orthodox Church, he assailed its sexual puritanism and hailed the supposed earthiness of the pagan and Jewish traditions in vividly sensual, unorthodox prose. When the meager political results of the 1905 revolution created widespread disappointment, intellectuals, professionals, and the reading public focused on sexual themes, in compensation for lost civic hopes and as a challenge to the puritanical anti-individualism of the radical left. Young people turned to Mikhail Artsybashev's brazen *Sanin* (1907) and Anastasiia Verbitskaia's melodramatic *Keys to Happiness* (1910–13) for tales celebrating the power of sexual desire.

The Russian secular elites drew their cultural vocabulary from the

I

Western repertoire and tested their values against a Western standard. It is therefore not surprising to find the question of sex deeply embedded in social and political concerns. Recent scholarship has shown that Western societies in the modern era have made sex and gender norms central to the structure of fundamental power relations and to the organization of cultural categories. Historians of Russia have begun to examine the position of women and the ideology of gender in the last half-century of tsarist rule; they have begun to investigate the family's role in the transformation of economic life and class categories; but they have not asked what role sexual ideology might have played in the struggle for public power and cultural influence waged between the old regime and the new social forces unleashed by the state's own program of modernization.[1]

The Great Reforms of the 1860s liberated the serfs, reconstituted the judicial system, laid the basis for local self-government, and alleviated censorship restraints. These changes permitted the emergence of a civic community resembling the Western type but deprived of access to political power. Within this amorphous public domain, members of the professional disciplines constituted a kind of civic establishment. While some of them cooperated with progressive bureaucrats in the interests of continuing reform (more than a few holding official positions at some point in their careers), most competed with the still vigorous administrative state for authority to shape the course of a rapidly but unevenly modernizing society.[2]

---

[1]On the social history of women and the family, see primarily David L. Ransel, ed., *The Family in Imperial Russia* (Urbana, Ill., 1978); Barbara Alpern Engel, *Mothers and Daughters: Women of the Intelligentsia in Nineteenth-Century Russia* (Cambridge, 1983); Rose L. Glickman, *Russian Factory Women: Workplace and Society, 1880–1914* (Berkeley, Calif., 1984); and Barbara Evans Clements, Barbara Alpern Engel, and Christine D. Worobec, eds., *Russia's Women: Accommodation, Resistance, Transformation* (Berkeley, Calif., 1991). On prostitution, see Richard Stites, "Prostitute and Society in Pre-Revolutionary Russia," *Jahrbücher für Geschichte Osteuropas* 31:3 (1983): 348–64; Barbara Alpern Engel, "St. Petersburg Prostitutes in the Late Nineteenth Century: A Personal and Social Profile," *Russian Review* 48 (1989): 21–44; and Laurie Bernstein, "Yellow Tickets and State-Licensed Brothels: The Tsarist Government and the Regulation of Urban Prostitution," in *Health and Society in Revolutionary Russia*, ed. Susan Gross Solomon and John F. Hutchinson (Bloomington, Ind., 1990). On political and social movements, see Richard Stites, *The Women's Liberation Movement in Russia: Feminism, Nihilism, and Bolshevism, 1860–1930* (Princeton, N.J., 1978); and Linda Harriet Edmondson, *Feminism in Russia, 1900–1917* (Stanford, Calif., 1984). For the role of sexual ideology in political and social life, see Stites, *Women's Liberation*; William G. Wagner, "The Trojan Mare: Women's Rights and Civil Rights in Late Imperial Russia," in *Civil Rights in Imperial Russia*, ed. Olga Crisp and Linda Edmondson (Oxford, 1989); and Wagner's *Marriage, Property, and the Struggle for Legality in Late Imperial Russia* (Oxford University Press, forthcoming).

[2]The politically moderate center of educated society has been studied from the point of view of the literary and cultural elite, the economic bourgeoisie, and political activism: see, e.g., Alfred J. Rieber, *Merchants and Entrepreneurs in Imperial Russia* (Chapel Hill,

The question of sexuality as it entered into that contest forms the subject of this book, which takes as its starting point the hypothesis elaborated by Michel Foucault in relation to Western Europe.[3] Foucault insists that sexual categories and norms constitute at once a system of power relations configuring the social body and a way of thinking about and organizing power through the medium of actual bodies. In the transition from traditional old regimes to the bourgeois capitalist order, he argues, the regulatory and punitive functions exercised by the absolutist state passed into the hands of professionals trained in such fields as medicine, criminology, pedagogy, and the law. Both power and personhood emerged transformed, and with them the uses and meanings of sexuality.

In Russia, as in Europe, the intrusion of the capitalist marketplace, the emergence of commercial culture, and the institutional consolidation of professional expertise generated a contest over the authority to regulate sexual conduct, to determine the boundaries of individual autonomy, and to demarcate public from private life. But the local context in Russia was different. Not only political expression but access to political power was severely restricted in the tsarist empire, even for those who stood at the top of the formal social hierarchy and enjoyed the privileges of Westernized culture. The transition from administrative to legal principles of governance occurred more hesitantly than it had in continental states. Urbanization and industrialization took forms that diverged in various degrees from the experience of Western countries, where the modern socioeconomic revolution was already well under way. European cultural developments continued to shape the outlook of the professional and intellectual elite, but Russia produced its own version of the Western tradition, in the sexual arena as well as in other cultural domains.

Just as the critique of capitalism preceded the full appearance of capitalism itself on the Russian scene, so Victorian notions of sexual re-

---

N.C., 1982); and Terence Emmons, *The Formation of Political Parties and the First National Elections in Russia* (Cambridge, Mass., 1983). On the professions, see V. R. Leikina-Svirskaia, *Intelligentsiia v Rossii vo vtoroi polovine XIX veka* (Moscow, 1971); idem, *Russkaia intelligentsiia v 1900–1917 godakh* (Moscow, 1981); Nancy Mandelker Frieden, *Russian Physicians in an Era of Reform and Revolution, 1856–1905* (Princeton, N.J., 1981); Julie Vail Brown, "The Professionalization of Russian Psychiatry, 1857–1911" (Ph.D. diss., University of Pennsylvania, 1981); John F. Hutchinson, *Politics and Public Health in Revolutionary Russia, 1890–1918* (Baltimore, 1990); Edith W. Clowes, Samuel D. Kassow, and James L. West, eds., *Between Tsar and People: Educated Society and the Quest for Public Identity in Late Imperial Russia* (Princeton, N.J., 1991), pt. 5; and Wagner, *Marriage.*
[3]Michel Foucault, *The History of Sexuality,* vol. 1, *An Introduction,* trans. Robert Hurley (New York, 1978); idem, *Discipline and Punish: The Birth of the Prison,* trans. Alan Sheridan (New York, 1977).

spectability and danger were questioned before they had a chance to take root. None of the interrelated protagonists of the Victorian sexual drama made a wholly successful transition to the Russian stage: neither the self-disciplined bourgeois male, nor his erotically unresponsive, homebound wife, nor the sexually undisciplined working-class male, nor the diseased and promiscuous prostitute. None survived intact as a social archetype. As part of their precocious acquaintance with the Western critique of bourgeois culture, Russians (having read Charles Fourier and John Stuart Mill) distrusted bourgeois values of sexual propriety well before anything resembling an economic middle class had emerged. They were likewise slow to imitate Western fears of lower-class sexual disorder, although Russian cities already harbored masses of the unwashed poor and of budding proletarians.

Such skepticism reflected not only a theoretical precocity but also the precarious status of Russian professional men, which distinguished them from the Western spokesmen for bourgeois respectability. Themselves excluded from power, Russian professionals were dependent on and resentful of the state, drawn into alliance with disgruntled groups from below, yet culturally related to those above them. Most were enemies of the traditional patriarchal order and all it stood for—old-style family life, police rule, human servitude. Many recognized the plight of women as analogous to their own disenfranchised state. They were fervent partisans of "enlightenment," enlisting education and social reform in the battle against communal control and the tyranny of accepted custom. But though they adopted the liberal ideal of the autonomous subject, they often rejected the Western bourgeois regard for self-interest and the goal of self-fulfillment. As a result, neither their feminism, their concepts of manhood, nor their understanding of class precisely fitted the Western standard.

Of course, the liberal ideal did not apply in equal measure to the two sexes in the West, where the principles of legal equality and respect for individual autonomy did not eliminate male domination. In Russia, precept and practice were more consistent: there patriarchy had barely been touched by the winds of change. Even after the Great Reforms, imperial laws governing family relations and gender rights reflected the patriarchal values that underlay both the autocratic regime and the peasant community. In all social classes, Russian wives and daughters depended on their husbands' and fathers' permission to take paid work, even to move about freely; their access to education was limited, and they were barred from many occupations, including state service. Divorce was nearly impossible, even legal separations hard to come by, though women did retain control of the property they brought into marriage. Abortion was a criminal offense; prostitution was subject to police regulation, on the continental European model, in some cities.

Western notions of privacy, private property, and domesticity were not entirely absent, but they competed with values and social patterns attached to precapitalist culture.

The traditional social order survived the impact of reforms and the demographic effects of economic change, especially in the rural parts of the empire. The state had freed the peasants from servitude yet continued to exclude them from civic life. The emerging urban working class, which lacked the right to organize until 1906, still retained its ties to the land. Male workers and peasants exercised patriarchal authority in their households but had only limited autonomy in their communities and the wider social world. Strict moral controls and the primacy of kin relations did not prevent women from earning wages outside the home but kept them tied to their menfolk. The peasant community did not recognize a divide between domestic and public affairs, and most peasant villages rejected private claims to landed property. Conjugal privacy was an exception, individual nonconformity rare. Once workers established themselves in town, family size began to shrink and the independent couple to emerge as a real possibility, but before 1905 the majority of workers lived apart from their families in collective or public arrangements.

It was among educated urban groups that the conjugal family—centered on the relationship between freely chosen partners, shielded from the intrusions of a larger community, and geared to the production of carefully nurtured offspring—began to emerge as a cultural norm. It was distinguished both from the extended peasant household and from the aristocratic town or country manor house, in both of which personal need, private satisfaction, and indeed privacy itself were subordinated to the family's public interests. Yet Russian defenders of individual rights in sexual and gender relations did not embrace middle-class notions of domesticity as a social ideal. Rather, traditional patriarchal relations functioned as the benchmark for elite values, in both a negative and a positive sense. Critics of the post-Reform order denounced the peasants' gender system as a vestige of serfdom: in their eyes, domestic patriarchy perpetuated the absolute power formerly exercised by landlords and still inherent in the autocracy, and female subordination symbolized the defective nature of male civic autonomy even after the Reforms. But these same critics were loath to celebrate the radical individualism latent in modern urban families and in women's claims to sexual autonomy. Rather, they seized on the peasant woman's reproductive prowess, a product of female subordination and cultural dependency, as a Rousseauian alternative to the self-indulgent, sterile urban world to which they themselves belonged but which they could not justify in moral terms.

After 1905 the nature of class relations shifted, a new political arena opened up, the rules of censorship softened, and elite men were both freer to act and more threatened by action from the lower ranks. Only then did the Russian professional establishment begin to adopt more "Victorian" sexual attitudes. The laboring classes, once seen as volatile yet unformed, now began to resemble the dangerous classes of European ill repute, criminal and debauched. Once deplored as martyrs to economic need and men's moral obtuseness, prostitutes began to represent the threat of biological as well as social pathology, as theories of organic determinism developed a more plausible ring. Further, the upper-class guardians of moral order and cultural hierarchy, having gained access to institutionalized power, feared that they themselves might succumb to the force of desire and the lure of personal gratification and began to focus on the problem of male self-control.

The pre–World War I crisis of the Russian elites thus expressed itself partly in a preoccupation with sexual pleasure and disorder. The frail plant of individualism they had been nursing all along now seemed a morbid growth. Yet even this belated adoption of a quasi-Victorian sexual ethic did not begin to match the original in force of conviction and power of stereotype, for two reasons. First, the resistance to Western attitudes shown by Russian professionals before 1905 was not merely a sign of cultural backwardness; rather, it indicated a conscious rejection of certain aspects of liberal society. Second, by the time the social and political changes inaugurated by the 1905 revolution began to erode this resistance, Western societies themselves had begun to challenge the nineteenth-century sexual system; thus, the Russian adoption of a (modified) Victorianism coincided with the revolt against Victorianism in the West. Russia's moral conversion in the sexual domain was therefore doubly ambiguous, and these contradictions of combined development reflected the complexities and ambiguities of Russian liberalism itself, a political and philosophical position that was at once behind and ahead of its time.

This book, as its subtitle suggests, is not only about sex but also about liberalism: the quest for a world in which "happiness" might become the goal of individual lives. By "liberalism" I mean not the ideas and endeavors of political activists in a narrow sense but a configuration of attitudes and values dispersed among different social groups and in various cultural locations. This liberalism was organizationally weak yet philosophically pervasive, at least in educated society. Russia had not only a radical but also a liberal intelligentsia, composed of the thousands of educated men and women who did not devote themselves to politics as a profession but who lived and thought politics in the course of their professional and civic lives. They constituted the core of a burgeoning "bourgeois" culture, not in the socioeconomic sense that they owned or controlled the means of production or were involved in productive activity, or in the sociocultural sense of incarnating Western-style individualism and Victorian propriety, but in the philosophical-cultural sense of inhabiting a partly realized, partly imagined public space in which they exhibited liberal values in the exercise of their professional skills.

Central among these values were personal autonomy, the protection of privacy, individual equality, and the rule of law. Even in nineteenth-century Europe, such ideals were often modified or compromised—out of fear of popular democracy, from a concern with the brutal social consequences of laissez-faire, in deference to patriarchal privilege.[4] The bourgeois public sphere that emerged out of the French Revolution was not accessible to all citizens alike; indeed, it was defined in such a way as to exclude the participation of women.[5] In Russia, the abstract civic values of the Enlightenment were tempered by two indigenous cultural strains: a powerful and persistent model of custodial statehood and a pervasive ethos of socialist collectivism that affected even the privileged elite. As the psychiatrist Nikolai Bazhenov remarked in 1906, Russia was obliged to reach beyond liberalism:

> The task of renovating Russia will be incomparably more complicated and harder than the task of liberation was in the West, for there is no other country as fundamentally democratic—by its historical, economic, anthropological, and sociocultural [kul'turno-bytovoi] essence—as our homeland. Moreover, the time for purely political revolutions has long passed. Russia will witness the struggle for land as well as for legality [pravo], and not for land alone but for the establishment of real social and economic justice [spravedlivost'].[6]

Bazhenov was among those educated Russians who, aware of the limits of the Western liberal order and susceptible to socialism's moral claims, felt the inadequacy of mere constitutionalism. This double consciousness had characterized the outlook of Alexander Herzen and continued to affect the thinking of many established professionals who, while sympathizing with its motivations, did not devote themselves to the revolutionary cause. Their skepticism about the West did not pre-

---

[4]See Anthony Arblaster, *The Rise and Decline of Western Liberalism* (Oxford, 1984), chaps. 15–16.

[5]Joan B. Landes, *Women and the Public Sphere in the Age of the French Revolution* (Ithaca, N.Y., 1988); Carla Hesse, "Reading Signatures: Female Authorship and Revolutionary Law in France, 1750–1850," *Eighteenth Century Studies* 22:3 (1989): 469–87.

[6]N. N. Bazhenov, *Psikhologiia i politika* (Moscow, 1906), 6. (Unless otherwise identified, all translations are mine.)

vent them from embracing many of the civic postulates of bourgeois society, including the respect for individuality and the importance of participatory politics. If populist sympathies engendered a marked ambivalence toward the prospects of modernity, they did not necessarily impede, though they often complicated, the work of Russian professionals and public figures dedicated to remaking their nation's civic life. The bearers of these liberal political aspirations determined to establish the conditions for what they thought of as modern civil society, either through officially sanctioned projects of reform (as in the case of the legal codifiers) or by the exercise of public and professional responsibility (as in the case of physicians, pedagogues, and criminologists) or in the molding of public opinion (in the case of journalists, writers, and critics).

Whether one believes that Russian liberalism ought to have triumphed, had any reasonable chance of succeeding, or was inevitably doomed, the liberal project has an unavoidable pathos in the social and political environment of late imperial Russia. The pathos resides first of all in the frustration that liberals experienced in confronting the archaic and recalcitrant political system that constrained their lives. It emerges also, however, from the tensions that affected the liberals' own intellectual universe—tensions that reflected, first, the disjuncture between the ideas imported from the West and the countervailing cultural assumptions of the local environment; and second, the Russian liberals' (understandable) failure to master the contradictions inherent in the borrowed discourse itself. But the ultimate drama of their situation rests perhaps in the nation's failure to enter the difficult and flawed terrain of postabsolutist public life, its failure to create a polity in which citizens might have struggled with the imperfections of the civic condition in terms supplied by the arsenal of liberal thought.

Battling the all too obvious constraints of a shamelessly repressive order, Russian reformers were spared the ironies of imperfect liberation that so focus Michel Foucault's iconoclastic ire. Foucault rejects the Enlightenment claim that the ideal civic regime consists in the rationally motivated activity of fully autonomous subjects. He argues that the liberation promised in the emergence of subjectivity from the shackles of absolute rule and from the limits of the ascriptive social hierarchy was seriously compromised—indeed, intrinsically obstructed—by the new power mechanisms brought into being by the liberal order. He describes the proliferation of professional discourse, the emergence of the free disciplines, and indeed the vindication of individual autonomy as techniques of political domination that translated the traditional forms of state-administered custodianship into self-administering instruments of rule.

Foucault is not the first to have identified the insidious mechanisms of constraint intrinsic in the new social order; they were recognized and deplored by some of the West's most fervent partisans of nineteenth-century liberalism—John Stuart Mill, for example.[7] With the experience of Western socialism and native populism already behind them, Russian liberals were even less naive in their enthusiasm for the promised order. Yet they persisted in their craving for the attributes of civic and cultural modernity, which were impeded by structures of absolutist governance that remained in place longer and with greater vigor in the Russian empire than anywhere else in Europe. In Russia the shift to political modernity failed to occur before the possibility of liberalism was pre-empted by another kind of transition.

I use the term "modernity" because it was a concept employed by the historical subjects who interest me—and who interest me precisely because it was modernity that enticed them—and because the mecha-nisms historically connected with the eighteenth- and nineteenth-cen-tury European societies that generated this idea served as a benchmark against which cultures on the European margin judged themselves. The term need not designate an identifiable stage of historical development which Western societies presumably had attained and toward which Russia was objectively striving. Russians did have that model in mind, however, and the emergence of a new kind of sexual regime was part of this desired, imagined, imperfectly actualized transformation. The very concept of sexuality as we use it today, Foucault argues, is a product of that pivotal moment. Sexual identity as an attribute of personality, sex-ual conduct as the focus of social anxiety and intervention, and (as Foucault might have added, though he did not) distinctions of gender all functioned as important elements in the social self-disciplining char-acteristic of the modern age.

The word "sex" in this book's subtitle is meant, by its somewhat dated vagueness, to encompass two different terms. On the one hand, it refers to the complex of ideas about sexual conduct and sexual catego-ries which Foucault includes in the concept of "sexuality." On the other, it engages the culturally variable understanding of the meaning and nature of sex difference, and in particular the political purposes accomplished by the creation of "man" and "woman" as social catego-ries—the burden of the concept "gender." What "sex" does not encom-pass here is the story of how historical subjects lived their intimate

---

[7]Mill feared the "social tyranny" that was "more formidable than many kinds of polit-ical oppression, since . . . it leaves fewer means of escape, penetrating much more deeply into the details of life, and enslaving the soul itself" (*On Liberty*, quoted in Arblaster, *Rise and Decline*, 278).

lives. It refers rather to the ideas articulated by members of the trained professions and other shapers of civic culture as part of an exercise in the appropriation and redefinition of political power—power embedded in social relations, public institutions, and intellectual authority.

The book is divided into two sections, distinguished both chronologically and conceptually. Part One goes back before the time of the Great Reforms of the 1860s, a revolutionary attempt at modernization accomplished from above; it ends on the eve of the 1905 revolution. Because the radical changes enacted by the Great Reforms deliberately exempted the institutions of absolute rule, the rudiments of a modern civic order were established within the confines of an unmodified political frame. The contradiction played itself out in the ensuing four decades: the social forces liberated from antiquated constraints but not yet in possession of the public field grew in extent, self-consciousness, and discontent until finally almost the entire spectrum of social groups arrayed themselves against the regime in a violent confrontation—a revolt of society against the state.

The particular story traced in this first section concerns the activities of two groups: the legal and medical professions. Within the wider cultural elite that persistently strove to broaden the changes initiated by the Great Reforms and to undermine the principles of autocratic rule, these professions used their specialized training and disciplinary expertise to legitimize their social activism. They are the focus of attention here because together they defined the sexual dimension of that civic revolution they were trying to effect—at first by peaceful, incremental means. Law and medicine represent the two disciplines most central to the constitution of the modern sexual regime in the West and to the steps taken in that direction in Russia: the first by redefining the status of individual persons in relation to political authority, articulating and implementing the state's intervention in social relations and creating the distinction between private and public space; the second by sharing in the job of regulating the private disposition of bodies (and later psyches) and the physical welfare of the community. The conflict between the principles represented by these professional communities and the state they both served and resisted provides the drama of these early years.

Part Two covers a period that is more overtly dramatic, from the 1890s to the eve of World War I, a time that includes the mobilization leading up to 1905, the revolutionary crisis itself, and its aftermath of continued social agitation and intellectual doubt. It is the period in which Russia first acquired a semblance of institutionalized political life in the form of the State Duma, created by the October Manifesto of 1905 and accompanied by other guarantees of civic freedom. If the gestation of the revolution is the story of society pitted against the state, its postpartum period reveals the conflicts latent within the social coalition. Thanks to the liberalization of censorship laws, the flourishing of commercial culture, and the expansion of urban life, the field of public discourse widened and its fissures spread. The highly politicized issues of culture and civic values were debated in a freer and more polyphonic context than ever before. In this atmosphere sexual themes often crosswired discursive fields occupied by newly enfranchised and emboldened groups (men free to express themselves in print and in parliament; the New Women), who successfully—if only temporarily—enacted a new kind of public life, even within the stubborn limits of the tsarist regime.

Trained professionals now found themselves in converse with a variety of looser groups, including journalists, commercial and serious writers, public activists (for example, the charity–social reform types), and political ideologues. Indeed, the transformed meaning of sexual themes in this period occurred precisely at the intersection of diverse cultural spheres. "Boulevard" fiction, for example, both constitutes and represents a fractured and all-inclusive cultural location: the mixing of high and low, public and private, professional and popular, respectable and transgressive, male-gendered and female-gendered. The boulevard, the undesired product of the new public world the liberals were trying to construct, confronted the intelligentsia with the limits of its cultural grandiosity. Violating both literary and gender convention even as it mirrored the sensibilities it shocked, pulp fiction, like other popular entertainments, freed itself from the tutelage of cultural elites and relied instead on the wiles of the marketplace, arousing and thriving on desire.

The narrower, more specialized concerns of Part One, focusing on the activity of physicians and legal experts, thus reflect the relative stability of the post-Reform years and the relatively self-contained nature of the debate over modernity. The professionals' confrontation with the problem of change and their interest in claiming a role in the workings of an imagined future order exhibited a certain decorum, however passionate the beliefs at stake. Part Two describes a more hectic, confused terrain in which the prestige of professional training and the authority of intellectual status did not always dominate the clamor of cultural contention.

In capturing the difficulty of these final years, no voice is more acute or troubling than that of the gifted and objectionable Vasilii Rozanov. By all rights he does not belong in this book. A philosopher by education, a journalist by profession, and certainly no liberal, Rozanov had an all-embracing command of the cultural landscape. As befitted the complexity of the contemporary scene, his prose is fractured and difficult, his thinking is intuitive and fragmentary, and his texts provide an uncanny reflection of what was in the cultural air—notably of the moment's salient contradictions. He was, above all, obsessed with the problem of sex, about which he was fearlessly idiosyncratic and of which he wrote with an immediacy, prolixity, and perversity that remain original to this day. A fierce opponent of modernity while a brilliant practitioner of modernism, he belongs neither to the company of earnest professionals nor to the ranks of cultural popularizers (his prose is too difficult and he does not entertain) nor to the lists of political ideologues but rather to the diffuse chorus of contending voices and invented genres that characterized the regime's last ten years. That sex was Rozanov's obsession is not to be taken lightly, just as his anti-Semitism cannot be dismissed; his sensibility was too sharp and his inhibitions were too lax for his meaning to be mistaken. Disturbing as he is, Rozanov provides the ominous red thread in a tragic story of contested modernity.

He does not have the last word, however, for his gleeful embrace of contradiction misses precisely the painful tension felt by those of his contemporaries who strove for consistency but failed to achieve it. It is perhaps not surprising that when liberalism spoke in a female voice, these tensions were at their most extreme. The emergence of the vocal "New Woman"—that Western gender rebel—on the cultural scene was facilitated by the spread of the market in fiction, entertainment, and female work. As a social type, literary stereotype, and cultural agent, the New Woman is represented in my discussion of the turbulent prewar sexuality debates by the author and the heroine of the period's best-selling boulevard novel, Anastasiia Verbitskaia's romantic potboiler *The Keys to Happiness*. Its six wordy volumes offered an eager audience of unsophisticated readers a compendium of modish themes, ranging from revolution to free love, in the adventures of an ambitious and sexually defiant young woman. Like the uncensored products of Rozanov's iconoclastic pen, this work scanned the cultural horizon, but its humble, indeed vulgar, aesthetic embraced a sexual vision too radical for the ideological company it wished to keep.

The same liberal critics who deplored Rozanov's political nihilism and patriarchal views mocked both Verbitskaia's progressive pretensions and the novel's popular success. The entire enterprise—its author, its protagonist, its careless and voluble text—represented the challenge posed at once by the appeal of cheap entertainment and by the emergence of women and female sexuality into respectable public space, thanks to the power of the modern commercial economy. Books that sold; women who earned their own living, boldly displaying their charms and seizing the instruments of culture; female authors who advertised themselves shamelessly in the press: these disturbing features of the post-1905 years called into question the resiliently traditional aspects of the sexual and cultural system to which even liberals subscribed. Yet, much like the conflicted response of its high-minded critics, striving for both individual autonomy and cultural control, the novel's message is itself ambiguous: in the end the keys are not found and happiness does not materialize, although the expectation is both liberating and erotic. Neither the boldest of narcissistic authors nor the most daring of heroines could transcend the ambivalence that plagued Russia's liberals. Avid for modern subjecthood yet unable to establish the kind of disciplinary authority with which to ensure preeminence in a world of unstable hierarchies and competing individualities, they feared the disorder—and even the pleasures—that desire can bring.

PART ONE

DISCIPLINING CHANGE:
LAW AND MEDICINE

Chapter One

Revising the Old Moral Order:
Family Relations and
Reproductive Sex

The regulation of family and sexual life constituted a battleground on which the Orthodox Church, the tsarist state, and professional elites struggled to define the basic principles governing the social, political, and civic order of late imperial Russia. Defenders of traditional social and institutional arrangements confronted reformers, both inside and outside the government, who sought to bring existing values and institutions into conformity with Western models, on the one hand, and with the changing character of Russian society, on the other. Such reformers thought of themselves as modernizers. In working to bring the Russian social and political system up to date, they felt they were continuing the work of Alexander II, who had not only liberated the serfs in 1861 but followed this profound revolution in social relations with an equally radical reconstitution of the nation's legal system in the judicial reforms of 1864.

These reforms represented a significant incursion into the absolute basis of autocratic rule. The institution of an independent judiciary, the introduction of trial by jury, the systematization of criminal procedure, the creation of justices of the peace—all these innovations constituted elements of a rule-of-law state in which the regularization of process, the autonomy of institutions, and the uniformity of rules protected the nation against the brunt of arbitrary power. Together, the Emancipation and the judicial reforms also reshaped the disposition of power within society itself. The formerly servile class now entered the body politic, in which the opportunity to participate in the exercise of power and authority was significantly enhanced. At the same time, however, the post-Reform autocracy continued to reinforce the principle of social hierarchy and to inhibit the development of autonomous instruments of social regulation. Consequently, the liberated serfs continued to suffer certain restrictions peculiar to their class position; both the range of their personal liberty and their access to formal judicial protection were limited by the authority of their local communities. Statutary law—itself a jumble of incoherent principles inherited from the past—did not apply in equal measure to all members of the national community.

Indeed, despite the progress achieved under Alexander II, institutional complexity and theoretical incoherence continued to characterize the tsarist judicial regime until its downfall in 1917. For example, four separate legal jurisdictions competed in the management and delineation of the sexual system: civil, criminal, ecclesiastical, and administrative law. Although such duplication was not unique to the tsarist regime, the wide extent of overlapping authority impeded the implementation of laws and regulations, which themselves predated the judicial reforms and therefore often failed to conform to the principles inherent in the post-1864 system. These deficiencies troubled the legal profession, which grew in authority and self-confidence as a result of the reforms and whose members often articulated a vigorous critique of autocracy as essentially incompatible with the rule of law.

Not all critics of existing legal institutions, however, considered themselves autocracy's opponents. The system's defects also worried men closely identified with the regime, who wished to save it from its own worst features. Recognizing the need to modernize the statutes upon which the courts relied, the Ministry of Justice itself authorized the revision of both the civil and the criminal codes in the last decades of the nineteenth century. It appointed reform commissions consisting of eminent statesmen and legal experts, who culled opinions from all corners of the legal and administrative establishment. Each commission generated a rich commentary on the law and aroused widespread public debate over the political and social issues entailed in its deliberations. Because conservative elements continued to block the implementation of reform, these debates yielded meager practical results; nevertheless, they amply document the attitudes of the country's political and professional elites toward questions of morality, citizenship, and political authority and testify to the presence and strength of a liberal judicial ethos.[1]

In undertaking the revision of the statutory law, the reformers confronted the maze of regulations and conflicting jurisdictions that gov-

---

[1] A detailed discussion of the civil code reform appears in William G. Wagner, *Marriage, Property, and the Struggle for Legality in Late Imperial Russia* (Oxford University Press, forthcoming) (I thank Bill Wagner for allowing me to read his manuscript). For juridical liberalism in Russian intellectual life, see Andrzej Walicki, *Legal Philosophies of Russian Liberalism* (London, 1987).

erned the contours of private life. They considered these matters as politically significant as any of the broadly constitutional matters connected with the modernization of the codes. The commission on criminal code reform described the family as "the indispensable precondition of social life [*obshchezhitie*], the primary and most solid foundation of the political order [*gosudarstvennost'*]," and insisted that protecting the family "serve[d] not merely the interests of the private persons [*chastnye litsa*] who are its members but the general interests of the entire state [*vse gosudarstvo*]."[2]

But what constituted protection in the era of rapid social, cultural, and economic change that succeeded the Great Reforms? What principles ought to govern sexual behavior outside the bounds of family life, at a time when traditional values had begun to lose their moral force and their power to shape collective and personal behavior? While conservatives defended the religious foundation of family life and the prohibition on extrafamilial sexual relations, reformers wished to replace traditional norms with the principles associated with a modern civic and judicial order: the primacy of individual rights over hierarchical precedence; equality before the law; secular standards of transgression and correction. But like their liberal analogues in Western Europe, the Russian reformers were swayed by their social values (whether openly acknowledged or un-self-conscious) and often defaulted on the consistent application of their own standards. The constitution of the ideal liberal subject was nowhere complete. What is interesting is the way in which these self-proclaimed modernizers adapted their abstract goals to the peculiarities of their particular political position and cultural setting.

The Institutional Context

The great theoretical and institutional conflict that animated nineteenth-century Russian legal discourse involved the contrast between two competing principles of governance: one, the custodial *Polizeistaat* model, in which the government exercises authority by administrative fiat in the spirit (optimally) of benevolent paternalism; the other, the *Rechtsstaat* model, under which political power is constrained within limits set by positive law.[3] In Europe as well as in Russia these two principles were embodied in separate institutional fields. Like those of

---

[2]*Ugolovnoe ulozhenie: Proekt redaktsionnoi komissii i ob"iasneniia k nemu* (St. Petersburg, 1895–97), 4:152; henceforth *UU*, vol. 4 (1895).
[3]See Walicki, *Legal Philosophies*, introduction; also Marc Raeff, *The Well-Ordered Police State: Social and Institutional Change through Law in the Germanies and Russia, 1600–1800* (New Haven, Conn., 1983).

the continental European states, the Russian courts operated on the
basis of formal statutes enumerated in civil and criminal codes.[4] The
administrative side of the law was not subject to the same formal constraints. The police, for example, fell under the authority of the Ministry of Internal Affairs, not the Ministry of Justice. They enforced the
statutory law by arresting those suspected of crimes or violations defined in the codes, but they also enforced their own set of regulations
intended for "the prevention and suppression of crimes."[5] Their powers
of enforcement, however, were not equivalent to those exercised by the
courts at any level. The Statute on Criminal Procedure declared that
"measures taken by the police and other administrative authorities for
the prevention and suppression of crimes in the manner established by
laws do not belong to judicial proceedings."[6]

The ideal distinction between judicial and police functions was nowhere fully realized, however, least of all in the Russian empire, where
the rule of law was so slow to take root. Not only did the monarch's
will take precedence over formal procedure; in addition, the guarantees
of due process introduced in 1864 failed to apply to the actions of the
police, whose measures "for the prevention and suppression of crimes"
were independent of the decisions and rules of the courts.[7] Indeed, the
confusion between police (so-called administrative) measures and judicial procedure was a feature of the autocracy's policy of deliberate mis-
rule.[8] Critics might denounce administrative regulations for their illegality when they contradicted written law, but technically speaking, the
Russian system allowed for competing legalities: statutes, regulations,
imperial decrees, and ministerial instructions jointly occupied the disciplinary field.[9]

The official distrust of legal procedure and uniform standards of law
was reflected in the history of codification. Because the great European

[4]The civil and criminal codes constituted vols. 10 and 15 of the Digest of the Laws.
For the organization of Russian statutory law, see "Kodifikatsiia," in E. N. Trubetskoi,
Entsiklopediia prava (Kiev, 1901), 143–50.

[5]Vol. 14 of the Digest of the Laws contains the Digest of Regulations on Public Decency (*Svod ustavov blagochiniia*), enforced by the police.

[6]Quoted from vol. 16, Digest of the Laws, in Marc Szeftel, "Personal Inviolability in
the Legislation of the Russian Absolute Monarchy," *American Slavic and East European
Review* 17 (1958): 2.

[7]On the precedence of the emperor's will over formal procedure, see Marc Szeftel,
"The Form of Government of the Russian Empire Prior to the Constitutional Reforms of
1905–06," in *Essays in Russian and Soviet History*, ed. John Shelton Curtiss (Leiden, 1963),
111.

[8]See George L. Yaney, *The Systematization of Russian Government: Social Evolution in
the Domestic Administration of Imperial Russia, 1711–1905* (Urbana, Ill., 1973), 260–65.

[9]For an example of complaints about the illegality of administrative regulations, see
A. L. Rubinovskii, "Povinnost' razvrata," *Vestnik prava*, no. 8 (1905): 156.

powers codified their laws and Russian rulers coveted the symbols of national and cultural prestige, the nineteenth-century tsars authorized the production of codes as an attribute of modern statehood. Wary, however, of any check on their own undivided sovereignty, the monarchs limited the extent to which legal principles could be systematically articulated and applied, as if mere codification might limit their absolute freedom to act.[10] Conversely, critics of absolutism associated the possibility of a modern political community with the principles of the rule-of-law state, challenging arbitrary power in the name of legal rationality.

The fate of Russia's criminal code illustrates the confrontation between the ambivalently modernizing regime and important forces within the legal establishment eager to deinstitutionalize that ambivalence. Constantly under scrutiny during the nineteenth century, the code was never brought into conformity with contemporary legal standards. The first codification project after Tsar Alexis's code of 1649 was instituted by Alexander I, under the direction of Count Mikhail Speranskii, an energetic partisan of judicial reform. A draft produced in 1813 with the help of European experts was rejected, however, in favor of the less radical text finally adopted in 1835.[11] The conservative

[10]Nicholas I, e.g., rejecting the idea that codes should embody abstract legal principles, thought of them as compilations of existing laws representing the monarch's will: Richard S. Wortman, *The Development of a Russian Legal Consciousness* (Chicago, 1976), 43. The complete code of Russian laws (*Polnoe sobranie zakonov Rossiiskoi Imperii;* henceforth *PSZ*), the chronological register of all laws ever enacted, most completely embodied this model. By contrast, the Digest of the Laws (*Svod zakonov Rossiiskoi Imperii*) indexed currently active laws by topic, thus introducing a measure of systematization and conceptual generality. I am grateful to Reginald Zelnik for remarking on this distinction, which exemplifies the ambivalence inherent in the monarchy's desire to conserve its unrestricted power while exercising it with greater effectiveness.

[11]Four attempts were made to reorder and modify the existing statutes, but only the codes of 1835 and 1845 went into effect; the drafts of 1813 and 1903 were rejected. The first code adopted (*Svod zakonov ugolovnykh* [St. Petersburg, 1835]; henceforth *Svod* [1835]) went through a second edition in 1842 and acquired eleven supplements: N. S. Tagantsev, *Russkoe ugolovnoe pravo: Lektsii,* 2d ed., rev. (St. Petersburg, 1902), 217–19. This code formed vol. 15 of the Digest of the Laws, which was finished in 1832 and took effect on January 1, 1835: Friedrich Barthold Kaiser, *Die russische Justizreform von 1864: Zur Geschichte der russischen Justiz von Katherina II bis 1917* (Leiden, 1972), 142. For the text of the 1813 draft, see *Proekt ugolovnogo ulozheniia Rossiiskoi Imperii* (St. Petersburg, 1813), in *Arkhiv Gosudarstvennogo Soveta,* vol. 4 (St. Petersburg, 1874); henceforth *Proekt* (1813). For its history, see [M. M. Speranskii], *Obozrenie istoricheskikh svedenii o svode zakonov* (St. Petersburg, 1833), 34–38; *Trudy komissii sostavleniia zakonov,* vol. 1, *Postanovleniia ob obrazovanii komissii,* 2d ed. (St. Petersburg, 1822); and I. I. Solodkin, *Ocherki po istorii russkogo ugolovnogo prava (pervaia chetvert' XIX v.)* (Leningrad, 1961), 8–17. The draft was supervised by Ludwig Heinrich von Jakob (1759–1827), professor of law at the University of Halle; see his *Entwurf eines Criminal-Gesetzbuches für das russische Reich. Mit Anmerkungen über die bestehenden russischen Criminalgesetze. Nebst einem Anhange, welcher enthält: Kritische Bemerkungen über den von der Gesetzgebungs-Commission zu St. Petersburg*

Nicholas I in turn sponsored revisions of the code that embodied extreme resistance to the very principles of positive law: the 1845 text, framed under the supervision of Count Dmitrii Bludov, head of the Second Section of His Imperial Majesty's Chancellery, remained on the books until 1917, variously amended but never fundamentally changed. It was rife with contradiction and illogic, commentators complained, and mired in a paralyzing specificity, avoiding the clear exposition of general principles in favor of the enumeration of particular cases.[12] One disgruntled scholar remarked that the criminal statutes' "aversion to theoretical definitions" and "fear of abstraction" recalled "a medieval collection of laws."[13]

The great turning point should have come in 1864, when Alexander II and his enlightened bureaucrats set the stage for judicial modernization. Indeed, the progressive forces responsible for these institutional innovations intended to refashion the content of the law as well. Despite top-level resistance, the reform impulse survived at the highest reaches of the state apparatus, even under the last two tsars' conservative regimes. The credit for the reformers' limited success goes to Dmitrii Nabokov, the minister of justice appointed by Alexander II in 1878, who had spent his formative years in the company of the progressive Grand Duke Konstantin Nikolaevich. Before being removed from office by conservatives in 1885, Nabokov managed to sponsor statutory revisions of a thoroughgoing and aggressively modernizing kind.[14] Indeed, the efforts that in 1895 produced the text of a new criminal code (officially approved in 1903 though never enacted) engaged the energies of the nation's foremost legal scholars and highest-ranking statesmen,

---

herausgegebenen Criminal-Codex (Halle, 1818). Jakob's reform, with its secular thrust and theoretical distinctions, proved too radical for the more conservative Russian members of the editorial committee to accept: Solodkin, Ocherki, 15–16.

[12]Ulozhenie o nakazaniiakh ugolovnykh i ispravitel'nykh (St. Petersburg, 1845); henceforth Ulozhenie (1845), replaced the 1835 code as vol. 15 of the Digest of the Laws. The draft version and editorial commentary are contained in Proekt ulozheniia o nakazaniiakh ugolovnykh i ispravitel'nykh, vnesennyi v 1844 godu v Gosudarstvennyi Sovet, s podrobnym oznacheniem osnovanii kazhdogo iz vnesennykh v sei proekt postanovlenii (St. Petersburg, 1871); henceforth Proekt (1844). Without its structure or basic premises being affected, the 1845 code appeared in three subsequent editions (1857, 1866, 1885), to which 34 supplements were appended between 1845 and 1895. See Tagantsev, Russkoe ugolovnoe pravo, 217–19, 222–23.

[13]V. D. Spasovich, quoted in Tagantsev, Russkoe ugolovnoe pravo, 222.

[14]On Nabokov's initiation of criminal reform, see Theodore Taranovski, "The Politics of Counterreform: Autocracy and Bureaucracy in the Reign of Alexander III, 1881–1894" (Ph.D. diss., Harvard University, 1976), 695–96 (I am grateful to the author for providing me with a copy). On progressive trends in the 1880s, see Heide W. Whelan, Alexander III and the State Council: Bureaucracy and Counter-Reform in Late Imperial Russia (New Brunswick, N.J., 1982); Nabokov and the civil code are discussed in Wagner, Marriage, chap. 4.

who were dedicated to modernizing and secularizing the law.[15] "All the political and social turning points of national life are reflected clearly in the concepts of crime and punishment," wrote leading commission member Nikolai Tagantsev, "and the more rapidly this life evolves, the more quickly do reforms occur."[16] Initiated, paradoxically, during the reign of the archreactionary Alexander III, the projects for civil and criminal code reform were definitively obstructed during Nicholas II's tenure.[17]

In addition to impeding the proper definition and implementation of positive law, the tsars used the administrative apparatus to circumvent the law altogether. Alexander III's exceptional regulations are the most

[15]The editorial committee consisted of eight men: I. Ia. Foinitskii, E. V. Frishch, N. A. Nekliudov, E. I. Nol'de, E. N. Rozin, V. K. Sabler, V. K. Sluchevskii, and N. S. Tagantsev. All were state senators, and four were professors of law; two were connected with the Second Section of His Majesty's Imperial Chancellery, two with the Ministry of Justice, one with the Committee of Ministers; and one was the legal counselor to the Holy Synod. The committees that reviewed the early drafts (one with 14, the other 48 members) consisted of equally highly placed men, including ministers and officials associated neither with the Ministry of Justice nor with reforming trends in the bureaucracy. On the composition and role of the committees, see "Obozrenie khoda rabot po sostavleniiu ugolovnogo ulozheniia" (St. Petersburg, 1903), rpt. in *Ugolovnoe ulozhenie (stat'i vvedennye v deistvie)*, ed. D. A. Koptev and S. M. Latyshev (St. Petersburg, 1912), 27, 31–33, 35–36, 40–41. On initiation of the reform, see ibid., 18–26, 29; also Tagantsev, *Russkoe ugolovnoe pravo*, 224–25. Career information is from *Gosudarstvennyi Sovet, 1801–1901* (St. Petersburg, 1901), 208 and app.; *Dnevnik gosudarstvennogo sekretaria A. A. Polovtsova* (Moscow, 1966), 2: index; *Zhurnal osobogo prisutstviia Gosudarstvennogo Soveta Vysochaishe uchrezhdennogo dlia obsuzhdeniia proekta ugolovnogo ulozheniia* (St. Petersburg, 1902); V. I. Gurko, *Features and Figures of the Past: Government and Opinion in the Reign of Nicholas II* (Stanford, Calif., 1939), notes; *Spisok g.g. chlenam Gosudarstvennogo Soveta (k 22 fevralia 1910 g.)* (St. Petersburg, 1910); *Spisok chinam vedomstva Ministerstva Iustitsii* (St. Petersburg, 1910–16); *Spisok vysshim chinam gosudarstvennogo, gubernskogo i eparkhial'nogo upravleniia* (St. Petersburg, 1891, 1903); *Ministerstvo Iustitsii za sto let, 1802–1902: Istoricheskii ocherk* (St. Petersburg, 1902).

[16]Tagantsev, *Russkoe ugolovnoe pravo*, 221.

[17]The commission to reform the civil code was dissolved in 1906. The conservative forces that blocked the reform are discussed in Wagner, *Marriage*. A revised criminal code completed in 1895, endorsed by the State Council, the Ministry of Justice, and the tsar himself in 1903, was never formally enacted. My interpretation of the reform process is based on *Ugolovnoe ulozhenie, Vysochaishe utverzhdennoe 22 marta 1903 g.* (St. Petersburg, 1903); henceforth *Ugolovnoe ulozhenie* (1903). The editorial committee's 1895 draft appears with commentary in *Ugolovnoe ulozhenie: Proekt redaktsionnoi komissii i ob"iasneniia k nemu* (St. Petersburg, 1895–97), vol. 6 (1897), ed. N. S. Tagantsev; henceforth *UU*, vol. 6 (1897). This volume contains most of the statutes on sexual crime; see Tagantsev, *Russkoe ugolovnoe pravo*, 227. The opinions of reviewers appear in *Ugolovnoe ulozhenie: Proekt, izmennyi ministrom iustitsii po soglasheniiu s predsedatelem Vysochaishe uchrezhdennoi redaktsionnoi komissii* (St. Petersburg, 1898); Ministerstvo Iustitsii, Pervyi departament, Chast' iuriskonsul'tskaia (marta 14 dnia 1898 goda), no. 8228, *Po proektu novogo ugolovnogo ulozheniia* (St. Petersburg, 1898); *Zhurnal osobogo prisutstviia; Zhurnal obshchego sobraniia Gosudarstvennogo Soveta po proektu ugolovnogo ulozheniia* (St. Petersburg, 1903). Texts that summarize the whole process and include the final version with commentary: N. S. Tagantsev, ed., *Ugolovnoe ulozhenie 22 marta 1903 g.* (St. Petersburg, 1904); and Koptev and Latyshev, *Ugolovnoe ulozhenie*.

famous example of this disruptive strategy. Strictly speaking, however, it was not an illegal course, for the tsar had the right to supersede laws and regulations of any kind.[18] Furthermore, exemption from the general principles embodied in the law was the defining feature of social status. The tsar's superiority to legality constituted his supreme power, and the delegation of specific immunities created the hierarchy of power over which he exercised final command. The right to exemption was, ironically, embodied in a statute: "Privileges granted by the Supreme Autocratic Power to private persons or associations exempt them from the action of general laws in matters concerning which there are precise provisions in those privileges."[19] The kind of society that nineteenth-century Russian monarchs thus wished to perpetuate was one in which power was embedded in a hierarchy of authority that could not be challenged by appeal to universal principles or institutionalized rules.

But even as the autocracy clung to its traditional principles, traditional cultural assumptions were breaking down in the cities and among the educated population. Throughout the nineteenth century, the intelligentsia's acute self-consciousness and intense preoccupation with the literary word had accompanied the emergence of a free-floating individuality, cut loose from the ties of community and the obligations of family that had bound alike the peasant and the aristocratic milieu—the two faces of the traditional rural order. As William G. Wagner has remarked in reference to the experts who revised the civil code, the legal profession was itself composed of highly trained men, with an acute sense of their own individual importance, whose own social experience and civic interests were reflected in the model of legality they tried to introduce into Russia's outdated system.[20]

These privileged yet disenfranchised men found themselves caught between the weight of unconstrained authority exercised from above and the weight of folk tradition that rooted the popular mass in its communal ways—or, as a liberal psychiatrist expressed it in 1906, between "the power of darkness below and the darkness of power above."[21] The abolition of serfdom had not eliminated the principle of arbitrary authority or the unequal distribution of civic rights. Indeed, insofar as the judicial reforms that came in the wake of emancipation made invidious distinctions of status, they failed to guarantee the prin-

---

[18]*Svod zakonov Rossiiskoi Imperii* (St. Petersburg, 1842), vol. 1, pt. 1, art. 70: "A ukase of the Emperor granted in a private case, or especially taking place with regard to any (special) kind of cases, abrogates the action of general laws concerning this very case or kind of cases" (quoted in Szeftel, "Form of Government," 111).

[19]Art. 71, quoted in Szeftel, "Form of Government," 111; see also 105–12.

[20]Wagner, *Marriage.*

[21]He concluded that the intelligentsia must defeat the regime in order to establish the "kingdom of light" among the popular masses: see N. N. Bazhenov, *Psikhologiia i politika* (Moscow, 1906), 7.

ciple of personal freedom for any member of Russian society. Rather, by establishing separate and parallel legal systems to serve distinct social communities, they perpetuated the central cleavage between the subordinate and privileged orders. Released from servitude to their masters and from subjection to manorial justice in 1861, peasants were thenceforth liable to prosecution in the regular courts for transgressions included in the criminal code and for less serious offenses committed against, or with the complicity of, nonpeasants. Peasants were, however, the only social group to be governed in some cases by rules and legal institutions peculiar to themselves. Minor offenses involving members of the peasant estate came under the jurisdiction of special courts, elected by the village communities and guided by local custom rather than positive law.[22]

Thus, the top and bottom of the tsarist order reflected the same conservative regard for particularity, the same distrust of the universal standards and binding principles embodied in the underlying structure and intention of the statutory system of law. Unlike the regular courts, which were governed by statute, the township courts offered no procedural guarantees or consistent standards of judgment. Instead, their decisions rested on the character and personal standing of the parties involved and on communal norms. Further, they were barred from connection with the rest of the system by denial of the right of appeal.[23] These local institutions might in practice demonstrate consistent patterns of response, but consistency was not essential to their function. They were intended instead to accommodate the variety and specificity of cultural practice. It was no accident of history, moreover, that this judicial structure should have so closely resembled the organic model of traditional social relations which nineteenth-century conservatives, including the Slavophiles, counterposed to the rule of contract and statute in modern civil society.[24] The Slavophiles' brand of romantic conservatism, in the Russian context, caused them to oppose the political in-

---

[22]M. I. Zarudnyi, *Zakony i zhizn': Itogi issledovaniia krest'ianskikh sudov* (St. Petersburg, 1874), 65–66; and Peter Czap, Jr., "Peasant-Class Courts and Peasant Customary Justice in Russia, 1861–1912," *Journal of Social History* 1:2 (1967): 152–53. In 1889 the jurisdiction of these courts was extended to other rural inhabitants: C. A. Frierson, "Rural Justice in Public Opinion: The Volost' Court Debate, 1861–1912," *Slavonic and East European Review* 64:4 (1986): 539.

[23]Zarudnyi, *Zakony*, 80; Aleksandra Efimenko, *Issledovaniia narodnoi zhizni*, vol. 1, *Obychnoe pravo* (Moscow, 1884), 176 (*gliadia po cheloveku*). Czap, "Peasant-Class Courts," emphasizes the peasants' juridical segregation.

[24]On the conservative paradigm and Slavophile thinking, see Andrzej Walicki, "Personality and Society in the Ideology of the Russian Slavophiles: A Study in the Sociology of Knowledge," *California Slavic Studies* 2 (1963): 1–20; idem, *The Controversy over Capitalism: Studies in the Social Philosophy of the Russian Populists* (Oxford, 1969), 29–80; and idem, *The Slavophile Controversy: History of a Conservative Utopia in Nineteenth Century Russian Thought*, trans. Hilda Andrews-Rusiecka (Oxford, 1975).

vasiveness of the autocratic state and the delegated absolutism of serf-dom even as it led them to participate in drafting the reforms that shored up the old social order.[25]

The juridical reinforcement of collective authority suited both the Slavophile ideal of "organic" social relations and the state's interests in maintaining rural stability in the absence of patrimonial ties. Critics objected that the separate system of justice maintained the peasants in a state of civil minority incompatible with the status of citizen.[26] Ultimately, the arbitrary nature of the peasant courts' authority and their exclusion from the jurisdiction of positive law became an obstacle to the enactment of the 1903 code. As the tsar and the Council of Ministers argued in refusing to put the new version into effect, the fragmented judicial structure established by the Great Reforms was not compatible with the principle of equality before the law embodied by the revised statutes.[27]

What the reform commission rejected in the existing code was precisely the foundation of civil status in exemption from general rules. For privilege conferred by special dispensation, the commission substituted the principle of equality before the law.[28] By removing distinctions based on civil status (which also underlay the very existence of the

[25]See Peter Czap, Jr., "The Influence of Slavophile Ideology on the Formation of the Volost' Court of 1861 and the Practice of Peasant Self-Justice between 1861 and 1889" (Ph.D. diss. Cornell University, 1959), chap. 2. For the way liberal hostility to old-regime absolutism might coincide with a conservative-sounding defense of corporate rights and particular interests, see James J. Sheehan, "Some Reflections on Liberalism in Comparative Perspective," in *Deutschland und der Westen*, ed. Henning Köhler (Berlin, 1984), 47.

[26]Frierson, "Rural Justice," 541, quoting a report from 1904.

[27]Whether or not this was the single most important objection to enacting the code, it was the one underscored by Koptev and Latyshev: Extension of "Obozrenie khoda," 44–47. While the Ministry of Justice wanted to implement the code as soon as possible, despite the fragmentation of the judicial system, members of the State Council objected on the grounds that only a unified system (without peasant courts and land captains) designed to afford "legal equality to persons of all social stations" (ukase of December 12, 1904) could operate under the new code. Other members recognized the structural problem but wanted to apply the new code in any case. See "Ministerstvo Iustitsii: O vvedenii v deistvie novogo ugolovnogo ulozheniia," Gosudarstvennyi Sovet v soedinennykh departamentakh (4 i 27 aprelia 1905 goda), no. 23, 1–7; "Osoboe mnenie chlenov Gosudarstvennogo Soveta . . . po delu o vvedenii v deistvie ugolovnogo ulozheniia," in *Obshchee sobranie Gosudarstvennogo Soveta* (30 maia 1905 g.); and N. S. Timashev, "Ugolovnoe ulozhenie i volostnoi sud," pts. 1–2, *Pravo*, nos. 19–20 (1914): 1529–36, 1618–25. By accepting the council's minority opinion and by countermanding his own endorsement of March 22, 1903, Nicholas exercised precisely those monarchical prerogatives that undermined the rule of law—prerogatives that were, of course, his *by* law. Legally enshrined absolutism was that contradiction in terms which so exasperated the partisans of modern statehood.

[28]For growing sentiment against hereditary privilege, see Gregory L. Freeze, "The *Soslovie* (Estate) Paradigm and Russian Social History," *American Historical Review* 91:1 (1986): 31.

township courts), the 1903 project exemplified the radical implications inherent in the formal system of law. The more coherent such a system became, the more it subsumed exceptional cases in comprehensive categories hard to reconcile (in theory) with the maintenance of social privilege and the exercise of absolute political discretion. In contrast to the projected reform, the existing statutes minimized this potential challenge to autocratic rule not only by their organizational fragmentation (their "medieval" interest in particular cases and specific detail) but also by their actual content, which endorsed the absolute nature of the monarch's power.

The existing codes also reflected the ambiguous relationship between secular and religious authorities, a confusion that aroused the displeasure of progressive legal reformers.[29] The imperial state had undermined the material and administrative autonomy of the Orthodox Church but continued to foster the church's moral preeminence and its privileged status in relation to other faiths. Peter the Great, as is well known, enhanced the power of the state apparatus and subordinated traditional customs to his own rational designs. Yet although Peter limited the church's independent judicial role, he did not challenge its doctrinal authority. The state defined the competence of the ecclesiastical courts but did not influence the substance of church law.[30] The other side of the church's incorporation into the secular apparatus consisted of the powerful secular support accorded religious principles. Moreover, the church retained a number of institutions whose functions paralleled those of the state—censorship, schools, courts, bureaucracy—and in many cases church and state shared both jurisdiction and disciplinary rights.[31] The criminal code sometimes equated religious offenses with

[29]On the relation between ecclesiastical and secular power, see Marc Szeftel, "Church and State in Imperial Russia," in *Russian Orthodoxy under the Old Regime*, ed. Robert L. Nichols and Theofanis George Stavrou (Minneapolis, 1978), 127–41.

[30]On Peter's limitation of church jurisdiction and its current extent, see M. E. Krasnozhen, *Tserkovnoe pravo*, 2d ed. (Iuriev, 1906), 200–201, 206–7. On the autonomy of church law, see "Doklad predsedatelia komissii Gosudarstvennoi Dumy III sozyva po delam pravoslavnoi tserkvi V. N. L'vova o predpolozhennykh v proekte ministra iustitsii 'O podsudnosti i poriadke proizvodstva del o rastorzhenii brakov lits pravoslavnogo ispovedaniia vsledstvie preliubodeianiia ili nesposobnosti k brachnomu sozhitiiu' izmeneniiakh v ustave dukhovnykh konsistorii," Komissiia Gosudarstvennoi Dumy III sozyva po delam pravoslavnoi tserkvi, sessiia pervaia, zhurnal no. 19, zasedanie 11 iunia 1908 g., in *Spravka po voprosu ob otnoshenii tserkovnogo zakonodatel'stva k gosudarstvennomu* (St. Petersburg, 1914), 80–82. For a contrasting opinion that emphasizes church law's dependence on state sanction, see "Vyderzhki iz ob"iasnitel'noi zapiski k proektu osnovnykh polozhenii preobrazovaniia dukhovno-sudebnoi chasti," Komissiia Gosudarstvennoi Dumy III sozyva po delam pravoslavnoi tserkvi, sessiia pervaia, zasedanie 14 fevralia 1913 g., in *Spravka*, 91.

[31]Bestiality, e.g., incurred loss of civil rights, deportation, and—for Christians—church penance: art. 997, in N. S. Tagantsev, *Ulozhenie o nakazaniiakh ugolovnykh i ispra-*

crimes against the state and prescribed secular penalties for religious offenses and religious penalties for transgressions for which no secular punishments were incurred.[32] The civil statutes also contained numerous prescriptions based entirely on religious principles. As for the church, the restriction placed on its role in worldly affairs seems to have encouraged its interest in strictly spiritual matters, in particular those bearing on the contours of private life.[33]

The family and sexual relations constituted a ground on which the moral interests of the church in reinforcing sacred values coexisted with the political interests of the state in defending old-style patriarchal forms of rule (domestic and public); both competed with the aspirations of a trained professional class eager to fashion a modern polity based on secular values and inhabited by rational self-governing subjects. All interpreted the family as the foundation stone of civil society and as a mirror of the principles governing national life.

Conjugal Black Mass: Russia's Sexual Outlaws

The marriage laws applied to all members in good standing of the national community; those to whom they did not apply constituted a stigmatized margin. The community was limited, in the first place, to adherents of the officially recognized religious faiths (Orthodox, Protestant, Catholic, Muslim, and Jewish), whose specific ritual requirements were endorsed by the statutory code.[34] The law did not provide for civil marriage; neither did the state recognize the legitimacy of dissenting religious sects, whose unions were therefore legally invalid. Only in 1874 did the Old Believers acquire the right to register at least

---

vitel'nykh 1885 goda, 11th ed., rev. (St. Petersburg, 1901), 522; henceforth *Ulozhenie* (1885).

[32]Although cohabitation, e.g., was a criminal offense (art. 994, *Ulozhenie* [1885], 515), the code called only for church-administered penance and a monetary fine (i.e., a civil penalty) until the law of June 3, 1902, transferred cohabitation to the civil jurisdiction, where it obviously belonged: A. Golubtsov, "Po povodu zakona 3 iunia 1902 goda," *Zhurnal Ministerstva Iustitsii*, no. 2, sec. 2 (1903): 194. A representative of the Warsaw circuit court condemned the retention of strictly religious sanctions in the imperial code as "an exclusively medieval concept" of law: see N. A. Nekliudov, ed., *Materialy dlia peresmotra nashego ugolovnogo zakonodatel'stva* (St. Petersburg, 1880–83), 3:306.

[33]Gregory L. Freeze, "Handmaiden of the State? The Church in Imperial Russia Reconsidered," *Journal of Ecclesiastical History* 36:1 (1985): 89–90. For a guide to nineteenth-century canon law, see S. V. Kalashnikov, *Alfavitnyi ukazatel' deistvuiushchikh i rukovodstvennykh kanonicheskikh postanovlenii, ukazov, opredelenii i rasporiazhenii Sviateishego Pravitel'stvuiushchego Sinoda (1721–1901 gg. vkliuchitel'no) i grazhdanskikh zakonov, otnosiashchikhsia k dukhovnomu vedomstvu pravoslavnogo ispovedaniia*, 3d ed., rev. (St. Petersburg, 1902), 46–66 (on marriage).

[34]E.g., art. 1568, *Ulozhenie* (1885), 749.

some of their marriages, a concession without which they and the Christian sectarians lived as conjugal outlaws. They had no legal recourse against change of partners or desertion, and their children could assert no legitimate claims.[35] The inability to abide by the laws regulating family life or to benefit from the advantages of legal protection reinforced the religious dissenters' status as social exiles. Even the 1874 law did not normalize the Old Believers' situation, because their own rites still had no legal force. According to a technical provision of the law, not all Old Believers qualified for the opportunity to inscribe their marriages in the state registry books (a de facto form of civil marriage), and few who did took advantage of the chance, since the heart of the Old Belief was loyalty to authentic spiritual rituals.[36] The very access to a form of civil marriage and the state's persistent refusal even after 1874 to acknowledge the legitimacy of their religious rites continued to distinguish the Old Believers from adherents of the officially sanctioned faiths.

The world of actual outlaws exiled to the penal colonies of Siberia constituted the second element in the empire's conjugal demimonde. For exiled convicts the administrative regime produced a universe in which the values defining the national community were stood on their heads, repeating in the sexual realm the inherent contradiction between law and administrative order that permeated the imperial system as a whole. Siberian exiles had, by definition, been deprived of their civil rights and their standing in the juridical order. As a sign of their consequent vulnerability to arbitrary and abusive treatment, they remained subject to corporal punishment for misbehavior in captivity even after it had been abolished for all other classes of the civilian population and after the crimes for which they had been condemned no longer incurred such punishment. The entire experience of transport, resettlement, and forced labor produced a brutalized population, more violent and vengeful than its members had been at the start.[37]

All the worst characteristics of the exile system were concentrated in the penal colony on Sakhalin Island, which Anton Chekhov visited in

---

[35]On the problems resulting from unregistered unions and the change introduced by the law of April 19, 1874, see Vereshchagin, "K voprosu o registratsii brakov raskol'nikov," *Iuridicheskii vestnik* 20:2 (October 1885): 288–303.

[36]Ibid., 297. For more on the Old Believers' marital and civil status, see N. S. Suvorov, *Grazhdanskii brak* (St. Petersburg, 1896), 117–26.

[37]See Alan Wood, "Sex and Violence in Siberia: Aspects of the Tsarist Exile System," in John Massey Stewart and Alan Wood, *Siberia: Two Historical Perspectives* (London, 1984), 23–42; and idem, "Crime and Punishment in the House of the Dead," in *Civil Rights in Russia*, ed. Olga Crisp and Linda Edmondson (Oxford, 1989), 215–33. For more detail on the sexual morality of Siberian exiles, see S. V. Maksimov, *Sibir' i katorga*, 3d ed. (St. Petersburg, 1900), 311–13.

1891. His account of what he observed, published in 1895, included a description of the island's sexual system, wherein everything illegal in normal society had the force of custom. Despite the laws against fornication and adultery, the colony provided no separate housing for female convicts; the penal authorities distributed newly arrived women among the male inhabitants as domestic partners, taking the youngest and most attractive for their own use. With men outnumbering women two to one, only the best-behaved and most industrious men got the chance to select a helpmate, and the lucky ones, Chekhov reported, looked forward to acquiring an extra hand at domestic and agricultural labor as they might anticipate the addition of a horse or cow. Allowed to examine the newcomers in an awkward mass encounter, they tried to estimate the women's capacity for work and reproduction.[38]

Though women were sentenced to the colony on an equal basis under the law and were even liable to corporal punishment once they got there, their place in the forced labor regime was unique. "When [the women] are transported to the island," Chekhov wrote,

> there is no thought of punishment or correction but only of their ability to bear children and do agricultural work. The women sentenced to hard labor [katorzhnye] are divided among the male settlers [poselentsy] supposedly in the capacity of workers. This arrangement is based on article 345 of the Exiles' Charter [Ustav o ssyl'nykh], which allows unmarried exiled women "to earn a living as servants in the nearest settlement until they get married." But this article exists only as a screen to cover violations of the laws against fornication and adultery. The female convict or exile living with an exiled man is not primarily his hired hand [batrachka] but his concubine [sozhitel'nitsa], an illegal wife whom he has acquired with the knowledge and consent of the administration.[39]

Paradoxically, the women considered themselves better off than they would have been in normal marriages, since on Sakhalin they were a scarce resource and had the freedom to change partners at will. No law constrained them to remain forever with men who beat or mistreated them, although Chekhov observed that physical abuse often accompanied otherwise stable relationships. While some couples evidently lacked any feeling for each other, he found, others showed enduring affection. Again paradoxically, wives who had accompanied their husbands into exile found themselves in a more difficult situation than the

---

[38]A. P. Chekhov, Ostrov Sakhalin (1895; Moscow, 1984), chaps. 16–17.
[39]Ibid., 218–19.

convict concubines, since each of the latter received an official allowance for the duration of her term. Without such economic support, legal wives sometimes turned to prostitution to earn their keep.

Converting their household arrangements into legitimate marriages was not easy, even for couples who wanted to do so. The sentence to exile and hard labor deprived the convict of all civil rights, including those connected with marital status, but the spouse who remained behind was still legally married. Exiles who wished to remarry were therefore obliged to obtain divorces from abandoned partners, who—out of religious scruple, resentment, or fear and ignorance of complex legal procedures—did not readily agree to provide them. And even if the obstacle of divorce could be overcome or was nonexistent, marriage was not easy to achieve: prospective couples had to offer documentary proof of their single state—a difficult feat, given their circumstances. Therefore, free unions and illegitimate progeny abounded.

In this world of moral inversion, where, as Chekhov put it, "cohabitation ha[d] become . . . the traditional order"[40] and personal relations were, if possible, even less private and individualized than in the patriarchal community, nonreproductive sex also put in a casual appearance. Physicians who worked in the colony noted that some women chose sexual partners among themselves, even though they did not escape the universal practice of cohabitation and often bore children. Some of them flaunted their defiance of gender norms by dressing in male attire, but it was impossible for their mutual love affairs to remain secret in any case.[41] It appears that homosexuality also flourished among the colony's many unattached men.[42] The regime of coercive lawlessness that governed Russia's penal universe thus made no attempt to regulate sexual conduct. With no respect for individuality or privacy, the result was not freedom of sexual expression but a condition of sexual impunity that mirrored the impunity of administrative rule.

Reform and Resistance: The Family

Just as sexual relations outside the civic community reflected the disregard for legality typical of the penal regime, so the constitution of

[40]Ibid., 222.

[41]A. D. Davidov, "Meditsinskii ocherk Sakhalinskoi katorgi," *Ezhenedel'nik*, no. 1 (1895), quoted in I. M. Tarnovskii, *Izvrashchenie polovogo chuvstva u zhenshchin* (St. Petersburg, 1895), 126–28.

[42]Wood, "Sex and Violence," 41. On male homosexuality and bestiality among exiles, see also Maksimov, *Sibir' i katorga*, 313. Chekhov's report on Sakhalin does not mention homosexuality.

sexual legitimacy within the community reflected the political principles that informed the body politic. Imperial Russian law established a
system of power within the family at least as autocratic as the one governing the operation of the state: the husband wielded absolute authority over the wife, and the father entirely dominated the children.
Women could not leave their households or undertake paid employment without the formal permission of father or husband, who controlled their access to the necessary official papers. No law protected
women against physical abuse short of severe bodily injury. No formal
grounds existed for legal separation; divorce, for which adultery constituted one of the few legitimate reasons, could be obtained only after
elaborate and humiliating (or duplicitous) procedures; annulment was a
rare and arduous attainment. No one of any age, male or female, could
marry without the permission of parents or other appropriate authorities. By ancient custom, women had the legal right to maintain their
own property after marriage, but they suffered severe disadvantages
when it came to inheriting family wealth.[43]

   The difficulty of obtaining a divorce or even physical separation
from an undesirable spouse became a central theme in scholarly and
journalistic descriptions of peasant life and in writings concerned with
the moral condition of educated society. Invocation of the peasant
woman's hard and immutable lot was a standard rhetorical flourish employed in legal treatments of marital issues and in such literary texts as
Anton Chekhov's "Peasants" (1897) and "Peasant Women" (1891).[44] His
eloquent "Lady with a Dog" (1898) and Leo Tolstoy's Anna Karenina
(1877) illustrate the impact of this legal stranglehold on the upper
classes. The private life of one particular literary figure came to exemplify the system's double binds. Vasilii Rozanov was a deeply religious,
if anticlerical, Christian philosopher who wrote regularly for the influential conservative newspaper Novoe vremia, where he made a specialty
of the "family question."[45] His preoccupation with this theme stemmed
from his own situation. As a young man he had married the considerably older former mistress of Fedor Dostoevsky, his literary idol. When

   [43]On the family and women's position in it, see William G. Wagner, "The Trojan
Mare: Women's Rights and Civil Rights in Late Imperial Russia," in Crisp and Edmondson, Civil Rights in Imperial Russia, 65–84. For Russian marriage law, see V. Ne-
chaev, "Brak," in Slovar' iuridicheskikh i gosudarstvennykh nauk, ed. A. F. Volkov and Iu.
D. Filipov, 1:1193–1224 (St. Petersburg, 1901).
   [44]For an example in the legal literature, see I. V. Gessen, Razdel'noe zhitel'stvo suprugov
(St. Petersburg, 1914), intro., 1–2. The literature on peasant culture is discussed in chap.
3.
   [45]Rozanov's columns were collected in Semeinyi vopros v Rossii (St. Petersburg, 1903).
For the journalistic debate on marriage, including Rozanov's contributions, see S. F.
Sharapov, ed., Sushchnost' braka (Moscow, 1901).

the romance soured, Rozanov was unable to obtain a divorce or to marry the woman he eventually considered his true wife, with whom he lived in unsanctified union and fathered four (illegitimate) children. An ardent partisan of patriarchal domesticity, hostile to secular cultural trends and to the abstract claims of the law and political ideology, Rozanov nevertheless joined with liberal critics of patriarchy in denouncing the existing system as unfair.[46]

Unlike most conservatives, who opposed divorce reform as a threat to the traditional social order—which reformers intended it to be[47]— Rozanov believed the law should accommodate the complex needs of individual lives. In this regard, he was no more inconsistent than the officials responsible for enforcing and interpreting the existing statutes, for in relation to family matters the regime played an ambiguous role consistent with its essentially arbitrary nature. Various powerful figures opposed any change in the formal rules governing spousal relations and, in particular, the right to legal separation (that is, they opposed provisions that would guarantee the wife's ability to live apart from her husband and adjudicate the guardianship of children and the allocation of material support), yet unhappy parties could obtain special dispensation from certain offices within the government, which arbitrated marital conflicts even in the absence of appropriate law. This extrajudicial discretion was exercised first by the Third Section of His Imperial Majesty's Chancellery (until its abolition under Alexander II) and then by the Ministry of Internal Affairs. Between 1889 and 1896, officials heard an average of 2,300 petitions a year (the vast majority presented by women). This strictly administrative procedure had the disadvantage, critics objected, of being inaccessible to most of the population living outside the capitals. Even when benevolent, progressives maintained, such resolutions remained entirely capricious. Administrative intervention could not dispense evenhanded and evenly distributed justice.[48]

---

[46]See, e.g., V. V. Rozanov, "A. L. Borovikovskii o brake i razvode," *Novoe vremia*, no. 9604 (November 28/December 11, 1902). Rozanov hailed domestic patriarchy as the bedrock of autocracy: A. Siniavskii, "*Opavshie list'ia*" *V. V. Rozanova* (Paris, 1982), 106.

[47]Wagner, "Trojan Mare," 66, 75, 80–81; and idem, *Marriage*, chap. 5. For an example of the liberal argument for divorce reform, see M. Filipov, "Vzgliad na russkie grazhdanskie zakony," pt. 1, *Sovremennik*, no. 2 (1861): 536–57. For the conservative defense of the existing family order, see, e.g., A. A. Bronzov (theology professor), *O khristianskoi sem'e i sviazannykh s neiu voprosakh* (St. Petersburg, 1901); and the popular opinions in "*Kak smotrit obshchestvo na tserkovnyi ili grazhdanskii brak?*" *Otvety na anketu, postavlennuiu knigoizdatel'stvom, so stat'eiu I. Tertychnogo "Ot chego ne prochnye nashi braki?*" (Kiev, 1908). Since Wagner, *Marriage*, discusses the debate in detail, I touch on it only briefly.

[48]See the historical and critical exposition in D. L., "Proekt pravil o razreshenii razdel'nogo zhitel'stva suprugov," *Vestnik prava*, no. 9 (1899): 141–53; Gessen, *Razdel'noe zhitel'stvo*, intro.; and Wagner, *Marriage*.

Not only did the government offer redress in cases of marital discord; it also provided the impetus for legal reform: while defending the letter of the law, the decisions of the Senate Civil Cassation Department opened the door to many exceptions.[49] In 1884 the State Council instructed the Ministry of Justice to revise the law on marital separation, and in 1897 the commission on civil code reform presented the council with a draft. Though the new formulation did not entirely eliminate the husband's dominant position, as critics complained, it did undermine the absolute nature of his authority and guarantee protection to subordinate family members.[50] All the ministries approved the proposal, but its acceptance was blocked by the stubborn opposition of Konstantin Pobedonostsev, the conservative overprocurator of the Holy Synod, who was also an expert in civil law. In the opinion of the liberal jurist Iosif Gessen, Pobedonostsev preferred to maintain the prevailing situation of prohibition and privileged exception rather than accept a principled resolution of the existing contradictions.[51] Thus the overprocurator not only defended the underlying religious injunctions embodied in the restrictive law but perpetuated the operative principle of autocratic governance with even more fervor than did some other high-ranking officials.

Pobedonostsev's obstruction was the last episode in the church's resistance to state-initiated proposals for marital reform.[52] Since the late eighteenth century, Gregory L. Freeze argues, the church had grown increasingly doctrinaire on matters of private sexual comportment as its organizational ability to enforce its prescriptions improved.[53] Yet whether because of Pobedonostsev's disappearance from the scene (his tenure as overprocurator ended in 1905, and he died in 1907) or of the

[49]Wagner, *Marriage*, chap. 5, notes that the department not only bent but violated statutory law in allowing marital separations. On the juridical authority of Cassation Department decisions, see K. Chikhachev, "O iuridicheskoi sile i prakticheskom znachenii reshenii kassatsionnykh departamentov Pravitel'stvuiushchego Senata," *Zhurnal iuridicheskogo obshchestva pri Imperatorskom S.-Peterburgskom Universitete*, no. 7, pt. 2 (1896): 40–56. On the Senate's role in granting peasant wives the right to separate passports (on a case-by-case basis), as well as conservative objections to this practice, see S. Kozhukhov, "O praktike Pravitel'stvuiushchego Senata po voprosu o vydache krest'ianskim zhenam otdel'nykh vidov na zhitel'stvo," *Zhurnal Ministerstvo Iustitsii*, no. 3 (1901): 158–68.

[50]For an example of the complaint that the new version still protected, even reinforced, male authority, see comments by A. L. Borovikovskii (appeals court judge, instructor at Novorossiisk University, and later senator in the Civil Cassation Department), "Konstitutsiia sem'i po proektu grazhdanskogo ulozheniia," *Zhurnal Ministerstva Iustitsii*, no. 9, sec. 2 (1902): 3–9. Information on Borovikovskii is in Wagner, *Marriage*, chap. 5.

[51]See Gessen, *Razdel'noe zhitel'stvo*, 8–9.

[52]Wagner, *Marriage*, chap. 4.

[53]For ecclesiastical court practice and increasing severity in the church's attitude toward family life, see Gregory L. Freeze, "Bringing Order to the Russian Family: Marriage and Divorce in Imperial Russia, 1760–1860," *Journal of Modern History* 62:4 (1990): 709–46. The church made annulments, separations, and divorces extremely difficult to obtain.

changed political climate after 1905, in 1910 the Synod finally approved the project for reform. The text had by then been amended through a process of political negotiation between the Synod, the Council of Ministers, and the State Duma. Its eventual adoption was facilitated by organized feminist pressure in the Duma. As a result of the compromise entailed in getting the bill passed, however, the final law did not contain the phrase "spousal separation," thus avoiding the implication that a right had been conceded. Reflecting the lawmakers' enduring distrust of abstract statement, it was officially said to concern "several changes and supplements to the existing laws on the mutual relationship between spouses and between parents and children." In substance, the law of March 12, 1914, consisted of amendments to the original restrictive civil statute, which remained in effect.[54]

Despite the liberal intentions of the civil code reformers, then—their desire to limit the range of public and domestic authority on the *Rechtsstaat* model; their attempts to liberate the secular law from the ideological influence and institutional constraints imposed by the church—they accepted the religious basis of family law and, indeed, upheld the importance of religious faith as the foundation of secular authority. The existing statutes endowed the specific ritual practices of the empire's various creeds (except for the Old Believers and the Christian sectarians) with the force of law, though these practices differed from one another and from the rites of the Orthodox Church. With a concern for detail that vitiated their larger political project, the reformers subjected each community's rules to professional scrutiny, rejecting some customs and endorsing others. In so doing, critics objected, they not only violated the universal principles represented by secular jurisprudence but infringed on the spiritual prerogatives of the ecclesiastical authorities whose judgment they intended to sustain.[55]

The same unsatisfactory compromise between secular and religious principles also characterized the 1903 text of the reformed criminal code. On the one hand, its authors believed that a model code should restrict the number of crimes classified under family law to those specific to family relations (such as adultery and child neglect). They argued that misdeeds involving violence, abuse of authority, or personal injury, which violated principles or rights inherent in general social re-

---

[54]Gessen, *Razdel'noe zhitel'stvo*, 10–11, 15; see also 40–47, on arts. 103 and 103[1] (law of March 12, 1914).

[55]A. I. Zagorovskii, "O proekte semeistvennogo prava," *Zhurnal Ministerstva Iustitsii*, no. 2, sec. 2 (1903): 55–109. An even more forceful criticism of the religious aspect of the new code appears in A. L. Borovikovskii, "Brak i razvod po proektu grazhdanskogo ulozheniia," *Zhurnal Ministerstva Iustitsii*, no. 8, sec. 2 (1902): 1–62.

lations, ought to be reclassified under the appropriate legal rubrics. On the other hand, the reformers insisted that beyond a concern with the violation of personal rights (such as the use of compulsion or fraud), the law must also protect those "religious-moral principles" (such as monogamy and the incest taboo) essential to the collective welfare. "These religious-moral principles infuse the entire system of state decrees on the marital union," the editorial commission approvingly wrote.[56]

The awkward combination of religious values and secular legal principles also characterized contemporary European legislation, which served Russians as a guide. Like their colleagues abroad, the Russians retained certain transgressions as violations of family law but decided to limit their number. The revised statutes covered marriage by force or fraud, polygamy, incestuous unions, marriages between Christians and the unconverted, marriages to persons outside the prescribed minimum and maximum age limits, and abduction for the purpose of marriage.[57] The reformers did eliminate the religious penalties (penance and monastic seclusion) still present in the existing code. Other provisions that fell by the wayside included the laws mandating the permission of parents and guardians; laws against abduction of married women as a special category; the prohibition against fourth marriages (a rule of the Orthodox Church); and the detailed regulation of correct religious ritual.[58]

Michel Foucault has suggested that the path to sexual modernity proceeds from a fixation on family and kinship to a concern with personal relations and the constitution of the self.[59] If this is so, then the 1903 code represented a step in the direction of modernity, as indeed its authors intended it to do. In all four criminal codes produced in the nineteenth century (the rejected codes of 1813 and 1903 and the actual codes of 1835 and 1845), the marriage laws rated a chapter of their own. In all except the 1903 text, this chapter contained as many statutes as those affecting all sexual transgressions combined: incest and abduction unconnected with marriage, seduction, rape, cohabitation, prostitution, procuring, and erotic practices other than intercourse.[60] In the 1903 version, by contrast, the statutes on the family constituted less than two-fifths of the total, reflecting the reformers' increased preoccupation with the interpersonal rather than the institutional aspect of sexual life.[61]

---

[56]UU, vol. 4 (1895), 154, 158–60.

[57]Ibid., 163.

[58]Cf. arts. 1549–79, Ulozhenie (1885), 741–53; and arts. 408–26, Ugolovnoe ulozhenie (1903), 146–50.

[59]Michel Foucault, The History of Sexuality, vol. 1, An Introduction, trans. Robert Hurley (New York, 1978), 106–8.

[60]Following contemporary practice, I use the term "intercourse" (polovoe snoshenie, sovokuplenie) to mean only penetration of the vagina by the penis.

[61]Sexual and family crimes occupied less than 5 percent of all articles defining specific crimes in the active codes, no more than 6 percent in the drafts (calculated by comparing

Progressive lawmakers thus attempted to reorient the law's attention (in relative terms) away from the hierarchy and rules of association defined by the family system, in the direction of individual behavior. It was not only the increased attention to interaction rather than combination, however, that distinguished the 1903 code, but also the way in which it classified and conceptualized the field of sexual crime. No previous nineteenth-century code had insisted so resolutely on the subjective and personal nature of the rights protected by the laws on sexual crime as did the 1903 version. Indeed, the history of classification reveals successive lawmakers' changing interpretations of the meaning of sexual regulation in relation to the maintenance of social order and to the codes' underlying legal values.

The story begins with a modest attempt to separate the different kinds of rights and interests implicated in the protection of family law as against the prosecution of sexual crime. Thus, the 1813 draft had drawn a principled distinction between public and private law, separating crimes affecting the political and social order from those impinging on the rights and welfare of individual persons.[62] It included the statutes on family and marriage (with the exception of adultery) in the first section; those dealing with what it called "offenses against good morals, or shameful crimes," were contained in the second section, in a chapter devoted to insults.[63] The concept of insult, however, worked to soften the contrast between the two categories. By classing sexual crime as a violation of personal rights but only in the sense that it injured one's public reputation (as an insult to one's honor), the 1813 draft failed to recognize a privacy distinct from the social context.[64]

---

the statutes on family and sexual crimes—excluding abortion and infanticide, classified as "special forms" of murder—with the total number of criminal statutes beyond the general introductory sections). The proportion holds, whatever the total number of articles in the various texts.

[62]*Proekt* (1813), tables of contents. Solodkin (*Ocherki*, 14) quotes from definitions in archival texts. D. N. Bludov, "Obshchaia ob''iasnitel'naia zapiska," in *Proekt* (1844), xiv, xliii, cites these distinctions with approval, despite his professed hostility to foreign example.

[63]Arts. 479–91, in pt. 3, chap. 3, "Ob obidakh," sec. 6, "O nakazaniiakh za obidy protiv dobrykh nravov, ili o stydnykh prestupleniiakh," *Proekt* (1813), 411–12. Similar language persisted in European legal texts throughout the century: the French "outrages aux bonnes moeurs," the German "Vergehen gegen die öffentliche Sittlichkeit," the English "crimes against public decency and good morals." The Germans seem to have abandoned the notion of modesty as the object of legal protection (*attentats à la pudeur; posiagatel'stva na tselomudrie*), and thus of sexual crime as "shameful," earlier than the French or Russians. See Wolfgang Mittermaier, "Verbrechen und Vergehen wider die Sittlichkeit. Entführung. Gewerbsmässige Unzucht," in *Vergleichende Darstellung des deutschen und ausländischen Strafrechts: Vorarbeiten zur deutschen Strafrechtsreform*, ed. Karl von Birkmeyer et al., Besonderer Teil (Berlin, 1906), 4:1–3, 9, 81; and I. Ia. Foinitskii, *Kurs ugolovnogo prava: Chast' osobennaia*, 4th ed. (St. Petersburg, 1901), 131.

[64]On the legal treatment of insult and its relation to social standing, see N. N. Rozin,

The two actual nineteenth-century codes (1835 and 1845) ignored even this hesitant step toward privatizing the status of sexual crime. The 1835 code abandoned all pretense at systematization, including the public-private distinction. It simply grouped the laws on sex in two equivalent sections, one devoted to family and kin (including adultery) and the other to "the illegal satisfaction of carnal desires" (*protivozakonnoe udovletvorenie plotskikh strastei*).[65] The 1845 version neglected the subjective dimension to an even greater extent by construing sexual crime as a species of public disorder. In this spirit, it too retained adultery as a disruption of family relations, not a personal insult. Sexual offenses unconnected to the institution of marriage it included in a grab bag of threats to "public order and decorum," along with slander, vagrancy, and violations of medical quarantine, sanitation rules, fire control, transport and postal regulations, and factory legislation.[66] It thus definitively separated these transgressions from the issue of personal rights and associated them closely with the interests of the community conceived as a whole. The enforcement of public order as it was construed in this section of the criminal code was a function intrinsic to the police (which, as we have noted, operated under their own separate statute). In reinforcing police authority with the power of the regular courts, the authors of the 1845 code may have reflected Nicholas I's preference for administrative measures, rather than legal procedures, as instruments of domestic policy.[67]

The 1845 treatment of prostitution strengthens this impression. Commercial sex had previously been penalized not in the 1835 code but only in the police regulations contained in volume 14 of the Digest of the Laws, which also included such transgressions as disorderly conduct, "luxury and prodigality," public drunkenness, and begging.[68] The

---

Ob oskorblenii chesti: Ugolovno-iuridicheskoe issledovanie, 2d ed., rev. (Tomsk, 1910), 94, 97–98, 105, 110–12.

[65]Bk. I, sec. 8, arts. 653–65 (the family), and sec. 9, arts. 666–78 (incest; illicit sexual intercourse [*blud*]; defloration and rape; and male homosexuality and bestiality); *Svod* (1835), 207–13. On this code's inconsistencies and lack of system, see Tagantsev, *Russkoe ugolovnoe pravo*, 222.

[66]Div. 8 ("O prestupleniiakh i prostupkakh protiv obshchestvennogo blagoustroistva i blagochiniia"), chap. 4 ("O prestupleniiakh protiv obshchestvennoi nravstvennosti i narushenii ograzhdaiushchikh onuiu postanovlenii"), arts. 1281–1305, *Ulozhenie* (1845), 344–49.

[67]In the view of the German scholar Mittermaier, Russian sexual legislation of this period bore "a marked moralistic police stamp": "Verbrechen," 74. On Nicholas's preference for police law, see Walicki, *Legal Philosophies*, 28.

[68]Sec. 3 ("O preduprezhdenii i presechenii besporiadkov pri uveseleniiakh i zabavakh, takzhe o presechenii iavnogo soblazna i razvrata v povedenii"), chap. 6 ("O nepotrebstve"), arts. 223–30, *Svod ustavov blagochiniia*, vol. 14 of *Svod zakonov Rossiiskoi Imperii* (St. Petersburg, 1836), 244–45.

1845 code, however, as if to underscore the primacy of the state's police function, made prostitution a crime. Indeed, the only sexual crime it excluded from the category of disorderly conduct was rape, which it treated in the section on personal health, safety, liberty, and honor, in a special subsection on "crimes against female honor and chastity."[69] It thus reintroduced the 1813 conception of sexual transgression as a kind of personal insult, but this time in relation to women alone.

The 1903 code rejected the existing classification of sexual misconduct as a "disturbance of the peace" and affirmed its character as a matter of interpersonal conflict, returning to the 1835 scheme in which sexual offenses occupied their own separate section. Questions of public morality and public standards of decorum were now invoked only in regard to immodest display and obscene expression.[70] Prostitution had disappeared from the criminal statutes with the 1866 edition and did not return.

What did these variations mean? The 1813 draft represented an attempt to bring Russian law up to European standards; the 1835 code marked the abandonment of that aim; the 1845 code embodied Nicholas I's preference for police over judicial authority; the 1903 code broke away from the existing mold. In terms of sexual crime and its relation to broader legal issues, the 1813 draft and the 1845 code represented opposite interpretations of the nature and extent of public regulation of private life: the former situated sexual conduct in the domain of interpersonal relations and individual rights; the latter went to the opposite extreme in conflating sexual misconduct with other forms of public misbehavior subject to administrative constraint. The scheme shared by the 1835 code (drafted under Alexander I but implemented by the repressive Nicholas I) and the innovative 1903 text, in which sexual crimes stood on their own, had more ambiguous political implications. Though the authors of the 1903 code thought of themselves as modernizers, critics argued that to isolate crimes according to their sexual component, rather than class them as impingements on more general social claims, was implicitly to rely on traditional or moral, rather than modern and therefore secular, conceptions of the law.[71] But the very notion

[69]On prostitution: sec. 8, chap. 4, arts. 128–88, *Ulozhenie* (1845), 344–45; on rape: div. 10 ("O prestupleniiakh protiv zhizni, zdraviia, svobody i chesti chastnykh lits"), chap. 6 ("O [*sic*] oskorbleniiakh chesti"), sec. 1 ("O prestupleniiakh protiv chesti i tselomudriia zhenshchin"), arts. 1998–2007, ibid., 514–17.

[70]Sexual crime: chap. 27 ("Nepotrebstvo"), arts. 513–29, *Ugolovnoe ulozhenie* (1903), 180–86; public decorum: chap. 13 ("O narushenii postanovlenii o nadzore za obshchestvennoiu nravstvennost'iu"), arts. 280–81, ibid., 107.

[71]See opinions of Salomon Mayer of Vienna and N. F. Vinogradskii in *UU*, vol. 6 (1897), 567–69; also V. D. Nabokov, "Plotskie prestupleniia, po proektu ugolovnogo ulozheniia," *Vestnik prava*, no. 9–10 (1902), rpt. in V. D. Nabokov, *Sbornik statei po*

of modernity was unstable: each side in this debate could cite European precedent for its own position, since the status and treatment of sexual crime was far from clear or uniform in contemporary Western law.[72]

Those who debated the question of what it meant to make sex a privileged object of legal scrutiny devoted special attention to the wording of the codes, for the status of sexual offenses was established as much in linguistic as in organizational terms. But vocabulary proved an uncertain guide to political intention. Alone among the codes and drafts, the 1835 version, oddly enough, used strictly secular language: it referred to "the illegal satisfaction of carnal desires,"[73] which implied that the rights and interests injured by the criminal act (its "object," in contemporary legal parlance) were not external to positive law but limited by published statute. Both the 1813 and 1845 texts, by contrast, used language with religious and moral connotations: "shameful crimes" (*stydnye prestupleniia*) and "offenses against good morals" (*obidy protiv dobrykh nravov*) in the first; "depraved behavior" (*razvratnoe povedenie*) and "unnatural vices" (*protivoestestvennye poroki*) in the second.[74] Surprisingly, the 1903 code did not accent the formal nature of the transgression, as the 1835 code had done, but preserved the vocabulary of moral opprobrium.

This vocabulary was at odds not only with the reformers' professed secular standards but also with their aim of introducing greater conceptual clarity into currently muddled laws. True, the confusion they could not seem to banish was not exclusively of their making. The term *nepotrebstvo*, used to mark the totality of illicit sexual acts in the 1903 code, was the Russian equivalent of the German *Unzucht*, which commentators interpreted to include any behavior—from the "pettiest [misdeed] to the most monumental perversion"—which violated accepted standards of "sexual propriety." The Russian version was equally vague.[75] The 1845 statutes, with their police approach to enforcing decorum, had used "vice" and "depraved behavior" to designate a wide range of offensive conduct, including not only particular sexual acts but also such "petty misdeeds" as being drunk in public and entering a pub-

---

ugolovnomu pravu (St. Petersburg, 1904), 90; and idem, *Elementarnyi uchebnik osobennoi chasti russkogo ugolovnogo prava*, vyp. 1 (St. Petersburg, 1903), 74–75.

[72]See Mittermaier, "Verbrechen."

[73]Title of sec. 9, bk. 1, *Svod* (1835), 210.

[74]Title of pt. 3, chap. 3, div. 6, *Proekt* (1813); title of pt. 1, sec. 1, chap. 4, div. 8, *Ulozhenie* (1845), 344.

[75]Mittermaier, "Verbrechen," 80–81, 83. Vladimir Dal', in his annotated dictionary *Tolkovyi slovar' velikorusskogo iazyka*, 4th ed., rev. (St. Petersburg, 1912), 3:1487, considered *nepotrebstvo* the equivalent of dissolute behavior, the result of "false teachings that turn a person from the true path" and lead to crime, debauch, and vice.

lic bath in the presence of the opposite sex. In that context, *nepotrebstvo* referred only to prostitution and illicit sexual intercourse.[76]

What the 1903 code accomplished was, on the one hand, the separation of general immorality from specifically sexual misconduct, and on the other, the broadening of the term *nepotrebstvo* to include all forms of illicit sex. The first step in this direction had already been taken in the 1885 edition of the code, which eliminated disorderly conduct of a general nature from the section on immorality, leaving that section's juridical construction a more narrowly sexual one.[77] It was the same complex of misdeeds, shorn of their nonsexual component, that appeared in the 1903 text under the umbrella of *nepotrebstvo*; at the same time, however, the word appeared in a restricted sense as a technical euphemism for prostitution.[78] Though the content of the new statutes stressed the subjective motives and effects of sexual crimes, the reinforced association between prohibited sexual conduct in general and the activity of public women in particular, as expressed by the term *nepotrebstvo*, suggests that the interest in public decorum had not entirely vanished from the reformed draft.

Prostitution represented women's escape from patriarchal authority and from reproductive sex. As a form of sexual activity that the family system did not control, it represented the very principle of illicit sexuality, which it was the state's direct responsibility to oversee. But the terms of sexual legitimacy within the family structure also constituted an object of public intervention. The law took a hand in defining the contours of legitimate conjugal sex and in policing the combinations that constituted the family as a social institution.

Sex within Marriage: Excess and Dearth

Sex could either reinforce the marriage bond or threaten its existence. In recognizing sexual incapacity as grounds for divorce, the church acknowledged sexual connection as a legitimate, indeed indispensable, part of married life. Sex with a person other than one's spouse (adultery) violated the exclusive sexual rights conferred by marriage and therefore also justified the dissolution of the bond. The prohibition

---

[76]See choice of terms and editorial comments on the drafts for arts. 1241–44, 1246 (1281–82, 1286 in the code), where *nepotrebnoe povedenie* is interpreted as *liubodeianie* (i.e., prohibited sexual intercourse), *nepotrebstvo* indicates prostitution, and prostitutes are called *nepotrebnye zhenshchiny* (*Proekt* [1844], 440–42).

[77]Chap. 4, *Ulozhenie* (1885), 515–24.

[78]E.g., in arts. 524, 526, *Ugolovnoe ulozhenie* (1903), 184–85.

against sexual contact with certain members of one's kinship network (incest) invalidated marriage with such partners. In all three areas, the state either reinforced church writ (providing criminal penalties for adultery and for infringements of the incest taboo) or supplied technical assistance (in determining sexual incapacity). In their joint jurisdiction over sexual violations of the marriage bond, church and state therefore engaged in complicated interactions on a shared but also contested terrain.

Incest

The criminal code and ecclesiastical law agreed in their treatment of incest as both an impediment to marriage and, outside marriage, a crime in its own right.[79] As a violation of the marriage rules, its legal meaning was clear. The law did not define the specific behavior that incest entailed as a separate transgression, but the term (krovosmeshchenie) was interpreted by legal scholars and the courts to mean the act of intercourse with a person known to be related by blood or marriage in certain specific degrees established by the church.[80] The criminal code penalized extramarital incestuous relations more harshly than the contracting of an illegal marriage but defined them more narrowly. In the most serious cases (between blood relations in direct descent), offenders were condemned to perpetual exclusion from civil life, exiled to remotest Siberia without the possibility of rejoining even the humblest juridical estate.[81]

Not all nineteenth-century European states penalized incestuous relations: the offense was absent from the codes of the Romance countries, and others provided criminal penalties only for a narrow range of prohibited ties. England, which first imposed secular sanctions for incest in 1908, saw the problem of disordered sexual relations within the family

[79]On incest as grounds for annulment in church law, see Krasnozhen, *Tserkovnoe pravo*, 219–24; also arts. 1559–62, *Ulozhenie* (1885), 748, in sec. 11 (crimes against family law), chap. 1 (crimes against the marriage union), div. 1 (illegal marriage), imposing religious as well as secular penalties. On incest as a crime in its own right, see arts. 1593–97, *Ulozhenie* (1885), 760–63, under crimes against the kinship bond, which also bore religious as well as secular penalties.

[80]N. A. Nekliudov, *Rukovodstvo osobennoi chasti russkogo ugolovnogo prava* (St. Petersburg, 1887), 498.

[81]Art. 1593 on incest in the direct line, *Ulozhenie* (1885), 760. In the Military Articles, beheading was the penalty for incest in the closest degree: arts. 173, 174, "Ustav voennyi: Voinskie artikuly" (1716), in *PSZ*, ser. 1 (St. Petersburg, 1830), 5:373; henceforth "Ustav voennyi" (1716). In a general revision of the punishment scale, later eighteenth-century laws substituted deportation for death; see legal precedent cited in art. 667, *Svod* (1835), 210.

as a consequence of poverty and ignorance, which the state, rather than the ecclesiastical authorities, must correct.[82] Mid-century Russian officials, by contrast, considered incest a particularly grave offense. The authors of the 1845 code declared that it "violates not only the laws of our faith and the positive law of all civil societies, but the very feelings instilled by nature. It is so vile that we have set a very special punishment [lifetime solitary confinement], which prohibits any direct communication with other people, even with those who have been banished for their own wicked lives or criminal acts."[83] In 1876, the Senate remarked that "incest with close blood relations not only constitutes a crime but also flouts the most basic principles of moral and religious duty; therefore those guilty of this crime not only deserve regular criminal punishment but must be removed forever from society, as persons who have definitively lost all understanding of morality and religion."[84]

Many crimes were punished with deportation; indeed, exile could result from misconduct considered not criminal at all but only in violation of communal standards. Various communities had the power—without recourse to the courts—to expel persons deemed "harmful and depraved"; vagrants could be deported by the police (or, conversely, forcibly returned to their native villages); and certain urban corporate organizations could invite the authorities to banish members said to be "depraved and corrupt."[85] Historians have estimated that before 1900, when exile was replaced in most cases by correctional labor, fully half

---

[82]Mittermaier, "Verbrechen," 145–46. See also Jacques Poumarède, "L'Inceste et le droit bourgeois au XIXe siècle," in *Droit, histoire et sexualité*, ed. Jacques Poumarède and Jean-Pierre Royer (Paris, 1987), 213–28. On England, see Victor Bailey and Sheila Blackburn, "The Punishment of Incest Act 1908: A Case Study of Law Creation," *Criminal Law Review*, 1979, 708–18. For the very reason that incest stemmed from social conditions, Mittermaier argued ("Verbrechen," 145), the criminal law was powerless to curtail it.

[83]Note to draft art. 2025, *Proekt* (1844), 672 (art. 2087 in the final code, art. 1593 in the 1885 edition). The draft version of the 1845 code had suggested lifetime solitary confinement, but the final text modified this punishment to life in Siberian exile; in 1885 offenders were still condemned to a period of solitary confinement and barred from reentry into the civil estates. Cf. *Ulozhenie* (1845), 540, and *Ulozhenie* (1885), 760.

[84]Decision 309 (1876), re Petrov, *Resheniia ugolovnogo kassatsionnogo departamenta Pravitel'stvuiushchego Senata za 1876 god* (Ekaterinoslav, 1911), 436; ruling reaffirmed in decision 23 (1888), re peasant X, *Resheniia . . . za 1888 god* (Ekaterinoslav, 1911), 60.

[85]"As a result, exactly the same severe penalty (i.e., exile to Siberia) could be administratively applied to drunkards, wastrels, tax-dodgers, fornicators, hooligans, and other social misfits as was passed by the higher judicial organs, after due process of investigation and trial, on murderers, arsonists, rapists, and thieves"; or communities could refuse to readmit criminals who had served their terms, forcing them into an exile to which they had not been sentenced: Alan Wood, "The Use and Abuse of Administrative Exile to Siberia," *Irish Slavonic Studies*, no. 6 (1985): 67–68, and 72 (on the continuation of the practice between 1864 and 1900).

the deported population of Siberia consisted not of criminals but of persons expelled by their communities for misbehavior.[86] Sexual crimes were obviously viewed as an extreme form of moral dereliction, and penalties sometimes included Siberian exile.[87] In contrast, however, to the simple term of deportation (*poselenie*) indicated in the statute on sodomy (another crime defined without reference to violence or coercion but merely in terms of the identity of adult partners), the language of the incest statute emphasized the element of permanent isolation. For incestuous relations in the closest degrees it prescribed "exile to remotest Siberia, not for settlement [the penalty for consensual sodomy] but for six years and eight months of solitary confinement in a local prison, followed by lifetime confinement at hard labor in a monastery."[88]

Despite the gravity of the offense, the number of legally penalized connections diminished over the years. Though the statutes rested on the church's definition of incest, the state nevertheless tried to narrow the range of prohibited relations, which had been set by eighteenth-century ecclesiastical law as far as the sixth degree in the collateral line. (Gregory L. Freeze argues, however, that the church had little interest in imposing strict incest taboos, since its overriding concern was to preserve existing unions.)[89] In 1810 the Holy Synod itself softened its rules to allow marriage in the fifth and sixth degree by special dispensation and loosened its control over marriage between in-laws. Secular law continued to enforce the church's absolute prohibition on marriage between collateral blood relations to the fourth degree; that is, up to and including first cousins.[90]

Because incest was defined by the church, the first nineteenth-century criminal code had not bothered to specify exactly which relations the incest statutes had in mind. But the 1845 code was particularly sensitive to hierarchy and careful to write all particulars into the law,

[86]Temporary rule no. 7, edict of June 10, 1900, no. 18777, *PSZ*, ser. 3, vol. 20, sec. 1 (St. Petersburg, 1902), 635. On expulsion, see Szeftel, "Personal Inviolability," 12. Wood ("Use and Abuse," 72–73) notes the irony that while the spouses of delinquents exiled by judicial order were not legally obliged to accompany them, the partners of administrative exiles were.

[87]E.g., a person convicted of sodomy might be deprived of civil rights and condemned to an unspecified term in Siberia; rape and nonconsensual sodomy were punished by loss of rights and exile to hard labor for specified terms. See arts. 995–96, 1523–25, *Ulozhenie* (1885), 521–22, 704–10.

[88]Art. 1593, *Ulozhenie* (1885), 760.

[89]Freeze, "Bringing Order," 727–28.

[90]See Krasnozhen, *Tserkovnoe pravo*, 220–23; the church also prohibited marriage between godparents and widowed parents of their godchildren (224). Also *O rodstve i svoistve, kak prepiatstviiakh k zakliucheniiu brakov, po deistvuiushchim zakonopolozheniiam* (Moscow, 1908), 9, 19–20, 22.

avoiding the articulation of general principles that might have left room for judicial discretion. Its authors were therefore anxious "to remove all confusion" in the case of incest by incorporating the proscribed relationships into the legal text.[91] Such specificity reflected the character of the Russian kinship system, which supplied a unique term for each particular link in the family chain. Detailed affinal terminology provided parallel, gender-marked sets for men and women which reflected the sex of the people to whom they were related, as well as their own. This folk culture suited a peasant society composed of extended, patrilocal families, in which kinship mapped a wide network of obligation and authority.[92]

The reform commission that produced the 1903 code wished to eliminate these archaic features of the law, in the interests of both modernization and secularization. European example offered no clear guidance in this case. Looking for new secular motivations to justify the remnants of old religious prescriptions, European legal scholars frequently ended in confusion about how to conceptualize the crime of incest in modern terms. Some thought incest as an independent crime had no place in the criminal code at all but ought to be broken into its component parts: abuse of authority, violation of trust, corruption of minors. Others disagreed.[93]

The Russian reformers took important steps toward reconceptualizing the crime and reassessing the gravity of the transgression. In the 1845 code and its 1885 edition, the incest statutes were included in a special section protecting kinship ties, under the broader rubric of family rights.[94] The authors of the 1903 reform dispensed with the kinship category altogether and eliminated the civil imposition of church penance. Instead, they reclassified incest among the sexual crimes unrelated to family law and supposed to injure personal, or private, rights and

[91]Note to art. 2026, *Proekt* (1844), 673. Of course, not all confusion was removed. E.g., the Senate ruled that illegitimate children must be considered the equivalent of legitimate ones in incest cases: decision 41 (1881), re Davydov, *Resheniia . . . za 1881 god* (Ekaterinoslav, 1911), 70–73. The Senate seems to have been divided, however, as to whether stepchildren were also protected. In Davydov it ruled that they were; in a later case involving both affinal and step relations, it ruled that the wife of a stepson was not the equivalent of a daughter-in-law, because her relation to her stepfather-in-law was tertiary, not secondary, and thus not legally prohibited: decision 21 (1895), re Kozlov, *Resheniia . . . za 1895 god* (Ekaterinoslav, 1911), 87–89.

[92]Paul Friedrich, "Semantic Structure and Social Structure: An Instance from Russian," in *Explorations in Cultural Anthropology*, ed. Ward H. Goodenough (New York, 1964), 140, 146, 151–52, 161.

[93]Mittermaier, "Verbrechen," 143–44, 147. For a Russian liberal's argument in favor of minimizing criminal penalties against incest, see Filipov, "Vzgliad na russkie grazhdanskie zakony," 531–33.

[94]Sec. 11, "O prestupleniiakh protiv prav semeistvennykh," chap. 3, "O prestupleniiakh protiv soiuza rodstvennogo," *Ulozhenie* (1845), 540–41.

interests. Reducing the penalties to a minimum, they also narrowed the definition of incest (following the contemporary European tendency), making intercourse illegal only between blood relations in direct descent, siblings, and a small circle of in-laws.[95] The need to maintain any sort of criminal sanction for offenses defined by the church and covered in provisions of the civil code was reaffirmed by the Ministry of Justice and the State Council, which wanted to avoid a "discordance" (razlad) between the moral positions of church and state and to provide additional deterrence for the "backward" part of the population (v srede malo razvitoi).[96]

Sexual relations between the head of the peasant household and his daughter-in-law (snokha) were in fact a common feature of the traditional extended family.[97] "Nowhere, it seems, except Russia," wrote the liberal jurist Vladimir Dmitrievich Nabokov (son of the reforming minister of justice) in dismay, "has at least one form of incest assumed the character of an almost normal everyday occurrence, designated by the appropriate technical term—snokhachestvo."[98] Russians perceived this problem as a vestige of social arrangements in older times when husbands had not yet acquired exclusive sexual access to their wives but shared them with other males in the family.[99] Aleksandr Radishchev, the outspoken critic of serfdom in Catherine the Great's reign, had invoked snokhachestvo to illustrate the abuse of power by the men in patriarchal peasant families.[100] Late nineteenth-century observers believed the custom was still alive, encouraged by the seasonal departure of young men for work outside the village. Though this form of incest was increasingly censured by the community, peasants did not consider it a

[95]On the European example, see commentary to draft arts. 440, 441 (518, 519 in code), UU, vol. 6 (1897), 613–14; on reduced limits, 619–20.

[96]Ministerstvo Iustitsii, Po proektu novogo ugolovnogo ulozheniia, 149; Zhurnal osobogo prisutstviia, 211–12; Zhurnal obshchego sobraniia, 72–74. Only a minority (3 of 48) on the council's special review commission thought the secular penalty superfluous: Zhurnal osobogo prisutstviia, 210–11.

[97]See Mikhail Kuznetsov, "Istoriko-statisticheskii ocherk prostitutsii i razvitiia sifilisa v Moskve," Arkhiv sudebnoi meditsiny i gigieny, no. 4 (1870): 120; S. Ponomarev, "Semeinaia obshchina na Urale," Severnyi vestnik, no. 1, pt. 2 (1887): 12–13; E. T. Solov'ev, Grazhdanskoe pravo: Ocherk narodnogo iuridicheskogo byta, vyp. 1 (Kazan, 1888), 10; Maksimov, Sibir' i katorga, 310–11. See also Friedrich, "Semantic Structure," 152; E. P. Busygin et al., Obshchestvennyi i semeinyi byt russkogo sel'skogo naseleniia Srednego Povolzh'ia: Istoriko-etnograficheskoe issledovanie (seredina XIX–nachalo XX v.) (Kazan, 1973), 101–2; M. O. Kosven, Semeinaia obshchina i patronimiia (Moscow, 1963), 75–76.

[98]V. D. Nabokov, "Plotskie prestupleniia," 129.

[99]Aleksandr Smirnov, "Ocherki semeinykh otnoshenii po obychnomu pravu russkogo naroda," Iuridicheskii vestnik, no. 3–4 (1877): 107. The anthropologist Paul Friedrich ("Semantic Structure," 153) notes that "the Russian patrilocal family often bore a striking resemblance to an extended polygynous one."

[100]Aleksandr Radishchev, A Journey from St. Petersburg to Moscow, trans. Leo Wiener, ed. Roderick Page Thaler (Cambridge, Mass., 1958), 135.

very serious offense, reports noted. Sons might occasionally take revenge, but peasant judges normally did not apply strict sanctions.[101]

Bearing an ancient pedigree that linked its condemnation to the critique of absolute power and feudal social relations, this particular practice obviously held potent symbolic value in the Russian context. "The
stricter the patriarchal principle," wrote an observer of peasant families
in 1887, commenting on such examples of "family despotism" (*semeinyi
proizvol*) as *snokhachestvo*, "the more powerfully was individuality suppressed."[102] By the time of the criminal code reform, however, Russians
were surely aware of the contemporary Western discourse on incest.
While Europeans and Russians at this time made similar claims for the
prevalence of incestuous habits among the uneducated population, the
Europeans ascribed them to changing social conditions, which had
eroded the traditional moral foundations of family life.[103] The crowding
and lack of privacy that supposedly encouraged such abuses in European city slums were also noted by students of urban poverty in Russia,
but it was not until after the revolution of 1905 that observers began to
blame modern life, rather than old-fashioned custom, for the scourge of
incest.[104] Once they did so, the proposed solutions involved social policy, not legal reform.

In the earlier legal debate, however, tradition remained the dominant
concern. Discussants saw themselves not as imposing order on an unruly
modern world but as modernizing an old order. They emphasized the
need to curb the exercise of unlimited power and to protect the dependent members of the family hierarchy. In considering the reform of the
incest laws, commentators typically emphasized the helplessness of
young female victims: "Who is not aware of the sadly subordinate position of the daughter-in-law in our peasant society?" asked a spokesman
for the St. Petersburg Juridical Society.[105] It was to encourage such victims to appeal for protection to the courts that the 1903 draft code ex-

---

[101]Smirnov, "Ocherki semeinykh otnoshenii," 110–12; see also Iakob Ludmer, "Bab'i
stony," pt. 2, *Iuridicheskii vestnik*, no. 12 (1884): 673–74. Incest is clearly one of the contradictions produced by the dual legal system: peasant courts sentenced offenders guilty
under the criminal statutes, which should have applied only to the operation of the regular courts.

[102]Ponomarev, "Semeinaia obshchina," 10–11.

[103]Mittermaier, "Verbrechen," 145. William Booth commented in 1890 on working-
class life: "Incest is so familiar as hardly to call for remark" (quoted in Bailey and Blackburn, "Punishment of Incest Act," 710). An indication that Russians were aware of the
European literature is found in S. S. Shashkov, "Detoubiistvo," *Delo*, no. 5 (1868): 31.

[104]Criminal statistics before the turn of the century showed that rural women were
more frequently convicted of incest than were city women: see I. Ia. Foinitskii,
"Zhenshchina-prestupnitsa," *Severnyi vestnik*, no. 3 (1893): 121–22. For the post-1905 situation, see chap. 8 below.

[105]*UU*, vol. 6 (1897), 623.

empted children under seventeen from punishment for incestuous acts. The State Council, however, changed the limit to fourteen.[106] The difference was more than a matter of minor detail; it represented a disagreement over the basic intention of the law.

What did it mean to hold girls between fourteen and seventeen (this was the archetypal case, although the child in consideration might also be a boy) criminally responsible for prohibited sexual relations with male relatives? In relation to most other sexual crimes, the State Council amended the text of the draft code to extend the period during which children were not held accountable for their own sexual conduct; when it came to incest, however, the council restricted their eligibility for special protection. By shielding the girl from criminal prosecution, the editorial commission had defined her as victim, not accomplice, in the act. The council's limitation on her immunity, by contrast, may have reflected its understanding of incest as a form of prohibited association rather than a violation of personal rights. In that sense, the council took a more conservative, or traditional, view of the juridical principle involved than did the authors of the reform. The case of incest shows how the reformers strove to reconceptualize traditional transgressions in modern terms even as they hesitated (like many of their European colleagues and, indeed, like many lawmakers today) to free themselves from essentially extralegal considerations. In part, the reformers had attempted to shift the rationale behind the incest prohibition from the protection of absolute moral principles (enshrined in religious writ) to the protection of individual welfare (exemplified by vulnerable family members who needed public safeguards against the abuse of private power).

Sexual Incapacity

The issue of sexual incapacity as grounds for divorce also entailed a conflict between personal interests and absolute moral standards but aroused greater conflict between secular and religious authorities than the problem of incest did. Here the church defined the performance of the conjugal duty—what could be said to constitute adequate sexual relations—in minimal terms so as to narrow the grounds for marital dissolution. The medical profession took a more generous position, allowing greater latitude for release from the marriage vows. The legal bureaucracy, for its part, tended to support the conservative view, defending the doctrinal over the scientific position.

[106]For the objection of the Osoboe Soveshchanie, see V. D. Nabokov, "Plotskie prestupleniia," 130. For commission discussion, see *UU*, vol. 6 (1897), 622–24.

Ambiguities in the existing law allowed free play to conflicting interpretations and authority claims. The civil code specified that sexual incapacity might constitute grounds for divorce if the condition was innate (*prirodnaia*) and already manifest before the marriage; it was the job
of the local medical officers to establish the spouse's claim.[107] But the
law did not specify what was meant by incapacity, and here the forensic
medical experts and the church disagreed. Whereas physicians considered the possibility of conception essential to the sexual purpose of marriage, the Synod focused more narrowly on the ability to perform the
act of sexual intercourse, regardless of whether children might ensue.[108]
By taking a "biological" view of the purpose of sex within marriage,
the medical establishment broadened the legitimate opportunities for
divorce and extended its own diagnostic role. The church's reluctance
to include infertility among the grievances legitimating dissolution of
the marriage bond did not constitute a more subtle understanding of the
importance of erotic satisfaction in conjugal life than did the physicians'
insistence on the centrality of producing children; rather, the clergy
were simply indisposed to enlarge the opportunities for divorce. Even
the impossibility of sexual relations did not necessarily constitute an
obstacle to continued marriage, in the eyes of the church, if disease or
mental derangement rather than organic malfunction could be identified
as its cause.[109]

On the issue of what constituted impotence, the Synod again took a
restrictive position. Only visible anatomical deficiencies, in its view,
qualified as legitimate incapacity, whereas the medical council insisted
that sexual malfunction might result from psychological causes in the
absence of organic defects, a syndrome it denominated by the technical
term "sexual neurasthenia." The Synod, for its part, objected that such
problems amounted to "an evasion of the Christian feelings of love and
respect for the other spouse essential to an Orthodox marriage." The
clergymen even suggested that medical treatment or self-discipline
might restore the sexual impulse in such cases, which therefore did not
satisfy the absolute requirements of the law. The medical council's

---

[107]See description of the investigative process in Sergei Grigorovskii, *Sbornik tserkovnykh i grazhdanskikh zakonov o brake i razvode i sudoproizvodstvo po delam brachnym* (St.
Petersburg, 1896), 170–71.

[108]D. P. Kosorotov, "O nesposobnosti k brachnomu sozhitiiu," *Zhurnal Ministerstva
Iustitsii*, no. 5 (1916): 76, 85; also Lev Bertenson, *Fizicheskie povody k prekrashcheniiu brachnogo soiuza* (Petrograd, 1917), 33–35.

[109]On the church's strict interpretation of sexual incapacity, see Freeze, "Bringing Order," 740–42. On the civil statute, see Grigorovskii, *Sbornik*, 167–68: "The law defines as
sexual incapacity only the spouse's physical inability to complete the sexual act. Therefore
any other physical or moral defect, such as insanity, madness, venereal or other diseases,
etc., de facto preventing intercourse, cannot serve as grounds for divorce."

greater flexibility also showed in its recommendation that people granted divorce for sexual incapacity be allowed to remarry once their disability had been overcome; that the mandated three-year waiting period before a suit could be brought be suspended in the case of obvious anatomical defects; and that the requirement that impotence precede the marriage be ignored when accidental or medical causes prevented a still childless couple from producing children.[110]

Discord between the ecclesiastical and medical authorities clearly led to confusion in the processing of cases in which their cooperation was at stake. In the spring of 1907 a gathering of high-ranking clergymen and physicians met to discuss their differences, and the clergymen met on their own once again in the winter of 1908–9 to review their position on the matter.[111] By the time of the joint meeting, the Synod had come around to accepting the medical interpretation of incapacity to include infertility as well as impotence. Some diehard proponents of canonical purity, including the Ministry of Justice, nevertheless continued to insist on the validity of the church's original stance. The Synod itself never relinquished its objection to the psychological definition of impotence.[112]

The clergy's concession on fertility, however, demonstrated an increased regard for the biological conditions of conjugal life, a new attitude also evident in the church's readiness to soften its uncompromising position on the question of physical disease. Whereas in 1896 the theologian Sergei Grigorovskii might insist that no illness of any kind could justify divorce, in 1909 he accepted the reasoning of the Russian Syphilological and Dermatological Society, endorsed by the state medical council, in arguing that syphilis made sexual contact impossible and threatened the welfare of the family.[113] The Syphilological Society underscored the medical, rather than moral, basis of its position by noting that divorce should be granted "regardless of whether the afflicted spouse had acquired the disease before or after the marriage or how it had been acquired (by venereal or nonvenereal means). The sad conse-

---

[110]Bertenson, *Fizicheskie povody*, 45–47, 69–76.

[111]The Ministry of Internal Affairs' medical council was represented by three members: Lev Bertenson, a physician; the psychiatrist Vladimir Bekhterev; and the court gynecologist Dmitrii Ott. Bertenson, chairman of the delegation, left a detailed account of these deliberations and an elaboration of his profession's point of view: ibid., 31–89.

[112]See, e.g., Kosorotov, "O nesposobnosti." Formerly a professor at the Academy of Military Medicine, Kosorotov was an advisory member of the medical council when he wrote this article in 1916. Bertenson, who denounced his position, called Kosorotov "more Catholic than the Pope" (*Fizicheskie povody*, 50; also 48, 69).

[113]Cf. Grigorovskii, *Sbornik*, 167–68, with the citation in Bertenson, *Fizicheskie povody*, 104 (and for medical opinions, 102–4). On the church's earlier resistance to syphilis as an excuse, see Freeze, "Bringing Order," 742.

quences for the entire family—husband, wife, and children—for neighbors, society, and future generations," the society insisted, "are one and the same."[114] In the end, the Synod recognized a medical diagnosis of syphilis as legitimate grounds for divorce, ceding another bit of ground to scientific authority in the resolution of moral plights. The guardians of scientific standards, for their part, asserted that only experts in venereal disease—not ordinary practitioners—should be granted official credence in this matter, thus reserving the right to establish a hierarchy of their own in the rendering of technical judgments. As for the legal authorities, neither the Ministry of Justice nor the commission to reform the civil code admitted syphilis as grounds for divorce, even though syphilis had been included in the law of March 12, 1914, as cause for spousal separation.[115]

Sex outside Marriage: Adultery and Fornication

Like incest, adultery was both a religious and a criminal offense in Russian law. Peter the Great included it in the Military Articles of 1716, which served as the basis for subsequent versions of the criminal code. Based on contemporary Swedish and Saxon law, these statutes did not distinguish between male and female wrongdoers and prescribed strictly secular penalties for the act. The penalties were reduced in cases in which the injured spouse forgave the offender or in which the adulterer claimed that the marriage had not provided sexual satisfaction.[116]

At the same time that the civil law established its jurisdiction over adultery, however, the church continued to exercise the authority to grant divorce. Its competence to establish adultery remained uncertain until the 1835 code, which retained adultery in the criminal statutes only in order to specify that complaints should be directed to the church, eliminating any secular penalty.[117] The authors of the 1845 code, in their typically compulsive way, insisted on reinforcing religious with secular punishment, citing existing European law as precedent.[118] Guilty

---

[114]Quoted in Bertenson, *Fizicheskie povody*, 104. On the venereal-nonvenereal distinction, see chap. 5 below.

[115]Bertenson, *Fizicheskie povody*, 101, 106–7.

[116]Arts. 169–70, "Ustav voennyi" (1716), 372. On the sources of the Military Articles, see the definitive P. O. Bobrovskii, *Voennoe pravo v Rossii pri Petre Velikom*, pt. 2, *Artikul voinskii*, vol. 1, *Vvedenie: Manifest, prisiaga i pervye chetyre glavy* (St. Petersburg, 1882), iv–vi, 4, 40–41; and idem, *Proiskhozhdenie artikula voinskogo i izobrazheniia protsessov Petra Velikogo po ustavu voinskomu 1716 g.*, 2d ed., rev. (St. Petersburg, 1881), 40–41. For their influence on later laws, see Bludov, "Obshchaia ob"iasnitel'naia zapiska," viii.

[117]Art. 663–65, *Svod* (1835), 209–10. See also M. M. Abrashkevich, *Preliubodeianie s tochki zreniia ugolovnogo prava: Istoriko-dogmaticheskoe issledovanie* (Odessa, 1904), 514–15, 525–26, 529–30.

[118]See comment to draft art. 2016, *Proekt* (1844), 669–70.

parties might now be sentenced to seclusion in a monastery or prison in addition to church-imposed penance. In two other innovations, adultery became a crime that could be prosecuted only upon complaint of the "spouse whose honor had been injured" (*oskorblennyi v chesti svoei suprug*), and a guilty spouse's unmarried partner was also liable to prison and penance.[119]

The procedural changes implemented along with the judicial reforms of 1864 finally redefined the division of labor between church and state on this matter. From then on, the offended spouse was obliged to choose between bringing the adulterer to task in criminal court (and thereby renouncing the right to divorce) and suing for divorce on grounds of adultery in ecclesiastical court (and thereby forfeiting the chance to bring criminal charges). In the first case, the satisfaction of legal redress was obviously alloyed by the need to spend the rest of one's days in indissoluble matrimony with the convicted partner. Most spouses chose the option of divorce, although few divorces were granted even on these grounds, the procedure for demonstrating infidelity being complex and humiliating for all concerned.[120]

The reform commission produced two versions of the adultery law: one in 1885 and another ten years later. In the first draft, adultery was repositioned in the chapter concerning illicit sexual acts as an offense to the individual person, not a violation of family law. Although the commission debated the option of decriminalizing adultery altogether in order to avoid continuing marriages between convicted spouses and their aggrieved (though vindicated) partners, it went only so far as to spare the nonmarried member of an adulterous pair the possibility of criminal prosecution. A number of legal experts objected, however, that adultery did not violate a person's right to sexual inviolability but, like incest, damaged values essential to the entire social body (*pravovye blaga tselogo*). For the same reason, some insisted that both adulterous partners should be penalized, since it was not merely the personal relation between husband and wife that had been violated but the very principle of marital fidelity, which society itself was obliged to uphold.[121]

A majority of the reform commission in fact favored decriminalization, which would have left adultery in the hands of the church. Both the Moscow and St. Petersburg juridical societies and a number of high-ranking judges supported this move, but the reformers ultimately decided to leave adultery in the criminal code, noting that almost all

[119]Art. 2077, *Ulozhenie* (1845), 537. See also Abrashkevich, *Preliubodeianie*, 534–35.

[120]Abrashkevich, *Preliubodeianie*, 536, 565–66; *UU*, vol. 4 (1895), 208; Grigorovskii, *Sbornik*, 149–52; and Freeze, "Bringing Order," 738–39.

[121]Abrashkevich, *Preliubodeianie*, 593–97, 599–601.

European systems considered it a crime in secular law. In the definitive 1895 draft, as well as the 1903 text, a social rather than personal interpretation of the crime triumphed. Neither draft retained the provision that adultery was actionable only upon complaint of the injured spouse. In this spirit, the statute was finally restored to its place under family law, and (despite some opposition in the commission's ranks) unmarried partners regained their vulnerability to prosecution. The convicted parties, however, risked nothing more serious than arrest, a considerable retreat from the monastery terms or prison sentences specified in the existing code.[122] When the State Council came to review the draft in 1901 and 1903, a number of its members voted in favor of decriminalization, but the majority approved the revised statute. Only a handful of councilmen held out for the stricter prison sentence.[123]

The 1903 reform effort to alleviate the problems inherent in the existing law had no effect on the fate of unhappy Russian families, since few of the 1903 statutes were ever put into effect. The outcome was different with the other crime of extramarital heterosexual love: fornication. The existing criminal code included "illegal cohabitation of an unmarried man and an unmarried woman by their mutual consent" under "provocative and depraved behavior and unnatural sins" (*soblaznitel'noe i razvratnoe povedenie i protivoestestvennye poroki*) in the chapter concerning "crimes against public morality."[124] Christians convicted of fornication in the secular courts were compelled by statute to submit to penance as imposed by the appropriate ecclesiastical authorities (the language included no mention of non-Christians). If the "sinful" union (*porochnaia zhizn'*) had produced a child, the law further stipulated, the father was compelled to support both mother and child according to his means. But a law of June 3, 1902, replaced this statute with an amendment to the civil code bearing on the status of illegitimate children, thereby decriminalizing fornication. Adultery and incest remained the only criminal forms of consenting intercourse (penis into vagina) between adults.

The shift in emphasis from illicit sex to illegitimate reproduction in fact restored the original intention of the fornication laws, which did not focus on the sexual activity itself. The purpose of Peter the Great's 1716 statute on fornication had been to place a twofold obligation

---

[122]Ibid., 603–4. See art. 359, *UU*, vol. 4 (1895), 207; and art. 418, *Ugolovnoe ulozhenie* (1903), 148. Cf. art. 1585, *Ulozhenie* (1885), 755.

[123]Abrashkevich, *Preliubodeianie*, 605–8. Among statesmen involved in this debate were Count Sergei Witte (at first against decriminalization, then for it), K. P. Pobedonostsev, V. K. Plehve, I. L. Goremykin, and Count V. N. Lambsdorff (all against it).

[124]Art. 994, *Ulozhenie* (1885), 515.

on fathers: to support their illegitimate offspring and to fulfill their promises to marry. The law addressed the father's behavior, not the mother's: "If an unmarried man sleeps with an unmarried woman," it began. In addition to providing financial support, the culprit was subject to imprisonment and church penance unless he subsequently married the mother of his child. The 1835 code incorporated this law unchanged. The 1845 code, however, altered its language and meaning by penalizing both partners in the illicit sexual connection and eliminating the man's option of marriage as an escape from punishment. The penalty itself was reduced to penance without a prison term.[125]

Despite the change in language, the statute was still interpreted as an instrument for the protection of the unwed mother's welfare. The Senate Criminal Cassation Department ruled, for example, that even a single act of sexual intercourse might legally constitute "fornication" (liubodeianie), even if it did not strictly qualify as "cohabitation" (sozhitie), so long as pregnancy had ensued. The Senate did, however, limit this protection to women of decent character who had succumbed to the power of seduction (ne razvratnaia, no obol'shchennaia zhenshchina); the benefits of the law did not extend to prostitutes or to women whose "loose conduct" (legkoe povedenie) indicated moral depravity. It did not penalize the prenuptial sexual activity of couples who later wed, nor the sexual relations of Old Believers, whose marriage rites were not recognized by the state.[126]

The criminal statute was intended to supplement the provisions of the civil code, which did not adequately provide for the support of illegitimate children.[127] When the law was changed in 1902, it therefore took the form of an amendment to the civil code. The new statute extended the rights of illegitimate children by granting legitimate status to the children of annulled marriages and compelling the married fathers of illegitimate children to provide material support (the existing law covered only the offspring of two unmarried partners). At least one commentator argued, however, that the new law made the woman's position worse by limiting the financial obligations of the father. It also made winning her case harder for the mother: in the criminal courts the plaintiff's role had been confined to bringing the suit, while the court itself summoned witnesses and presented the argument; in civil court the woman was obliged to gather witnesses and evidence on her own behalf and was also responsible for legal fees. For his part, the father,

[125]Art. 176, "Ustav voennyi" (1716), 373; arts. 669–74, Svod (1835), 211–12; art. 1289, Ulozhenie (1845), 345.

[126]Senate decisions are cited in Ulozhenie (1885), 515–17.

[127]Ibid., 518.

no longer threatened by criminal prosecution, could more easily evade the obligations imposed by the court.[128]

The history of the fornication statute does not represent a straight line of progress from archaic to modern principles. Essentially a welfare measure for the protection of mothers and at the same time for placing the burden of material support in private rather than public hands (on fathers rather than orphanages and foundling homes), the law adopted an explicitly religious formulation only in the mid-nineteenth century, at which time it also extended its punitive range to cover women as well as men. The 1902 reform once again eliminated the religious element and the inappropriate role of the criminal courts in regulating consensual moral conduct, but at the same time the new law obstructed the mother's ability to protect her own interests through the courts.

The way in which legal reformers treated the female partner in criminal intercourse between consenting adults shows the limits to the strict application of their liberal juridical principles. Of the three cases in this category—incest, adultery, and fornication—only fornication was decriminalized. By enlarging the range of sanctioned sexual conduct, this change promoted personal autonomy in private life, thus promoting the reformers' underlying goals. Yet by retaining the statutes against incest and adultery, which were intended to reinforce the solidity of social institutions (marriage and the family), while eliminating the one designed to protect the welfare of individual women, the actual changes effected in the statutory law shored up exactly those principles—respect for hierarchy and convention—challenged by the progressive 1903 code.

---

[128]Golubtsov, "Po povodu," 193–95. Wagner discusses the civil statutes on illegitimacy in *Marriage*, chap. 4.

Chapter Two

Gender and the Juridical Subject:
Sodomy, Prostitution, and Rape

Whereas the established code looked at sexual offenses primarily as violations of accepted patterns of association and hierarchy and retained certain categories justified only on religious grounds, the authors of the 1903 reformed code viewed sexual crime as an area of personal interaction, which they delineated in secular terms. The new emphasis on individual rights and bodily integrity, however, led reformers to expand rather than restrict the range of culpable sexual behavior. Nor did questions of status disappear in the new statutes on sex; instead, status came increasingly to be defined in terms of biological attributes rather than as a function of social position. The asymmetrical relationship of men and women to public life and to agencies of the state therefore emerged in sharp relief, as sexuality itself became a defining feature of civil stature.

Changing legal attitudes toward prostitution and male homosexuality underscore the gender differences that helped shape public responses to the problems of constructing a model civic arena in late nineteenth-century Russia. Insofar as the two transgressions represent the extremes of the public-private continuum, they both raised issues of public intervention and the limits of privacy in the rule-of-law state dreamed of by legal reformers. In the lawmakers' view, homosexuality tested the limits of the permissible in private conduct, prostitution its limits on public ground.

Despite these contrasts, the two activities share certain traits. Both involve sexual connection between persons who have abandoned (if only temporarily) the patriarchal framework of reproductive sexuality. Both also entail inversions of expected gender attributes, enacting, respectively, the possibility of female sexual aggression and male sexual submissiveness, thus also refusing (or ignoring) the established sex-gender hierarchy. Both practices were thought to disturb the moral foundation of social life, though on different grounds: the legal prohibition against sodomy rested on a religious base, while prostitution constituted a problem of public order (though it technically violated the law and precept against adultery and fornication). Both raised questions about the moral role of the law, about the transformation of traditional into modern forms of welfare and custodianship, about the qualifications for citizenship and the criteria for wardship. Though their legal status remained asymmetrical until the end of the old regime (sodomy subject to criminal prosecution, prostitution to administrative regulation), in both cases critics pressing for change insisted on greater formal recognition of individual rights and of personal standing before the law. In both cases the pressure failed, leaving men and women in their respective traditional places: the first compelled to adopt the role of (sexually dominant) public agent, the second compelled not to.

Sodomy: The Refusal of the Private

Proposals for reform of the antisodomy statutes, considered as part of the criminal code revision, led to a debate over the juridical limits of privacy and individual autonomy as well as to continued discussion of the place of religious injunctions in the secular law. Despite its controversial nature, however, the issue of homosexuality (specified as male) never aroused as widespread and impassioned a controversy as the reform of family and marriage law. Conservatives continued to invoke the traditional religious objections to this practice, but the tradition itself did not encourage excessive vigilance. Though Russia shared the Christian and legal heritage of Western culture, Orthodox canon law had taken a milder view of deviant sexual practices than had the Byzantine church.[1] The fact that Russian secular laws penalizing homosexuality were less severe than the European statutes upon which they were based may reflect the impact of Orthodox attitudes.[2]

Russian professionals also failed to seize on the question of homosexuality as the focus of therapeutic intervention or ideological harassment. Until the end of the nineteenth century, liberals in the legal community

[1] On the relative leniency toward homosexual activity in medieval Russian law, see Eve Rebecca Levin, *Sex and Society in the World of the Orthodox Slavs, 900–1700* (Ithaca, N.Y., 1989), 199–204.

[2] This connection has not, however, been established in the existing scholarship. On the Saxon legal authority Benedict Carpzow, on whom Peter's articles drew, see Louis Crompton, "The Myth of Lesbian Impunity: Capital Laws from 1270 to 1791," *Journal of Homosexuality* 6:1–2 (Fall–Winter 1980–81): 22.

argued for the insignificance of homosexual behavior as a marker of social deviance. The medical interpretations of homosexuality, which Foucault sees as crucial to the development of modern sexual identity in the West, had little effect on Russian legal thinking. Even among Russian physicians, as we shall see, the Western interest in the subject failed to arouse a correspondingly intense concern.[3] On the public stage, homosexuality never served as a vehicle for symbolic politics, as it did in England and Germany during the same period.[4]

The legal debates focused on the constitutional implications of the crime. The reformers' primary goal was to reconceptualize homosexuality in secular terms as a physically or psychologically injurious practice, rather than an act proscribed in religious writ. As a crime that did not involve the use of violence or coercion (homosexual rape was penalized in all the codes as analogous with heterosexual rape, but sodomy was criminal even when consensual), sodomy also challenged the limits of personal autonomy under the law. Although the statutes did not specify the gender of the partners in anal intercourse, the original context of the law revealed that the state's interest in repressing it concerned precisely the regulation of intimate relations between men. The meaning and status of the sodomy statutes therefore entered into the definition of maleness as an attribute of subject-status before the law.

Though based on biblical injunctions (filtered through the text of contemporary European codes), the earliest secular legislation on sodomy appeared as part of Peter the Great's Westernizing enterprise and served to regulate a social network, rather than to instill moral virtues of a purely personal kind. It was, moreover, originally intended to control the behavior not of civilians but of soldiers: that is, of men deprived of full civil status. Today we think of antisodomy legislation as an invasion of privacy, but the statutes that appeared in the Military Articles of 1716 applied to men in a public rather than a private capacity.[5] Soldiers belong to a closed society of males, linked by ties of hier-

---

[3]For discussion of the Western medical literature, see Robert Nye, "The History of Sexuality in Context: National Sexological Traditions" (paper presented at Princeton University, 1990).

[4]For a political interpretation of the Eulenburg affair, see Isabel V. Hull, *The Entourage of Kaiser Wilhelm II, 1888–1918* (Cambridge, 1982), chap. 5. One obvious candidate to become the Russian Eulenburg was Prince Vladimir Meshcherskii, a noted homosexual who was close to both Alexander III and Nicholas II but never became the object of a political-sexual scandal. See W. E. Mosse, "Imperial Favourite: V. P. Meshchersky and the *Grazhdanin*," *Slavonic and East European Review* 59:4 (1981), 529–47.

[5]Arts. 165–66, "Ustav voennyi: Voinskie artikuly" (1716), in *Polnoe sobranie zakonov Rossiiskoi Imperii* (hereafter *PSZ*), ser. 1 (St. Petersburg, 1830), 5:370; henceforth "Ustav voennyi" (1716). Homosexuality in pre-Petrine law is discussed in Ardalion Popov, *Sud nakazaniia za prestupleniia protiv very i nravstvennosti po russkomu pravu* (Kazan, 1904), 195–96; legal history in B. I. Piatnitskii, *Polovye izvrashcheniia i ugolovnoe pravo* (Mogilev,

archy and fraternity. In the army, sex between men affects the official system of ordered relationships; it is not simply a matter of private choice.

At first applied only to the military, Peter's code was the basis for subsequent criminal statutes that regulated conduct throughout society. As the laws were modified or criticized in the nineteenth century, the terms in which they were couched and their legal interpretation gradually altered. This development illustrates the changing conception of homosexuality among the Russian juridical elite and the evolving social implications of the practice.

Though language is often an important marker of such conceptual and social shifts, no clear movement emerges in the legal vocabulary before 1903. Except for its title, which invokes "the sin of sodomy," the chapter on sex in Peter's code contains no religious language. It also avoids proper nouns.[6] Assuming a male subject, the first article condemns "congress" (the verb *smeshat'sia*) with a "nonrational creature" (*bezumnaia tvar'*); the second proscribes the defiling (the verb *oskvernit'*) of a young man and sex between men. The accompanying German text describes the latter with the phrase *es treibet sonst Mann mit Mann Unzucht* ("two men commit illicit sex"), for which the Russian offers the even more awkward *muzh s muzhem muzhelozhstvuet* (literally, "two men engage in men-lying-with-men"). The crimes are thus depicted as actions, not fixed categories, and the acts themselves are not defined. The nineteenth-century codes substituted nouns for verbs but were no more precise in their definitions. The 1813 draft code spoke in biblical tones of unnamed but "shameful acts contrary to nature" (*protivoestestvennoe stydodeianie*). The 1835 code penalized the crimes of *muzhelozhstvo* and *skotolozhstvo* ("lying with a beast"); in the 1845 version these became "vices contrary to nature" (*protivoestestvennye poroki*), which they remained until 1917.

In contrast to the statutory language, which did not move toward greater secularization, the treatment of sodomy in succeeding texts shows the effect of modernizing juridical practices. Whereas the 1835 code followed the wording of the Military Articles with little change,[7] the 1845 statutes reflected the impact of new directions in legal thinking. As part of an interest in juridical accountability, defined according to biological standards of competence, the 1845 code added a provision

---

1910), 25–27; first modern appearance noted in *Ugolovnoe ulozhenie: Proekt redaktsionnoi komissii i ob"iasneniia k nemu* (St. Petersburg, 1895–97), 6:587; henceforth *UU*, vol. 6, (1897).

[6]Chap. 20, "O sodomskom grekhe, o nasilii i blude," in "Ustav voennyi" (1716), 370.

[7]Arts. 677–78, *Svod zakonov ugolovnykh* (St. Petersburg, 1835), 213; henceforth *Svod* (1835). The nature of the punishments altered with overall penal reforms.

that equated consensual homosexual relations with a boy or feeble-minded man with the homosexual rape of an adult.[8] Committed to a secular vision of the law, the authors of the 1903 code dropped the bestiality statute. Though they retained consensual homosexual intercourse between adult males as a crime in its own right, they eliminated the imposition of penance and stressed those aspects of the crime that showed its analogy with other forms of sexual contact rather than its peculiar status as a sinful or depraved act, historically associated with the uniquely "unnatural" act of intercourse with a beast. The 1903 code thus emphasized elements common to both heterosexual and homosexual rape: the use of physical violence or other forms of coercion; statutory rape; and sex with an unconscious or incompetent partner.[9]

But given this secular orientation, why should homosexuality remain in the law at all? Western practice provided no clear guidance. Consensual homosexuality was not criminal in the Romance countries; in others the definition of the prohibited act varied, as did the justification for its inclusion in the criminal codes.[10] Faced with such variety, the Russians defended their choices in local terms that reveal their understanding of the nature of privacy and of individual autonomy.

The vagueness of the law left ample room for creative interpretation. Even the meaning of the term *muzhelozhstvo*, which the 1903 code retained, was open to debate. Though the label seemed to designate the activity of men (given the prefix *muzhe*), the statute did not specify whether the prohibition was intended to cover only sex between men or anal sex with a woman, too. The confusion resulted from the absence of any law penalizing heterosexual contact that did not involve penetration of the vagina by the penis. Thus, the Senate ruled that the rape statutes could not be applied to cases of forcible anal intercourse between a man and a woman, which should instead be prosecuted under the statute on *muzhelozhstvo*.[11] This decision was not as strained an interpretation as it might seem, since contemporary medical texts often

---

[8]Arts. 1293–94, *Ulozhenie o nakazaniiakh ugolovnykh i ispravitel'nykh* (St. Petersburg, 1845), 346; henceforth *Ulozhenie* (1845); appearing as arts. 995–96 in N. S. Tagantsev, ed., *Ulozhenie o nakazaniiakh ugolovnykh i ispravitel'nykh 1885 goda*, 11th ed., rev. (St. Petersburg, 1901), 521–22; henceforth *Ulozhenie* (1885).

[9]Art. 516 in code (438 in draft), *Ugolovnoe ulozhenie, Vysochaishe utverzhdennoe 22 marta 1903 g.* (St. Petersburg, 1903), 181–82; henceforth *Ugolovnoe ulozhenie* (1903).

[10]Wolfgang Mittermaier, "Verbrechen und Vergehen wider die Sittlichkeit. Entführung. Gewerbsmässige Unzucht," in *Vergleichende Darstellung des deutschen und ausländischen Strafrechts: Vorarbeiten zur deutschen Strafrechtsreform*, ed. Karl von Birkmeyer et al., Besonderer Teil (Berlin, 1906), 4:147–57.

[11]Decision 642 (1869), re Mikirtumov, *Resheniia . . . za 1869 god*, pt. 2 (St. Petersburg, n.d.), 982–84. See also V. D. Nabokov, "Plotskie prestupleniia, po proektu ugolovnogo ulozheniia," *Vestnik prava*, no. 9–10 (1902), rpt. in V. D. Nabokov, *Sbornik statei po ugolovnomu pravu* (St. Petersburg, 1904), 105.

used the terms sodomy and pederasty interchangeably to indicate the act of anal intercourse, without regard to the gender of the partners involved.[12] Existing statutes also failed to specify whether the term "vice," as applied both to sex with animals and sex between men, indicated a pattern of habitual behavior or a single instance of prohibited conduct. That omission had been remedied in 1872 when the Senate rejected the argument that bestiality could be prosecuted only as a repeated performance.[13]

The editorial commission decided that tradition, as well as etymology, supported only the narrow construction of *muzhelozhstvo* to mean "anal intercourse between men" (*sovokuplenie muzhchiny s muzhchinoiu v zadnii prokhod*), excluding even other forms of same-sex pleasure.[14] To fill the remaining gap in the law, the new code introduced penalties for sexual acts other than intercourse, when accomplished without the consent of an adult woman or with underage partners of either sex.[15] It did not, however, penalize consenting sex between adult women or unwanted sexual contact with an adult man that did not include anal intercourse.

There was nothing controversial about the reformers' decision to exclude bestiality from the code. Indeed, an eighteenth-century Senate ruling, though ignored in the subsequent statutes, had recommended leniency for an act it considered depraved but not malicious.[16] Members of the post-1864 judiciary had for decades recommended decriminalization, arguing that bestiality was in itself a sign of serious mental impairment, extreme ignorance, or overwhelming physiological compulsion—all conditions that by definition deprived the perpetrator of responsibility before the law. Mild penalties should be imposed, these judges and prosecutors maintained, only in cases of public exposure. The very severity of current penalties led to frequent acquittals. In fact,

[12]Russia's leading forensic textbook on sexual crime thought women might be considered "passive pederasts": Vladislav Merzheevskii, *Sudebnaia ginekologiia: Rukovodstvo dlia vrachei i iuristov* (St. Petersburg, 1878), 203. So did Louis Martineau in *Leçons sur les déformations vulvaires et anales produites par la masturbation, le saphisme, la défloration et la sodomie,* 2d ed., rev. (Paris, 1886), 124.

[13]Decision 1362 (1872), re Kosterin, *Resheniia . . . za 1872 god* (St. Petersburg, n.d.), 1464–65. Aleksandr Lokhvitskii, in his *Kurs russkogo ugolovnogo prava,* 2d ed. (St. Petersburg, 1871), 465, was one legal authority who interpreted "vice" to imply habitual rather than single commission.

[14]The Senate's application of the sodomy statute to heterosexual partners had not been popular with judges either: e.g., the Kharkov judicial chamber asserted that "sodomy has a far less corrupting effect on female than on male victims," and therefore that the penalty was too severe to be applied when women were the victims; see N. A. Nekliudov, ed., *Materialy dlia peresmotra nashego ugolovnogo zakonodatel'stva* (St. Petersburg, 1880–83), 2:220.

[15]Arts. 513–14, *Ugolovnoe ulozhenie* (1903), 180–81.

[16]Case of Bantse-Kaspar (1786), cited in Nekliudov, *Materialy,* 2:528–31.

they argued, no criminal sanction could deter culprits benighted enough to commit such an act in the first place; only education, not punishment, these practitioners believed, could produce the socially desirable effect.[17]

The problem whether to penalize anal intercourse between consenting men was more difficult. It vexed European as well as Russian jurists.[18] The latter offered two possible reasons (or conditions) for keeping *muzhelozhstvo* in the criminal code: as an "offense to public sensibility" (*oskorblenie obshchestvennoi stydlivosti*) when accomplished before witnesses; and as an inherently "unnatural" act (*protivoestestvennoe deianie*) or vice (*porok*) incompatible with public standards of virtue even when it occurred out of sight. No matter how discreet the partners, the practice might easily become a public menace, the editorial commission argued in 1895, citing the case of European cities that had developed "a particular class of prostitutes" who specialized in the act. Their analogues could already be found in St. Petersburg, the commission warned, though as yet in less impressive numbers.[19]

Reactions to the commission's decision and reasoning were divided. Some commentators thought the draft overly severe; others thought it too lenient; all found it inconsistent. An official from the Plotsk circuit court, whose opinion was solicited by the editorial commission, questioned whether men needed protection against sexual assault. The apparent subjection involved in the act was misleading, the official noted; the "passive pederast" could not be thought a victim solely by virtue of his role, since he too was capable of seduction. Nor, he went on, was the damage that anal intercourse might inflict, such as syphilitic infection, unique; that was a risk incurred in any form of sexual contact.[20] Most members of the Moscow Juridical Society believed that homosexuality should remain in the revised code, but a sizable minority advocated decriminalization.[21] The St. Petersburg Juridical Society argued that modern jurisprudence did not punish vicious acts as such but only those that "violate the legally protected interests of individual persons or of society as a whole." Formally speaking, society members pointed

---

[17]Nekliudov, *Materialy*, 2:221–22, 516–19, 521, 523; 3:307–8.

[18]Mittermaier, "Verbrechen," 147–57.

[19]*UU*, vol. 6 (1897), 589, citing the European experts Tardieu, Legrand Du Saulle, and Casper. For more on organized male homosexuality, citing local examples, see Merzheevskii, *Sudebnaia*, 207–9.

[20]*UU*, vol. 6 (1897), 591.

[21]The vote was 23 to 9 against decriminalization: "Proekt osobennoi chasti ugolovnogo ulozheniia v obsuzhdenii Moskovskogo iuridicheskogo obshchestva," *Iuridicheskii vestnik*, no. 10 (1886): 375. For argument in favor of decriminalization, see the opinion of the Samara circuit court, cited in Nekliudov, *Materialy*, 3:307–8; V. D. Nabokov, *Elementarnyi uchebnik osobennoi chasti russkogo ugolovnogo prava*, vyp. 1 (St. Petersburg, 1903), 81; and idem, "Plotskie prestupleniia," 117, 119, 125.

out, consenting anal intercourse, like bestiality, harmed no one and should therefore not be considered a crime. They argued that homosexual intercourse between adult men should be penalized only when it took the form of male prostitution (*doma kinedov*), which should not enjoy the tolerance granted the commercialization of "natural" sex.[22]

Liberals thus criticized the commission's decision to retain homosexuality in the criminal code, on the grounds that private conduct between willing adults did not merit prosecution. Conservatives, by contrast, rejected the argument for decriminalization because they refused to recognize the distinction between public and private. The most uninhibited expression of the conservative position came from the Lutheran procurator of Archangel Province, Richard Kraus.[23] As guardian of the general interest, Kraus argued, the state had the duty to prosecute socially undesirable acts (such as the disregard of medical quarantine) even if those acts did not violate the rights of particular persons. Such reasoning applied to sexual behavior might sound out of date by modern legal standards, Kraus realized, but Russian reality, he wrote, was itself out of date: "The criminality of such acts is deeply rooted in the popular consciousness. In Russia we have not yet reached the degree of depravity [*razvrashchennost'*] of the big European centers, where acts of unnatural sexual gratification are considered an acceptable diversion [*dozvolennaia zabava*]. In narrowing the scope of criminality regarding acts of this kind, the law merely yields to the demands of changing times and customs, but it must not anticipate these demands, by legitimizing what is still considered a crime."[24]

This argument echoed the commission's own justification for retaining the sodomy statute as a defense against a corrupt modern culture responsible for such abominations as male prostitution. Less consistently, in view of the reformers' professed legal principles, they too invoked the standards of folk culture reflected in an old-fashioned way of life. The unique status of homosexuality was enshrined, so the commission claimed, in "the popular legal mind" (*v nashikh narodno-iuridicheskikh vozzreniiakh*), which distinguished it sharply from other lascivious acts.[25]

Kraus objected to the singling out of anal intercourse between males only because the law had neglected the other forms of sexual perversion. In startling contrast to the staid language, dotted with strategic Latin turns of phrase, employed by the commission and the St. Peters-

[22]*UU*, vol. 6 (1897), 590, 594.
[23]Born in 1848, of gentry background, and educated at St. Petersburg University, Kraus had served in the Ministry of Justice since 1873: *Spisok chinam vedomstva Ministerstva Iustitsii* (St. Petersburg, 1900), 406.
[24]*UU*, vol. 6 (1897), 595.
[25]Ibid., 592.

burg Juridical Society, the vocabulary of this moral purist is rich in
proper nouns and simple Russian words for despicable practices: no
"act *in os*" for Kraus, but rather the homely "*minet*," a colloquialism
missing even from the pages of Dal'.[26] Vladimir Dmitrievich Nabokov
called him "naively pornographic."[27] "Surely," asserts Kraus, "anal intercourse [*sovokuplenie v zadnii prokhod*], not with a man, for example,
but with a woman; or satisfaction obtained in a manner analogous to
intercourse, by rubbing one's member between the other person's
thighs or breasts [*sis'ki*]; or so-called oral sex [*minet*] does not deviate
any less from the natural instincts, is no less depraved, than anal intercourse between men [*muzhelozhstvo*]."[28] Though more explicit than the
ordinary legal text, Kraus's language hews to the established structure
of discourse. The subject is assumed to be male, and the male gaze is
directed outward to rest on the female object: the target of action and
desire.

Such use of language also illustrates the social and juridical distinctions embedded in contrasting linguistic fields. In all Kraus's plain-
speaking, only the penis hides in euphemism; indeed, nowhere in the
entire text of the editorial commission's deliberations is the male sexual
organ called anything but "member" (*chlen*), "genital organ" (*detorodnyi
chlen*), or "penis" in Latin script—whereas the vagina, like the thighs
and nipples of "the other person," enters proper discourse without disguise as *vlagalishche*. Linguistically, male sexual agency thus belongs to
that realm of speech which, like statutory law itself, deals in general
categories and abstract notions of individuality. By contrast, the lexicon
of female sexuality, as deployed by Kraus and other lawmakers, is specific and colloquial, appropriate in its particularity and regional variation, to the folk culture that follows custom, rather than to principles
applicable in all places and to all people. Women, like the peasantry as a
whole, embodied (literally) the conservative ideal enshrined in customary law, which reflected local rather than universal standards. The systematic prohibitions of a written code better applied to the control of
male sexual activity. At the same time, however, the content of the
code endorsed extralegal values and juridically unjustified norms that
suited its function as the guardian of female virtue and the traditional
way of life.[29]

---

[26]The St. Petersburg Juridicial Society called oral sex "akt *in os*"; see ibid. On *minet*,
see D. A. Drummond and G. Perkins, eds., *Dictionary of Russian Obscenities*, 2d ed., rev.
(Berkeley, Calif., 1980), 35.

[27]V. D. Nabokov, "Plotskie prestupleniia," 118.

[28]*UU*, vol. 6 (1897), 594.

[29]On the way language systems help establish hierarchies of cultural and social domination, as well as class-marked practices of individuation, see Basil Bernstein, "Social

According to Kraus, the ordinary Russian community did not generate the conditions under which the right to personal integrity might have replaced traditional moral categories in setting the limits to permitted sexual conduct. Not content with naming as many names as he could attach to the female body, Kraus provided an equally detailed and intimate view of corruption in everyday life. A physician of homosexual inclination takes advantage of his young wife's ignorance to engage exclusively in anal intercourse for the first two years of married life. A retired lieutenant scandalizes the provincial town in which he lives by keeping a young girl and rubbing "his member" against her thighs. Perversion, in Kraus's view, did not emerge from the crowded or benighted circumstances of rural life but intruded from the outside. His culprits are not simple peasants but men of education and urban ways.

Fraudulent intercourse, Kraus believed, was most common in the big city, whose artificial culture, he and many contemporaries thought, was destroying the natural, organic relationships of country life. Virgin girls, for example, specialized in the prostitution of their thighs, he claimed, adding the parody of true sex to the parody of love inherent in their trade. This was all very well in St. Petersburg, commented Kraus, but the distinction between public and private, a product of new-fangled ways, was foreign to the small provincial town, where even the most private act "is soon known by all and cannot remain unpunished without constituting the most obvious temptation and violating the moral feelings of others." The right to control one's own body did not include the right to use it in ways that constituted "outrageous violations of the laws of nature," Kraus declared, for these undermined "the basic principles of human existence and community."[30] Deviations must therefore be repressed by the state.

The issue at stake in these discussions was the relationship between collective welfare and individual behavior, the issue in fact central to all aspects of the criminal law. Though moral principles might be offended by a particular act, legal experts agreed, a crime must have as its target an actual object.[31] But what qualified as an "object" under the law? The editorial commission argued, for example, that bestiality injured no legally recognized rights and therefore should not be prosecuted, except as an offense to public decency. In that case the "object" protected by

Class, Language, and Socialization," in his *Class, Codes, and Control*, vol. 1, *Theoretical Studies towards a Sociology of Language* (London, 1971).

[30]Quoted in *UU*, vol. 6 (1897), 594–95.

[31]Mittermaier, "Verbrechen," 10; N. D. Sergeevskii, *Russkoe ugolovnoe pravo: Posobie k lekstiiam, chast' osobennaia*, 8th ed. (St. Petersburg, 1910), 234; I. Ia. Foinitskii, *Kurs ugolovnogo prava: Chast' osobennaia*, 4th ed. (St. Petersburg, 1901), 131. See also "Pravo i nravstvennost'," in E. N. Trubetskoi, *Entsiklopediia prava* (Kiev, 1901), 35.

law was the moral sensibility of those who witnessed the act.[32] Kraus, by contrast, assailed the legal immunity of private conduct on the grounds that no behavior could avoid having social effects. True privacy was virtually impossible in the ordinary (nonurban) Russian community, he maintained; therefore, even the potential to give offense must be subject to criminal sanction.

The equally conservative legal commentator Leonid Vladimirov took the opposite tack, arguing that all conduct, whether private or public, must be judged by its effect on personal well-being. An action might damage only its perpetrator yet still be considered a crime: "Drunkenness, professional prostitution, male homosexuality, and bestiality may be viewed either as vices that do not impinge on the rights of others or as crimes directed against the actor himself. The concept of a crime against the self depends on the existence of an interest within the self that can be violated and therefore must be protected by the laws of the state. The object capable of being violated . . . consists of the self-respect essential to every person." The "basic task of jurisprudence" was therefore individual "moral training," Vladimirov believed: "The ignorant and weak-willed person has the right to punishment, as a form of rehabilitation [perevospitanie]!"[33]

Witnesses, in this view, were beside the point. The state confronted the individual directly, not as a social being but, unmediated by the collective interest, as a being "created in the image of God," beholden to the highest authority alone.[34] The state, in Vladimirov's religious conception, functioned as an agent of the transcendent, not the secular, order. A similarly moral construction of the state's role, based on quite different premises, was presented before the St. Petersburg Juridical Society in 1902 by the eminent jurist Anatolii Koni, known for having presided at the trial of Vera Zasulich, would-be assassin of St. Petersburg's municipal governor. For Koni, secular law did not represent but replaced the traditional moral code. Formerly, as he argued in debate with his junior colleague Vladimir Dmitrievich Nabokov, people had avoided the "unnatural vices" from a knowledge of sin; later they did so from a sense of shame; now, only fear of penal sanctions kept these impulses in check. Like the heretical Skoptsy, who defied the laws of God by self-castration and undermined the social order by challenging the true faith, homosexuals, Koni asserted, were active propagandists

---

[32]UU, vol. 6 (1897), 595.

[33]L. E. Vladimirov, Ugolovnyi zakonodatel', kak vospitatel' naroda (Moscow, 1903), 124, 210–11.

[34]Ibid., 132.

for the nonprocreative life. Just as the state outlawed the Skoptsy, he concluded, so it should outlaw homosexuals.[35]

It was against both the religious and the moral arguments in the law that Vladimir Dmitrievich Nabokov argued, invoking the latest in European legal and medical thought. Even though he characterized homosexuality as socially undesirable and "deeply repugnant" to the "healthy and normal" person, Nabokov did not think it should be classified as a crime. To decriminalize such behavior did not constitute an endorsement, he pointed out, but merely acknowledged that neither moral standards nor medical norms could be imposed by legislative means. First, they were ineffective, as the case of England, cited by Havelock Ellis, amply proved; there the number of homosexuals remained unaffected by the existence of harsh laws. Second, such means were inconsistent, for consistency was in fact unthinkable; if homosexuality harmed the family, Nabokov asked, should not bachelorhood be criminal as well?

> From the moral point of view, satisfaction of the sexual urge is permissible only within marriage, and even then, only by natural means. Thus *every* extramarital satisfaction and every unnatural one within marriage should be penalized. To claim that pederasty is more repugnant than other perversions is, in the first place, a matter of subjective judgment. In the second place, such *distinguo* [*sic*] are hardly possible or permissible in this particular domain. Finally, the claim that sodomy is extremely widespread is entirely doubtful, especially if it is defined strictly as *coitus per anum*.

Scientific research showed that anal penetration was in fact relatively rare, Nabokov claimed, in comparison with other practices, such as lesbianism, that were more common yet ignored by the law.[36]

Nabokov was a staunch partisan of privacy; homosexuality should be kept out of public sight. "But the question of public morality is irrelevant to sodomy in secret, *intra muros*, so to speak. The 'public disclosure' feared by the editorial commmittee occurs only against the will and without the knowledge of the participants (rarely will anyone brag of such a 'peculiarity'). They cannot therefore be held responsible for such publicity themselves." There were additional formal obstacles

[35]Nabokov's presentation is summarized in the proceedings of the St. Petersburg Juridical Society, *Pravo*, no. 1 (1903): 50–56, Koni's reply in "Khronika: Iz deiatel'nosti iuridicheskikh obshchestv. Ugolovnoe otdelenie S.-Peterburgskogo obshchestva, 7 dekabria 1902," *Zhurnal Ministerstva Iustitsii*, no. 1 (1903): 235.

[36]V. D. Nabokov, "Plotskie prestupleniia," 102, 119–20, 125.

Caricature of Vladimir Dmitrievich Nabokov and group portrait of March 1912 congress of criminologists. Nabokov is seated third from left in the front row. *Iskry*, no. 9 (1909), no. 14 (1912). Helsinki University Slavonic Library.

to legal prosecution, in Nabokov's view. How should the act or the actor be defined? Was penetration the sine qua non? Should the law target the person who struggled helplessly against his own inclination or the one who indulged in a cold-hearted fling? Nabokov dismissed as absurd Vladimirov's notion of the subject-as-object: if it were criminal to behave without dignity or self-respect, then masturbation, too, should be a crime, not to speak of dishonesty or greed. If the state must protect the citizen's good health, then it ought logically to punish smoking and outlaw dangerous trades. The problem of intention or legal responsibility was another hornet's nest; European doctors now distinguished, at least in theory, between homosexuality as a pathology present from birth and a practice adopted at will. Either way, the legal issue remained the same: how to protect victims against abuse, not participants against their own desires.[37]

Nabokov thus distanced himself from colleagues who saw the problem in either religious or moral terms, and in so doing he made the strictest possible case for secular law based on abstract and universal principles. The culture-bound specificity of traditional moral beliefs, which conservatives hailed, served Nabokov as an argument against them. Laws on homosexuality, he noted, had changed over time and still varied across cultural traditions; there was no single, absolute standard common even to the Christian world upon which to base consistent legal norms.[38] Though he did not make the argument himself, Nabokov must have been aware of the testimony of judges from the Muslim areas of the Russian empire, who explained the impossibility of enforcing the antisodomy laws among peoples who did not disapprove of homosexual behavior.[39] The cultural relativity of moral standards in regard to sexual comportment was no secret to anyone who sat on the imperial bench.

Nabokov argued in the European liberal tradition, following the example of Paul Johann Anselm von Feuerbach, who had championed the strict separation of the moral and legal domains and had therefore excluded homosexuality, as a purely private matter, from the 1813 Bavarian criminal code.[40] Though he decorously announced his disap-

[37]Ibid., 107, 115–22.

[38]Ibid., 110–13.

[39]E.g., the president of the Irkutsk provincial court, cited in Nekliudov, *Materialy*, 3:307. For a similar argument, see Grigorii Iokhved, "Pederastiia, zhizn' i zakon," *Prakticheskii vrach*, no. 33 (1904): 872–73.

[40]V. D. Nabokov, "Plotskie prestupleniia," 111; see also Ludwig von Bar, *History of Continental Criminal Law* (1918; rpt. New York, 1969), 332. Although Feuerbach dropped homosexuality from his 1813 criminal code, he later wanted to make it subject to police regulation and reintroduced it into the 1824 revision. These reversals were not, however, well known to nineteenth-century jurists, and Feuerbach retained his reputation as a lib-

proval of homosexual behavior, Nabokov went so far as to support the decriminalization campaign of the German physician and homosexual rights activist Magnus Hirschfeld.[41] He approved of the sentiments voiced in 1901 at the international congress on criminal anthropology by a Dutch physician, who argued that homosexuality was neither inherently abnormal nor morally reprehensible. Since modern society no longer justified sex only for purposes of procreation, Nabokov agreed, there was no reason to condemn the homosexual in his search for pleasure and love.[42]

Legal prosecution, Nabokov concluded, was an exercise in futility, open only to abuse. The problem of homosexuality, if it was one, could be neither prevented nor cured, and attempts to gather evidence and bring charges in courts of law would lead only to the undesirable extension of police power.[43] To be consistent, one later commentator pointed out, the law would have to penalize all forms of vice, which "would end with the complete enserfment [zakreposhchenie] of the individual by the state."[44] This evocation of the defunct principles of pre-Emancipation Russia was no hyperbole. Liberals opposed the moralistic conception of the laws on private behavior as a vestige of the kind of political authority they opposed. The defense of privacy and secular standards of criminal law accompanied a demand for individual autonomy in the civic arena, itself still in rudimentary form before 1905.

Though there is no doubt that such political issues were embedded in the legal debates, it is harder to evaluate the product of the 1903 reform. Old-fashioned moral values have to this day not disappeared from the books in the most "enlightened" of nations. In using the laws on homosexuality and other sexual practices to protect the integrity of persons vulnerable to abuse, the reform did represent progress from the juridical point of view. In the case of heterosexual rape, for example, the new code replaced the older concern with female chastity, as an attribute central to the perpetuation of the patriarchal family system, with an interest in the physical and psychic integrity vital to personal well-being. Insofar as the revised antisodomy laws focused on the harm inher-

---

eral: Isabel V. Hull, "The Early 19th-Century German State(s) and the Criminalization of Homosexuality" (paper presented to the International Scientific Conference on Gay and Lesbian Studies, Amsterdam, December 1987), 4–6.

[41]Nabokov translated the section of his article on homosexuality, minus the moralistic disclaimer, for Hirschfeld's journal: Vladimir v. Nabokoff, "Die Homosexualität im russischen Strafgesetzbuch," *Jahrbuch für sexuelle Zwischenstufen*, no. 2 (1903): 1159–71.

[42]A. Aletrino, "La Situation sociale de l'uraniste," in Congrès international d'anthropologie criminelle, *Compte rendu des travaux de la cinquième session tenue à Amsterdam du 9 au 14 septembre 1901* (Amsterdam, 1901), 25–36; cited in V. D. Nabokov, "Plotskie prestupleniia," 113–14.

[43]V. D. Nabokov, "Plotskie prestupleniia," 124.

[44]Piatnitskii, *Polovye izvrashcheniia*, 29.

ent in certain relations of power, rather than on the moral character of particular acts, they revealed the progressive influence of secular liberalism. The reformers refused, however, to endorse a notion of privacy and individual self-determination that would have allowed adult males freely to deviate from public norms of reproductive sexuality and gender definition. Their refusal took the form of continued support for criminal sanctions on the grounds that in this case private interests could not be distinguished from social ones. Despite the existing medical literature on the pathology of male homosexuality, the lawmakers did not choose the alternative of subjecting deviant males to the extralegal constraints, administrative or therapeutic, which they favored in the case of prostitution and abortion, when female sexuality was at stake.

Prostitution: The Refusal of the Public

The moral supervision exercised by public authorities in relation to the family and male homosexuals, a concern reflected even in the reformed code of 1903, entailed the redefinition of religious norms in social welfare terms. The translation was neither complete nor consistent with a strictly liberal vision of civil society, however. Not surprisingly, the reforming lawmakers did not solve the knotty problem of how to recast or reappropriate the state's custodial role to accommodate the needs and values of a society that increasingly rejected the paternalistic tutelage of the absolutist regime. The question how to treat the subordinate members of the sexual hierarchy posed similar challenges to the modern principles of jurisprudence that inspired the 1903 reforms. Even in the new code, for example, women remained the objects of a custodial solicitude that deprived them of certain advantages which might reasonably have accrued from the weakening of the old-style patriarchal system. Like children and the mentally incompetent, women continued to be marked by special disabilities in relation to the law, as the case of prostitution graphically demonstrates.

They were not, of course, the only group in tsarist Russia to be denied full legal stature. The Emancipation had set in train a process that only gradually began to free the peasantry—the least autonomous of estates—from traditional, communal bonds (endorsed by the state in the form of peasant courts, passport regulations, and the institution of collective responsibility) which subordinated the young to the old and the individual to the whole. In addition, peasants were still governed by administrative regulation operating from outside the community.[45] Just

---

[45]On the power of the collective, see M. I. Zarudnyi, *Zakony i zhizn': Itogi issledovaniia krest'ianskikh sudov* (St. Petersburg, 1874), 55–56; B. N. Mironov, "Traditsionnoe

as the regime combined social discipline (collective and household norms) with public control (police power) to constrain peasant autonomy, so women in particular were doubly constrained in relation to their sexual position: subject to male authority within the family and to police surveillance if, as prostitutes, they left the familial and reproductive context behind. Like peasants in general, women were not entirely excluded from the general rule of law: they, too, might come before the regular courts on criminal charges. But insofar as women of any class escaped or transgressed domestic controls, they were subject to the authority of the police, which exercised the same kind of absolute, uncontestable power over them as did their husbands and fathers.[46] Indeed, police power and administrative regulation, as we have noted, constituted the autocracy's bulwark against principles of formal justice.[47]

The existing code classified sexual crimes as a species of public disorder, stressing their affinity with the kinds of transgressions usually relegated to police control. The 1903 reform, by contrast, penalized sexual crimes not as public offenses but insofar as they damaged the physical and psychological integrity or infringed the personal rights of their victims. This newer notion carried the democratic assumption that bodily inviolability was intrinsic to the constitution of the self. The code nevertheless retained the one term used to characterize the sexual crime most closely associated with public misconduct and the intervention of the police: *nepotrebstvo*—prostitution—the archetypal female sexual transgression. This choice suggests that conflicting attitudes toward female sexual agency were responsible for some of the confusion embedded in the new statutory scheme.

In Russia as well as Europe, groups deficient in the attributes of personal autonomy became the objects of special disciplinary regimes.

---

demograficheskoe povedenie krest'ian v XIX–nachale XX v.," in *Brachnost', rozhdaemost', smertnost' v Rossii i v SSSR: Sbornik statei*, ed. A. G. Vishnevskii (Moscow, 1977), 83–84. The dismantling of the peasants' peculiar legal status occurred piecemeal: universal military conscription (1874) and the abolition of the poll tax (1885) eliminated two essential markers of inferior status; collective responsibility ended with the State Council ruling of March 12, 1903; corporal punishment was finally eliminated in connection with the cancellation of redemption arrears (Manifesto of August 11, 1904); the decree of October 5, 1906, relaxed the commune's control over passports and ended the peasants' liability to prosecution for actions not defined as crimes in the statutes of the justices of the peace. See Alexander Gerschenkron, "Agrarian Policies and Industrialization: Russia 1861–1917," in *Cambridge Economic History of Europe* (London, 1966), 6:769, 785–88. Thanks to Reginald Zelnik for commenting on these points.

[46]On male domestic authority as the analogue of the administrative regime, see Pavel Kokhmanskii, *Polovoi vopros: Razbor sovremennykh form polovykh otnoshenii* (Moscow, 1912), 9.

[47]On the conflict between the "bureaucratic-administrative" concept of law and abstract legal principles, see Andrzej Walicki, *Legal Philosophies of Russian Liberalism* (London, 1987), 4.

Both exclusion from the arena of formal justice and subjection to physical violence were hallmarks of the peasants' former servile status—male and female alike. Before 1861 the rule of law and exemption from corporal punishment applied only to those members of society not subject to private authority that superseded that of the state. The freeing of the servile population from this delegated jurisdiction was the necessary precondition for penal reform.[48]

With the end of the serf-owners' authority, loss of freedom and status could for the first time replace bodily pain as a principal instrument of official disciplinary action against subordinate social groups, who now enjoyed a freedom and status of which they could be deprived. Serious restrictions on the use of physical force were introduced in 1863, as part of the Emancipation process.[49] But a paradox arose: in that year, immunity from corporal punishment, which in the eighteenth century had come to be associated with privileged status, was extended to include women of all social ranks.[50] The reform commission explained its decision as a concession to the moral vulnerability of women, for whom the shame of a public beating was thought to have particularly grave moral (and implicitly sexual) consequences harmful to the family and to the community as a whole. "A disgraced woman is a ruined creature," the reformers wrote, in terms that might have described the results of public—or illicit—sex rather than a public beating. "She usually turns to depravity, destroys family happiness, and corrupts the morals of her female intimates and friends."[51] If exemption thus marked women as in need of special care, it marked men as potentially capable of self-governance. Among peasants, the only males ex-

---

[48]D. N. Zhbankov and V. I. Iakovenko, *Telesnye nakazaniia v Rossii v nastoiashchee vremia* (Moscow, 1899), 22.

[49]Edict of April 17, 1863, no. 39504, *PSZ*, ser. 2, vol. 38, sec. 1 (St. Petersburg, 1866), 352–55. For its main provisions see N. S. Tagantsev, *Russkoe ugolovnoe pravo: Lektsii*, 2d ed., rev. (St. Petersburg, 1902), 218; also Zhbankov and Iakovenko, *Telesnye nakazaniia*, 27. On the reforms, see Bruce Friend Adams, "Criminology, Penology, and Prison Administration in Russia, 1863–1917" (Ph.D. diss., University of Maryland, 1981); and idem, "Progress of an Idea: The Mitigation of Corporal Punishment in Russia to 1863," *Maryland Historian* 17:1 (1986): 57–74.

[50]Exemption according to social rank was indicated by the phrase "iz″iatye po pravam sostoianiia ot telesnykh nakazanii"; formula taken from the edict of June 10, 1900, no. 18777, *PSZ*, ser. 3, vol. 20, sec. 1 (St. Petersburg, 1902), 631. On the origins of the class distinction, see Sergeevskii, *Russkoe ugolovnoe pravo*, 200–202; also Zhbankov and Iakovenko, *Telesnye nakazaniia*, 6–8. Women were exempted not on the basis of juridical status (*po pravam sostoianiia*) but by special dispensation (*osoboe postanovlenie*): see Senate decision 319 (1867), re Lebedeva, *Resheniia . . . za 1867 god* (Ekaterinoslav, 1910), 282–83. Until 1893, this exemption did not cover female convicts guilty of crimes committed while they served their terms (Zhbankov and Iakovenko, *Telesnye nakazaniia*, 27, 29).

[51]*Okonchatel'noe zakliuchenie komiteta dlia rassmotreniia proekta voinskogo ustava o nakazaniiakh, po voprosu ob otmene telesnykh nakazanii* (n.p., [1862]), 20. I thank Allan Pollard for providing a copy of this document.

empt from corporal punishment were those elected to public office,
which elevated them to exceptional civic standing. The rest remained
subject until 1886 to beatings of various kinds by decree of the regular
courts, and until 1904 to birching by order of the township courts or
(for soldiers) of the military authorities.[52]

The edict of 1863 did not mean that peasant women had been admitted to a higher status than men of their class but that they were
henceforth subject to physical discipline only by their husbands, not by
agents of public authority.[53] Women's exemption functioned, in fact, as
a mark of the peasant male's improved standing, constituting the family
as his inviolable domain and reinforcing the wife's "private" status. The
irony did not go unremarked: Russian women, one commentary declared in 1899, were by then "completely free from any kind of corporal punishment, *at least under the law* [*po kraine mere, po zakonu*]."[54] A
similar principle had been at work in the eighteenth-century edict that
prescribed the penalties for delinquent minors: fifteen- to seventeen-
year-olds might be whipped, ten- to fifteen-year-olds merely birched,
but children younger still were to be chastised only by their parents or
the lord of the estate. The law thus renounced jurisdiction in cases of
the greatest physical and psychic immaturity. Such offenders did not
qualify as members of the civic community but remained subordinate
to the patriarchal or private authorities in whose hands they were left.[55]

The other primary case of backhanded judicial leniency for women
in the nineteenth century was that of prostitution. Under a system of
medical control adopted from the French in 1843, prostitution was subject only to administrative regulation, not criminal prosecution (except
for the years between 1845 and 1866, when, however, the relevant statutes were largely ignored). But if prostitutes did not rate as criminals
under the law, the men who recruited, managed, and maintained them
did. The laws regulating sexual conduct cast such males (even peasant
males) almost exclusively as culprits and women as victims, granting
men the privilege of agency that women were denied. The prostitute's
clients might be detained by the police for short periods, but they rarely
were. They were never subjected to medical intervention.[56] It was the

---

[52]Zhbankov and Iakovenko, *Telesnye nakazaniia*, 28; Sergeevskii, *Russkoe ugolovnoe
pravo*, 130; even after 1904, birching continued in some penal institutions. Samuel
Kucherov, *Courts, Lawyers, and Trials under the Last Three Tsars* (New York, 1953), 103,
erroneously asserts that corporal punishment ended altogether in 1863.

[53]On the husband's prerogative to take family discipline into his own hands, see N.
Lazovskii, "Lichnye otnosheniia suprugov po russkomu obychnomu pravu," *Iuridicheskii
vestnik*, no. 6–7 (1883): 383.

[54]Zhbankov and Iakovenko, *Telesnye nakazaniia*, 29 (my emphasis).

[55]M. F. Vladimirskii-Budanov, *Obzor istorii russkogo prava*, 7th ed. (Petrograd, 1915),
364.

[56]Art. 227, *Svod ustavov blagochiniia*, vol. 14 of *Svod zakonov Rossiiskoi Imperii* (St.

sexual activity not of men but of women that the state wished to control; males were granted the option of self-regulation, in recognition of their stronger sense of self. In sum, the criminal system focused on the pimp, not the customer, while the prostitute came under police supervision.

Admission to the system of criminal justice, the ability to qualify for criminal status, was in fact a mark of acceptance into civil society, a sign of inclusion, not marginalization. As the jurist Nikolai Sergeevskii remarked in connection with the general principle of criminal liability, persons to whom the law did not apply for reasons of incompetence or immaturity were "not really members of the national community [*gosudarstvennyi soiuz*]." Freedom from punishment, he wrote, "is not some sort of privilege [*l'gota*]; it is the greatest restriction of civil rights—exclusion from the community, as subjects unfit for civic life [*grazhdanskaia zhizn'*]."[57] Insofar as sexuality acted as a defining feature of female subjectivity under the law, the legal limits on women's accountability for their own sexual choices also limited their recognition as autonomous civic agents. On the one hand, the continuing (or even extended) restrictions on women's right to sexual self-determination reflected the lawmakers' desire to protect them from the abuse of power and authority to which they were subject in the domestic and sexual realm. On the other hand, the reluctance to acknowledge their sexual agency kept them in the position of subordination that had established that very vulnerability in the first place.

Rape, Children, and the Legal Constitution of Sexual Desire

The law defined rape as sexual intercourse achieved through violence or other forms of coercion. Much nineteenth-century discussion of what constituted rape thus hinged on what constituted either legally valid resistance or consent. Since the crime was construed to apply only to women thus victimized by men (not to the objects of female sexual aggression or to the male victims of men), the discussion also focused on the question of women's responsibility for their sexual behavior or (the other side of the coin) their ability to express and satisfy sexual desire.

Eighteenth-century Russian law considered all women, no matter

---

Petersburg, 1836), 244. The only men subject by law to medical examination for venereal disease were factory workers. Here we see proof that such exams were inflicted only on persons of subordinate status and also that the brothel client's role as sexual agent is what saved him from this humiliation.

[57]Sergeevskii, *Russkoe ugolovnoe pravo*, 209.

what their physical or social condition, able to consent to sexual relations and also worthy of protection from assault. Peter the Great's Military Articles penalized rape regardless of the victim's status: "a person
of the female sex, whether old or young, married or single, on hostile
or friendly soil." However, the statutes' moral neutrality was undercut
by the procedural conditions that hemmed it in. Torn between the desire to protect female sexual integrity—an attribute central to maintaining the traditional family system—and the reluctance to elevate the interests of an individual woman over those of a man, the law at once set
severe penalties for rape and made it difficult to prove.[58] Though she
had a right to legal redress, a woman of dubious character was unlikely
to make her charges stick. Honor was not the juridical object infringed,
but it was a factor in establishing a woman's claim.

The first nineteenth-century code (1835) likewise prescribed a uniform penalty for all cases of rape but described victims according to
marital status, noting that unwanted sexual intrusion might count as
rape when committed either against a married woman or widow or
against an unmarried woman regardless of age (*devitsa sovershennoletniaia
ili maloletniaia*).[59] Although it had no legal consequences, the distinction
implied a difference in sexual experience, since the rape of an unmarried
woman was equated with defloration (*nasil'noe rastlenie*, in the statutory
text).[60]

The 1845 code altered the existing scheme in two ways: by focusing
on the injury to "female honor and chastity" (*chest' i tselomudrie
zhenshchin*), and by invoking for the first time in Russian law the notion
of noncoercive, or statutory, rape if the victim was under the "age of
consent."[61] Both innovations deflected attention from the element of

---

[58]"Ustav voennyi" (1716), 370–71, concerning arts. 167–68.

[59]Art. 675, *Svod* (1835), 212. Though the statute does not make this exception explicit,
a married woman could be the object of rape only by a man who was not her husband.

[60]*Rastlenie* is defined as intercourse that ruptures the hymen in O. A. Filippov,
"Vzgliad na ugolovnoe pravo po predmetu oskorbleniia chesti zhenshchin," *Iuridicheskii
vestnik*, no. 2 (1862): 25. Same definition supported by the Senate: see decision 1018
(1869), re Grigor'ev, *Resheniia . . . za 1869 god*, pt. 2 (St. Petersburg, n.d.), 1553–54;
decision 386 (1870), re Skirda, *Resheniia . . . za 1870 god*, pt. 1 (St. Petersburg, n.d.), 457.
See also Zuk, "O protivozakonnom udovletvorenii polovogo pobuzhdeniia i o sudebno-meditsinskoi zadache pri prestupleniiakh etoi kategorii," *Arkhiv sudebnoi meditsiny
i obshchestvennoi gigieny*, no. 2, pt. 5 (1870), 9. For debate on juridical meaning of the
term, see *UU*, vol. 6 (1897), 626–27.

[61]See commentary to arts. 1933–34 (1868–69 in actual code), *Proekt ulozheniia o
nakazaniiakh ugolovnykh i ispravitel'nykh, vnesennyi v 1844 godu v Gosudarstvennyi Sovet, s
podrobnym oznacheniem osnovanii kazhdogo iz vnesennykh v sei proekt postanovlenii* (St. Petersburg, 1871), 641–42; henceforth *Proekt* (1844). Statutory rape was first introduced into
the French penal code in 1832, and the age of consent raised from eleven to thirteen in
1863: Jacques Poumarède, "L'Inceste et le droit bourgeois au XIXe siècle," in *Droit, histoire et sexualité*, ed. Jacques Poumarède and Jean-Pierre Royer (Paris, 1987), 223.

violence to the circumstances of the victim, but they represented contradictory ways of classifying sexual crime: the first stressed social convention as the object of harm; the second, physiological well-being. By substituting age boundaries for distinctions in civil status to mark the difference between innocence and experience, and endowing the difference with practical significance, the code centered the very definition of the crime on the victim's *capacity* to consent, rather than on the circumstance of whether she had in fact done so. The status-blind eighteenth-century statutes had considered any female, "young or old," to have agreed to sexual intercourse unless it could be (laboriously) established that she had tried loudly and forcibly to ward the aggressor off. The 1845 code reversed the presumption in relation to females thought to be sexually immature, for whom failure to resist did not imply consent.[62]

The key term, which united the social and biological elements of the new approach, was chastity. It was also responsible for the considerable confusion generated by the 1845 statutes, which obliged the Senate to create a series of legal precedents that became a subject of debate and confusion in their own right.[63] Although the code defined sexual responsibility as a function of physical development, it interpreted chastity not as a verifiable physical condition but as a socially constituted attribute, like honor (with which it was paired), since its juridical meaning varied automatically with a woman's age and marital status.

The 1845 code thus failed to imagine that a girl under fourteen could be anything but a virgin and therefore described all cases of sexual intercourse (forcible or not) with girls under fourteen as the crime of "defloration." It provided no category specifically penalizing intercourse, whether consensual or coercive, with an experienced underage partner.[64] To supplement the letter of the law, the State Council ruled in 1848 that the girl's sexual experience was irrelevant in cases where rape proper was at issue.[65] In another case, the Senate chose rather to invoke the letter of the law in order to circumvent its actual wording, ruling in 1870 that "a girl under fourteen is always innocent"; thus, for her to

---

[62]Note to art. 167, "Ustav voennyi" (1716), 371; art. 1999, *Ulozhenie o nakazaniiakh ugolovnykh i ispravitel'nykh* (St. Petersburg, 1845), 514; henceforth *Ulozhenie* (1845).

[63]See the Senate's review of its own history of decisions in decision 6 (1897), re Litvin and Zhuk, *Resheniia . . . za 1897 god* (Ekaterinoslav, 1911), 12–14; and "Obozrenie ugolovnoi kassatsionnoi praktiki za 1897 g.: K voprosu o rastlenii," *Pravo*, no. 9 (February 27, 1899): 436: "In relation to Articles 1523 and 1524 [on sexual contact with underage girls], the Senate performed not only an interpretive but an entirely creative role."

[64]Art. 1523, *Ulozhenie* (1885), 704.

[65]State Council decision 32 (1848), re Bogdanovye, cited in Senate decision 96 (1876), re Vedilin, *Resheniia . . . za 1876 god* (Ekaterinoslav, 1911), 125, and Senate decision 6 (1897), re Litvin and Zhuk, *Resheniia*, 12; see also "Obozrenie ugolovnoi kassatsionnoi praktiki," 435.

have intercourse without losing her virginity was "unthinkable" (*nemy-slimo*).[66] In yet another case the Senate found a different solution for the dilemma, applying the statute on the rape of adult women to cases of forcible intercourse with girls whose experience technically deprived them of protection under the statute on coercive rape of minors (which defined such victims as virgins).[67] In this rendition, sexual experience rather than age became the mark of adulthood. But chastity in the absolute sense was not uniformly protected by law: defloration of a mature unmarried woman (older than fourteen) was not a crime in itself unless her partner or seducer was her guardian, tutor, or teacher.[68] Abuse of authority, rather than damage to the victim, was the statute's true concern.

The nation's leading law journal concluded that "the meaning of defloration is highly indeterminate and varies in relation to age; it is without doubt an unreliable juridical concept."[69] So was the notion of psychic innocence. In contrast to the statute on rape proper—which assumed that underage victims lacked sexual experience and were thus by definition deprived of virginity in the act—the statute on noncoercive intercourse with underage girls (statutory rape) defined the action as a crime only if the offender had "taken advantage" of the child's "innocence and ignorance" (*po upotrebleniiu vo zlo ee nevinnosti i nevedeniia*).[70] This qualification, however, threatened to undermine the entire purpose of the age-linked statutes by making it possible for minors to consent to sex. To avoid such an outcome, the Senate ruled that girls under ten must always be presumed innocent; only for those between ten and fourteen need the jury ask whether they knew what they were doing.[71] In the case of statutory rape, the Senate thereby introduced a

---

[66]Senate decision 386 (1870), re Skirda, *Resheniia*, 456.

[67]Decision 958 (1869), re Igorev, *Resheniia . . . za 1869 god* (St. Petersburg, n.d.), 1457–58.

[68]Senate decision 1490 (1872), re Bakanov, *Resheniia . . . za 1872 god* (St. Petersburg, n.d.), pt. 2, 1615–16. Defloration figured only as an aggravating circumstance for adult rape: see art. 1528, *Ulozhenie* (1885), 710.

[69]"Obozrenie ugolovnoi kassatsionnoi praktiki," 440.

[70]Art. 1524, *Ulozhenie* (1885), 707.

[71]In 1870 the Senate ruled that a thirteen-year-old girl might be a responsible partner in sexual intercourse because she was beyond the age of "unconditional nonaccountability" (*bezuslovnaia nevmeniaemost'*), which was ten and under: decision 1167 (1870), re Vialikov, *Resheniia . . . za 1870 god* (St. Petersburg, n.d.), 1553. But in 1875 the Senate ruled that no girl under fourteen could "understand the full consequences of the act perpetrated upon her [and] therefore defloration, even with her consent, constitutes an abuse of her innocence": decision 356 (1875), re Usachev, *Resheniia . . . za 1875 god* (Ekaterinoslav, 1911), 533. In 1876 the Senate reaffirmed the Vialikov decision and attempted to resolve the conflict in its own rulings: "[As in the case of girls under ten] intercourse with a girl [between the ages of ten and fourteen] is likewise considered defloration, since the law assumes that every girl under fourteen is virgin . . . , but the law does not recognize as a

distinction between technical virginity and the psychological maturity necessary to grasp the significance of sexual acts: a girl who had once had sexual intercourse might never have grasped the meaning of what had occurred, while a girl still physically intact might show enough sophistication to make sexual choices on her own behalf.[72] One loophole remained: consenting sex with a knowledgeable girl between ten and fourteen was not rape at all but merely illegal cohabitation.[73]

In addition to complicating the distinctions of age in relation to prohibited intercourse, the 1845 code also broke new ground in calibrating the punishment for rape to the circumstances of the crime, reserving the severest sanctions for cases resulting in the victim's death, or in the defloration of a girl under fourteen using force (rape of minors) or without the use of force but "taking advantage of her innocence and ignorance" (statutory rape).[74] The code also detailed a series of variations, in descending order of gravity, depending on the victim's age, marital status, and relation to the aggressor. What might be interpreted as consideration for the personal features of the parties to the crime in actuality reflected the same concern with position and hierarchy that permeated all other aspects of the code. Unequal treatment was, in fact, key to the system of privilege. Indeed, penalties that still varied with the offender's juridical status came under attack in the 1860s by advocates of liberal penal reform. An 1862 article on "crimes against female honor," for example, argued that "every rational system of positive law" should rest on a foundation of "universal human truth" (*obshche-chelovecheskaia pravda*), not on the distinctions of social standing and privilege (*soslovnye nachala i drugie preimushchestva*) central to Russian law.[75]

The authors of the 1903 code abandoned the notion of honor and chastity as the juridical objects of rape, but their interest in biological development led them to identify women's legal responsibility with sexual maturity and to focus on children (mostly female, but also male) as the objects of sexual crime.[76] In particular, they redefined the nature

---

general rule that girls between ten and fourteen always lack sufficient understanding of their own actions; rather the court must determine in each case whether the girl acted with awareness [*razumenie*] or not": decision 96 (1876), re Vedilin, *Resheniia*, 125; reaffirmed in decision 6 (1897), re Litvin and Zhuk, *Resheniia*, 13. Some European codes also defined degrees of maturity: see Mittermaier, "Verbrechen," 117.

[72]"Obozrenie ugolovnoi kassatsionnoi praktiki," 435.

[73]Art. 994 (1885 ed.): decision 6 (1897), re Litvin and Zhuk, *Resheniia*, 14.

[74]Arts. 1998–2007, *Ulozhenie* (1845), 514–17; see commentary on corresponding arts. 1933–44, *Proekt* (1844), 641–46.

[75]Filippov, "Vzgliad na ugolovnoe pravo," 20, 23.

[76]The question of children as subjects rather than victims of crime also attracted legal attention in this period; penalties for underage offenders were thoroughly revised in the

of the damage incurred by sexual contact with children from loss of virginity, narrowly conceived as the rupture of a specific membrane in the course of a specific sexual act, to impairment of the sexual function itself, conceived as a complex physiological system governing emotional as well as bodily health. For the first time, therefore, boys could logically qualify as victims of molestation without rape.[77] This reconception widened the scope of illegal acts to include a greater variety of erotic behavior and to cover victims already deprived of physical integrity but still susceptible to psychological abuse. Unlike the existing law, however, the new code did not penalize statutory rape more seriously than the actual rape of an adult woman but instead stressed their equivalence.[78]

The move to strengthen the protection of children or, seen from the other standpoint, to limit their autonomy had wide support in the legal community. Even those members of the judiciary who considered the state's intervention in matters of private sexual conduct among adults to be an unwarranted and "arbitrary [proizvol'noe] intrusion of the police and other authorities into the privacy of family life" endorsed the legal defense of minors.[79] The St. Petersburg Juridical Society argued for absolute protection up to the age of fourteen, when menstruation was presumed to begin for most peasant women. The authors of the 1903 reform, however, while accepting the Senate's distinction between childhood and adolescence, decided that the possibility of informed consent should be recognized only at the age of twelve, not ten. They also raised the limit of judicial discretion from fourteen to sixteen, after which, in the absence of physical force, a woman's ignorance of the sexual act could not serve as grounds for the charge of rape.[80] In reviewing the draft, the State Council extended the presumption of sexual innocence even further, from age twelve to fourteen. The final version of the 1903 code thus widened the range of state tutelage over sexual conduct and limited the span of adult responsibility.[81]

State Council ruling of June 2, 1897, no. 14233, *PSZ*, ser. 3, vol. 17 (St. Petersburg, 1900), 357–64. See also Sergeevskii, *Russkoe ugolovnoe pravo*, 224–25.

[77]See discussion in *UU*, vol. 6 (1897), 631–32.

[78]Art. 522, text and commentary in ibid., 629–30.

[79]Opinion of the procurator of the Samara circuit court, in Nekliudov, *Materialy*, 3:304.

[80]*UU*, vol. 6 (1897), 629–30; conversely, sexual experience did not disqualify her from bringing charges of rape when force had been applied (651–52).

[81]The age of consent in European law ranged from thirteen to sixteen: ibid., 574; Mittermaier, "Verbrechen," 117. Preference for a lower age of consent was not always a mark of "liberalism": Bludov's 1844 draft had suggested a cutoff at twelve, which the State Council raised to fourteen in the final 1845 code. Cf. arts. 1934 (draft) and 1999 (code), *Proekt* (1844), 642, and *Ulozhenie* (1845), 514. Mittermaier ("Verbrechen," 116)

Increased concern for the effects of age on responsibility before the courts was part of a general tendency in jurisprudence to elaborate the concept of legal accountability (*vmeniaemost'*) in biological terms. In the case of rape, it was the victim's state of mind that was decisive. Childhood, or immaturity, was the functional equivalent of unconsciousness or insanity in rendering a woman unable to show the resistance essential to proving she had been raped or to claim she had not been. Intercourse with a woman deliberately rendered unconscious, through drugs or other means, was considered an aggravated form of rape in the 1845 code.[82] The 1903 code for the first time considered the victim's insanity and other organic limitations valid grounds for the charge of rape in the absence of physical compulsion.[83] The statutory language showed the imprint of biological conceptions of personal autonomy. Even without drugs, lack of consent could be inferred when "a woman [had been] deprived of the ability to understand the nature and meaning of the deed or to govern her own actions, as a consequence of the pathological disturbance of her mental faculties [*vsledstvie boleznennogo rasstroistva dushevnoi deiatel'nosti*], unconsciousness, or mental retardation resulting from physical defects or illness."[84]

Likewise, the consequences of rape came to be construed in terms of organic rather than social effects. The same thinking explains the greatest innovation of the 1903 sexual code: its inclusion of a broad range of nonreproductive sexual practices not affecting a woman's marriageability or public standing, which were targeted not so much for their inherent immorality as for the psychological damage they caused. The new interest in the varieties of possible sexual experience accompanied the intensified concern for the welfare of children, who represented the formative, and hence most vulnerable, period of human development.[85]

---

notes a tendency for the age of consent in European law to rise over the course of the nineteenth century, reflecting a growing concern for the special vulnerability of adolescents, linked precisely to the process of sexual maturation.

[82] Art. 2001, *Ulozhenie* (1845), 515.

[83] The issue of what forms of compulsion other than violence qualified the offender's behavior as rape is outside this discussion, but it should be noted that judges were sensitive to the vulnerability of peasant wives to sexual harassment by their powerful fathers-in-law; they could not bring charges of rape in the absence of physical coercion, though they could theoretically find redress under the statutes on incest. See the comments of the procurator of the Moscow judicial chamber in Nekliudov, *Materialy*, 2:289–90.

[84] Art. 520, *Ugolovnoe ulozhenie* (1903), 183. See also commentary on draft art. 442 in *UU*, vol. 6 (1897), 633–35.

[85] V. D. Nabokov, among others, stressed the long-term consequences of early sexual experience: "Lewd acts [*bludnye deistviia*] often cause permanent damage and sometimes lead to the [child's] complete and irrevocable degradation [*bespovorotnoe padenie*]" ("Plotskie prestupleniia," 95). "Degradation" is used in the same context by Lokhvitskii,

The only forms of "unnatural" sex penalized in existing Russian law were anal intercourse between males of any age and intercourse between man and beast. Before the changes introduced in 1845, the statutes singled out the use of coercion or force for special attention but did not provide additional sanctions for adults who involved young boys in the act. The 1845 code, as we have seen, introduced distinctions based on the status of the passive partner, penalizing nonviolent anal intercourse with a feebleminded man or an underage (*maloletnii*) boy on a par with the use of force in the same act between grown men and with the most serious forms of rape.[86] The authors of the 1903 text proceeded further in modifying the laws on intercourse between men to conform to the model established by the laws on heterosexual contact. Though the code retained the statute against homosexuality, it was reconceived as one of a number of proscribed sexual acts supposed to inflict damage on victims of both sexes.[87] Thus, distinctions formerly relevant only to heterosexual rape now applied to rape between males as well. The penalty for forcible anal intercourse with adolescent boys between fourteen and sixteen, or for seducing them into the act for the first time, was exactly the same as that for heterosexual intercourse with females under the same conditions. The parallel was not complete, however; the abduction for sexual purposes of a willing but inexperienced adolescent between fourteen and sixteen was a special crime only in the case of girls. Likewise, intercourse with a girl under fourteen, regardless of her state of mind, earned a harsher sentence than comparable behavior in relation to a male.[88]

The 1903 text also surpassed the 1845 code in the variety of sexual behavior it subjected to criminal sanctions. The only instance of hetero-

---

Kurs, 593; *sittlich verdorben* is Mittermaier's phrase ("Verbrechen," 115). On the absolute protection accorded "chastity" in children, see Nabokov, *Elementarnyi uchebnik*, 77. Mittermaier, "Verbrechen," 114–15, 117, notes a European tendency to penalize "unnatural" acts in connection with greater attention to child sexual abuse.

[86]Art. 1294 made statutory and actual homosexual anal rape equal to the violent defloration of a girl under fourteen (art. 1998) or the nonviolent defloration of a girl under fourteen by a man under whose power or authority she fell (art. 1999, para. 2) or rape followed by the death of the victim (art. 2002), each of which incurred ten to twelve years' hard labor; its exact heterosexual equivalent, intercourse with an inexperienced girl under fourteen accomplished by an unrelated male who used seduction rather than brute force (art. 1999, para. 1) incurred only eight to ten years: *Ulozhenie* (1845), 346, 514–15.

[87]When the Moscow Juridical Society debated the issue of sexual protection for children, A. M. Fal'kovskii noted that society placed greater demands on the sexual purity of brides than of grooms and that girls therefore needed more protection; V. M. Przheval'skii (supported by a majority of those present) countered that such a distinction held for intercourse but not for other forms of sexual contact, whose effects were moral and psychological. See "Proekt osobennoi chasti ugolovnogo ulozheniia v obsuzhdenii Moskovskogo iuridicheskogo obshchestva," 374.

[88]V. D. Nabokov, *Elementarnyi uchebnik*, 77–78. Cf. art. 516 on male homosexuality with arts. 520 and 522 on heterosexual rape: *Ugolovnoe ulozhenie* (1903), 181–84.

sexual conduct penalized in existing law (whether incest, adultery, fornication, or rape) involved the intrusion of a man's penis into the vagina of a woman or girl who was not his wife.[89] All other forms of
sexual contact between men and women were legal, unless they occurred in a public setting. This state of affairs, the 1903 reformers complained, left Russia the only European nation without penalties for the
majority of sexual acts ("indecent acts," *attentats à la pudeur, unzüchtige
Handlungen*).[90] The draft therefore established a sequence of offenses
parallel to the rape sequence but lower on the punishment scale: "lewd
acts" (*liubostrastnye deistviia*) with children under fourteen, with or without consent; with adolescents between fourteen and sixteen against their
expressed desire or with their uninformed, naive compliance. These offenses were aggravated in all cases by the perpetrator's position of authority or by the use of force or threats of force. Special penalties were
also set for deflowering a girl or woman without intercourse and for
unwanted sexual contact of any kind with women over sixteen.[91] These
provisions covered female as well as male offenders.

Though anal intercourse between consenting adult men constituted
its own special crime, other kinds of erotic contact with an *unwilling*
adult male did not, in the drafters' view, merit criminal sanction.[92]
Thus, the reformers did not extend the analogy between boys and girls
to men and women. If unwanted sexual contact other than anal rape
between grown men should involve either "dishonor or coercion," the
reformers believed, it ought to be penalized under those categories for
the nonsexual damage sustained. If, however, the object of such advances did not feel himself besmirched, then no harm could be said to
have been done.[93] The act was objectionable, if at all, because it was
degrading or unsolicited, not because it was sexual. The law insisted,

[89]On the impossibility of charging marital rape, see Lokhvitskii, *Kurs*, 591; also *UU*,
vol. 6 (1897), 642–43.

[90]*UU*, vol. 6 (1897), 566. The absence of such legal protection had been the subject of
commentary among jurists and members of the judiciary since the 1860s; see Nekliudov,
Materialy, 2:219; 3:372–73. For terms, see Mittermaier, "Verbrechen," 81.

[91]On defloration without intercourse, see art. 515, pt. 3, *Ugolovnoe ulozhenie* (1903),
181. On manual defloration, see V. D. Nabokov, "Plotskie prestupleniia," 100; idem,
Elementarnyi uchebnik, 80; *UU*, vol. 6 (1897), 562–64; and Foinitskii, *Kurs*, 145. The problem of how to classify and penalize acts that destroyed virginity without intercourse gave
rise to an enormous amount of discussion; see, e.g., Nekliudov, *Materialy*, 2:288–89. The
relevant Senate decisions are 1018 (1869), re Grigor'ev, *Resheniia*, 1551–56; 386 (1870), re
Skirda, *Resheniia*, 456–57; 330 (1872), re Sukhino, *Resheniia . . . za 1871 god*
(Ekaterinoslav, 1910), 287–91. On unwanted contact with adult women, see arts. 513–15,
Ugolovnoe ulozhenie (1903), 180–81. On the new code's use of the concept of *liubostrastnye
deistviia*, see O. Portugalov, "Prestuplenie nepotrebstva po novomu ugolovnomu ulozheniiu," *Iuridicheskaia gazeta*, no. 62 (Sept. 14, 1903): 2.

[92]Consensual lesbian sex, however, did not constitute a special crime; see V. D.
Nabokov, "Plotskie prestupleniia," 110.

[93]*UU*, vol. 6 (1897), 577.

however, that women must be protected in their capacity as sexual beings. It placed the crime of subjecting a woman to unwanted bodily
contact among the statutes on child sexual abuse.[94] This arrangement
underscored the extent to which the adult woman was likened to the
child of either sex, the immature of the species, for in both cases issues
of health and physical integrity aroused the lawmakers' concern. Passage to adult status elevated the male but not the female to a different
category. On the analogy with statutory rape, the antisodomy statute
made it impossible for the passive partner to consent to penetration. In
assuming a female sexual posture, the "victim" thus acquired a female
subject-position before the law, as the object of judicial solicitude. Except when engaged in conduct that mimicked vaginal penetration, however, grown men were allowed to define their own relation to sexual
experience in a way that grown women were not.

Furthermore, adult men did not have to endure public supervision of
their sexual conduct outside the framework of the law. Along with
children, women monopolized the interest of these lawmakers, who
now invoked public health—the administrative regulation of bodies—
instead of religious precepts or patriarchal values to justify the custodial
function of the law. The old police orientation of the 1845 code, designed to sanitize the public arena, thus crept back into the reformed
statutes under new ideological auspices. The purpose of the law, the St.
Petersburg Juridical Society argued, was to "shield [the child's] immature organism . . . [not only] against direct [and] lascivious manipulation but also in general against whatever influence might prematurely
arouse the sexual instinct and artificially encourage sexual desire." The
threats included all sources of moral corruption—pornography and
cynical or brazen language as well as public sexual behavior—which
must be regulated, the society believed, not only in the interests of the
child but in the national interest (v interesakh obshchegosudarstvennykh).[95]
And the quintessential lewd and corrupting public influence, traditionally the target of state intervention, was prostitution—the quintessentially female sexual crime.

Prostitution: The Crime Which Is Not One

More congenial to the absolutist Russian regime than the rule of law
(Rechtsstaat) was the eighteenth-century notion of the police, or custodial state (Polizeistaat), whose role was to promote the general well-

[94]Literally, between arts. 513 and 514.
[95]Quoted in UU, vol. 6 (1897), 571–72.

being through surveillance and active social intervention. These ambitious "police" functions, as initially conceived, were to be not so much repressive as educative and formative. In the classic tradition of enlightened despotism and the domestic tradition of paternalistic rule, the Russian state had been quick to claim jurisdiction over the medical terrain. Following the European example, Catherine the Great established Russia's first public health institutions and laid the legislative foundation for the "policing" of public behavior and moral standards. She opened the first venereal disease wards for women in 1763, for example, under the auspices of the St. Petersburg Admiralty Hospital.[96]

By the nineteenth century, however, European states had relinquished much of the burden of initiative and regulation to society itself, restricting the notion of police to the limited, technical agency by which the state enforced the laws.[97] Despite this shift, continental nations continued to use the police as an instrument of social regulation in some cases, enlisting the authority of medicine and social science in defense of this traditional custodial role. The prime example of such an adaptation was the system of state-regulated prostitution—opposed by European social reformers on the grounds that it violated the civil rights of the supervised persons while defeating the public health goals it claimed to achieve.[98]

In Russia the regulatory functions of the police were still extensively employed. In principle, legal experts distinguished between the function of the courts, which penalized actions that violated written laws,

[96]On the notion of the "police" state, see Alice Erh-Soon Tay and Eugene Kamenka, "Public Law–Private Law," in *Public and Private in Social Life*, ed. S. I. Benn and G. F. Gaus (London, 1983), 75; Reinhold August Dorwart, *The Prussian Welfare State before 1740* (Cambridge, Mass., 1971), 3; Hans Maier, *Die ältere deutsche Staats- und Verwaltungslehre (Polizeiwissenschaft): Ein Beitrag zur Geschichte der politischen Wissenschaft in Deutschland* (Neuwied am Rhein/Berlin, 1966), 116–30. I thank Isabel V. Hull for these references. In the Russian context, see Marc Raeff, *The Well-Ordered Police State: Social and Institutional Change through Law in the Germanies and Russia, 1600–1800* (New Haven, Conn., 1983), pt. 3; and V. F. Deriuzhinskii, *Politseiskoe pravo: Posobie dlia studentov*, 3d ed. (St. Petersburg, 1911), 6–8. For the *Polizeistaat*'s role in public health, see George Rosen, "The Evolution of Social Medicine," in *Handbook of Medical Sociology*, ed. Howard E. Freeman, Sol Levine, and Leo G. Reeder (Englewood Cliffs, N.J., 1963), 24–25; John T. Alexander, "Catherine the Great and Public Health," *Journal of the History of Medicine* 36:2 (1981): 185–204; and Ia. A. Chistovich, *Istoriia pervykh meditsinskikh shkol v Rossii* (St. Petersburg, 1883), 452.

[97]This interpretation is outlined in Deriuzhinskii, *Politseiskoe pravo*, 9–11, 15. See also Lucie Luig, *Zur Geschichte des russischen Innenministeriums unter Nikolaus I* (Wiesbaden, 1968), 50. The "polite" citizen (*der polizierte [oder polite] Mensch*) of the modern polity was one who "policed" himself, whose manners showed he had internalized behavioral norms formerly promoted by the state to ensure public order: Maier, *Die ältere deutsche Staats- und Verwaltungslehre*, 120, 127–29.

[98]Alain Corbin, *Les Filles de noce: Misère sexuelle et prostitution (19ᵉ et 20ᵉ siècles)* (Paris, 1978).

and that of the police, which disciplined behavior thought to promote criminal activity but not criminal in itself.[99] In relation to sexual conduct, one such authority explained, "criminal retribution . . . follows upon the commission of an immoral action forbidden by law," whereas police interventions "suppress behavior and remove conditions that exert a negative influence on moral development." Police targets were "enticements that disturb the moral order, interfere with the moral development of the masses, and often constitute the main cause for the moral downfall of individual persons."[100]

Prostitution was one of those "enticements" against which the police were empowered to act by the eighteenth-century laws still embodied in the Regulations on Public Order. These regulations—which Vladimir Dmitrievich Nabokov, a prominent enemy of administrative rule, termed a "monstrous fossil"[101]—allowed the police to punish both parties in the commercial exchange of sex by imposing fines or brief terms of detention, because the trade, like public drunkenness, was considered a noncriminal activity that created an atmosphere conducive to immorality and crime. The police were also enjoined to round up any "vagrant girls" of "dubious character" from the poorest and most disreputable classes (*brodiachie, podlye i podozritel'nye devki*) who might be suspected of harboring venereal disease.[102] Beginning in 1843, prostitutes were subject to registration and medical inspection according to rules enforced by the police or specifically empowered local bodies, under the authority of the Ministry of Internal Affairs. Russia thereby acquired the system of medically supervised brothels (*doma terpimosti*) already familiar in France (*maisons de tolérance*).[103] Wherever it was employed, the use of discipline to promote public welfare and public health in particular was a legacy of the eighteenth-century concept of police: the curative and the punitive were considered not contradictory but mutually reinforcing impulses.[104]

[99]Deriuzhinskii, *Politseiskoe pravo*, 310.

[100]I. E. Andreevskii, *Politseiskoe pravo*, 2d ed., rev. (St. Petersburg, 1876), 2:7, 9.

[101]V. D. Nabokov, "Plotskie prestupleniia," 137.

[102]Arts. 223–30, *Svod ustavov blagochiniia*, 244–45; see also A. I. Elistratov, "Prostitutsiia v Rossii do 1917 goda," in *Prostitutsiia v Rossii*, ed. V. M. Bronner and A. I. Elistratov (Moscow, 1927), 19.

[103]See G. M. G[ertsenshtein], "Prostitutsiia," in *Entsiklopedicheskii slovar' Brokgauz-Efron*, 25A:479–86 (St. Petersburg, 1898). On France as the model for the Russian system, see N. B—skii, "Ocherk prostitutsii v Peterburge," *Arkhiv sudebnoi meditsiny i obshchestvennoi gigieny*, no. 4, sec. 3 (1868): 67, 82, 88. On regulation in Russia, see Laurie Bernstein, "Yellow Tickets and State-Licensed Brothels: The Tsarist Government and the Regulation of Urban Prostitution," in *Health and Society in Revolutionary Russia*, ed. Susan Gross Solomon and John F. Hutchinson (Bloomington, Ind., 1990), 45–65.

[104]See S. K. Gogel', "Iuridicheskaia storona voprosa o torgovle belymi zhenshchinami v tseliakh razvrata," *Vestnik prava*, no. 5 (1899): 116. Minister of Internal Affairs L. A. Perovskii, who initiated the system of regulation, also reorganized the entire medical bureaucracy; see Luig, *Zur Geschichte*, 53–54.

Despite the existence of regulation on the continent, its Russian enemies associated the practice with the persistence of premodern forms of
rule. In 1907, when Russia's most eloquent feminist critic of regulation,
the public health physician Mariia Pokrovskaia, denounced the system,
she represented it as a reflection of Russia's enduring political backwardness:

> The police-medical surveillance of prostitution conforms to the feudal
> [*krepostnicheskie*, a term specifically associated with serfdom] inclina-
> tions of the male sex and to their contempt for the human dignity of
> women. This connection is demonstrated by the fact that the Russia
> of serfdom [*krepostnaia Rossiia*] was among the first countries to copy
> the French morals police. . . . [This system] has flourished in the
> country of lawlessness [*bespravie*] and arbitrary power [*proizvol*],
> while in England, a country of freedom and respect for the person,
> police regulation did not take root.[105]

Indeed, the eighteenth-century police legacy had been infused with
new legal blood precisely in the conservative reign of Nicholas I, when
Count Bludov had incorporated prostitution directly in the 1845 criminal code. Theoretically, to be sure, criminalization was incompatible
with regulation, which tolerated the conduct of prostitution itself and
penalized only the infringement of specific rules governing the trade. In
practice, however, the new criminal sanctions did not render such toler-
ation illegal, for in fact they aimed simply to enhance the kind of public
discipline exercised by the police. Bludov's justification reads very
much like an extract from a textbook on police law. Admitting that
"local authorities cannot keep track of the bad behavior of private persons when unaccompanied by provocative actions," he felt obliged to
concentrate instead on "illicit sexual behavior [*nepotrebnoe povedenie*],
when it takes the form of brazen or provocative actions in a public
place." These were "more criminal" than similar acts performed in private, he explained, "because of the harmful influence they exercise on
others and because of the violation of what might be called moral propriety [*nravstvennost' prilichii*]."[106] In addition to making prostitution itself a crime, the code, somewhat contradictorily, also set a small fine
for patrons whose "provocative behavior" made their transgression impossible to ignore.[107]

---

[105]M. I. Pokrovskaia, "Prostitutsiia i bespravie zhenshchin," *Zhenskii vestnik*, no. 10
(1907): 226.

[106]Note to draft art. 1241 (1281 in code) and commentary on draft art. 1246 (1286 in
code), *Proekt* (1844), 440, 442; and art. 1281, *Ulozhenie* (1845), 344.

[107]Art. 1286, *Ulozhenie* (1845), 345. A leading commentator believed the code penalized only disorderly conduct, not commercial sex as such: N. A. Nekliudov, *Ruko-*

It was, in fact, the province of the police to control the external or public manifestations of moral disarray. Because of their reliance on compulsion, an expert admitted, the police could not "fight immoral attitudes or principles as such but only insofar as they became evident in external actions, harmful to the interests of individual persons or society as a whole."[108] This construction of administrative discipline led to the legal paradox of tolerated prostitution. "Victorian" moralists, Russians among them, stressed the horror of exchanging money for sex, but from the legal point of view it was the commercial basis of prostituted sex that legitimized the otherwise criminal acts of fornication and adultery in which the customer and professional engaged. The law, explained the Senate in 1892, "strictly distinguishes extramarital carnal congress, whether permanent or even temporary, from depravity [razvrat] or prostitution [nepotrebstvo], defined as commerce in one's own body, the offer or acceptance of an offer to engage in carnal congress for a stipulated fee with anyone who asks."[109]

The exchange of money took the sexual act out of the private domain and into the public arena patrolled by the police. But money was not the only distinction the authorities invoked. From 1864 to 1909 (when the prostitution articles of the 1903 code went into effect), women inscribed on the official lists and thereby termed "public women" (publichnye zhenshchiny) were subject to penalty for insubordination to administrative rules under Article 44 of the charter of the justices of the peace. The Senate drew an explicit analogy between the managers of "public establishments" and public servants such as government bureaucrats, who were equally subject to penalty for noncompliance with administrative rules.[110] Women who traded in sex on their own, without the official stamp, risked prosecution for "secret depravity" (tainyi razvrat) under the Regulations on Public Order and, between 1845 and 1866, under the criminal code as well. During those twenty years, the conflict between police and judicial authority was particularly acute, prompting the State Council to rule in 1853 that women could not be tried for the crime of prostitution unless they came before the courts on other, unrelated charges.[111]

---

vodstvo osobennoi chasti russkogo ugolovnogo prava (St. Petersburg, 1887), 477. Nekliudov interpreted the antiprostitution statute as "a salve to the lawmaker's conscience" (quoted in V. D. Nabokov, "Plotskie prestupleniia," 136).

[108]Deriuzhinskii, Politseiskoe pravo, 309–11.

[109]"Vypiska iz 'reshenii obshchego sobraniia pervogo i kassatsionnykh departamentov [Prav. Senata]' 3 fevr. 1892 g.," in Svod uzakonenii pravitel'stva po vrachebnoi i sanitarnoi chasti v Imperii, ed. L. F. Ragozin (St. Petersburg, 1895–96), 1:67–68.

[110]Ibid., 68.

[111]Note to art. 44 in N. S. Tagantsev, ed., Ustav o nakazaniiakh, nalagaemykh mirovymi

In 1892, however, the Senate seemed to imply that even police measures must in fact be judged by legal standards; administrative regulations, it ruled, were valid only if they did not "destroy or restrict the legally protected rights of private persons, because no regulation of the subordinate organs can invalidate or restrict the law emanating from the Supreme Power." Yet the ruling went on to specify the conditions under which regulation should be applied and enforced under Article 44. Within that context, the Senate did try to strengthen the principle of due process by insisting that the police exercise caution in entering women of dubious character on the official rolls, a procedure that had been a target of criticism in both the legal and the medical press.[112] The Senate thus showed itself sensitive to the growing demand for judicial controls over police power. In the spirit of the 1892 Senate decision, the authors of the 1903 reform underlined the importance of due process by incorporating Article 44 directly into the criminal code.[113] They did not thereby eliminate police jurisdiction over prostitution but made the violation of police rules a criminal offense. This change was consistent with the current movement in European legal and medical circles calling for the reestablishment of regulation on a proper legal footing.[114]

The objection that police control deprived prostitutes of their civil rights ought to have posed no obstacle to the registration of girls who had not achieved the status that adulthood theoretically conferred; indeed, partisans of regulation argued that underage prostitutes would be better protected if inscribed on the official rolls.[115] But registration, while limiting sexual activity, also confirmed that such activity was taking place and thereby endowed girls with the defining feature of adult women, a consequence the 1903 reformers wished to avoid. When the editorial commission submitted its proposals in 1895, the minimum age

---

sud'iami. Izdanie 1885 goda. S dopolneniiami po svodnomu prodolzheniiu 1912 goda, s prilozheniem motivov i izvlechenii iz reshenii kassatsionnykh departamentov Senata, 21st ed., rev. (St. Petersburg, 1913), 191–92; also *UU*, vol. 6 (1897), 599–601.

[112]"Vypiska iz 'reshenii obshchego sobraniia pervogo i kassatsionnykh departamentov,'" 67–68. See the much-cited and eloquent denunciation in P. E. Oboznenko, "Po povodu novogo proekta nadzora za prostitutsieiu v Peterburge, vyrabotannogo komissieiu Russkogo sifilidologicheskogo obshchestva," *Vrach*, no. 12 (1899): 347–50. On the legal side, see V. D. Nabokov, "Plotskie prestupleniia," 138; and A. L. Rubinovskii, "Povinnost' razvrata," *Vestnik prava*, no. 8 (1905):161.

[113]Art. 528, *Ugolovnoe ulozhenie* (1903), 186.

[114]Corbin, *Les Filles de noce*, 372–73. The position was popular among a number of Russian specialists in venereal disease: see Oboznenko, "Po povodu," 349; and idem (reporting European opinion), "Vopros ob uporiadochenii prostitutsii i o bor'be s neiu na dvukh mezhdunarodnykh soveshchaniiakh 1899 goda," *Vrach*, no. 30 (1900): 910–12.

[115]For debate at the 1897 conference on syphilis sponsored by the Medical Department of the Ministry of Internal Affairs, see *Trudy Vysochaishe razreshennogo s"ezda po obsuzhdeniiu mer protiv sifilisa v Rossii, byvshego pri Meditsinskom Departamente s 15 po 22 ianvaria 1897 goda* (St. Petersburg, 1897), 2:95–99.

set by the police for registering prostitutes was sixteen, which the commission raised to seventeen in accord with its goal of retarding the judicially recognized onset of female sexual autonomy. In 1901, however, the Ministry of Internal Affairs itself raised the limit to twenty-one, the age of legal majority, which it remained in the final version of the code.[116] Under the new statute, admitting a girl under twenty-one to work in a brothel incurred the same penalty as forcing an adult woman to remain in a brothel after she had expressed the desire to leave.[117] On the model of statutory rape, which excluded the possibility of consent on the grounds of age alone, this legislation made it impossible for young women legally to agree to work in brothels.

Lawmakers who pictured women as vulnerable to manipulation and sexual exploitation also directed their attention to procuring—defined in the Regulations on Public Order (where it figured as a police infraction) as the act of providing the conditions for others to engage in commercial sex.[118] The 1845 code, in its zeal to institutionalize the disciplinary function of the state, had elevated procuring (svodnichestvo) to a criminal offense in order to protect "those superficial and inexperienced people who might otherwise retain their moral purity were it not for seduction and depravity."[119] True to its authors' obsession with social status and their compulsion to specify all possible variations of a crime, the 1845 code established a hierarchy of aggravating circumstances geared to the relative position of perpetrator and victim. Their choice marked the problem as one of protecting authority relations within the family, not of securing the rights of women.[120] The most serious cases thus involved parents who enticed their own children into "prostitution and other vices" and husbands who did the same with their wives, followed by guardians and other custodial figures who exploited their underage charges, and finally those who engaged in procuring as a regular occupation or trade.[121]

The framers of the 1903 reform followed this model by specifying the degree of exploitation inherent in certain relationships and in the age and sexual status of the woman whose cooperation might be obtained, even without violence to her own wishes. They also filled a gap left in

---

[116]Ministry of Internal Affairs Circular 1314 (June 6, 1901), confirmed in *Zhurnal osobogo prisutstviia Gosudarstvennogo Soveta Vysochaishe uchrezhdennogo dlia obsuzhdeniia proekta ugolovnogo ulozheniia* (St. Petersburg, 1902), 256–57.

[117]Art. 529, *Ugolovnoe ulozhenie* (1903), 186; see also comments on draft art. 292, *UU*, vol. 6 (1897), 351.

[118]*UU*, vol. 6 (1897), 665.

[119]Comment on draft arts. 1256–60 (1296–1300 in code), *Proekt* (1844), 445.

[120]As we have seen, the same principle was embedded in the statute on defloration of dependent adult women.

[121]Arts. 1296–98, 1300, *Ulozhenie* (1845), 347–48.

the 1845 edition, which had neglected to penalize the procuring of girls who were not members of the culprit's family or under his (or her) charge.[122] The 1903 statutes, in fact, penalized offending parents or guardians less severely than strangers who solicited girls between the ages of fourteen and sixteen or virgins up to the age of twenty-one.[123] In the code's final version the State Council strengthened the sanctions by prohibiting the use of violence or fraud to recruit potential prostitutes and also to transport them abroad. A particular (though lesser) penalty was indicated for "a person of the male sex" who traded in the sexual services of women under his influence or authority, or who turned a profit from supplying regulated brothels with new recruits.[124] Of the code's statutes on sexual conduct, only those on prostitution were enacted into law.[125] Their conception, as well as their adoption in 1909, reflected the influence of the international campaign against the so-called white slave trade, defined as the abduction by gangsters and procurers (said to operate on a worldwide scale) of innocent young women for purposes of prostitution.[126]

[122]For complaints about this gap, see, e.g., Nekliukov, *Materialy*, 2:223.

[123]Art. 524, *Ugolovnoe ulozhenie* (1903), 184–85.

[124]Art. 527, ibid., 185–86.

[125]Arts. 524–29 (chap. 27) and 500 (chap. 26) on personal freedom. See D. A. Koptev and S. M. Latyshev, eds., *Ugolovnoe ulozhenie (stat'i vvedennye v deistvie)* (St. Petersburg, 1912), 3. For a discussion of the law of December 25, 1909, enacting these statutes, see A. A. Shchipillo, "Sostav prestuplenii, predusmotrennykh zakonom 25 dekabria 1909 goda o merakh k presecheniiu torga zhenshchinami v tseliakh razvrata," *Zhurnal Ministerstva Iustitsii*, no. 10 (1911): 56–102. For the text of the law, see "Odobrennyi Gosudarstvennym Sovetom i Gosudarstvennoi Dumoiu Vysochaishe utverzhdennyi zakon: O merakh k presecheniiu torga zhenshchinami v tseliakh razvrata," *Sobranie uzakonenii i rasporiazhenii pravitel'stva*, no. 10, sec. 1 (January 12, 1910):91–94.

[126]On the influence of international congresses on the Council and Ministry of Justice, see commentary to arts. 526–27 in N. S. Tagantsev, ed., *Ugolovnoe ulozhenie 22 marta 1903 g.* (St. Petersburg, 1904), 716. The law of December 25, 1909, enacting the prostitution and procuring statutes of the 1903 code, had been promoted by the antiregulationist Russian Society for the Protection of Women, itself an offshoot of the 1899 London congress on the international prostitution trade: Deriuzhinskii, *Politseiskoe pravo*, 380–83; Elistratov, "Prostitutsiia v Rossii," 34; V. D. Nabokov, "Plotskie prestupleniia," 134–35; I. M. Faingar, "Detskaia prostitutsiia," *Vestnik psikhologii, kriminal'noi antropologii i pedologii*, no. 3 (1913): 42. The councilman and former education minister A. A. Saburov, who sat on both council committees reviewing the new code, was vice chairman of the Russian Society for the Protection of Women and attended the congresses of 1899 (London) and 1900 (Paris): Oboznenko, "Vopros ob uporiadochenii," 913; V. F. Deriuzhinskii, "Mezhdunarodnaia bor'ba s torgovleiu zhenshchinami," *Zhurnal Ministerstva Iustitsii*, no. 8, sec. 2 (1902): 185. For the archival record of the bill's presentation to the Duma, see TsGIA, f. 1405, op. 543, d. 512, including a letter from the Society for the Protection of Women (l. 1) and correspondence between the Ministries of Justice and Internal Affairs, with references to international congresses and foreign laws. For the justification of the new statutes as presented to the Duma, see Report no. 266: S. V. Andronov, "Doklad po vnesennomu ministrom iustitsii zakonoproektu o merakh k presecheniiu torga zhenshchinami v tseliakh razvrata," in *Prilozheniia k stenograficheskim*

Lawmakers decried procurers and pimps, who were largely though not exclusively male, as unscrupulous villains who preyed on the virtue of innocent girls and therefore deserved to feel the full force of the criminal law, from which the prostitutes (all female) were shielded.[127] But in fact, to target the men and leave the women to the vagaries of police regulation was to acknowledge the male's independence of action while keeping the female subject to the custodial discretion of the state. Despite the greater stigma attached to criminal trials and the heavier penalties they entailed, it was better from the point of view of due process and respect for personal rights to have committed an offense that fell under the criminal code than one that came under the jurisdiction of the peasant courts or the police. This distinction was not lost on contemporaries: "In the present day and age," wrote jurist Sergei Gogel' in 1899, "when all are free and what Montesquieu called civic freedom exists in every state; when . . . even criminals have their 'magna charta libertatum' in the form of the criminal code and the rules of criminal procedure, which determine down to the smallest details for what and under what circumstances they may be punished, only prostitutes remain without legal defense, without specific rights and duties."[128]

The severity of laws against the abuse of male authority and power served ultimately to bolster the legitimacy of male control. Police power, as one legal text frankly admitted, was there to make up for the defects of family life, to provide moral support when the domestic system failed.[129] If a woman's father and husband controlled her freedom of movement, the police took over when she entered the "public" ranks, confiscating her passport in exchange for the notorious yellow card designating registered public women. When married women were inscribed on the police rolls, it was their husbands who lost by the exchange, for the women themselves had not been free agents to begin with.[130]

The Juridical and the Sexual Subject

The criminal satisfies individual desire or private need at the community's expense. The sexual offender is a kind of moral entrepreneur who

---

otchetam Gosudarstvennoi Dumy, tretii sozyv, sessiia vtoraia (St. Petersburg, 1909), 2: 1–2, followed by the text of the proposed statutes.

[127]Male prostitutes existed but were not covered in the police rules or the criminal statutes; homosexual acts were criminal, whether committed for commercial gain or not (although so was fornication)—another gender-related inconsistency in the law.

[128]Gogel', "Iuridicheskaia storona," 115. The prostitutes' lack of rights was emphasized by many writers, including V. D. Nabokov in "Plotskie prestupleniia," 137.

[129]Andreevskii, Politseiskoe pravo, 2:5, 9.

[130]"Polozhenie vrachebno-politseiskogo komiteta v S.-Peterburge (utv. M-rom V.D. 28 iiulia 1861 g.: Pravila dlia publichnykh zhenshchin," in Ragozin, Svod uzakonenii, 102. Rubinovskii ("Povinnost' razvrata," 167) claimed that the registration of married women was a violation of their husbands' legal rights.

achieves personal gratification in defiance of collective norms. "The role of criminal law," wrote the assistant procurator of the Odessa circuit court, ". . . is to sanction the existing order, by protecting it from arbitrary incursions of the individual will."[131] The structure of the law on sexual crime makes it clear that individualism of this type was perceived as overwhelmingly male. Given the organization of family life and the legal position of women, this perception did not distort actual social experience. The purpose of the law, however, was not merely to reflect but also to shape the terms of social interaction, and in this sense the lawmakers helped to ensure that individual autonomy remained a male preserve.

The juridical subject of sexual crime could be female in three cases in the nineteenth-century criminal codes—adultery, incest, and cohabitation—although circumstances in each case mitigated the active female role and minimized its public dimension.[132] Adultery was actionable only upon the complaint of the wronged spouse; on the principle that "my house is my castle," it was by definition a private affair. Even though an erring wife was considered a more serious threat to family life than a guilty husband, she was nevertheless viewed either as the hapless victim of her own desire or the object of seduction by her partner in crime.[133] Adult women might be charged with incest, but in fact the statutes were designed to protect young girls from the depredation of males. And the statutes on fornication and cohabitation, while applicable to both parties, were principally intended to impose paternal responsibility for the illegitimate child.[134]

Prostitution would seem unequivocally to entail a female agent, yet here too the law managed to construe the real subject as male. Because the practice was technically legal, prostitutes were not criminal offenders. Instead, public women and brothel managers (male or female) were subject to police surveillance, which deprived them of due proc-

---

[131]M. M. Abrashkevich, *Preliubodeianie s tochki zreniia ugolovnogo prava: Istoriko-dogmaticheskoe issledovanie* (Odessa, 1904), 619.

[132]Even in relation to infanticide, an overwhelmingly female crime, jurists tended to exculpate the guilty women on the grounds of physiological or mental incapacity at the time of the act. See, e.g., Ivan Ozerov, "Sravnitel'naia prestupnost' polov v zavisimosti ot nekotorykh faktorov," *Zhurnal iuridicheskogo obshchestva pri Imperatorskom S.-Peterburgskom Universitete*, no. 4 (1896): 60; M. G., "O detoubiistve," *Arkhiv sudebnoi meditsiny i obshchestvennoi gigieny*, no. 1, sec. 2 (1868): 55; A. A. Zhukovskii, "Detoubiistvo v Poltavskoi gubernii i predotvrashchenie ego," *Arkhiv sudebnoi meditsiny i obshchestvennoi gigieny*, no. 3, sec. 2 (1870), 10; and chap. 3 below.

[133]Art. 1585, *Ulozhenie* (1885), 755; English phrase in Abrashkevich, *Preliubodeianie*, 638. O. A. Filippov believed, for example, that married women, whose sexual passion was "already developed," were easily seduced because their urges were "deeper and stronger" (i.e., less under control) than those of women who had not known sex ("Vzgliad na ugolovnoe pravo," 29). Abrashkevich concluded: "Often aggressive men make women the victims, rather than the accomplices in the crime" (*Preliubodeianie*, 632).

[134]*UU*, vol. 6 (1897), 623; art. 994, *Ulozhenie* (1885), 515–21. On the 1902 decriminalization of cohabitation, see chap. 1 above.

ess. Their clients (all male) were subject neither to administrative discipline nor to judicial sanctions (though they might, in principle, be put in jail for a few days). In rendering the prostitutes "noncriminal," the system negated a delinquency that implied self-assertion and would otherwise have commanded consideration in the courts. Indeed, the only participants in the prostitution trade who risked criminal prosecution under the statutory law were the men engaged in recruiting and exploiting women. The women who directly supplied sexual services were neither entirely free nor criminally liable.

By contrast, anal intercourse between men remained a criminal act, even though, like prostitution, sodomy did not necessarily violate either partner's personal rights. The reformed code of 1903 also enlarged the range of criminal sanctions against men who subjected women to unwanted sexual contact and who engaged in noncoercive sexual acts with children of either sex. Progressive lawmakers thus chose to extend legal protection for women as victims of male sexual aggression (or simply as objects of male sexual desire) but limited the cases in which women were formally acknowledged as sexual actors. Insofar as the male constituted the archetypal public figure, the state accentuated the public consequences of his personal behavior, refusing to accept the sexual subjugation of men as a protected private practice or to categorize men so subjugated as candidates for custodial care and therapeutic marginalization. Conversely, insofar as women moved their sexual activity onto the public stage (in the form of prostitution), they found themselves under the control and supervision of an administrative regime that deprived them of autonomy and civil status.

What do these contrasts and choices tell us about the level of Russia's civic development at the turn of the twentieth century or about the nature of the revised social order about which legal reformers dreamed? The content of Russian criminal legislation owed a great deal to the European model, and laws on sexual crime are notoriously unreliable guides to levels of cultural and political development: England, after all, had no civil law on incest until 1908 but executed male homosexuals into the nineteenth century.[135] Nevertheless, one need make no claim for the originality of Russian laws to interpret them in relation to the national context. The nineteenth-century criminal codes reinforced the principle of ascriptive distinction, even as they reflected the continuous modification of actual social categories. The 1903 version reveals, how-

---

[135]See Victor Bailey and Sheila Blackburn, "The Punishment of Incest Act 1908: A Case Study of Law Creation," *Criminal Law Review*, 1979, 708–18; and Louis Crompton, *Byron and Greek Love: Homophobia in 19th-Century England* (Berkeley, Calif., 1985), 38—not to mention the June 1986 antisodomy decision of the U.S. Supreme Court.

ever, that Russia's highest statesmen had begun to accept the erosion of this principle and the emergence of individual autonomy as the central dynamic of modern life.

These men accepted the emphasis on female dependence and vulnerability, on women's prolonged immaturity and need for protection, as a sign of cultural progress and sensitivity to personal need.[136] Yet such attitudes may just as easily be said to mark the persistence of their attachment to the traditional social structures upon which the old regime was based.[137] The weight of domestic patriarchal constraints that limited the autonomy of all women did not diminish when patriarchy in the larger sense was shaken by the freeing of the serfs. Indeed, as peasants gradually moved out of their villages into the towns, out of their politically imposed cultural isolation, women remained as the emblem of the old ascriptive order, the guarantee of stability and submission to patriarchal rule. Special treatment for women based on the peculiarities of their physical and, in particular, sexual constitution endorsed their civil subordination at the very moment when the weakening of formal status barriers challenged the legal subordination of lower-class men. The European liberal tradition, of course, extended its own individualism only up to the family door, and the problem of how to protect the rights and serve the needs of less powerful social groups while promoting their civic equality is one that plagues feminists and social activists to this day.[138] In this sense, it might be said, the legal reconstitution of women's dependent status on a modern basis, within the general context of greater individual autonomy, was consistent with Russia's move toward the acceptance of contemporary legal ideas.

[136]E.g., in granting women exemption from corporal punishment, some argued that respect for women, in particular distinctive treatment of their physical selves, was the mark of "every educated society" (quoted in Adams, "Criminology," 34).

[137]William G. Wagner, *Marriage, Property, and the Struggle for Legality in Late Imperial Russia* (Oxford University Press, forthcoming), shows the same limitations in outlook among reformers of the civil code.

[138]On European liberals' attitudes toward gender, see Susan Moller Okin, *Women in Western Political Thought* (Princeton, N.J., 1979), 199–200. A notable exception was John Stuart Mill, who was cited by the Russian feminist M. I. Pokrovskaia, *O polovom vospitanii i samovospitanii* (St. Petersburg, 1913), 24, both for his ideas and for his relationship with Harriet Taylor.

Chapter Three

Power and Crime in
the Domestic Order

The familiarity with Western legal thinking that caused reformers to embrace the principles of juridical modernity also led them to follow contemporary criminological trends that focused on the relationship between crime and the sexual attributes attached to gender. The study of crime in Russia, however, was complicated by the duality of the post-Reform legal system. This duality, in turn, complicated the way in which observers interpreted the role of gender, for the duality coincided with differences of class and social environment, which strongly affected women's position in the household and community and limited their resources in different ways.

In charting the contours of crime, experts relied on official records, which covered only those transgressions prosecuted under the formal system of statutory law. It was another matter to discover the prevalence and variety of crime according to the standards applied by the peasant township courts. The search for the principles of popular legal practice constituted an ethnography of peasant culture. Indeed, a number of studies were sponsored by the ethnographic division of the Imperial Russian Geographical Society.[1] Europeanized Russians thus approached their own native culture, to which the majority of the

[1] I. Ia. Foinitskii, "Programma dlia sobiraniia narodnykh iuridicheskikh obychaev: Ugolovnoe pravo," and A. F. Kistiakovskii, "K voprosu o tsenzure nravov u naroda," both in *Zapiski Imperatorskogo russkogo geograficheskogo obshchestva po otdeleniiu etnografii*, vol. 8 (St. Petersburg, 1878). Similarly, F. Pokrovskii, "O semeinom polozhenii krest'-ianskoi zhenshchiny v odnoi iz mestnostei Kostromskoi gubernii po dannym volostnogo suda," *Zhivaia starina: Periodicheskoe izdanie otdeleniia etnografii Imperatorskogo russkogo geograficheskogo obshchestva*, no. 3–4 (1896), and M. P. Chubinskii's work on customary law in his *Trudy etnograficheskoi statisticheskoi ekspeditsii v zapadno-russkii krai*, vol. 6 (St. Petersburg, 1872).

population belonged, as anthropologists confront an alien society. "We know as little about the dark forest," wrote a rural justice of the peace a decade after Emancipation, "as we know about the dark mass of the common folk."[2]

If the peasants were a foreign country, women were a foreign race. Followers of the Italian school of so-called criminal anthropology, which focused on the biological determinants of crime, naturally concentrated on sex differences as a key to patterns of criminal behavior. But the interest in women as criminal subjects antedated the vogue for biological explanation. Statistical sociologists such as the Belgian Adolphe Quételet, who left his mark on a generation of Russian scholars, investigated the influence of gender along with that of age, marital status, and education.[3] To students of popular culture, the status and treatment of women served as a measure of moral development and a key to the ethical basis of customary law. This chapter considers the picture of criminality generated by the two coexisting legal jurisdictions—the criminal courts and the institutions of peasant justice—in order to reconstruct the way in which contemporaries imagined the relationship between female sexuality, criminality in general, and sexual crime in particular.

For educated Russians, as well as their European contemporaries, female crime posed a logical challenge. The psychosexual traits associated with the use of violence and defiance of the law were thought to apply only to males. Thus, with men clearly in mind, the Moscow professor and political economist Ivan Ozerov noted that murder and assault were more frequent in the spring and summer months, when "the blood boils more passionately, the organism's vital forces swell, its physiological need for sex makes itself more powerfully felt, [and] awareness of one's individuality grows."[4] By contrast, the aggression, premeditation, and self-assertion necessary to crime were exactly the traits women were supposed to lack. How was the submissive, irrational, self-abnegating female to muster the drive and egotism responsible for crime? Women's sexual passion, in the contemporary view, should have led them to self-sacrifice, maternal and romantic devotion,

---

[2]V. N. Nazar'ev, "Sovremennaia glush': Iz vospominanii mirovogo sud'i," *Vestnik Evropy*, no. 2 (1872): 611.

[3]Quételet cited in E. N. Anuchin, *Issledovaniia o protsente soslannykh v Sibir' v period 1827–1846 godov* (St. Petersburg, 1866), 34; P. N. Tarnovskaia, *Zhenshchiny-ubiitsy: Antropologicheskoe issledovanie* (St. Petersburg, 1902), 96; N. A. Nekliudov, *Ugolovno-statisticheskie etiudy*, I: *Statisticheskii opyt issledovaniia fiziologicheskogo znacheniia razlichnykh vozrastov chelovecheskogo organizma po otnosheniiu k prestupleniiu* (St. Petersburg, 1865), 1.

[4]Ivan Ozerov, "Sravnitel'naia prestupnost' polov v zavisimosti ot nekotorykh faktorov," *Zhurnal iuridicheskogo obshchestva pri Imperatorskom S.-Peterburgskom Universitete*, no. 4 (1896): 66. Thanks to Reginald Zelnik for identifying Ozerov.

or impulsive behavior lacking evil intent. "The habit of carnal indulgence *easily leads to further excesses*," a Kiev criminologist emphatically warned, "which undermine the character and prepare the ground on which crime so easily grows." Because women "refrained . . . from sensual gratification," he insisted, they therefore possessed a series of virtues, including industry, modesty, patience, charity, and piety, which prevented them from committing crimes.[5]

Such assumptions might explain why women in fact committed fewer crimes than men, but they could not explain why the crimes they did commit focused on precisely the domestic and sexual sphere that ought to have kept them in check. The inability of prevailing gender stereotypes to account for female crime became all the more obvious when differences of class and culture complicated the picture. Russian peasant society endorsed behavioral and sexual norms that did not always coincide with those of the educated elite. Violence itself played a different role in the peasant community than it did in the town, sustaining village cohesion rather than threatening the stability of village life.[6] Observers realized that crime did not mean the same thing in urban and rural Russia, that families served different purposes, that relations between the sexes were of a different order.

Though imbued with Western values, most educated Russians nevertheless thought of folk culture as the core of national identity. Many persisted in holding up the lower strata as standards of virtue and unspoiled moral integrity, despite their own revulsion against certain popular practices and the evidence of social change. Yet Western ideas shaped the terms in which the educated framed their interpretation of social reality, if only as a foil against which to measure national authenticity. In thinking about crime and sexual delinquency, Russians were taking the measure of their cultural distinctiveness and expounding a critique of the political system under which they lived.

The Legal Norm: Crime by Gender

In 1874 the Ministry of Justice began publishing empirewide data on trials and convictions by the justices of the peace and the circuit courts.[7] These reports were analyzed in the ministry's own journal by the indefatigable Evgenii Tarnovskii, who summarized his conclusions about

---

[5]Nikolai Zeland, *Zhenskaia prestupnost'* (St. Petersburg, 1899), 35, 37–39.

[6]See Stephen P. Frank, "Popular Justice, Community, and Culture among the Russian Peasantry, 1870–1900," *Russian Review* 46:3 (1987): 239–65.

[7]S. S. Ostroumov, *Prestupnost' i ee prichiny v dorevoliutsionnoi Rossii* (Moscow, 1960), 85.

the first twenty years of recordkeeping on post-Reform justice in a monograph issued in 1899.[8] Tarnovskii's figures showed that prosecutions for sexual crime rose dramatically between 1874 and 1894. Indeed, convictions for moral offenses considered as a single category increased at over four times the rate of the next most rapidly expanding group, crimes against family and kinship rules. His evidence has led a Soviet historian to decry the "moral decay" into which tsarist society had fallen.[9]

Modern calculations support the impression that in these twenty years the number of convictions for sexual crimes in the circuit courts increased at a dramatically higher rate than convictions for any other category of crime, and ten times more sharply than the number of criminal convictions as a whole. However, even at its peak in 1894, the category of sexual and moral offenses accounted for no more than 7.5 percent of all criminal convictions. Fully three-quarters of this total, moreover, consisted of convictions for cohabitation. This crime's meteoric rise as an object of legal prosecution accounts for two-thirds of the rise in convictions in the sexual category as a whole.[10] Once cohabitation was decriminalized, in 1902, the number of convictions for the totality of sexual crimes fell sharply.[11]

Even at their maximum, the absolute number of persons convicted of sexual crimes remained small: only ten men, for example, were found guilty of homosexuality in 1894, and a mere 120 persons of incest.[12] The perception generated by Tarnovskii's figures was nevertheless the one that dominated contemporary discourse on the subject of crime. In particular, what struck most observers in the quarter-century preceding 1905 was the concentration of female crime in just those "moral" offenses apparently so sharply on the rise.[13] In other words, the supposed deterioration of morals seemed to represent a change in female sexual behavior or in women's behavior connected to their sexual role. "Woman's moral physiognomy," wrote Ozerov in 1896, "has

[8]E. N. Tarnovskii, *Itogi russkoi ugolovnoi statistiki za 20 let (1874–1894 gg.)* (St. Petersburg, 1899).

[9]Ostroumov, *Prestupnost'*, 174.

[10]Richard Cummer Sutton, "Crime and Social Change in Russia after the Great Reforms: Laws, Courts, and Criminals, 1874–1894" (Ph.D. diss., Indiana University, 1984), 64, 232–33.

[11]D. N. Zhbankov, "Polovaia prestupnost'," *Sovremennyi mir*, no. 7, pt. 2 (1909): 56.

[12]Sutton, "Crime," 232.

[13]In addition, modern analysis of the data available at the time shows that between 1874 and 1894, female convictions in the sexual-crime category rose from about 40 to 45 percent of the total and that sexual crimes as a percentage of all female crimes rose from 4 to 25 percent (ibid., 232, 238). If one eliminates the single category of cohabitation, however, then the contribution of women to the overall sexual crime rate appears to have diminished, falling from almost 34 percent of all other sexual crimes in 1874 to 30 percent in 1894.

enormous significance for every society. . . . Woman is the measure of the social condition [*Kakova zhenshchina—takovo obshchestvo*]."[14]

Overall crime rates for women were lower in Russia than in Western Europe, all agreed, with Germany considered highest of all. Women were overrepresented, however, in the commission of certain crimes. In all societies, for example, infanticide was by definition an almost exclusively female offense. Russian statisticians found that the proportion of women among people convicted of incest and adultery was four to six times the proportion of women among people convicted of the full range of crimes. The experts also noticed that family and kin constituted a larger proportion of the victims of female than of male murderers. Though convicted of only 9 percent of all crimes and a mere 7.6 percent of murders against unknown or unrelated victims, women were responsible for killing 85 percent of murdered parents, 54 percent of murdered stepparents, 38 percent of murdered spouses, 35 percent of murdered parents-in-law, and 26 percent of murdered servants.[15]

Female domestic killers were not unique to the Russian scene, and they raised the same paradoxical questions for all students of crime. Women's position in the family ought to have kept them from murder, not driven them to it. "Women are less given to crime than men," wrote Quételet in 1835, "because they are restrained by the feeling of shame and modesty . . . , by their position of dependence and seclusion . . . , and by physical weakness." Nevertheless, he admitted, "compared to men, [women] sooner kill within the family than without."[16] The Russian literature echoed these conflicting views. Evgenii Anuchin, an admirer of Quételet, repeated the explanation for women's lower propensity toward crime in his often cited 1866 study of felons exiled to Siberia during the 1830s and 1840s. While documenting the high level of women's domestic criminality, Anuchin nevertheless maintained that women's restricted household role narrowed the range of female crime: "Limited in her activity by the vicious circle of family life, confined within the four walls of the household, women have fewer experiences that stimulate the passions than do men."[17]

---

[14]Ozerov, "Sravnitel'naia prestupnost'," 45.

[15]Adolphe Quételet, *Sur l'homme et le développement de ses facultés; ou Essai de physique sociale* (Paris, 1835), 216; I. Ia. Foinitskii, "Zhenshchina-prestupnitsa," pt. 1, *Severnyi vestnik*, no. 2 (1893), 132; Ozerov, "Sravnitel'naia prestupnost'," 57–58, 61; E. N. Tarnovskii, *Itogi*, 142; Anuchin, *Issledovaniia*, 30–33, 37–38 (showing family killings as 7 percent of murders by men, 23 percent of murders by women). See also Pauline Tarnowsky [P. N. Tarnovskaia], "Criminalité de la femme," in Congrès international d'anthropologie criminelle, *Compte-rendu des travaux de la quatrième session tenue à Genève du 24 au 29 août 1896* (Geneva, 1897), 232–33.

[16]Quételet, *Sur l'homme*, 216–18.

[17]Anuchin, *Issledovaniia*, 36.

Two distinct notions were combined in this argument. The first was that public life gave rise to conflicts and passions absent in the private sphere, which therefore constituted a shelter against crime. Thus Ozerov praised the family as "a marvelous bulwark against onslaughts of the criminal urge." The second notion, however, conceded that private life generated pressures of its own, which could also result in crime. The same Ozerov therefore attributed "the high level of female participation in crimes against sexual morality" to "the fact that woman's nonparticipation in public life confines her activity to the sphere of intimate relations."[18] The very unworldliness that restrained a woman from crime in general, these writers had to admit, inclined her to sexual misconduct (incest and adultery) and domestic violence. The family, Ozerov lamented, "not only fails to raise woman's moral level but causes her to react against it in the form of crime."[19]

Observers thus developed a gendered interpretation of crime that matched the public-private duality of social organization. The psychological motives for crime differed between women and men, Tarnovskii wrote: men were moved by "greed, the urge to acquire what belongs to someone else"; women were driven by "affect derived from the sphere of sexual life [*affekty, vytekaiushchie iz sfery polovoi zhizni*]." The sort of crime they committed reflected the private circumstances of their lives: "As a consequence of a woman's generally more restricted circle of activity, sexual feeling and the disturbances and affect associated with it occupy a much greater place in her inner world than they do in a man's." The particular crimes to which women were prone—fornication, adultery, incest, infanticide, child neglect, murder of husband and other relatives—"for the most part flow from one or another abnormality or complication of sexual as well as of family life. The burden of these complications or deviations from the ordinary, normal type have a greater impact on women," the statistician averred, "since the latter are more dependent and less capable of autonomous existence than men. These circumstances increase the probability that women will commit serious crimes such as infanticide or the murder of close relations, despite the fact that female criminality is in general lower than men's."[20]

It was not clear, moreover, that family life had to be "abnormal" to impose intolerable burdens on female members. From the standpoint of many educated observers, it was precisely the traditional peasant household that was least conducive to a woman's happiness. For the lower

[18]Ozerov, "Sravnitel'naia prestupnost'," 65, 59.
[19]Ibid., 61–62; Anuchin, *Issledovaniia*, 38.
[20]E. N. Tarnovskii, *Itogi*, 149–50, 143–44.

classes, these critics argued, domesticity did not represent a benign re-
spite from life's travails. Married life among the common folk, wrote
the Kiev criminologist Nikolai Zeland, was a site of sexual conflict and
violence, in which women usually figured as the unrecorded victims of
physical and sexual abuse and all too often as aggressors in their own
right. In no sense could the domesticity of the peasant household,
where women worked as mothers, housekeepers, fieldhands, and paid
laborers and were subject to the absolute authority of men, be considered a haven from the stress and provocations that led to crime. Rather,
the family perpetuated the kinds of violence and abuse associated with
serfdom or the penal regime, in which the powerless had no legal personality or legal recourse. "On the one hand," Zeland wrote, "the
physical abuse [*zhestokoe obrashchenie*] of women is entirely routine
among the common folk [*narod*]. Because it is indissoluble, marriage, in
these cases, becomes a form of enslavement for the wife [*tiazhkoe kre-
postnoe sostoianie*]. On the other hand, since women's individuality is
not recognized, they are often forced to marry decrepit or repulsive
men. Since the wife has no means of escape, she may experience such a
marriage as a kind of punishment at forced labor [*katorga*]." [21]

Nor, according to Zeland, could the clear division between public
and private, which might apply to educated urban women, explain the
configuration of peasant life. The wives of peasants, soldiers, and petty
urban tradesmen (*meshchanki*), who constitued four-fifths of the female
population, "are not at all confined within the 'four walls of the
home'"; working at various paid jobs outside the household, women in
these categories "come into contact with the extra-domestic sphere
[*vnedomashnaia sfera*] no less than men, only in somewhat different
form." The arduousness of such women's domestic existence, coupled
with their outside responsibilities, Zeland concluded, meant that provocations to crime were more, not less, abundant for women than for
men, who did not bear the same household and reproductive burdens
and were not the objects of physical and psychological abuse. Women's
relative restraint could not therefore be said to result from lack of opportunity, he asserted, but must reflect innate moral virtue. [22] Having
destroyed the icon of sheltered womanhood, Zeland nevertheless retained the image of woman as moral paragon.

This picture of the traditional household constituted at once an in-
dictment of the old system and a nostalgic longing for the values that
system was supposed to represent. It was more common, however, for
experts to blame the opportunities offered by modern life rather than

---

[21]For this argument, see Zeland, *Zhenskaia prestupnost'*, 8–10, 14.
[22]Ibid., 7–10.

the lack of opportunity offered by old-style arrangements for producing female crime. In comparing the offenses committed by rural and urban women, the eminent St. Petersburg law professor Ivan Foinitskii noted that "the criminality [of urban women already] has a more individual character, directed toward the fulfillment of needs derived not so much from family and communal relations as from personal demands and interests."[23] Greater individual autonomy, Foinitskii argued, led women to commit more and different kinds of crimes.

The association of rising criminality with the emergence of individuality from the bondage of communal life was one that Foinitskii shared with most of his colleagues, although he was careful to point out that rural crime was vastly underreported.[24] In the 1860s, for example, Anuchin had drawn the connection between the traditional family order and female virtue, noting that Protestant women enjoyed the greatest freedom and therefore committed the greatest number of crimes; Jewish and Muslim women appeared more virtuous because they were more severely restrained.[25] On balance, however, Foinitskii did not see the process of emancipation as a negative one. The ultimate effect of culture, he insisted, was positive. Education, salaried labor, and involvement in public life were the keys to moral fortitude, he believed, for men and women alike: "Education gives women new powers . . . with which to satisfy life's various demands, but it also gives them new concepts and ideas that restrain them from crime." Women who worked outside the home, Foinitskii found, had the lowest crime rates of their sex: "Women who perform all kinds of honest labor, not excepting factory women, who are usually depicted in uncomplimentary terms, are distinguished by a remarkably low rate of crime relative to other groups. The domestic hearth certainly does not protect women from crime; rather, the best recourse against female crime, at least under current conditions, is [paid] labor."[26]

Despite his celebration of individual autonomy, gender equality, the moral advantages of the labor market over the patriarchal community, and the value of education, Foinitskii, like other Russian liberals, seems to have been ambivalent about individualism itself. This unease may explain his defense of paid labor on the grounds that it counteracted the individualistic ethos of city life, making people into "more civic-minded [obshchestvennye] creature[s]" who were "less inclined to those capricious manifestations of personal will [proiavleniia proizvol'nogo i lich-

---

[23]Foinitskii, "Zhenshchina-prestupnitsa," pt. 1, 139.
[24]Ibid., 125.
[25]Anuchin, Issledovaniia, 36.
[26]I. Ia. Foinitskii, "Zhenshchina-prestupnitsa," pt. 2, Severnyi vestnik, no. 3 (1893): 128; pt. 1, 142.

nogo] of which crimes are made." In his view, it was social, not domestic, labor that nurtured the selflessness associated in most people's eyes with women's family role. Women active in the world, Foinitskii maintained, enjoyed the additional advantage of relief from the pressures of exclusive family life that drove them to the types of crimes with which they were commonly connected.[27]

All these various commentators on female crime offered different solutions to the domestic contradiction. Zeland attributed the overall low level of female crime to women's innate moral superiority, blaming actual crime levels on the provocations supplied by brutal family relations and the stresses of the work world. Foinitskii denied that women were constitutionally superior to men in moral terms and suggested that widening women's sphere would strengthen their ability to resist temptation. Ozerov too denied that women had innate moral strengths: "As a result of special social conditions," he wrote, "women's criminal instincts slumber for want of the force to bring them to life." But he did not agree that women could profit from wider opportunities. If family life had a baleful influence, he insisted, it was only because the "hardship" created by "present abnormal conditions" were driving women to extremes. The key to female, and by implication social, regeneration was therefore the reform of family life: "If we do not want the process of woman's moral decline to continue," Ozerov wrote, "and with it the breakdown of the next generation's civic moral instincts, we must pay special attention to the woman's sphere and somehow or other transform the present abnormal conditions of marriage and the family."[28]

But the supposedly traditional values embedded in familial and collective existence no longer seemed to satisfy conservatives either. Even Ozerov, who wished to reinforce the domestic hearth, acknowledged that modern life had distinct advantages. The same personal independence that led male criminals to turn against the community, Ozerov conceded, also prepared them to survive in the disorderly, competitive urban world. This strength was not altogether a negative feature. Ozerov believed the moral fiber to be made of tougher stuff in men than in women, despite man's stronger criminal urge. Crime, though deplorable, was a measure of personal fortitude: "To commit serious crimes one must have a greater strength of will," Ozerov wrote, "one must participate more in life." When women left the village for the big city, their crime rates soared, whereas men, who were more delinquent in the countryside to begin with, did not react as sharply to the oppor-

[27]Ibid., pt. 2, 123, 129; pt. 1, 143.
[28]Ozerov, "Sravnitel'naia prestupnost'," 52, 60, 62.

tunity presented by urban life. Ozerov recognized that women's economic disadvantages were greater than men's but refused to explain their turn to crime on this ground alone. Weakness of character was crucial: "The surrounding temptations of city life, the mass of new impressions, have a fatal effect on woman. She does not have as solid inner moral supports as man and yields more easily than he to external influence."[29]

The conservative argument for women's distinctive moral character thus could take opposite forms: it could be a brief either for her superiority, as in Zeland's argument, or for her inferiority, as in Ozerov's. Women's "moral physiognomy," in the latter's view, might be the index to society's moral health, but a woman's "moral constitution" (*nravstvennyi sklad*), he insisted, was more fragile than a man's. If *her* transgressions stemmed from weakness, *his* derived from strength. Tempered by the grim conflicts of public life, the human male emerged a superior being, Ozerov argued in Darwinian terms, just as animal species proved their biological advantages in the evolutionary wars. "Man," wrote Ozerov, "has hardened himself in the struggle for existence; for him the corrupting influence of the factory, the manufacturing center, the black days of unemployment are not so terrible. His moral principles have taken a more solid form and show greater resistance to life's temptations. True, men have more powerful criminal instincts, acquired in the struggle for existence, which hones their egotistical tendencies, but for this very reason their moral principles are more solid than women's prove to be."[30] Sense of self, or ego, was not the enemy of virtue but its base.

In this scheme, capitalism offered the occasion for the male's self-assertion, while women represented the premodern values shielded from the aggressive inroads of individualism by the remnants of patriarchy: the family regime. "Man has kept woman out of the arena of the struggle for existence," Ozerov wrote, "perhaps at the price of his own morality. He has barred her heart against the criminal instincts, so she might sustain and protect the high moral ideals he himself has squandered in the pursuit of material wealth." But the old values had lost their organic connection to life. The male had nurtured woman's moral character under conditions of social isolation, and this "female morality, cultivated far from real life, has a hothouse character. So long as woman sits under glass, she is a model of virtue, but no sooner is she touched by the cold wind than she perishes."[31]

[29]Ibid., 51, 54–55.
[30]Ibid., 75.
[31]Ibid., 75–76.

Such a vision of hothouse virtue certainly did not describe the lot of peasant women, who did not benefit from careful nurture, however patriarchal their circumstances. The pattern of rural crime suggested indeed that old-style household regimes generated far higher rates of female domestic crime than the more individualistic relationships in the towns. Among village women, the rate of infanticide and abortion was fourteen times higher than the average conviction rate for all crimes; among urban women, this rate was a mere four times higher than the average. The rate of conviction for the murder of spouses and family members was well below average among city women but six times above average for women in the countryside.[32] The sharpness of this contrast may have reflected the higher overall crime rate among urban women, but the statistical illusion impressed contemporaries nevertheless. Not only did the traditional institutions of sexual life seem to produce a relatively high level of sexual disorder and violence, but the most natural, organic, and timeless of female functions—motherhood—revealed a sinister dimension that threatened the dichotomous female-male, domestic-public, rural-urban, popular-elite scheme. The reputation of the common folk itself was at stake.

A Digression on the Archetypal Female Crime

In the contemporary view, woman was at once vulnerable to the criminal impulse and incapable of real or important crime. She was both the guardian of morality and likely to be led astray. The family kept her from transgression yet drove her to the most heinous of violent acts. Underlying these contradictions ran a unifying theme: the sexual impulse central to woman's existence was by definition an irrational force. Its domination left her incapable of the intentionality intrinsic to the notion of crime. If for no other reason than the nature of a woman's daily life, a disproportion of her criminal acts necessarily touched on sex, reproduction, and family relations. For the same reason, they could be seen as involuntary reactions to emotions beyond a person's control. Infanticide was a case in point.

Our own cultural bias inclines us to view the murder of newborn infants with special horror as an abuse of the child's vulnerability, precisely because it is not yet self-conscious and is physically helpless. Ancient Russian law, however, showed no concern for the value of children. Medieval laws did not list the murder of one's own children as a crime. The regard for parental rights expressed in this omission, cou-

[32]Foinitskii, "Zhenshchina-prestupnitsa," pt. 1, 138–39.

pled with the desire to control illicit sex, also explains the fact that the criminal code of 1649 exacted the death penalty for women who killed their illegitimate offspring but was more lenient toward the murder of legitimate children.[33] Inspired by a new pronatalist interest in population growth as a national resource, Peter the Great's Military Articles for the first time attempted to protect the existence of children, making the penalty for the murder of legitimate children as severe as that for those born out of wedlock. Further, the Military Articles differentiated for the first time between infants and grown children, setting a more severe penalty for the murder of the former. It was also in the eighteenth century, under Catherine the Great, that the state established Russia's first foundling homes for unwanted infants, in part to reduce the incidence of infanticide and foster the nation's demographic strength.[34]

If the early laws were designed to protect parental (specifically paternal) rights over their children and then to curb extramarital sexual activity, the increased severity of penal sanctions represented the state's new role as guardian of the people's biological welfare. Only in the nineteenth century did the custodianship extended to children begin to apply to mothers as well. In 1845 the illegitimacy of the murdered infant became a mitigating rather than an aggravating factor in the crime, if the murder occurred immediately after the birth and was motivated by the unwed mother's "shame or fear" (*styd ili strakh*).[35] The neglect or abandonment of newborns leading to their death was also treated with special consideration under the same circumstances.[36] A woman convicted of that crime faced the loss of her civil rights plus a prison term of one and a half to two and a half years; the lowest penalty for outright infanticide with mitigating circumstances included loss of rights and as long as six years of penal incarceration, while repeated or premeditated cases might incur ten years to a lifetime at hard labor. Most juries considered these penalties too harsh and preferred to return acquittals or to convict for the least possible offense, the concealment of a dead infant's body.[37] In order to make the law more effective, the 1903 code lowered

---

[33]David L. Ransel, *Mothers of Misery: Child Abandonment in Russia* (Princeton, N.J., 1988), 10–12; see also M. M. Borovitnikov, *Detoubiistvo v ugolovnom prave* (St. Petersburg, 1905), 13–14.

[34]See Ransel, *Mothers of Misery*, esp. chap. 3.

[35]Ibid., 17–19; Borovitnikov, *Detoubiistvo*, 14–15. See art. 1451 in N. S. Tagantsev, ed., *Ulozhenie o nakazaniiakh ugolovnykh i ispravitel'nykh 1885 goda*, 11th ed., rev. (St. Petersburg, 1901), 665; henceforth *Ulozhenie* (1885). This distinction was absent from the 1835 Digest; see art. 341, *Svod zakonov ugolovnykh* (St. Petersburg, 1835), 119.

[36]See art. 1460, *Ulozhenie* (1885), 670.

[37]Russian juries in the last quarter of the nineteenth century were almost twice as likely to dismiss charges of infanticide as of other forms of murder. See M. N. Gernet, "De-

the penalties radically in all cases to confinement in prison or a correctional institution for no more than eight years and as little as two weeks. The phrase "shame or fear" disappeared from the text, but the crime was still defined as the action of a mother murdering an illegitimate newborn infant.[38]

The nineteenth-century infanticide laws, like their European equivalents, thus created a special category of female offender, defined by a peculiar state of mind: the psychic and physical distress occasioned by childbirth in socially unfavorable circumstances.[39] The presence of such a condition was the defining feature of the crime, not grounds for acquittal. For that it was necessary to demonstrate that the accused had suffered from a kind of derangement even more severe that the one seen by the law as normally occasioned by unwanted births.[40] But writers on crime often invoked the impaired psychic state accompanying the act as a sign of diminished responsibility that ought to nullify the crime itself. An anonymous contributor to the journal of forensic medicine sponsored by the medical department of the Ministry of Internal Affairs argued that infanticide should not be considered a form of murder, as it was in the Russian penal code, but "a crime sui generis." Citing English, German, and French authorities, this physician argued that infanticide lacked the crucial element of premeditation, since it was committed under the irresistible pressure of two emotions: shame and physical suffering.[41] A physician from Poltava argued that infanticide should not be considered a crime at all, since by definition it lacked a responsible

toubiistvo," in *Entsiklopedicheskii slovar' T-va Br. A. i I. Granat i Ko.*, 7th ed., rev. (Moscow, 1910), 19:309–10; N. A. Nekliudov, ed., *Materialy dlia peresmotra nashego ugolovnogo zakonodatel'stva* (St. Petersburg, 1880–83), 2:247–48. English juries were similarly lenient; see George K. Behlmer, "Deadly Motherhood: Infanticide and Medical Opinion in Mid-Victorian England," *Journal of the History of Medicine* 34:4 (1979), 412.

[38]Borovitnikov, *Detoubiistvo*, 22–23; art. 461, *Ugolovnoe ulozhenie, Vysochaishe utverzhdennoe 22 marta 1903 g.* (St. Petersburg, 1903), 162. The designation of illegitimate changed from *nezakonnorozhdennyi* to *prizhityi vne braka*; that is, from the offspring of an unwed mother to the offspring of people unmarried to each other.

[39]Franz v. Liszt, "Die Kindestötung," in *Vergleichende Darstellung des deutschen und ausländischen Strafrechts: Vorarbeiten zur deutschen Strafrechtsreform*, ed. Karl von Birkmeyer et al., Besonderer Teil (Berlin, 1905), 5:108–10, 112–13.

[40]For acquittals on the basis of temporary insanity—one of a married woman who murdered a twelve-day-old child (not a newborn and thus not covered by the infanticide statute) and another of a married woman who bore the child of a man other than her husband—see N. I. Dobrotvorskii, "Detoubiistvo: Sudebno-psikhiatricheskaia ekspertiza," *Arkhiv psikhiatrii, neirologii i sudebnoi psikhopatologii*, no. 3 (1893): 91–96; and idem, "Ubiistvo mater'iu svoego nezakonnorozhdennogo rebenka vo vremia rodov," *Voprosy nervnopsikhicheskoi meditsiny*, no. 1 (1905): 139–43. Apparently such defenses were extremely rare: see Viktor Lindenberg, *Materialy k voprosu o detoubiistve i plodoizgnanii v Vitebskoi gubernii (Po dannym vitebskogo okruzhnogo suda za desiat' let, 1897–1906)* (Iuriev, 1910), 39.

[41]M. G., "O detoubiistve," *Arkhiv sudebnoi meditsiny i obshchestvennoi gigieny*, no. 1, sec. 2 (1868): 26, 41–45, 55.

agent. "What kind of crime is it," he asked, "in which there is no pre-meditation, no prepared plan, and no subsequent attempt to hide it?" Rather, he said, the woman reacted *"unwillingly, passively"* to forces beyond her control.[42]

The suffering allegedly experienced by women in such conditions seemed to exonerate them from moral taint. We must understand, wrote Nikolai Zeland in deploring the harshness of the law, "the state of despair into which the inexperienced woman's soul descends under the combined pressure of shame, solitude, poverty, and postpartum in-disposition."[43] The feminist writer Serafim Shashkov cited both the psychological disturbance attending childbirth under adverse conditions and the moral scruples expressed in the feelings of "shame and fear" as reasons to exempt infanticide from the force of the law.[44] The same Ozerov who found woman's moral fabric so threadbare and easily rent refused to consider infanticide a moral failing on the grounds that nei-ther agent nor victim was in possession of the mental resources neces-sary to the definition of self: "If [woman] commits a high percentage of infanticides," he wrote, "this hardly speaks against her moral character. The major role is played by her peculiar psychological state at the mo-ment of the crime, and moreover one must note that the crime is com-mitted against a still uncomprehending child."[45] Far from being im-moral, the desire to dispose of an illegitimate child had its admirable side, some even argued: "An immoral woman can never be brought to such a pitiable position," one physician affirmed, "because she is com-pletely indifferent to her disgrace; but one with a keen sense of her dishonor is often unable to bear such misfortune. . . . In other circum-stances, she could become a tender and morally irreproachable wife and a good mother to her children."[46]

The "shame and fear" that gave rise to uncontrolled impulses and impaired judgment had the happy ambiguity of both social and psycho-logical interpretation. Most commentators combined the two implica-tions, suggesting that social pressure contributed to psychic distress. It was the rare writer who refused this ambiguity, but the eminent Anu-chin did just that in asserting that under contemporary social circum-stances, infanticide was a sign of good judgment, a rational response to a difficult situation. He refused to connect the crime to "the physiologi-cal peculiarities of the female organization," blaming it instead on the

---

[42]A. A. Zhukovskii, "Detoubiistvo v Poltavskoi gubernii i predotvrashchenie ego," *Arkhiv sudebnoi meditsiny i obshchestvennoi gigieny*, no. 3, sec. 2 (1870): 6, 10 (original em-phasis).
[43]Zeland, *Zhenskaia prestupnost'*, 15.
[44]S. S. Shashkov, "Detoubiistvo," pt. 3, *Delo*, no. 6 (1868): 54.
[45]Ozerov, "Sravnitel'naia prestupnost'," 60.
[46]M. G., "O detoubiistve," 26–27.

"disadvantageous conditions of woman's social situation, conditions under which all the blame for illegitimate sexual congress falls on [her]."[47]

More commonly, however, the "shame and fear" resulting from social pressure were considered merely adjuncts to the psychic disturbance attendant on the physical trauma of birth.[48] Occasionally a writer went so far as to consider delivery itself the source of mental impairment. Childbirth under any circumstances was enough to drive a woman out of her mind, explained an anonymous contributor to the journal of forensic medicine. Many a well-bred, cultivated woman had been seen with face distorted in pain, a crazed look in her eye, sweat pouring from her brow, rage rising in her breast toward the loving husband and newborn child. Often a woman in first childbed, this writer claimed, lost complete consciousness or fell to raving senselessly. How much more traumatic must the experience be for the ignorant, unmarried peasant girl, alone in some drafty barn, driven literally mad by pain and fear![49]

Who in fact were the women who committed infanticide? Did the alleged psychological dynamics identified with the act apply to mothers throughout society? Studies spanning the years between 1870 and 1910 allow us to see how contemporaries understood the sociology of this crime.[50] The overall statistical picture drawn by Foinitskii in 1893 indicates that infanticide was more common among rural than among urban women; higher among young and illiterate women than among older or better educated ones; remarkably low among women working in factories or those with high levels of education. Fifteen or twenty years later this profile had not changed.[51]

---

[47]Anuchin, *Issledovaniia*, 38.

[48]For this elision, see Dobrotvorskii, "Ubiistvo mater'iu," 143; S. Glebovskii, "Detoubiistvo v Lifliandskoi gubernii," pt. 2, *Vestnik obshchestvennoi gigieny, sudebnoi i prakticheskoi meditsiny*, no. 10 (1904): 1428.

[49]M. G., "O detoubiistve," 41–43. On the English application of the "merciful legal fiction" of "temporary phrenzy," see Behlmer, "Deadly Motherhood," 413. In fact, European experts were divided in their evaluation of the mother's postpartum mental state: see Gernet, "Detoubiistvo," 307.

[50]For studies based on crime statistics, see Foinitskii, "Zhenshchina-prestupnitsa," and M. N. Gernet, "Detoubiistvo: Sotsiologicheskoe i sravnitel'no-iuridicheskoe issledovanie" (1911), in *Uchenye zapiski Imp. Moskovskogo Universiteta* (Moscow, 1912). For a critical review of Gernet's study, see S. V. Poznyshev, "Kritika i bibliografiia: M. N. Gernet, *Detoubiistvo* (M, 1911)," *Voprosy prava*, no. 1 (1912): 178–92. Local studies by physicians include Zhukovskii, "Detoubiistvo"; Glebovskii, "Detoubiistvo," pts. 1–2, *Vestnik obshchestvennoi gigieny, sudebnoi i prakticheskoi meditsiny*, nos. 9–10 (1904): 1269–83, 1397–1438; and Lindenberg, *Materialy*.

[51]Foinitskii, "Zhenshchina-prestupnitsa," pt. 1, 138–39; pt. 2, 115, 121–22, 125–27. See the study using data up to 1906 in Lindenberg, *Materialy*, 32–33, 53; also Gernet, "Detoubiistvo" (1910), 311–12.

Foinitskii could have surprised no one with his findings. Writers in the 1860s and 1870s had already emphasized the role of poverty and ignorance in driving women to this extreme. In 1868 Shashkov associated infanticide with the particular conditions of urban poverty, especially among the proletariat.[52] In the same year a physician observed that the "full force of the law" fell unfairly on women of the lower classes: "Upper-class women go serenely abroad," he wrote, "where they deliver their illegitimate children and leave them to be raised by poor families for a hefty fee. The lone servant woman, peasant widow, or soldier's wife is forced to give birth where the pains of labor find her."[53] A colleague from Poltava reported in 1870 that the number of cases involving both legitimate and illegitimate births varied inversely with the mother's social standing. The culprits were inexperienced and immature: the vast majority of infant victims were firstborns, and over half the guilty mothers were less than twenty years old.[54]

The Poltava physician, like his colleagues, emphasized the special hardship experienced by peasant and other working women. Deliveries occurred in yards, outhouses, gardens, attics, in the fields, and on riverbanks. At least two-thirds of mothers fell unconscious and remembered nothing. Many infants died naturally from lack of proper care. The conditions under which unmarried peasant women gave birth and committed their crime had not apparently changed by the early twentieth century.[55] Stories of women giving birth in the snowy woods or while milking cows or cutting hay emphasized their pitiable circumstances. Commenting on the mothers' desperation and lack of forethought, a physician from Lifland reported in 1904 that "the umbilical cord is either torn by hand, bitten off, or cut with the nearest knife, sickle, or scythe, but in most cases it is simply torn, since there is not always a sharp instrument at hand."[56]

The social and personality traits attributed to murdering mothers overlapped with cultural characteristics attributed to the different social classes. The disadvantages suffered by the uneducated were not only material but also emotional and intellectual, wrote the physician from Poltava in 1870. Infanticide in his province was, he declared, "in most

[52]Shashkov, "Detoubiistvo," pts. 1 and 2, *Delo*, 1868: no. 4, 78; no. 5, 1–3, 6.
[53]M. G., "O detoubiistve," 29–30.
[54]Zhukovskii, "Detoubiistvo," 5–6.
[55]Ibid., 7; see also Aleksandra Efimenko, *Issledovaniia narodnoi zhizni*, vol. 1, *Obychnoe pravo* (Moscow, 1884), 75. For remarkable similarity to cases in Bavarian judicial records, see Regina Schulte, "Infanticide in Rural Bavaria in the Nineteenth Century," in *Interest and Emotion: Essays on the Study of Family and Kinship*, ed. Hans Medick and David Warren Sabean (Cambridge, 1984), 300–316. For outdoor places, see Glebovskii, "Detoubiistvo," pt. 2, 1412; Lindenberg, *Materialy*, 29.
[56]Glebovskii, "Detoubiistvo," pt. 2, 1412.

cases committed by inexperienced adolescent women, whose thinking is still unformed, whose character and will are still weak, and who are caught in the period of most intense passionate desire and romantic attachments. Most of them are neither well brought up nor intellectually and morally mature." It was not merely their age, however, but their cultural environment that fostered their carelessness and violence. Infanticide, this physician argued, was but the final result of a more general moral disorder: "Day and night, during work and rest, sleep and recreation, they [the women] are constantly in the closest relation to men as coarse as themselves, who cannot restrain their desires and are not accustomed to respecting women's rights. In no other social stratum is there such a high incidence of rape by immature boys of underage girls as among the common folk [*prostoi narod*]."[57]

Such critical views of peasant morality were rare among contemporary observers (but note that women figure in this benighted tableau as the victims of male seduction). The writer was, nevertheless, typical of his generation in his faith in education and enlightenment as the key to civilized behavior. The peasants' negligent sexual habits, he felt, reflected the popular disregard for individual rights and responsibilities and for the principles of legality. "Raising the level of popular intellectual and moral development will diminish, on the one hand, the degree of prejudice, of false conceptions, and of wanton violence [*proizvol*]; on the other hand, it will increase respect for legality [*zakonnost'*] in general and for personal and civil rights [*lichnye i obshchestvennye prava*] in particular. This in turn will change the nature of current public attitudes to unmarried mothers and the mothers' attitude to their own illegitimate children."[58]

The literature shows an interesting ambiguity, however, precisely about the question of peasant morality. Certainly the national statistics demonstrate that infanticide was more common in the countryside than in the towns. This distribution might have suggested that peasant morality was stricter and communal sanctions more severe than the vague opprobrium an unwed pregnant woman might confront in the city. Yet Shashkov insisted in the late 1860s that peasant women rarely killed their newborns from motives of "shame and fear." Peasant communities, he claimed, did not condemn illegitimacy. Rather, husbands who returned from long absences looked with favor on children conceived "without their participation" as useful additions to the household labor force.[59] Other writers stressed the impossibility of generalizing about

[57]Zhukovskii, "Detoubiistvo," 10.
[58]Ibid., 12.
[59]Shashkov, "Detoubiistvo," pt. 2, 30–31.

the peasantry's attitude toward extramarital sex. As one physician had observed in the 1860s, "at one end of Russia the unmarried pregnant woman pulls in her stomach, trying to hide her state, while at another, she flaunts her belly. Having proved her ability to conceive, she is sure of marrying after the delivery."[60]

Whatever the variety of rural customs, most writers agreed that cities provided something of a haven for sexually delinquent women. "In the cities," wrote a physician from Lifland in 1904,

> the unwed pregnant mother encounters less ferocious disdain than in the countryside and is less troubled by the fear of losing her job. The adolescent mother's ties with family, relatives, and friends have weakened . . . [as] she leads a more independent life; she therefore feels the shame of her condition less acutely. Circumstances in the city do not limit a mother's personal life as sharply as those in the village. On the contrary, the city offers her the services of lying-in hospitals, mothers' shelters, and even foundling homes. Here the child's existence is merely an encumbrance to her continued work, not an obstacle. Having established her right to independent personal existence, the girl begins to ignore the opinions of those she has left behind.[61]

City life did not seem to make children more desirable, however. Rather, cities offered more effective, less violent and desperate ways of getting rid of them. Having cited the public institutions and social conditions that lessened the pressure to do away with newborn infants, the same observer noted that although urban living "reduce[d] infanticide, it increase[d] abortion." Having celebrated individualism as a liberation from coercive morality of communal life, he proceeded to denounce it as the source of an equally immoral habit. Abortion, he declaimed, was "the sad accompaniment of our [modern] development and the growth of personal needs." It did not represent progress but constituted "only a new method of preventing the birth of those 'excess' children, who, among the poor, and especially in the countryside, end up among the dead."[62]

These two forms of birth control, infanticide and abortion, came to represent opposing modes of action: the first, irrational, impulsive, violent; the second, rational and planned.[63] Infanticide was seen as a spontaneous deed committed by young, desperate, illiterate village girls; abor-

---

[60]M. G. "O detoubiistve," 49; also Glebovskii, "Detoubiistvo," pt. 2, 1408.
[61]Glebovskii, "Detoubiistvo," pt. 2, 1402.
[62]Ibid., 1403.
[63]See, e.g., ibid., 1419. For more extensive discussion, see chap. 6 below.

tion and especially contraception were favored by more educated women. Even female factory workers, observers claimed, exercised greater control over their reproductive lives than rural women did.[64] Foinitskii had emphasized that education and employment outside the home diminished the level of female crime in general. A later researcher noted the apparent paradox that in Lifland villages in the early twentieth century, most murdering mothers were not ignorant fieldhands but servants who had acquired at least the rudiments of literacy. Yet for such women, he argued, the conditions of employment that demanded literacy also precluded privacy or independence. Only the factory woman, he asserted, had attained the degree of personal autonomy and enjoyed access to the kinds of public resources that no longer made infanticide seem the only way out.[65]

In the years before the war, infanticide never ceased to figure as a sign of cultural deprivation: the hapless peasant girl who was caught unprepared in a barn and smothered her infant in the hay stood for the traditional woman's irrational, passive, victimized life. As we have seen, the very crime of infanticide presupposed the absence of intentionality and full mental competence. Indeed, the literature offers only the slightest suggestion that some peasant women may have tailored their actions to take advantage of the law, deliberately endangering the lives of their newborns rather than killing them outright so as to incur a lesser penalty in case of conviction.[66] The overwhelming numbers of infants taken to foundling homes and the impossibility of correlating illegitimacy rates with the incidence of infanticide may also suggest that the murder of newborns, though gruesome and distressing to the mother, was not the mental aberration nineteenth-century lawmakers and physicians wanted it to be.[67] Whatever the subjective experience of these desperate mothers, it is clear that observers were intent on seeing them as the victims rather than the mistresses of their fate.

The Cultural Norm: Customs of the Countryside

The wayward peasant woman, as she appears in the literature on crime, did not strike out on her own. Her crimes were reactive: re-

---

[64]On rational methods of birth control as a mark of cultural development, see N. A. Vigdorchik, "Detskaia smertnost' sredi peterburgskikh rabochikh," *Obshchestvennyi vrach*, no. 2 (1914): 212–53.

[65]Glebovskii, "Detoubiistvo," pt. 2, 1417–21, 1430–31.

[66]Ibid., pt. 1, 1273–74.

[67]On foundlings, see Ransel, *Mothers of Misery*; on lack of correlation, see Glebovskii, "Detoubiistvo," pt. 2, 1409.

sponse to the sexual seduction of a man, panic at the birth of an unwanted child, murder in revenge for domestic abuse.[68] She did not seek
wealth, opportunity, or a sense of power in relation to those with more
power than herself. She preferred stealth and cunning to violence and
open attack: her favorite means were poison and the torch.[69] Among her
most serious crimes was arson. She stood in relation to the male peasant, in this view, as the peasantry stood to society as a whole: instinctive,
unreasoning, inhabiting a world of organic growth and natural functions, despite her remarkable tendency to give motherhood short shrift.

To serve this supposedly organic, unselfconscious community, the
rural township courts introduced in 1861 were designed, as we have
said, to operate in accordance with peasant custom, not written law.
Part of the interest in evaluating this new institution in its first decades
of operation was therefore linked to the question of whether the system
could be said to fulfill in any recognizable way the requirements of
justice as understood by the cultural elite.[70] The answer clearly depended on one's evaluation of the practices on which the system drew,
to which there could be no better guide than the evidence of the cases in
these very courts. "The decisions of the township courts," wrote Aleksandra Efimenko, one of the first to examine them in depth, "are of
great interest both for students of the common people [narod] in general
and especially for students of folk concepts of justice [spravedlivost']."[71]

Many cases dealing with public problems concerned family affairs,
since family and community were tightly intertwined. Legal scholars
searching for the cultural basis of post-Reform peasant justice were
therefore led to consider the nature of family relations and, in particular, the role and experience of peasant women.[72] The relationship between men and women showed them how the peasant male—who

[68]On male responsibility for sex, see M. G., "O detoubiistve," 27. On murder resulting from forced marriage and the husband's violence, see Anuchin, Issledovaniia, 39; Zeland, Zhenskaia prestupnost', 14.

[69]Anuchin, Issledovaniia, 39; Foinitskii, "Zhenshchina-prestupnitsa," pt. 1, 133; Tarnovskaia, Zhenshchiny-ubiitsy, 345; Paul Kovalevsky [P. I. Kovalevskii], Psychopathologie
légale, vol. 1, La Psychologie criminelle (Paris, 1903), 182.

[70]See C. A. Frierson, "Rural Justice in Public Opinion: The Volost' Court Debate,
1861–1912," Slavonic and East European Review 64:4 (1986): 526–45; and Moshe Lewin,
"Customary Law and Russian Rural Society in the Post-Reform Era," Russian Review 44
(1985): 1–19.

[71]Efimenko, Issledovaniia, 99. Her status as a "pioneer" is noted in Lewin, "Customary
Law," 4.

[72]See Peter Czap, "The Perennial Multiple Family Household, Mishino, Russia 1782–
1858," Journal of Family History 7:1 (1982): 5; and Paul Friedrich, "Semantic Structure and
Social Structure: An Instance from Russian," in Explorations in Cultural Anthropology, ed.
Ward H. Goodenough (New York, 1964), 141. In her essays on customary law written in
the 1870s and early 1880s, Efimenko discusses marriage, the peasant woman, and family
subdivision; see her Issledovaniia.

headed the family, ran community affairs, and sat on the peasant courts—wielded his power. It revealed his sense of justice. If the investigators were not gender-blind, it was because the gender issue hit them squarely in the eye. It was also because the 1860s intelligentsia, influenced by Jeremy Bentham and John Stuart Mill, had seen the "woman question" as intrinsic to the problem of social reform.[73]

Educated observers of the rural scene recognized that peasant custom had the force of law in the sense that it determined the way the community ran its affairs, disposed of property, established and maintained family ties, and upheld standards of conduct. But they disagreed as to whether this popular culture could be said to reflect a sense of fairness or justice. By definition tied to local tradition, peasant customs certainly varied from place to place, but lack of uniformity did not necessarily imply the absence of coherent guiding principles.

Those who argued for the existence of an underlying legal sense explained it as a reflection of the peculiarities of peasant social organization. Folk values, they insisted, ought not to be judged by European standards appropriate to societies based on individual autonomy rather than on collective responsibility, as in the Russian case. In a Slavophile vein, Efimenko went so far as to assert that customary law was the key to "our real national character," whereas positive law was a foreign import suited only to the elite's "artificial" way of life.[74] Others felt that justice might once have prevailed at the popular level but that peasant society had been corrupted by modern ways of life associated with manufacturing or the city and had therefore succumbed to the rule of violence and fallen into moral disarray. Still a third argument held that violence and the arbitrary exercise of power were intrinsic to those hallowed ancient ways and that only modernity, in the shape of education and positive law, held out the promise of justice in any meaningful sense of the term. Some commentators combined more than one point of view, to the detriment of good logic and political consistency, though perhaps with the greatest respect for the complexity of popular experience.

The defense of the peasants' traditional moral integrity sometimes took ingenious turns, as in the case presented by Mitrofan Zarudnyi, a member of the official government commission that investigated the operation of the township courts. An ardent Anglophile and proponent of the Benthamite ideal of equal and accessible justice for all, Zarudnyi

---

[73]Citing Bentham is M. I. Zarudnyi, *Zakony i zhizn': Itogi issledovaniia krest'ianskikh sudov* (St. Petersburg, 1874), 13 and elsewhere. See also Richard Stites, *The Women's Liberation Movement in Russia: Feminism, Nihilism, and Bolshevism, 1860–1930* (Princeton, N.J., 1978), 73–75.

[74]Efimenko, *Issledovaniia*, 181, 179.

nevertheless argued in 1874 that a uniform system of justice was still inconceivable in the Russian empire because of the vast cultural division between its educated and traditional parts. The decree establishing the township courts had correctly intended them to serve the interests of the peasant estate as a whole and not of its individual members, Zarudnyi held, precisely because peasants made no real distinction between individual and community.[75]

Though the township courts did not meet Zarudnyi's standards of judicial performance, he did not attribute their inadquacy to defects in the popular understanding of justice. Rather he criticized them as arbitrary impositions of the central state, which in fact threatened to obliterate the instinct for justice inherent in the simple life but lacking in society's educated strata. This instinct, Zarudnyi felt, was revealed in the activity of folk institutions entirely independent of the state, such as village meetings and informal arbitration courts. Far from being an authentic peasant institution, Zarudnyi insisted, the township court was as artificial as any other construction of the bureaucratic mind. He nevertheless insisted that it served the peasant community better than did the rural justices of the peace, who offered the advantages of explicit judicial procedure and regularized standards of judgment but belonged to a social and mental world remote from the reality of village life.[76]

In locating authentic popular culture at a deep level of collective practice, where dominant values expressed themselves in consistency of action but not in clearly articulated abstract principles and formal rules, Zarudnyi represented a view widely though not universally held. Most other observers agreed that peasants settled their conflicts in a distinctive fashion that reflected the collective basis of their social life. Some, however, identified the distinctive feature as the absence of consistency and refused to acknowledge any virtue in the township courts, which they described as entirely arbitrary institutions—not because they imposed bureaucratic practices on self-regulating communities but because

[75]Though Zarudnyi insists on this point (*Zakony i zhizn'*, 55–56), he mentions that "minority" and individual interests were sometimes in need of special protection (59–60), implying that the identity of collective and private needs was not always perfect. Biographical data are in A. A. Polovtsov, *Russkii biograficheskii slovar'* (St. Petersburg/Petrograd, 1896–1918), 7:239–40.

[76]Zarudnyi, *Zakony i zhizn'*, 13–15, 75, 172, 184–85. Another advocate of informal peasant justice (below the volost' court level) was V. V. Tenishev; see his *Pravosudie v russkom krest'ianskom bytu* (St. Petersburg, 1907), 69. On the justices of the peace, see Joan Neuberger, "Popular Legal Cultures: The St. Petersburg *Mirovoi Sud*," in *The Great Reforms*, ed. John Bushnell and Ben Eklof (Indiana University Press, forthcoming). The justices operated under the rules contained in N. S. Tagantsev, ed., *Ustav o nakazaniiakh, nalagaemykh mirovymi sud'iami. Izdanie 1914 goda. S prilozheniem motivov i izvlechenii iz reshenii kassatsionnykh departamentov Pravitel'stvuiushchego Senata*, 22d ed., rev. (Petrograd, 1914).

they reflected the abuses of power endemic to the peasant world.[77] Because this world rested firmly on the back of family organization, observers turned their attention to the regulation of sexual behavior as a central aspect of village life.[78]

Nothing separated that life more sharply from upper-class experience than the way such regulation was carried out. European notions of privacy and individuality had no meaning in the "archaic" folk context, wrote Aleksandr Kistiakovskii, professor of law at Kiev University, in an 1878 essay documenting community control of sexual purity through the traditional marriage rite. The wedding itself and the confirmation of the bride's virginity, he reported, were in some parts of Russia an entirely public affair (*delo obshchestvennoe*).[79] Describing in detail the ceremony of defloration (in which an older man or the matchmaker's finger might substitute for the groom in case of need) and the rituals surrounding the announcment of the results, Kistiakovskii contrasted these practices with the mores of educated society. There

> the conclusion of marriage and its attendant events are more and more becoming the exclusive affair of individual life. The public ceremony of marriage is gradually falling into disuse. An educated person does not like and tries not to permit outside eyes to penetrate his strictly individual life. Although, on the one hand, marriage in European societies is more and more losing the character of a church sacrament [*tainstvo*] and acquiring the traits of a civil institution, on the other hand, it is also assuming the character of a secret [*taina*], in the sense that outside eyes are ever more removed from direct observation of this event.[80]

Even peasants who left the villages to work in towns and factories resisted the old practices of sexual publicity, Kistiakovskii found, and standards of virtue were themselves in the process of changing.[81] But peasants who remained saw nothing mysterious in sex; everything was natural, out in the open, and therefore also completely social. Dmitrii Zhbankov, himself of peasant origin, noted that "everything [sexual]

---

[77]N. Kalachov, *Ob otnoshenii iuridicheskikh obychaev k zakonodatel'stvu* (St. Petersburg, 1877), 5, 8–9. This speech on the relationship between law and custom, presented to the first congress of Russian jurists in 1875, was published by the Imperial Russian Geographical Society, Ethnographic Division.

[78]V. A. Aleksandrov, *Obychnoe pravo krepostnoi derevni Rossii XVIII-nachalo XIX v.* (Moscow, 1984), 17.

[79]Kistiakovskii, "K voprosu," 161–62. Biographical data for Kistiakovskii are in Polovtsov, *Russkii biograficheskii slovar'*, 8:720–25.

[80]Kistiakovskii, "K voprosu," 161–62.

[81]Ibid., 170, 167.

occurs in one hut, in front of everyone, and therefore belongs to custom and habit [*obychnyi i privychnyi*]." To him this meant that "sexual relations [were] simpler and more natural" in the village than in the city.[82]

Certainly, Kistiakovskii and Zhbankov agreed, sexual customs revealed the community's deepest values and organizing principles. Good morals in the peasant context thus involved public disclosure, and in particular the physical examination and display of the woman's body, whereas among the Europeanized classes purity entailed privacy and concealment. The element of exposure and subjection to the public regard, as these educated observers understood it, was in some way an essential aspect of peasant culture that transcended local differences in standards of behavior. Controlling access to one's body, whether physical or visual, their accounts suggest, was a symptom of the development of certain notions of personality. In the absence of such notions, it was impossible to imagine individual rights in the Western juridical sense. A woman's physical experience was therefore more than a folkloric curiosity. It expressed (or embodied) the culture's basic concept of itself.

The community's intrusion into the sexual lives and even the sexual organs of its members occurred in the context of general physical hardship, bodily neglect and abuse, and levels of interpersonal violence unacceptable to educated observers, some of whom saw absolutely no value in the old patriarchal ways.[83] Traditional folk culture, observed the journalist and justice of the peace Valerian Nazar'ev in 1872, was the remnant of a time when lawlessness prevailed at all levels of society. "Over the centuries," he wrote, "there emerged a unique and peculiar way of life, characterized by the use of unlimited violence [*bespredel'nyi proizvol*]. . . . Everywhere there reigned relations of an unimaginably patriarchal and primitive nature [*patriarkhal'nost' i pervobytnost'*]."[84] If the continuing hold of violence and corruption in the backwoods was to be destroyed, the rule of law must be imposed.

To post-Reform Russian ears, the term "patriarchy" evoked the era

---

[82]Zhbankov, "Polovaia prestupnost'," 82.

[83]For critical evocations of male brutality, see, e.g., E. T. Solov'ev, *Grazhdanskoe pravo: Ocherk narodnogo iuridicheskogo byta*, vyp. 1 (Kazan, 1888), 10–13. The monarchist newspaper *Moskovskie vedomosti*, by contrast, attacked critics of peasant domestic violence for imposing their own cultural standards on the people, thereby helping to undermine the legitimacy of popular family life; see S. Kozhukhov, "O praktike Pravitel'stvuiushchego Senata po voprosu o vydache kres'ianskim zhenam otdel'nykh vidov na zhitel'stvo," *Zhurnal Ministerstva Iustitsii*, no. 3 (1901): 159.

[84]Nazar'ev, "Sovremennaia glush'," 605–6. Biographical data for Nazar'ev are in S. A. Vengerov, *Istochniki slovaria russkikh pisatelei* (St. Petersburg/Petrograd, 1900–1917), 4:485.

of serfdom and thus the horror of unlimited authority and abuse of force.[85] Patriarchy was certainly a way of life: it had its rules, rituals, and assigned roles. But it was lawless because there was nothing to check the exercise of the patriarch's power on those under his control. Even Zarudnyi, despite his admiration for peasant self-regulation, deplored the existence of shaming rites directed particularly against theft and sexual misconduct. Because these rituals seemed to occur without the endorsement of the village assembly, Zarudnyi believed, they violated the standards of legality that generally prevailed in the village. To defend the integrity of peasant justice, Zarudnyi therefore tried to explain such events as the peculiar legacy of serfdom, symbol of all that was unfair, arbitrary, and violent in Russian culture—and furthermore, according to him, imposed by the state.[86] Another commentator singled out the area of family relations as one particularly open to abuse in peasant courts, which, in settling domestic matters, "frequently produce[d] entirely arbitrary decisions, bordering on vicious mob rule [*zhestokii samosud*]."[87]

In revealing the abuses of patriarchy in its domestic sense, writers also condemned the moral foundation of the "old dispensation" (*starye poriadki*) as it persisted among former victims of the abolished serf system.[88] Indeed, women were the last remaining victims of that old order. For all her Slavophile defense of peasant justice, Efimenko never equated the practice of the township courts with the principles that dominated the traditional, patriarchal family, which she condemned for disregarding the "human dignity" of its female members. "Nowhere is the continued vitality [of patriarchal assumptions] as sharply revealed as in the sphere of relations defining the woman's place in the family," Efimenko wrote. "In all the various forms of the peasant family, the patriarchal principle emerges as a powerful force, crushing woman's individuality and depriving her of autonomy."[89]

Given the force of patriarchal repression, observers found it remarkable that the downtrodden peasant wife had the courage and independence to defend her own interests on public ground. Such initiative did

[85]"Serfdom was not only an institution that determined the relationship between two social estates but one that cast its shadow on decidedly all aspects of Russian life, defining its structure and content. The serf-owner was not only the lord of his peasants but also the bearer of civic and state power [*nositel' obshchestvennoi i gosudarstvennoi vlasti*], administering justice and meting out punishment for the mass of the population": S. N. Igumnov, "Zemskaia meditsina i narodnichestvo," in *Trudy odinnadtsatogo Pirogovskogo s"ezda*, ed. P. N. Bulatov (St. Petersburg, 1911), 1:79.

[86]Zarudnyi, *Zakony i zhizn'*, 194.

[87]Tenishev, *Pravosudie*, 63.

[88]Nazar'ev, "Sovremennaia glush'," pt. 1, is headed "The old ways and the new court."

[89]Efimenko, *Issledovaniia*, 69, 123.

not fit with notions of traditional female behavior. Yet women did appeal to justices of the peace. "One after another," Nazar'ev wrote, "peasant women appeared before me with bloody stains instead of faces, sobbing violently, describing with swollen, shaking lips their husbands' beastly treatment." They came because they had realized that the newfangled law, in the person of the justice of the peace, could work to their advantage. "The women have taken heart," said Nazar'ev in reporting the complaints of peasant men, "twirling their tails like magpies." They had gotten out of hand, said the men. "She wants to stand higher than a man," Nazar'ev quoted one as saying. "You raise a stick to her, and she runs to the judge: 'I'll get my rights, you'll answer for the blow, and that's the last you'll see of me!'"[90] There was nothing harmonious, in this view, between the interests of the community and its individual members.

The criticism of folk custom was widely repeated in the 1880s. In an 1883 article on "personal relations between spouses according to Russian customary law," which appeared in the journal of the Moscow Juridical Society, N. Lazovskii followed Nazar'ev's lead in denouncing the patriarchal basis of peasant society. He too had noticed that women flocked to the justices, "complaining not of trifles but of serious blows and often injuries." The township courts did not help, Lazovskii explained, because they were guided by "ossified" patriarchal principles that were "still alive among the popular masses and [had] the greatest impact on the position of women." In crushing female autonomy, the senior male acted in the interests of the familial collectivity, which did not tolerate assertions of individuality by any of its subordinate members.[91]

Just as the peasantry as a whole often appeared to the ruling classes (and to hopeful revolutionaries) as the bitterest enemy of the old order, precisely because it was society's most oppressed group, so the long-suffering peasant wife might figure not only as the incarnation of traditional life but as its fiercest opponent. Was she not already engaged in sniper action, poisoning her husband and burning the barn, as the criminologists reported? Indeed, Aleksandra Efimenko remarked in the 1880s that women had ceased to submit patiently to "the yoke."

> Wherever the [large, patriarchal family] remains firm, women passively accept their bitter lot. But no sooner do they sense that life might be organized on different principles than consciousness of the

---

[90]Nazar'ev, "Sovremennaia glush'," 610, 612; Efimenko (*Issledovaniia*, 72) cites this article to prove the low esteem in which peasant wives were held.

[91]N. Lazovskii, "Lichnye otnosheniia suprugov po russkomu obychnomu pravu," *Iuridicheskii vestnik*, no. 6–7 (1883): 382, 359, 365.

entire burden of their actual situation and a burning desire to change it are immediately aroused. The already wavering principles of the patriarchal clan find in [the peasant woman] an opponent embittered by the age-old burden that has weighed upon her with such unbearable oppression. . . . Everywhere the people . . . recognize woman as the malicious opponent of the common joint clan-family life and consider her the main cause of the downfall of the old order.[92]

Adopting the position developed by Efimenko, Lazovskii argued that "the internal organization of the Russian folk family represents a conflict between the patriarchal system and new aspirations, between the family's devouring of individual members and its members' insistence on the recognition of their individuality. . . . The destruction of the clan family [rodovaia sem'ia]," he maintained, ". . . is the first step toward women's emancipation"—but only a first step. Better off though women were in the smaller family units that formed as the younger generation broke away, their subjection continued nevertheless.[93]

Efimenko derived her argument from a general view of the dynamics of social development in the countryside. It was true, in her opinion, that "the woman's position in the smaller [nuclear] peasant family was still to a significant degree determined by conceptions retained from the extended clan organization [rodovoi byt]." But these old notions did not have the same effect, Efimenko observed, in the absence of kin to enforce them. Though the wife still worked hard, perhaps even harder, in her own household, she was no longer subject to the authority of other women; she managed her own affairs, and her husband respected her authority in matters that concerned her. Now that she was obliged (or permitted) to exercise the initiative, intelligence, and ingenuity repressed in the old collective unit, her personality began to emerge: "The woman in the smaller family must of necessity throw off the dullness and apathy that were the products of her former slavish position." This development reflected the incursion of the "labor principle," which, Efimenko believed, was displacing the outmoded patriarchal basis of communal life and promoting the emergence of "autonomous individuality" (samostoiatel'naia lichnost') in peasant society.[94]

Although Zarudnyi deplored the township courts as recent innovations that failed to reflect the deeper impulses of the peasant commu-

---

[92]Efimenko, Issledovaniia, 84, 89.
[93]Lazovskii, "Lichnye otnosheniia," 358–60, 379–80, 393; cf. formulations in Efimenko, Issledovaniia, 122, 91.
[94]Efimenko, Issledovaniia, 68, 91–92, 95, 122. On labor principle, see Aleksandrov, Obychnoe pravo, 20–27.

nity, most observers indicted these institutions precisely for reinforcing the community's old-fashioned, patriarchal ways. This complicity was clearly revealed in the courts' manner of handling domestic conflict. According to customary law, as manifest in the decisions of the township court and endorsed by village opinion, the husband had absolute control over his wife's movements and the right to her labor power, her presence in the household, and her obedience, which he was justified in obtaining by physical means. "The peasant woman's position in everyday life," Lazovskii wrote, "is *contingent* [*sluchaino*], entirely dependent on her husband's *arbitrary power* [*proizvol*], because the norms of customary law on this question are so vague and give so much scope to the husband's abuse." Among the husband's powers, Lazovskii explained, custom acknowledged the right to apply physical discipline (*telesno nakazyvat'*), the traditional method used "to suppress the wife's 'will' and make her into some sort of creature without personality [*bezlichnoe sushchestvo*]." Her complaints of mistreatment were perceived as insubordination: "The people view almost any evidence of the wife's independence as an affront to her husband's dignity." If husbands, for their part, did not often turn to the township court to chastise recalcitrant wives, it was because they preferred to "teach their wives a lesson" on their own. In the case of adultery, for example, village leaders and neighbors did not disapprove of the husband's taking punishment into his own hands, short of murdering the allegedly faithless wife.[95]

In evaluating the role of the township courts, Efimenko tried to reconcile her populist regard for the value of folk authenticity with her indignation over the fate of women in this traditional order. Like everyone else, she recognized that popular attitudes did not favor women in cases of physical abuse. Since beating was considered a husband's prerogative, township courts settled such complaints in their own way, usually with leniency toward the man, rather than referring them to the regular courts, where, as criminal offenses, they belonged. But, Efimenko argued, the very fact that township courts agreed to hear the complaints of wives against husbands was a sign of their sensitivity to the demands of real life and of their superiority to the regular courts, which were forbidden to consider cases brought by one spouse against the other except in conflicts over property and in criminal cases (of which adultery was officially one). Their freedom from rigid rules, Efimenko believed, often allowed the peasant courts to show a "hu-

---

[95]Lazovskii, "Lichnye otnosheniia," 381–83, 392, 394 (original emphasis). On community tolerance for husbands' violence, see also Solov'ev, *Grazhdanskoe pravo*, 10. For another view that customary law reinforced patriarchy and did nothing to inhibit male violence, see Iakob Ludmer, "Bab'i stony," pt. 1, *Iuridicheskii vestnik*, no. 11 (1884): 462, 465.

manitarian" flexibility absent from the "cold callousness of positive law. . . . Difficulties arising from the marital union are resolved more simply, expediently, and fairly by the peasants than by other social estates. The interests of the weak party, that is, women, are considered more carefully by the township courts than by courts obliged to adhere strictly to the law."[96]

Rejecting Efimenko's view of customary law as "humanitarian," Lazovskii called it "inhuman" and outmoded,[97] but he agreed that it showed a practical flexibility in dealing with concrete cases—sometimes even to the benefit of women, whose rights it systematically denied. Indeed, he conceded that the peasant courts sometimes accommodated the woman's interests in a manner impossible for courts restricted by positive law. Despite the statutory insistence on the duty of marital cohabitation, for example, township courts occasionally granted a wife's request for separation on grounds of the husband's sexual incapacity, his failure to provide economic support, or his exile on criminal charges (but only if he had abused her in the past). Welcome as these exceptions may have been, however, they did not represent concessions to individual rights; the first two in fact derived from the traditional assumption that the family was justified only by procreation and could not function without a roof over its head. The third did acknowledge—though in a limited way, contingent on the husband's prior domestic misbehavior—that spouses were not responsible for each other's behavior. Lazovskii cited further instances in which the township courts had authorized separation in the case of violent physical abuse, without absolving the pair of marital obligations, by stipulating as a condition the husband's continuing economic support or, alternatively, the wife's payment of monetary compensation for the labor she had ceased to provide.[98]

If women fled the township courts for the justices of the peace despite these accommodations, according to Lazovskii, it was because the justices were not personally implicated in their immediate communities, as every peasant judge on a township court necessarily was. What Zarudnyi saw as the drawback of formal justice was precisely its virtue in women's eyes: the justices promised them a fair, because disinterested, hearing.[99] Even here, however, the women were often disap-

---

[96]Efimenko, *Issledovaniia*, 99–100, 103, 115–16. Spouses under Russian law retained independent property rights.

[97]Lazovskii, "Lichnye otnosheniia," 385–86.

[98]Ibid., 371–73. So restrictive was the law binding women to their husbands that even the Senate often intervened to grant peasant women their own passports; see Kozhukhov, "O praktike," 158–68.

[99]The same point was made by another justice of the peace (also a correspondent for

pointed, as Nazar'ev and Lazovskii both complained. One reason was that the statutory limits of the justices' power prevented them from intervening in a range of domestic and family matters reserved precisely for the local peasant courts. The justices were empowered to hear cases of abuse, for example, only when a medical examiner deemed the injuries sufficiently serious to qualify as criminal assaults. A second reason, Lazovskii argued, was that "the basic principles of customary law, deriving from the very nature of the patriarchal family order, entered whole into the Digest of the Laws." "How little attention," he lamented, "is paid to the wife's individuality both in customary and in positive law. Our legislation sees woman as a limited, weak creature who needs protection, and regards her with condescension, rather than offering her the rights that should belong to her as a member of society."[100] No woman could legally leave her husband, refuse him domestic and conjugal services, or deny him full obedience and respect.

In extending his criticism of peasant justice to the system of statutory law as well, Lazovskii showed an ambivalence typical of Russian liberals. Applying a Western standard against which to judge national institutions and avoiding the sentimental view of folk custom, they at the same time refused to idealize bourgeois social relations. Thus, Lazovskii maintained that the brutality of Russian peasant existence served only to underscore the defects of a moral system from which society as a whole was not free:

> Among the peasantry the arbitrary exercise of power [proizvol] and the habit of taking the law into one's own hands [svoevolie] are more visible, since they are expressed in more tangible form in the physical suffering of the wife from her husband's blows. Among the urban lower middle class [meshchanstvo] the wife's sufferings are even more terrible, since the husband often lives off his wife, beats her, and goes on the rampage, while she remains without any defense. Among the upper social classes imposition of the husband's will [svoevolie] is already more or less concealed, often almost indiscernible to the outsider's eye, expressed only in the wife's secret moral suffering. There is no recourse to the courts in such cases, since existing law does not touch on this question.[101]

---

the liberal newspaper *Russkie vedomosti*), who recounted fifty-four cases in which wives complained of physical abuse: Ludmer, "Bab'i stony," pt. 1, 447–48, 450. (On Ludmer, see Vengerov, *Istochniki slovaria*, 4:16.) Another article in the same vein, though less detailed and original, is Vereshchagin, "O bab'ikh stonakh," *Iuridicheskii vestnik*, no. 4 (1885): 750–61.

[100]Nazar'ev, "Sovremennaia glush'," 619; Lazovskii, "Lichnye otnosheniia," 399, 401.

[101]Lazovskii, "Lichnye otnosheniia," 404. Efimenko (*Issledovaniia*, 73) likewise argued

Iakob Ludmer, another liberal justice of the peace who wrote in the same period, also insisted that the cruelty of peasant husbands could not be blamed entirely on the defects of folk culture, for no part of society was isolated from the rest. Domestic tyranny, he too declared, originated in the ancient patriarchal family in which the old man ruled the roost, beat his wife, and slept with his sons' young wives in the hallowed tradition of *snokhachestvo*. But in his experience the disadvantages of olden times paled before those of his own. Unlike Efimenko, who sensed a turn for the better, Ludmer believed that woman's lot had deteriorated "along with the people's ruin, which has caused a decline in popular morals. Many observers of contemporary popular life confirm the growing cruelty, amounting at times to simple brutality, of the popular mass." This deterioration, however, only reflected the moral tone of society as a whole, Ludmer believed. The male peasant's own subjection to abuse caused him to mistreat those even more powerless than himself: "Peasants punished with birching most often later become despotic with their own wives and children." In a culture of general physical violence, it was not surprising that women suffered most. "After the [Russo-Turkish] war," Ludmer reported, "the level of family concord among the peasants fell markedly." Rudeness, conflict, and cruelty increased, especially in the families of returned soldiers.[102] Thus, violence administered in the public realm filtered down into personal relations.

Just as the state, in resorting to institutionalized violence and failing to respect the personal dignity of its subjects, created conditions that reinforced the worst aspects of traditional culture, so the state might help transform popular mores through the instrument of the law, which was in Ludmer's view the agency of progress and enlightenment. The only hope for women, he insisted, was the strict application of general laws based on the principle of individual rights. Divorce must become a legal possibility, or women would continue to suffer as they had always done. "To put one's hopes for the improvement of women's position only on the general softening of manners, on the spread of education and material well-being, is unthinkable," he warned, "for many years will pass before the softening of manners becomes an indisputable fact. To keep women outside the law for all this

---

that "civilized society" had contempt for women but covered it with hypocritical talk of women's rights.

[102]Ludmer, "Bab'i stony," pt. 1, 446, 455–58; "Bab'i stony," pt. 2, *Iuridicheskii vestnik*, no. 12 (1884): 673–74.

time, leaving them in slavery, would be inhuman and incompatible with the goals of the civil community [*gosudarstvennyi soiuz*]."[103]

As these reflections show, the discussion of family relations, sexual morality, and crime led criminologists, folklorists, justices, and physicians to elaborate their views on the underlying principles of social organization and the basic values of the national culture; hence, they promoted contending visions of the ideal state and the ideal conditions of political existence. Though many observers celebrated folk tradition, only archconservatives defended the patriarchal relations on which it was based.[104] And though many liberals, as well as populists, were ambivalent about the consequences of individualism (urban crime, for example), few glorified the absence of personal autonomy to which village conditions led. Biological theories suggested a resolution to such ambivalence, for they explained crime and violence as the products of uncontrollable forces, organic processes in which neither individuality nor culture played the central part. Such theories also offered the medical profession an active role in the identification and regulation of social deviance.

[103]Ibid., pt. 2, 678–79.
[104]E.g., *Moskovskie vedomosti*, cited in Kozhukhov, "O praktike," 158–59.

Chapter Four

Female Sexual Deviance and
the Western Medical Model

Prostitution offered physicians and criminologists choice terrain on which to exercise both therapeutic and theoretical expertise. Not a crime, and thus not the object of prosecution in the courts, prostitution was a form of social deviance subject to various kinds of public custody, from police surveillance to medical examination to charitable rehabilitation. As links in the spread of venereal disease, public women fell under the scrutiny of the public health establishment. Prostitution also constituted the archetype of female sexual deviance. Open to professional as well as sexual intrusion, prostitutes provided forensic experts, criminologists, and physicians with their primary access to the functioning of female sexuality, since respectable women were defined precisely by their domestic seclusion and purported indifference to the erotic urge.

Because it was linked to the spread of disease, prostitution came to embody a diseased condition in its own right. Over the course of the nineteenth century, European experts became less likely to regard public women as unfortunate creatures performing a socially stigmatized role and began to interpret their irregular sexual behavior as a symptom of underlying pathology.[1] The assumption of abnormality was not an

[1] For the evolution of nineteenth-century medical concepts of sexual deviation, cf. Alexandre Parent-Duchâtelet's enormously influential work first published in 1836, *De la prostitution dans la ville de Paris, considérée sous le rapport de l'hygiène publique, de la morale et de l'administration*, 3d ed. (Paris, 1857), with the equally influential 1893 study by Cesare Lombroso and Guglielmo Ferrero, *La Femme criminelle et la prostituée* (Paris, 1896). Viewed initially as the commission of morally reprehensible acts by persons otherwise normal, deviation came to be seen as an organic disturbance, which explained the commission of such acts by mentally and bodily defective persons.

initial component of the state's justification for the system of police-medical control, but it eventually helped to bolster the custodial rationale. The assumption reflected the burgeoning influence of biological theories of deviance, which were widely debated in late nineteenth-century Europe.[2]

In adopting Western models of public administration, professional organization, and scientific expertise, Russia also implemented a system of regulated prostitution and absorbed the sexual ideology elaborated in Western medical texts.[3] Official surveillance of public women comfortably suited the Russian *Polizeistaat* tradition, but biological theories of social and sexual deviance, which translated these state functions into scientific terms, seem to have encountered strong cultural resistance as they crossed the national divide.

Two structural factors may help explain the extent to which Russian physicians resisted this ideological pressure from the West. The first was the medical community's relation to the state. The foremost student of Russian medicine has argued that the profession sought autonomy and legitimation not by seeking official patronage and validation, as Western physicians usually did, but rather by resisting the state's attempt to monopolize control of public policy and technical skill.[4] Forensic obligations were particularly onerous to Russian doctors, who were compelled, for example, to supervise the execution of corporal punishment, a practice that symbolized everything progressive educated society repudiated in the autocratic regime. By facilitating the state's

[2]See Robert A. Nye, *Crime, Madness, and Politics in Modern France: The Medical Concept of National Decline* (Princeton, N.J., 1984); Daniel Pick, *Faces of Degeneration: A European Disorder, 1848–1918* (Cambridge, 1989); and Ruth Harris, *Murders and Madness: Medicine, Law, and Society in the Fin de Siècle* (Oxford, 1989), 80–120.

[3]For political and social context, see Jane Caplan, "Sexuality and Homosexuality," in *Women in Society: Interdisciplinary Essays*, ed. Cambridge Women's Studies Group, 149–67 (London, 1981). For the ideology's evolution, see Annemarie Wettley, with Werner Liebbrand, *Von der "Psychopathia sexualis" zur Sexualwissenschaft* (Stuttgart, 1959). An early example of the Russian translation of foreign texts is K. I. Babikov, *Prodazhnye zhenshchiny: Kartiny publichnogo razvrata (prostitutsiia) na vostoke, v antichnom mire, v srednye veka i v nastoiashchee vremia vo Frantsii, Anglii, Rossii i dr. gosudarstvakh Evropy, sostavleno po Paran-Diu-Shatele, Zhanneliu, Sherru, Lakrua i dr.* (Moscow, 1870).

[4]Nancy Mandelker Frieden, *Russian Physicians in an Era of Reform and Revolution, 1856–1905* (Princeton, N.J., 1981). This antistate ethos lasted until 1905; see John F. Hutchinson, "'Who Killed Cock Robin?': An Inquiry into the Death of Zemstvo Medicine," in *Health and Society in Revolutionary Russia*, ed. Susan Gross Solomon and John F. Hutchinson (Bloomington, Ind., 1990); and idem, *Politics and Public Health in Revolutionary Russia, 1890–1918* (Baltimore, 1990), chaps. 1–3. For the struggle between state-dominated and corporatist versions of medical professionalism in France, see Jan Goldstein, *Console and Classify: The French Psychiatric Profession in the Nineteenth Century* (Cambridge, 1987), 15, 40, 168. On the German profession, see Claudia Huerkamp, *Der Aufstieg der Ärzte im 19. Jahrhundert: Vom gelehrten Stand zum professionellen Experten: Das Beispiel Preussens* (Göttingen, 1985).

use of violence, physicians felt themselves implicated in the repressive mechanism that impeded their own chances for personal and professional freedom.[5] Participation in the system of regulated prostitution likewise aroused a high level of professional ambivalence.

The second factor that distinguished Russian physicians and criminologists from their Western colleagues was the class configuration of Russian society. At the top of the social scale, the educated few were largely excluded from political power and resented their own position as the objects of an oppressive custodial regime. Below them the elite confronted the mass of common folk, no less politically disenfranchised but also culturally and economically deprived. This popular mass aroused in its social superiors a feeling of apprehension mixed with a strong sense of moral obligation and collective guilt. The biological marginalization of subordinate groups, such as workers and women, which accompanied the consolidation of bourgeois public life in nineteenth-century Western nations did not begin to serve the cultural purposes of the Russian professional elite until the 1905 revolution significantly reconfigured the public terrain. It was only after workers, peasants, and professionals had jointly engaged in a common political venture and after the privileged groups had secured a measure of political responsibility for themselves that the biological determinism already current in the West began to exert a noticeable appeal.

Pathological Prostitutes

Of course, biological determinism was not universally accepted as an interpretation of social and sexual deviance even in Europe. The early public health tradition, exemplified in the work of Alexandre Parent-Duchâtelet, favored a sociological approach to explaining prostitution.[6] Though he deplored the commercial trade in sex on moral and sanitary grounds, Parent did not consider public women themselves to be abnormal. In recording the fact that prostitutes sometimes engaged in lesbian relationships with each other, however, Parent established a connection between prostitution and sexual perversion which allowed later observers to challenge his own normalizing claims.[7]

---

[5]Frieden, *Russian Physicians*, 189–92. For a physician's revulsion against corporal punishment, see Anton Chekhov's account in *Ostrov Sakhalin* (1895; Moscow, 1984), 298–99.
[6]Parent was heir to the eighteenth-century tradition of "medical ecology," which regarded disease as a response to the social environment; see Alain Corbin, "Présentation," in Alexandre Parent-Duchâtelet, *La Prostitution à Paris au XIXe siècle* (Paris, 1981), 10.
[7]Though Parent disapproved of lesbianism, he regarded it as an understandable response to emotional deprivation and denied that it represented a physical anomaly or that

Indeed, the tendency to explain deviant sexuality, whether in the form of homosexual practices or of prostitution, as the product and symptom of organic pathology was manifest primarily among the practitioners of forensic, not public health, medicine. Obliged to testify in courts of law, the forensic specialists began with the problem of anatomical visibility. Some believed that the commission of deviant acts left physical marks, offering scientific proof that the law had been broken. Others denied that the genitals and other body parts could speak of crime with such assurance. These same experts also disagreed as to whether persons who habitually indulged such tastes either came to exhibit distinguishing anatomical features or manifested organic faults that explained their predilections in the first place and that might serve as the basis for sexual classification. Far from generating a consensus in medical circles, the subject of sexual perversion provoked a heated quarrel between the leading midcentury forensic authorities: the eminent Frenchman Ambroise Tardieu and his equally august colleague from Berlin, Johann Ludwig Casper.[8] Tardieu and his disciples insisted that deviant acts and deviant propensities could be deciphered on the subject's physical self; Casper treated their assertions with scorn.[9]

Despite Casper's well-publicized skepticism, the German-language medical community helped conceptualize the distinctive features of sexual perversion. The physicians Albert Moll and Carl Westphal and the Austrian psychiatrist Richard von Krafft-Ebing elaborated a typology of sexual deviance from which the classification of sexual types easily emerged. The most uncompromising version of organic determinism, however, took the form of criminal anthropology, a theoretical position articulated by the Italian forensic psychiatrist Cesare Lombroso in the 1880s and 1890s. Though he devoted some attention to deviant sex-

---

its presence could be detected by a masculine appearance or other unusual bodily signs: Parent-Duchâtelet, De la prostitution, 1:150–61, 163–67, 211–12. Some later physicians accepted Parent's observations while rejecting his interpretations; see annotations added by the editors of the third edition; also Louis Martineau, Leçons sur les déformations vulvaires et anales produites par la masturbation, le saphisme, la défloration, et la sodomie, 2d ed., rev. (Paris, 1886), ii (on Parent's "clear and precise facts"); Lombroso and Ferrero, La Femme criminelle, 401: Parent "is not always as successful in his explanations as he is precise in his information."

[8]The relevant texts are Johann Ludwig Casper, Practisches Handbuch der gerichtlichen Medizin, nach eigenen Erfahrungen, 2 vols. (Berlin, 1857–58), trans. as A Handbook of the Practice of Forensic Medicine, Based upon Personal Experience, 3 vols. (London, 1861–64; idem, Klinische Novellen zur Gerichtlichen Medizin, nach eigenen Erfahrungen (Berlin, 1863); and Ambroise Tardieu, Etude médico-légale sur les attentats aux moeurs (1857), 7th ed. (Paris, 1878). On Tardieu and Casper, see Biographisches Lexikon der hervorragenden Ärzte aller Zeiten und Völker (vor 1880), ed. August Hirsch, 3d ed. (Munich, 1962), 1:848–49, 5:515–16. Tardieu was president of the French Academy of Medicine.

[9]Tardieu's most influential follower was Louis Martineau, whose two major works were widely read: La Prostitution clandestine (Paris, 1885) and Leçons.

ual practices, Lombroso focused mainly on the question of crime. Prostitution afforded him an opportunity to talk of both pathologies in conjunction. Indeed, in his view, prostitution exemplified all that was inherently abnormal about the female constitution.

When it came to taking sides in these scientific debates, which they followed both in the original and in translation, the Russians tended to resist organic explanations.[10] In so doing, they assumed a posture of theoretical modesty with respect to the subject at hand, for theories of biological determinism endowed the physician with wide powers of diagnosis and broad grounds for professional intervention. It was precisely such extensive claims that the Russians preferred to avoid. For example, Russian forensic specialists generally favored Casper's cautious approach to proving the existence of sexual perversion and seemed reluctant to explain it on organic grounds. On the question of anatomical evidence, an 1870 article in the forensic journal of the medical department of the Ministry of Internal Affairs endorsed Casper's opinion that "the signs of male homosexuality are extremely unreliable" and that "lesbian love rarely leaves any traces at all." The author showed a similar caution about medicine's ability to ascertain, after the alleged fact, whether defloration or even rape had actually occurred.[11]

Russia's leading textbook on sexual crime, published in 1876 by Vladislav Merzheevskii, former obstetrician for the city of St. Petersburg and member of the medical council of the Ministry of Internal Affairs, likewise shared Casper's doubt that the practices and inclinations of homosexual men could be confirmed by physical examination.[12]

[10]Among the forensic and sexology texts available in Russian translation before 1900 were Casper, *Handbook*; Havelock Ellis, *Man and Woman: A Study of Human Secondary Sexual Characters* (London, 1894) (trans. 1898); Richard von Krafft-Ebing, *Grundzüge der Criminalpsychologie auf Grundlage des Strafgesetzbuchs des deutschen Reichs für Ärzte und Juristen* (Erlangen, 1872) (trans. 1874); idem, *Psychopathia sexualis, mit besonderer Berücksichtigung der conträren Sexualempfindung: Eine klinisch-forensische Studie* (Stuttgart, 1886) (trans. 1887, 1909); idem, *Lehrbuch der gerichtlichen Psychopathologie, mit Berücksichtigung der Gesetzgebung von Österreich, Deutschland und Frankreich* (Stuttgart, 1886) (trans. 1895); Lombroso and Ferrero, *La Femme criminelle* (trans. 1898); Martineau, *La Prostitution clandestine* (trans. 1885, 1887). More appeared after 1900: Iwan Bloch, *Das Sexualleben unserer Zeit in seinen Beziehungen zur modernen Kultur* (Berlin, 1907) (trans. 1910, 1911); Havelock Ellis, *Sexual Inversion* (London, 1897) (trans. n.d.); Magnus Hirschfeld, *Berlins drittes Geschlecht* (Berlin, [1904]) (trans. 1908, 1909); Albert Moll, *Untersuchung über die Libido sexualis* (Berlin, 1897, 1898) (trans. 1910); Hermann Rohleder, *Vorlesungen über Sexualtrieb und Sexualleben des Menschen* (Berlin, 1901) (trans. 1907). Freud's work began to be translated in 1911, but commentaries on his ideas had appeared earlier in professional journals.

[11]Zuk, "O protivozakonnom udovletvorenii polovogo pobuzhdeniia i o sudebno-meditsinskoi zadache pri prestupleniiakh etoi kategorii," *Arkhiv sudebnoi meditsiny i obshchestvennoi gigieny*, no. 2, pt. 5 (1870): 9, 12–13. For the same kind of restraint, see Parent-Duchâtelet, *De la prostitution*, 1:203, 215.

[12]Vladislav Merzheevskii, *Sudebnaia ginekologiia: Rukovodstvo dlia vrachei i iuristov* (St. Petersburg, 1878), 216–19. Merzheevskii nevertheless accepted the notion (see Casper, *Hand-*

As for lesbian love, "consist[ing] of mutual onanistic stimulation," Merzheevskii followed Casper in believing that it offered even less material evidence than sexual contacts between men. He also believed the practice to be extremely rare: "Having paid special attention to this question," he was convinced that "onanism and the unnatural satisfaction of sexual passion between women are developed to an incomparably lesser extent than among men."[13]

It was only with the emergence of criminal anthropology as an influential school that the Russians began to adopt organic explanations. The case of prostitution demonstrates, however, that even the most ardent Russian partisans of this approach refused to accept the theory's full social implications. Their reluctance is all the more striking in its contrast to the active enthusiasm of these same converts for the school's leading ideas and institutional opportunities. In 1883 Pavel Kovalevskii, professor of psychiatry at Kharkov University, founded a journal modeled on the one started by Lombroso and his disciples three years before,[14] and Russians attended the six international congresses held in major European capitals between 1885 and 1906.[15] Indeed, Lombroso and his coauthor, Guglielmo Ferrero, were themselves indebted to a Russian physician, Praskov'ia Tarnovskaia, for much of the material analyzed in their 1893 book on female crime. Her "anthropometric" study of prostitutes and female thieves appeared in Paris in 1889, followed in 1902 by a study of female murderers.[16]

Tarnovskaia was the daughter of Nikolai Kozlov, the prominent advocate of women's medical education who had served as director of the St. Petersburg Academy of Military Medicine during the tenure of the

---

book, 3:331; Tardieu, *Etude*, 218) that men of such inclinations formed a "brotherhood," obvious to initiates even when invisible to others (204–6).

[13]Merzheevskii, *Sudebnaia ginekologiia*, 261–62.

[14]*Arkhiv psikhiatrii, neirologii i sudebnoi psikhopatologii.*

[15]On the Russian delegates, see *Actes du troisième congrès international d'anthropologie criminelle, Bruxelles, 1892* (Brussels, 1893), and the following publications issued by the Congress: *Compte-rendu des travaux de la quatrième session tenue à Genève du 24 au 29 août 1896* (Geneva, 1897), xxvii; *Compte-rendu des travaux de la cinquième session tenue à Amsterdam du 9 au 14 septembre 1901* (Amsterdam, 1901), xiv, xxxii; *Comptes-rendus du sixième congrès international d'anthropologie criminelle (Turin, 28 avril–3 mai 1906)* (Turin, 1908). Among the regular participants were D. A. Dril', V. F. Chizh, V. P. Serbskii, V. M. Bekhterev, V. M. Tarnovskii, and P. N. Tarnovskaia.

[16]Pauline Tarnowsky [P. N. Tarnovskaia], *Etude anthropométrique sur les prostituées et les voleuses* (Paris, 1889); P. N. Tarnovskaia, *Zhenshchiny-ubiitsy: Antropologicheskoe issledovanie* (St. Petersburg, 1902). For Tarnovskaia's career, see *Rossiiskii meditsinskii spisok* (St. Petersburg, 1891), and D. Nikol'skii, "Pamiati vrachei-antropologov: N. V. Gil'chenko i P. N. Tarnovskoi," *Prakticheskii vrach*, no. 13 (1911): 222–23. Though she was trained as a physician, her interest in criminology led her to join the St. Petersburg University Juridical Society; see *Iuridicheskoe obshchestvo pri Imperatorskom S.-Peterburgskom Universitete za dvadtsat' piat' let (1877–1902)* (St. Petersburg, 1902), 122.

reform-minded minister of war Dmitrii Miliutin.[17] She was married to Russia's distinguished authority on venereal disease, Veniamin Tarnovskii, professor at the Academy of Military Medicine. His expertise ranged from syphilis to prostitution to sexual perversion, and his works were widely translated and read abroad.[18] Husband and wife each achieved an international scholarly reputation and together came to exemplify Russia's contribution to the field of criminal anthropology.

Despite their general agreement, however, the couple did not always see eye to eye. An ardent partisan of regulation of prostitution, Tarnovskii used the arguments pioneered by the Italian school to attack the abolitionist position. Among the school's Russian adherents, he was the most faithful to its vision of the public woman as congenitally depraved and sexually predatory. He followed Lombroso in depicting the habitual prostitute as a retrograde type who behaved in a manner more appropriate to normal men—she was egotistical, self-assertive, and sexually aggressive.[19] Part criminal, part insane, the female prostitute was not, in Tarnovskii's view, the passive victim of male lust, male deception, and economic misfortune, as the enemies of regulation believed, but an enterprising and resourceful figure responding to commercial opportunity. Indeed, in his estimation, the male was her victim, not the other way around: "Without pity or remorse . . . ," Tarnovskii wrote, "she sows evil [seet zlo] and spreads syphilis unchecked, . . . certain to infect her generous and trusting clients," many of whom "accidentally" fall for her charms (sluchaino uvlech'sia prostitutkoiu).[20] Rather than reinforcing conventional gender roles (male as seducer, female as prey), as many feminists claimed it did, prostitution in Tarnovskii's disturbing picture reversed the expected relation between the sexes. Not only did the public woman leave the domestic confines to compete in the mar-

---

[17]On the bureaucratic push for female medical education, see Richard Stites, *The Women's Liberation Movement in Russia: Feminism, Nihilism, and Bolshevism, 1860–1930* (Princeton, N.J., 1978), 85; Christine Johanson, *Women's Struggle for Higher Education in Russia, 1855–1900* (Kingston, 1987), 78–81.

[18]For V. M. Tarnovskii's service record, see TsGVIA, f. 546, op. 2, d. 7945, ll. 57–69. On his career, see "Tarnovskii, V. M.," in *Entsiklopedicheskii slovar' Brokgauz-Efron*, 32A:650–51 (St. Petersburg, 1901); S. P. Arkhangel'skii, *V. M. Tarnovskii* (Leningrad, 1966); and *Biographisches Lexikon*, 5:519–20. He published work on regulated prostitution, *Prostitutsiia i abolitsionizm* (St. Petersburg, 1888) (trans. in German 1890); on male sexual perversion, *Izvrashchenie polovogo chuvstva: Sudebno-psikhiatricheskii ocherk dlia vrachei i iuristov* (St. Petersburg, 1885) (in German, 1886; in English, 1898; in French, 1904); on sex education, *Polovaia zrelost', ee techenie, otkloneniia i bolezni* (1886), 2d ed. (St. Petersburg, 1891).

[19]Lombroso and Ferrero, *La Femme criminelle*, 464.

[20]V. M. Tarnovskii, *Prostitutsiia i abolitsionizm*, viii, 133–34, 136, 159, 164. For a picture of the prostitute as a victim of male lust, see the pamphlet by the feminist antiregulationist M. I. Pokrovskaia, *O zhertvakh obshchestvennogo temperamenta* (St. Petersburg, 1902), 26, 30–31.

ketplace for gain, but she practiced the masculine art of sowing seed (*seiat'*), an activity that contrasted with the passivity of the hapless men who allowed themselves to be taken in (*uvlech'sia*).

Though Tarnovskii did not exclude the impact of the environment in stimulating or inhibiting crime, he stressed the overriding power of organic forces. "The real prostitute," he wrote, "is born with a predisposition to vice," which might or might not be realized, depending on circumstances, social class, and personal experience. The potential might be aroused by unfavorable conditions, but in the absence of "essential personality traits and peculiarities," Tarnovskii insisted, a defective environment by itself would have no such negative effect. True, a small percentage of prostitutes were victims of misfortune, he conceded; these women tried to escape from brothels, killed themselves in desperation, and welcomed the refuge of charitable homes for fallen women. But the vast majority of public women refused to reform, because they had found their true calling. "A woman who willingly and consciously engages in the prostitution trade," Tarnovskii affirmed, "is always a morally vicious and most often a physically abnormal being."[21]

Minimizing the role of social conditions in inducing women to trade in sex, Tarnovskii denied that prostitutes consisted largely of the proverbially innocent peasant maids corrupted by the city's evil ways who figured in most Russian physicians' accounts of prostitution. They were indeed peasant girls, he agreed, but they were those who were congenitally incapable of restraint, who had lost their virtue before leaving home, and who awaited only the excitement of the metropolis to fulfill their destiny. Such reasoning, however, did not account for the existing sociological distinctions: if the readiness for sexual indulgence were a biological trait, it could not be expected to favor one class over another, yet educated and well-bred women were rare in the brothels of the capital cities. Tarnovskii was therefore obliged to invoke the role of "circumstance" once more: "In the middle and upper social classes," he asserted, "the influence of family, the surrounding environment, upbringing, and intellectual development" prevented biology from taking its toll.[22]

In this scheme, neither the privileged nor the downtrodden woman logically bore more or less responsibility for her fate: one was the involuntary beneficiary of cultural advantages that fortified her powers of restraint; the other lacked similar strength of character because of the

---

[21]V. M. Tarnovskii, *Prostitutsiia i abolitsionizm*, 165–66, 183–89, 243.

[22]Ibid., 180–81. Lombroso himself wavered in the relative weight he assigned to biological and social factors; see Marvin E. Wolfgang, "Cesare Lombroso," in *Pioneers in Criminology*, ed. Hermann Mannheim, 2d ed. (Montclair, N.J., 1972), 187, 208.

equally fortuitous circumstance of cultural deprivation. By insisting on the primary role of the psychophysical constitution in driving prostitutes to their trade, Tarnovskii justified the caretaking role of the police and medical authorities. Yet the force of sexual compulsion and the masculine attributes associated with the exercise of sexual initiative, especially in the public world, gave Tarnovskii's archetypal prostitute an ambiguous twist: was she the passive instrument of biology (just as the unemployed seamstress was the victim of economic fate) or the responsible agent of her own (however morbid) desires? Was she a remnant of archaic organic forms (Lombroso's atavism) or an exemplar of modern capitalist initiative? Tarnovskii used the notion of inappropriate masculinity, extensively developed by Lombroso in his own work on public women, to reconcile (however inadequately) these divergent implications: those traits associated with modernity were "normal" only in biological males; strong sexual drive in the female sex could be understood only as a mark of arrested or retrograde development.[23]

Tarnovskii's insistence that the prostitute's self-assertiveness was a consequence of her genetic constitution allowed him to withhold the compassion accorded the victims of injustice. Convinced that such women actively elicited the contacts that opened them to abuse, he claimed that innate moral obtuseness shielded them from the full impact of exploitation and mistreatment. "No normal woman, even from the lowest, most ignorant social stratum," Tarnovskii declared, would tolerate the "insult and injury" (unizhenie i oskorblenie) a prostitute encountered every moment of her life. In invoking the title of Dostoevsky's 1861 novel, The Insulted and Injured (Unizhennye i oskorblennye), which depicted in melodramatic Dickensian terms the victimization of an impoverished and helpless young girl who fell prey to the vicious designs of unscrupulous adults, Tarnovskii did not evoke the novel's moral pathos. In contrast to Dostoevsky's proud and vulnerable waif, Tarnovskii's women thrived on vice—indeed, could not live without it. Morally impervious, they could not feel humiliation. Part of a "separate social element" by virtue of their "innate qualities," prostitutes should be deprived of personal freedom, Tarnovskii argued, "the concept of which applies only to the normal member of society"; in their case it could lead only to the spread of sin and disease.[24]

Although Tarnovskii's preference for the organic interpretation of crime led him to champion administrative measures that deprived those controlled by them of civil rights and personal dignity, such political

[23]On female hypersexuality and atavism, see Lombroso and Ferrero, La Femme criminelle, 402, 409–10.
[24]See V. M. Tarnovskii, Prostitutsiia i abolitsionizm, 136–39, 163–64, 189; F. M. Dostoevskii, Unizhennye i oskorblennye (1861), in Polnoe sobranie sochinenii (Leningrad, 1972), 3:169–442.

applications were not the inevitable consequences of Lombroso's theory. Lombroso cast himself as a champion of women's rights and considered the organic interpretation of crime a contribution to progressive penology in that it substituted treatment for the punishment of offenders unable to exercise freedom of the will.[25] If criminals were sick rather than oppressed or impoverished, then medical intervention rather than social reform was the key to reducing crime. With the exception of Veniamin Tarnovskii, however, Lombroso's Russian followers tended to emphasize the interaction between biology and society rather than the absolute priority of biology, and to stress social policy initiatives rather than the isolation and treatment of the delinquent as the most effective response to crime.[26]

The invocation of criminal anthropology in the name of social reform characterized the work of Praskov'ia Tarnovskaia, who valued what she saw as the flexibility of Lombroso's approach and soft-pedaled its deterministic implications. Lombroso, Tarnovskaia pointed out, did not claim that every defective was a criminal and every criminal sick. A physiologically normal person might stumble accidentally into crime or be driven to it by a burst of strong emotion, while a criminally inclined personality might never encounter conditions that brought the impulse to life. Tarnovskaia ranked the causes of crime in order as heredity, poverty, bad example, alcoholic parents, deprived childhood, and weak upbringing, thus conceding a substantial role to circumstance and the social environment in leading people to break the law. She interpreted Lombroso's ideas as a program for social rejuvenation, a formula for renewal rather than repression. "Prisons," she wrote, "when turned into places of breeding, correction, and training [*vospitanie, ispravlenie i obuchenia*], will help educate the common people, bringing knowledge to the milieu that needs it most of all, enlightening and humanizing the darkest and most deprived part of the population."[27]

---

[25]On women's rights, see Lombroso and Ferrero, *La Femme criminelle*, xiv–xv. On the progressive element in therapy for criminals, see P. N. Tarnovskaia, *Zhenshchiny-ubiitsy*, v, 91–92, 484, 488. Lombroso, a Jew, spoke out against anti-Semitism and rejected biological theories of racism: Cesare Lombroso, *L'Antisémitisme* (Paris, 1899), 33–38.

[26]E.g., the social activist D. A. Dril', trained in both law and medicine, attacked the absolute standards of classical legal thought and praised the Italian school for its attention to the individual offender and for its compatibility with sociological explanations of crime; see his "Antropologicheskaia shkola i ee kritiki (Zametki po povodu statei g. Obninskogo)," *Iuridicheskii vestnik*, no. 4 (1890): 582–83, 587–88; S. A. Shumakov's review of D. A. Dril', *Uchenie o prestupnosti i merakh bor'by s neiu* (St. Petersburg, 1912), in *Voprosy prava*, no. 3 (1912): 197–203; and S. S. Ostroumov, *Prestupnost' i ee prichiny v dorevoliutsionnoi Rossii* (Moscow, 1960), 288–92. Although V. D. Nabokov rejected the strictly biological interpretation of crime and insisted that the social factor was primary, he was grateful to Lombroso for directing attention to the personality of the criminal: "Chezare Lombrozo," *Pravo*, no. 43 (1909): 2292–97.

[27]Pauline Tarnowsky [P. N. Tarnovskaia], "Criminalité de la femme," in Congrès

Tarnovskaia gave a biological spin to the cultural distinction invoked by her husband to explain the observable distribution of crime by social class. Where he considered environment a counterweight to heredity, she explained organic differences as a product of culture, claiming that the common folk were less evolved, physiologically more primitive, and hence psychologically more obtuse than people with the advantage of breeding and education. Determined criminals, Tarnovskaia believed, belonged to the "social classes whose way of life and occupation indicate a more or less pronounced incompatibility with the moral sense inherent in the cultivated, healthy, and normal man." In her view, virtue was a function of biological development, itself the product of civilization, not an attribute of unspoiled folkloric simplicity. Thus she could argue that the group of prostitutes and thieves who claimed her attention were "the product of the lower depths, of the dregs of society, which will shrink in size as culture improves the conditions of biological evolution."[28] Thus, hidden in her language of biological determinism was a call for social reform.

For all her preference for anthropometric methods, Tarnovskaia in the end explained prostitution in social terms. Having measured public women's facial features and body size, noted their hair and eye coloring and the shape of their ears, Tarnovskaia concluded that biological deformity was the result of ills attributable to poor living conditions and that deformity alone, in the absence of unfavorable environmental conditions, did not produce crime. Thieves and prostitutes were bred by the same social context, she wrote: "the same crude and uncultivated surroundings, sexual promiscuity, and parents' poverty, illness, and vice; [circumstances in which] children are left entirely to themselves from the youngest age without any upbringing at all."[29] The solution to prostitution, therefore, was social intervention, not medical treatment or punishment:

> We must attack the evil at its very source: improve the conditions under which these abnormal women are born and live; reduce their poverty by widening access to honest and renumerative labor for women who want to stay honest; admit them to the many profes-

---

international d'anthropologie criminelle, *Compte-rendu des travaux de la quatrième session*, 235; Tarnovskaia, *Zhenshchiny-ubiitsy*, ii–iii, 89, 487. One Soviet scholar calls Tarnovskaia a monarchist who denied that social conditions cause crime: S. S. Ostroumov, "Levaia gruppa kriminalistov," *Pravovedenie*, no. 4 (1962), 145. Earlier, however, Ostroumov had called her liberal in spirit, despite the contradictions with her own basic premises: *Prestupnost'*, 307.
[28]Tarnowsky, *Etude anthropométrique*, 2, 203.
[29]Ibid., 195.

sions and trades exercised mainly by men. Every new opportunity for honest labor necessarily reduces women's difficulty in finding work, and therefore also diminishes their poverty—the evil counselor that sustains depravity and nourishes vice, regardless of inborn inclinations.

We should protect the young from the pernicious influence of vicious parents and incorrigible profligates by providing humanitarian means to raise abandoned children. A well-conceived and well-organized philanthrophy will reduce the damage to families caused by drunkenness, syphilis, and other debilitating parental illnesses; in short, it will increase the possibility of healthy parents who can produce a healthy new generation, unharmed by the effects of sickly heredity.[30]

Neither custodial regimes nor medical cures could eradicate "the source of evil," in Tarnovskaia's view. Rather, society must offer the material and cultural conditions essential to bodily and psychic health. Public health, in this sense, was the obligation of the public.

In their theoretical work, both Tarnovskiis invoked the influence of class and social environment to explain the distribution of traits supposedly bred in the genes. Though the husband focused on regulating the noxious effects of prostitution through administrative means and the wife on preventing prostitution in the first place, both belonged to the major civic organization dedicated to alleviating the plight of prostitutes through charitable efforts. Founded in 1901, the Russian Society for the Protection of Women eventually—but only after Tarnovskii's death—denounced the system of state regulation whose champion he had been.[31]

The attempt to reconcile biological and social causation reflected the contemporary belief that genetic material might be modified by the circumstances under which people lived.[32] Some of the couple's medical colleagues concluded, however, that such arguments undermined the

---

[30]Ibid., 203–4.

[31]The society's roster appears in TsGIA, f. 1335, op. 1, d. 24 (Alfavitnyi spisok chlenov Rossiiskogo obshchestva zashchity zhenshchin s 1901 goda), l. 69. Later in life Tarnovskii became more critical of the inadequacies of the regulatory system but never suggested it should be abolished: V. M. Tarnovskii, *Sifiliticheskaia sem'ia i ee niskhodiashchee polozhenie: Biologicheskii ocherk* (Kharkov, 1902), 71–72. For more on his doubts at this time, see Laurie Bernstein, "Yellow Tickets and State-Licensed Brothels: The Tsarist Government and the Regulation of Urban Prostitution," in Solomon and Hutchison, *Health and Society*, 50.

[32]See Elizabeth Lomax, "Infantile Syphilis as an Example of Nineteenth Century Belief in the Inheritance of Acquired Characteristics," *Journal of the History of Medicine* 34:1 (1979): 23–39.

basic premises of hereditary determination.[33] "There is no doubt," said psychiatrist Mikhail Nizhegorodtsev in 1887 at a conference at which Tarnovskaia presented her statistical data, "that the extreme depravity which leads to prostitution very frequently exists independently of any anthropological sign of degeneration."[34] The psychiatrist Vladimir Dekhterev objected that Tarnovskaia had mistaken differences of culture and education for the effects of biological disorder. "Often," he remarked at an 1886 conference, "what seems to be the result of degeneration or moral insensibility may in fact stem from inadequate upbringing and an absense of moral feeling in the environment in which the women have grown up. It is well known that peasant custom demonstrates completely different and less complicated attitudes toward sexual relations than those we consider appropriate. . . . [Such attitudes] do not represent the population's moral decline but simply reflect more primitive relations between the two sexes."[35]

Tarnovskii's distinguished contemporary Eduard Shperk, director of the St. Petersburg venereal disease hospital for women, objected to the organic theory as a denial of the realities of social class.[36] "One must not forget that there are no inborn prostitutes," he wrote. "Before resorting to this sad extremity, all such women once belonged to that part of the population in which labor ennobles the impulses of the heart." Shperk did not exclude the possibility that some among them might turn to prostitution as an expression of temperament rather than need, but such examples, he insisted, were not typical of Russian culture. "The social stratum that produces prostitutes who begin as young girls because they are incapable of any other work is still very limited in Russia,"

[33]See, e.g., Shumakov, review of Dril', 199. A professor of law at Kazan University denounced the notion of the biological criminal, expressing particular irritation at "the conclusion that woman and children are in general criminal": N. Gregorovich, *Voprosy tak nazyvaemogo ugolovnogo prava* (Kazan, 1897), 35.

[34]Quoted in P. N. Tarnovskaia, "Antropometricheskie issledovaniia prostitutok, vorovok i zdorovikh krest'ianok-polevykh rabotnits (zasedanie 21 noiabria 1887 g.)," in *Protokoly zasedanii obshchestva psikhiatrov v S.-Peterburge za 1887 god* (St. Petersburg, 1888); rpt. in appendix to *Vestnik klinicheskoi i sudebnoi psikhiatrii*, no. 2 (1889), 62, with criticism by other colleagues, 62–65. For similar objections at the 1886 meeting, see P. N. Tarnovskaia, "Klassy vyrozhdaiushchikhsia v sovremennom obshchestve," in *Protokoly zasedanii obshchestva psikhiatrov v S.-Peterburge za 1886 god* (St. Petersburg, 1887); rpt. in appendix to *Vestnik klinicheskoi i sudebnoi psikhiatrii*, no. 1 (1887): 60–63. Nizhegorodtsev was of a politically conservative disposition; see Hutchinson, *Politics and Public Health*, 42–43, 49.

[35]Quoted in Tarnovskaia, "Klassy vyrozhdaiushchikhsia," 63.

[36]On Shperk, see *Rossiiskii meditsinskii spisok* (St. Petersburg, 1890–93); E. I. Lotova, *Russkaia intelligentsia i voprosy obshchestvennoi gigieny: Pervoe gigienicheskoe obshchestvo v Rossii* (Moscow, 1962), 45; "Esquisse biographique," in Edouard Léonard Sperk [Eduard Shperk], *Oeuvres complètes: Syphilis, prostitution, études médicales diverses* (Paris, 1896), 1:xv–xxxviii. Shperk's international stature is indicated by his inclusion in *Biographisches Lexikon*, 5:361.

Shperk declared. ". . . The vast majority are girls and women whose calloused hands prove they led lives of hard labor before their fall. Most are peasants from provinces near the capital, whence they have come looking for work."[37]

Whereas Shperk believed that the common folk must embody the physical and moral virtues of the plain and simple rural life, the Tarnovskiis condemned poverty and ignorance for the lower classes' physical and moral corruption. Like Shperk, populist-minded opponents of the Italian school also blamed modern life for the problems the Tarnovskiis blamed on backwardness. If criminals were "mentally weak and undeveloped," wrote the Kharkov psychiatrist Isak Orshanskii, it was not because they were throwbacks to a more primitive era but because they were the casualties of progress, "the cultural refuse of civilization, victims of its negative sides. . . . Therefore, this group [of defectives] must be substantial in countries with old and complex cultures and insignificant in a country as young from the historical and cultural point of view as Russia, which is in reality what we observe."[38]

Shperk and Orshanskii both associated the emergence of organic pathology with the Western way of life. Tarnovskaia, by contrast, saw the organic interpretation of deviance as uniquely suited to Russia's particularities. For her, the theory's greatest attraction was its adaptability to Slavophile notions of the Russian national character. Despite its aggressively positivist methods (based on measurement, physical quantities, and statistics), with their claim to scientific exactitude and universal application, criminal anthropology served her as an alternative to the classical theory of law, with its insistence on abstract and universal principles. Such an absolutist approach, she believed, overlooked the subjective and particular dimensions of crime, whereas physicians and juries "introduce[d] the elements of public conscience and positive knowledge [to judicial practice], constantly moderating and clarifying the pointless cruelty involved in rigidly applying the abstract theory of law to real life." Absolute equality of rights and the classification of crime solely in terms of its objective consequences, she feared, did not do justice to the real complexity of human motivation. Far from constituting an abstraction in its own right, positive science in the guise of criminal anthropology better reflected the particularity of actual experi-

[37]Edouard Léonard Sperk, "Données scientifiques concernant la réforme des mesures de police sanitaire" (trans. from *Voenno-meditsinskii zhurnal*, 1887) in his *Oeuvres complètes*, 2:479–80.
[38]I. G. Orshanskii, "Uchenie Lombrozo o tipe prestupnika," in his *Sudebnaia psikhopatologiia dlia vrachei i iuristov*, pt. 1 (St. Petersburg, 1900), 168. For Orshanskii's populist view of peasant culture, see V. A. Aleksandrov, *Obychnoe pravo krepostnoi derevni Rossii, XVIII–nachalo XIX v.* (Moscow, 1984), 23.

ence. Moreover, the principle of cure rather than punishment, Tarnovskaia believed, accorded well with Russian culture, unspoiled by secularization and imbued with the religious idea that a person's fate was often not a matter of choice or subject to control. "The Christian view of criminals is entirely supported by criminal anthropology," she wrote, "and our Slavonic folk view of criminals as '*unfortunates*' expresses these attitudes best of all."[39]

This conception of the "subjective" or particular nature of popular justice was not peculiar to Tarnovskaia. Students of customary law such as Aleksandra Efimenko noted that peasant courts decided cases according to the character and circumstances of the accused, not in strict relation to the category of the offense. Efimenko contrasted the native Russian principle with the Roman law tradition, which formed the basis for Russian positive law and which she characterized as "objective," abstract, and formal. But even in the West, she noted, the virtues of the subjective approach were acknowledged in the institution of trial by jury and in the "individualistic" trend in modern penology and criminal law.[40] Anatolii Koni, the well-known progressive judge, likewise noted that absolute standards were no longer universally applied even in formal courtrooms. Attention to the individual criminal's character had already begun to influence penal practice, and judges often took account of the offender's motivation and personality by linking accountability to age and mental state.[41] The anthropologists attempted merely to put the judgment of criminal responsibility on a firm "scientific" base, he explained, in cases where they believed the organic component loomed large.

Despite such references to the weakening of classical penology even in the West, Lombroso's Russian adherents admired him precisely for diverging from or modifying what they considered essential but inadequate in the Western legal tradition. In so doing, Tarnovskaia and Efimenko, along with many of their educated contemporaries, counterposed two ideal types of social organization: the individualistic but depersonalized social order administered by remote, formalized institutions; and the supposedly more human, collective, or communal order operating on the basis of immediate face-to-face encounters.[42] Insofar as

[39]Tarnovskaia, *Zhenshchiny-ubiitsy*, 490, 82–86, 487 (original emphasis); idem, "Criminalité de la femme," 234.

[40]Aleksandra Efimenko, *Issledovaniia narodnoi zhizni*, vol. 1, *Obychnoe pravo* (Moscow, 1884), 176–79.

[41]A. F. Koni, "Antropologicheskaia shkola v ugolovnom prave," in *Poslednie gody: Sudebnye rechi (1888–1896), iuridicheskie soobshcheniia i zametki, vospominaniia i biograficheskie ocherki, prilozheniia*, 2d ed., rev. (St. Petersburg, 1898), 400.

[42]For this paradigm, see Andrzej Walicki, "Personality and Society in the Ideology of Russian Slavophiles: A Study in the Sociology of Knowledge," *California Slavic Studies* 2 (1963): 1–20.

Tarnovskaia celebrated the principle of trial by jury, respected the social conditions of crime, and believed in the power of science (in this case, psychiatric and medical expertise) to cure social ills, her argument constituted a "liberal" interpretation of the Lombroso school. But it also entailed a conservative defense of exceptionalism and particularity, based on organic notions of personality and the just social order, not unlike the thinking of the early Slavophiles and her populist-minded contemporaries. Tarnovskaia's contribution was to enlist the positivist ethos in the service of national distinctiveness. Like other Russian liberals, she also sought to translate old-style paternalism into the idiom of professional expertise. Given her hostility to the vision of a rule-of-law polity, the conflict between its abstract principles and the custodial strategies of therapeutic intervention did not trouble her peace of mind. Ambivalence about the salient features of the liberal ideal—individualism, conflict, universal standards—led progressives such as Anatolii Koni to accept the same contradiction. But its dangers did not go unremarked by others.

Lombroso himself, for all his enlightened disclaimers, had drawn from the organic theory of crime certain unabashedly conservative conclusions that his Russian critics, and even some of his Russian followers, rejected. The Italian might declare himself a partisan of female education and equal rights, for example, but he nevertheless viewed women as a primitive species. At best he thanked learned women for their limited but worthy labor in spreading ideas already developed by men—tipping his hat to Tarnovskaia, among others whose "masculinity" supposedly explained the unusual intelligence they displayed.[43] In stressing the importance of medical testimony in the courtroom, Lombroso called for an end to trial by jury, for restricted publicity, and for curbs on public association. "For the sake of progress," objected an unsympathetic Russian jurist, "it would seem one must return to the kind of social order that has already proved to be incompatible with progress." In the name of modern science, he argued, the theory relied on retrograde ascriptive principles that violated progressive social values; in its more radical form, he pointed out, the concept of the criminal type or born criminal as a hereditary abnormality—a throwback to more primitive levels of biological development—disqualified the offender from the rights of citizenship. In such cases, he continued, sounding a familiar Russian theme, punishment must be "an affair of state administration, not of criminal justice," since the essence of deviance thus defined was the violation not of socially constituted norms but of biological ones. Such a theory therefore had no bearing on questions of law, which dealt with choice, intention, and conventionally

[43]Lombroso and Ferrero, *La Femme criminelle*, xvi, 167, 172–73.

established rules.[44] It could not hope to find widespread favor among Russian professionals eager to replace a custodial with a rule-of-law regime.

The Lombrosians in fact reasoned backward: the inveterate criminal demonstrated his or her incorrigibility by declining to reform, and this persistence served as proof of organic compulsion. Anton Chekhov captured the tautology of the biological argument, as well as its sinister social implications, in his short story "The Duel" (1891), in which the zoologist Von Koren explains the latest in criminological theory to Deacon Pobedov: "As the consumptive and the scrofulous are known by their symptoms," Chekhov has the scientist declare, "so are the immoral and insane by their acts." Immorality is a function of mental disease. "Knowledge and common sense," Von Koren continues, "tell us that the morally and physically abnormal constitute a menace to mankind. If so, then you must wage war on these freaks. If you can't raise them to the norm, you at least have the strength and skill to neutralize them—exterminate them, in other words."[45]

Chekhov had met Dr. Tarnovskaia at a dinner party in 1888 and described her in a letter to his sister: "She's an obese, bloated chunk of flesh," he wrote. "If she were stripped naked and painted green, she'd be a swamp frog. After a chat with her, I mentally crossed her off my list of physicians."[46] In addition to Tarnovskaia's other highly placed family connections, her niece had married the liberal jurist Vladimir Dmitrievich Nabokov, son of Alexander II's last minister of justice and father of the novelist Vladimir Nabokov, who later commented on this incident in his memoirs. Concluding from Chekhov's letter that Tarnovskaia had "somehow offended [the writer] in the course of a medical conversation," Nabokov noted that "she was a very learned, very kind, very elegant lady, and it is hard to imagine how exactly she could have provoked the incredibly coarse outburst Chekhov permits himself."[47] Perhaps Chekhov, himself a physician, projected his distaste for her view of human nature onto her physiognomy.

Female Sex

Certainly, when it came to the question of female crime, adherents of the anthropological approach did their utmost to undermine their theory's claim that it respected extrabiological causation, for they iden-

[44]N. S[ergeevskii], "Antropologicheskoe napravlenie v issledovaniiakh o prestuplenii i nakazanii," *Iuridicheskii vestnik*, no. 2 (1882): 212, 218–20.

[45]Anton Chekhov, "The Duel," in *The Russian Master and Other Stories* (Oxford, 1984), 90–91.

[46]Quoted in Simon Karlinsky, ed., *Anton Chekhov's Life and Thought* (Berkeley, Calif., 1973), 311.

[47]Vladimir Nabokov, *Speak Memory: An Autobiography Revisited* (New York, 1966), 67–68.

tified precisely the feature that distinguished women from men—the biological distinction of sex—as the origin of women's criminal impulse. It was therefore easy for physicians of this theoretical disposition to elide the normal and the pathological.[48] In a study of menstruation, for example, the psychiatrist Pavel Kovalevskii explained how the ordinary manifestations of the female reproductive cycle might easily result in mental imbalance, even psychosis. Knowing that most women did not, as a result of their unfortunate constitution, commit an excessive number of crimes, Kovalevskii, like the Italians, was obliged to imagine a sexual condition more pathological still: the "precocious and excessive eroticism" that drove certain women to choose prostitution as a way of life.[49]

Tarnovskaia's point of difference with her husband and Kovalevskii, as well as with Lombroso himself, was her refusal to consider the sexual organization of women—whether reproductive or simply erotic—as organically deficient, or to condemn female sexual activity as pathological. Whereas Tarnovskii singled out the prostitute's aggressive, entrepreneurial behavior as a symptom of her disordered disposition, Tarnovskaia believed that self-abasement, not self-assertion, marked prostitutes as sick.[50] "It is logically inconceivable," she declared, "that a human being in possession of her mental faculties, healthy in body and mind, should be able to engage in the procreative act with the first man to come along, [who is] often drunk on wine, crude, brutal, and cynical, and rewards her only with contempt." Her indignation echoes the words of her husband, who saw the prostitute's tolerance for humiliation and insult as a sign of moral obtuseness. But whereas Tarnovskii described such women as actively soliciting the contacts that opened them to abuse, Tarnovskaia stressed the passivity behind their submission: "Can one truly imagine," she rhetorically asked, "the enslavement, self-abnegation [*le renoncement à sa personnalité*], and exploitation of all sorts to which a brothel prostitute submits?"[51]

Thus, although Tarnovskaia too described the prostitutes' condition as inherently pathological, she did not connect its origin with the peculiarities of female sexual organization. Rather, she disputed the claim that anything in the normal female character or sexual constitution ei-

---

[48]For the full extent of such thinking among medical men, see Esther Fischer-Homberger, *Krankheit Frau: Zur Geschichte der Einbildungen*, 2d ed. (Darmstadt, 1984).

[49]P. I. Kovalevskii, "Menstrual'noe sostoianie i menstrual'nye psikhozy," *Arkhiv psikhiatrii, neirologii i sudebnoi psikhopatologii*, no. 1 (1894): 73; Paul Kovalevsky [P. I. Kovalevskii], "La femme criminelle et la fille publique," in his *Psychopathologie légale*, vol. 1, *La Psychologie criminelle* (Paris, 1903), 196. Cf. Lombroso and Ferrero, *La Femme criminelle*, 420.

[50]See V. M. Tarnovskii, *Prostitutsiia i abolitsionizm*, 135–36; cf. Tarnowsky, *Etude anthropométrique*, 3–4.

[51]Tarnowsky, *Etude anthropométrique*, 5–6.

ther inhibited or encouraged crime. The moral sense, she maintained, was evenly distributed in men and women. Crimes too would be evenly distributed, she declared, if prostitution were counted as one: "By plying their trade, inveterate prostitutes, whom one cannot consider healthy, normal, or honest, fill the excessively large gap that criminal statistics create in favor of women. We believe that inveterate prostitutes supply the counterweight that balances the criminal scales and distributes crime more evenly and fairly between the two sexes."[52] Building her case for inherited organic damage among such women, Tarnovskaia compared 150 veteran prostitutes (chosen from the St. Petersburg hospital for women with venereal disease) with 100 illiterate peasant women from the same geographical areas. The prostitutes included a majority with alcoholic parents, many whose parents had tuberculosis, and some with family histories of epilepsy, mental illness, and syphilis. They differed from the "honest" controls, Tarnovskaia reported, in beginning to menstruate at an earlier age, having fewer children, and more often becoming sexually active while still physically immature.[53]

Although she insisted that her sample of prostitutes was constitutionally distinctive, she cited no evidence among them of the pathological intensity of sexual desire that Lombroso reported among public women.[54] Like Parent-Duchâtelet, who emphasized the anatomical normality of the prostitutes he had examined, but unlike later experts who claimed to have observed characteristic genital deformities, Tarnovskaia insisted that prostitutes' genitalia showed no unusual signs.[55] Her most striking divergence from Lombroso's model was the denial that the prostitutes' extreme sexual activity expressed virility of character and accompanied masculine physical features. Instead, she divided her "constitutionally predisposed" prostitutes into four personality types, which reproduced the full range of possible character traits: from hypersexuality to complete sexual indifference, from fat and lazy to charming and seductive.[56] She did not suggest that expressions of sexual appetite

---

[52]Ibid., 97.

[53]Ibid., 53–58.

[54]See Lombroso and Ferrero, *La Femme criminelle*, xii–xiii, 346–47 (women's passion as virile); 527 (prostitutes' sexual precosity); 53, 106–7, 115, 119–20 ("normal" female sexuality as self-denying, maternal).

[55]Tarnovskaia, "Antropometricheskie issledovaniia," 63. See Parent-Duchâtelet, *De la prostitution*, 1:188–89, 204–7, 209, 211. For the specific source that Parent refutes, see Marie-Jo Bonnet, *Un Choix sans équivoque: Recherches historiques sur les relations amoureuses entre les femmes, XVIe–XXe siècle* (Paris, 1981), 174–75. Martineau denied that prostitutes' genitals differed from other women's but asserted that many indulged in lesbian sex, which did alter the appearance of their vaginas: *Leçons*, 21–22, 39, 61–62, 74–77.

[56]Tarnovskaia, "Antropometricheskie issledovaniia," 67–69, 73–75, 79–86. One colleague (cited in Tarnovskaia, "Klassy vyrozhdaiushchikhsia," 63) objected that these psychological types, too various to belong only to prostitutes, could be found also among normal women.

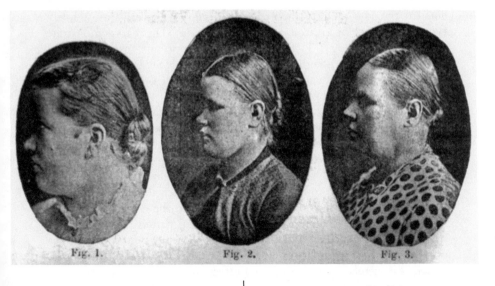

"Congenitally defective" prostitutes: photographs and statistical table. Pauline Tarnowsky, *Etude anthropométrique sur les prostitutées et les voleuses* (Paris, 1889), 36, 94–95. Firestone Library, Princeton University.

entailed the loss of feminine appeal; indeed, her sexually most lively types were those of greatest conventional charm. Eager to demonstrate the group's pathology, Tarnovskaia did not report any cases of lesbianism among them, though she did discuss female sexual perversion in other contexts and must have been familiar with the observations of both Parent and Lombroso in this regard.

It was the absence, not the presence, of passion that Tarnovskaia found the need to explain. Lombroso considered lesbians sexually the most voracious (hence virile) female type; he believed that normal women tolerated intercourse only for the sake of motherhood, whereas Tarnovskaia regarded such indifference as a form of revulsion that might eventually lead to lesbian preferences, which she identified as the lowest level of sexual desire. She also attributed the noted tendency of women to murder their husbands—a statistical pattern as common in Russia as in Europe, despite vast differences in culture and social formation—to the offenders' temporary or constitutional sexual indifference, which she viewed as the combined effect of organic malfunction and cultural factors relevant to specifically Russian conditions. That sort of aggression on the part of women in traditional roles, Tarnovskaia observed, reflected the impact of everyday experience: "Extreme ignorance and coarse manners," she wrote, "unfortunately still prevail among the majority of the rural population, especially in areas far from the big cities." Marriages arranged by peasant families for adolescent daughters unprepared for sex, the physical and sexual brutality of husbands, the incestuous depredation of the heads of households—these were the pressures that led some women to acts of criminal violence.[57]

Borrowing her sexual typology from Krafft-Ebing's well-known work, Tarnovskaia described a scale of increasingly organic sexual deviance with which she explained the pattern of husband murder. The least abnormal cases she identified with the experience of adolescent brides: the young wife's early response to her husband's advances might well be a desperate one, but at the onset of menses her sexual revulsion might abate. Most cases, however, revealed "various departures from the norm of sexual development." Some women, Tarnovskaia affirmed, never experienced sexual desire at all; submitting to intercourse only in the interests of children (the pattern Lombroso designated as

---

[57]Tarnovskaia, *Zhenshchiny-ubiitsy*, 345–46, 479. The assertion that many peasant girls married prematurely concealed a presumption about the normal process of female sexual maturation. E.g., a physician found that 65 percent of peasant women in Kostroma Province married after the onset of menses but almost all before twenty-five, which he considered the age of sexual maturity; he therefore concluded that the majority of peasant marriages were premature: S. A. Olikhov, "K voprosu o plodovitosti krest'ianok Kineshemskogo uezda Kostromskoi gubernii," *Zemskii vrach*, no. 52 (1890): 824.

normal), they might easily grow to hate a husband who insisted on his rights. In educated circles such hatred might lead to divorce; among peasants, to murder.[58]

Two other forms of sexual deviance that contributed to female criminality, according to Tarnovskaia, involved increasing degrees of attraction to other women, accompanied by permanent repulsion toward men. The most common variety, an "instinctive abnormal attraction to persons of the same sex," was most often asexual, she believed, a "vaguely conscious urge, short of literal perversion, . . . accompanied by indifference to the normal satisfaction of sexual desire." The passionate friendship of inseparable companions belonged to this type. The more developed variety, much rarer and rarely acknowledged, entailed full-fledged passion for members of their own sex on the part of women of obviously masculine appearance.[59] Tarnovskaia thus identified the masculine woman as someone who shunned men rather than as one who enticed and entrapped them, as she appeared in the imagination of Lombroso and Tarnovskii.

Though Tarnovskaia found it hard to believe that normal women would endure the indignities of brothel life, she avoided the parallel conclusion that submission to the ordinary conditions of traditional marriage might also be thought abnormal, even as she detailed the ways in which conjugal relations in these households might drive a wife to extremes. Because the vast majority of women did not strike back (most wives and daughters patiently "bore their cross"), she argued, those who did could not be responding in a reasonable fashion to pressures felt by many women in the same intolerable degree; rather, they must find sex itself unusually repugnant.[60]

In identifying sexual revulsion rather than hyperactivity as a motive for violent crime, Tarnovskaia reversed one of Lombroso's central postulates, but she was unable to relinquish the Italian's belief that disturbed, deviant, or uncontrolled sexuality was closely connected to defects of the moral sense. Despite her analysis of husband murderers as sexually repressed, she nevertheless described criminals in general as suffering from "a weakening of the moderating centers, which leads them to succumb without the least struggle to their frequently extreme sensuality. Their lack of moral equilibrium takes the form of passionate, impulsive desires, which they hasten to indulge, lacking the power of self-restraint." Ultimately—and even less satisfactorily from the logical standpoint—she returned to criminal anthropology's underlying

---

[58]Tarnovskaia, *Zhenshchiny-ubiitsy*, 345–48.
[59]Ibid., 347–48.
[60]Ibid., 479, 345.

tautology, that those who kill are those willing to do so. "Women who opt for crime," Tarnovskaia wrote, "are distinguished primarily by an absence of revulsion to the act of murder, the practical ability to carry it out, and a low regard for the value of another's life."[61]

However readily Lombroso and Tarnovskii equated female submissiveness with health and pathology with inappropriately aggressive female sexual conduct, Tarnovskaia seems to have had trouble reconciling her theoretical ideas with her dismay at women's painful subordination.[62] But dismayed though she may have been, the biological argument allowed Tarnovskaia to avoid the social implications of her own critical views. In the end, she joined the men in denouncing the prostitute's escape from family obligation as a mark of organic disturbance, asserting that peasant women who remained in brothels when they had the opportunity and means to leave did so for pathological reasons.[63]

What Tarnovskaia identified as pathological when measured against the standard of an educated person's self-esteem, however, was not incompatible with the customary behavior of the members of a different social class. To show that prostitution was not a response to material need but must satisfy some warped psychological inclination, Tarnovskaia cited the case of a village girl rescued from brothel life by a wealthy benefactor who gave her the money to return home and raise the child he thought was his. After six short months, the young mother consigned the infant to her parents' care and resumed her place in the brothel. When interviewed, she denied having abandoned the child— whom she visited regularly, if infrequently—and explained that work in the fields was too hard.[64]

Tarnovskaia might have interpreted this woman's choice of brothel life and economic independence as a sign of enterprise, a willingness to strike out on her own and take her chances in the big city. Or she might have recognized the decision as part of the peasantry's changing world of new opportunity framed by traditional expectations. After all, the young mother acted much like other peasant women who sent their offspring home to be raised in the village, while they themselves left the arduous work of the fields for more remunerative labor in the towns. Indeed, even unwed mothers of peasant origin sheltered by the Russian Society for the Protection of Women in a refuge outside St. Petersburg refused to perform the agricultural tasks designed for their rehabilitation, "so spoiled" were they, their protectors complained, by "even a

[61]Tarnowsky, "Criminalité de la femme," 235; Tarnovskaia, Zhenshchiny-ubiitsy, 480.
[62]V. M. Tarnovskii, Prostitutsiia i abolitsionizm, 135–36; see also Kovalevsky (following Lombroso), Psychopathologie légale, 196.
[63]Tarnowsky, Etude anthropométrique, 62.
[64]Ibid., 7–8.

single year of city life."[65] But to have endorsed the young woman's step as rational or, if not freely chosen, at least as following a legitimate social path would have been to acknowledge, in stronger terms than Tarnovskaia was willing to apply, the intolerable situation of lower-class Russian women as it appeared to educated outsiders.

In terms of brute physical suffering, such observers had to admit, the prostitute was perhaps better off than the overburdened village woman. As Parent had noted in his day: "With the exception . . . of the syphilitic diseases, the prostitutes' profession is not in itself unhealthy. If I compare the existence of these women to that of working-class women obliged to lead a sedentary life and exhaust themselves working to make ends meet, the latter deserve more pity than the former."[66] Russian peasant women were not sedentary; they worked hard, bore many children, aged early, and often endured contempt and ill treatment at home, as educated Russians knew. Contemporary portraits do not show such women in command of their own bodies, nor mistresses of their own desire.[67] Rather, they imply that the sexual basis of a woman's subjection was no less explicit in the village than in the brothel. Students of folk custom, for example, denounced what they perceived as the widespread practice of incest between fathers and their daughters-in-law in traditional households.[68] It is possible that country women sometimes turned sexual impositions to their own advantage; one Soviet historian has suggested that the young wives who attracted the patriarch's attention may sometimes have been willing partners in a bargain that strengthened their position in relation to the powerful old man.[69] Contemporaries, however, never saw the arrangement in this light. The peasant woman remained, in their eyes, a standard of virtue and an icon of oppression.

Contemporary descriptions of the poor woman's lot nevertheless suggest a parallel their authors themselves did not acknowledge: subject

[65]A. G. Borodina, "Tsel' i zadachi Obshchestva zashchity zhenshchin," in *Trudy pervogo vserossiiskogo zhenskogo s"ezda pri Russkom zhenskom obshchestve v S.-Peterburge 10–16 dekabria 1908 goda* (St. Petersburg, 1909), 59.

[66]Parent-Duchâtelet, *De la prostitution*, 1:258.

[67]On the abuse of peasant women, see, e.g., Iakob Ludmer, "Bab'i stony," pts. 1–2, *Iuridicheskii vestnik*, nos. 11–12 (1884). See chap. 3 above. On the peasant woman's lack of autonomy, see, e.g., Efimenko, *Issledovaniia*, 123.

[68]The practice was known as *snokhachestvo*. See Ludmer, "Bab'i stony," pt. 2, 673–74; V. D. Nabokov, "Polovye prestupleniia, po proektu ugolovnogo ulozheniia," *Vestnik prava*, no. 9–10 (1902): 129–89, rpt. in V. D. Nabokov, *Sbornik statei po ugolovnomu pravu* (St. Petersburg, 1904), 129; M. O. Kosven, *Semeinaia obshchina i patronimiia* (Moscow, 1963), 75–76. See chap. 1 above.

[69]E. P. Busygin et al., *Obshchestvennyi i semeinyi byt russkogo sel'skogo naseleniia Srednego Povolzh'ia: Istoriko-etnograficheskoe issledovanie (seredina XIX–nachalo XX v.)* (Kazan, 1973), 102.

to the brothel's rule and the compulsion of the market, the prostitute was perhaps no more constrained than the virtuous peasant wife. True, the public woman was more anonymous, more depersonalized, in the sense that her identity was unknown to the men she received, whereas everyone knew the peasant wife, where she had come from, where she belonged. But the wife too suffered the anonymity of subordination and incorporation in a system built upon the suppression of her will and her desire. This disturbing analogy was one that most students of prostitution resolutely failed to accept, resisting the possibility that prostitution might represent an attractive choice in light of the available alternative. Tarnovskaia claimed to demonstrate with scientific rigor that prostitutes and peasant wives were not the same, but her recognition of the perils of ordinary womanhood ended by dulling her theoretical commitment to biological anomaly as the source of female deviance from the social and sexual norm.

The Refusal of Sexual Perversion

In rejecting Veniamin Tarnovskii's claim that prostitutes were often genetically abnormal, his colleagues spurned the suggestion that peasants might be anything but sexually naive and organically sound. But even Tarnovskii and his wife, the most committed Lombrosians on the Russian medical scene, modified the master paradigm in significant ways. Though he viewed male sexual perversion and female prostitution as predominantly organic disorders and shared the notion that sexual agency was appropriate only to men, Tarnovskii refrained from exploring the possibilities of female sexual perversion. Tarnovskaia, for her part, did pursue the matter of deviance among women but insisted on viewing female self-assertion and sexual desire as signs of mental health, not morbid aberrations. Neither one bolstered the case for the prostitutes' supposed psychosexual pathology by reference to the homosexual propensities of public women.[70] There was a point beyond which even they would not go in pursuing the logic of their own theoretical position.

Tarnovskii's lack of interest in the reputed lesbianism of prostitutes is all the more striking in view of his scientific concern with male sexual perversion. Indeed, he shared Lombroso's view that criminality and male homosexuality were analogous physiological states. Both conditions, Lombroso claimed, involved "the same completely amoral" psy-

[70]On lesbianism among prostitutes, see Lombroso and Ferrero, *La Femme criminelle,* 403–6.

chology. "Thus among homosexuals one observes frivolity, egotism, jealousy, deceit, lying, loquacity, a vain delight in ornament, a certain aesthetic penchant. . . . Likewise, criminals show a penchant for orgies and for revenge, affection for animals, love of evil for its own sake, use of slang, of tattoos, of symbols, all of which indicate the most profound atavism." Tarnovskii for his part considered "innate pederasty," like crime, "one of the signs of psychological degeneracy."[71] He regarded male homosexuality with the same moral revulsion disguised as scientific detachment that Lombroso showed for the female sex. Distinguishing between innate and acquired forms, as Krafft-Ebing had done in the case of homosexuality and Lombroso of crime, Tarnovskii denounced the use of penal sanctions in organically defective cases. Yet he described the effeminate, congenital male homosexual—a "monstrous type," "repugnant to men and despised by women"—in the same terms Lombroso used in speaking of the female sex: capricious, hysterical, jealous, cowardly, petty, vengeful, impulsive—and mysterious.[72]

If Tarnovskii accepted the tendency to see gender inversion as characteristic of deviance in men, he did not apply the same rule to women. Although he saw the congenital prostitute and the innate homosexual male as similar degenerate types, he did not apply the stigma of lesbianism to set the women apart. Having described the typical brothel inhabitants in terms borrowed from Parent-Duchâtelet, dwelling on their laziness, slovenliness, dishonesty, and greed, he neglected Parent's mention of their mutual love affairs. Even in citing the authority of Tardieu's disciple Louis Martineau, who was obsessed with precisely that detail of Parent's account, Tarnovskii did not echo Martineau's concern.[73] Another Russian partisan of the Italian school, the psychiatrist Pavel Kovalevskii, left the same gap. Writing in 1903, Kovalevskii compiled from the work of Parent and Lombroso a portrait of the typical prostitute, whom he considered an organically defective breed, but like Tarnovskii (and Tarnovskaia), he ignored any reference to her canonical lesbian tastes.[74]

[71]Cesare Lombroso, "Du parallélisme entre l'homosexualité et la criminalité innée," in *Comptes-rendus du sixième congrès international d'anthropologie criminelle*, 7–8; V. M. Tarnovskii, *Izvrashchenie*, 23–24. On the relation of Tarnovskii's ideas to current European thinking about male homosexuality, see Wettley and Liebbrand, *Von der "Psychopathia sexualis,"* 49, 55, 62, 66–67. A Soviet biographer, who lauds Tarnovskii's role in the study of venereal disease, blames his adherence to the "erroneous" doctrine of the born criminal type on the unfortunate influence of his wife: Arkhangel'skii, *V. M. Tarnovskii*, 69.

[72]V. M. Tarnovskii, *Izvrashchenie*, 10, 20.

[73]The absence of this theme is particularly evident in V. M. Tarnovskii, *Prostitutsiia i abolitsionizm*, chap. 14, on the prostitutes' moral constitution. Cf. Martineau, *Leçons*, 21–22.

[74]Kovalevsky, "La Femme criminelle."

The conviction that nature and virtue had survived among even the urban lower classes, still so closely tied to village life, may help explain why nineteenth-century Russian texts so largely ignored the possibility of sexual perversion among prostitutes drawn from the common folk. Across the political spectrum, Russians were more comfortable explaining sexual turpitude as a function of class privilege than of social and cultural deprivation.[75] It took some ingenuity, however, to explain the evidence that sexual crime and sexual variation were not in fact a monopoly of the cultivated elite. Vladislav Merzheevskii, for example, reported in the 1870s that St. Petersburg, like other great capitals, harbored men who "made pederasty their special trade," extorting money from gentlemen clients, and bathhouse attendants who operated a "completely organized perverts' guild" (*artel' razvratnikov*), offering sexual services in exchange for tips. The blackmailers belonged to "the very lowest social stratum: lackeys, idle children of petty shopkeepers," Merzheevskii claimed, as though to distinguish them from the noble poor. Yet even they, he insisted, were merely responding to their well-heeled clients' demands.[76] Thirty years later the populist-minded physician Dmitrii Zhbankov imagined that rapes committed by educated men were distinguished by cold-blooded calculation, whereas those perpetrated by hot-blooded lower-class youths were passionate and impulsive expressions of the natural sexual urge, and therefore less repugnant.[77]

With the exception of the two Tarnovskiis' work, Russian medical treatments did not usually present prostitution as a form of sexual pathology. In the 1890s, when professional psychiatric journals finally came into their own, most largely ignored the subject of sexual deviance. When they tackled it at all, they focused on case histories drawn from private clinical practice. The journal that devoted the greatest attention to the issue confined itself to the discussion of homosexuality in educated men.[78] One exception to the general neglect of female homo-

[75]On upper-class depravity, see, e.g., the populist democrat D. N. Zhbankov, "Polovaia prestupnost'," *Sovremennyi mir*, no. 7, pt. 2 (1909): 59–60; the reactionary jurist A. Likhachev, "Novye raboty v oblasti ugolovnoi statistiki i antropologii," *Zhurnal grazhdanskogo i ugolovnogo prava*, no. 3 (1883): 2; and the feminist physician I. I. Kankarovich, *Prostitutsiia i obshchestvennyi razvrat: K istorii nravov nashego vremeni* (St. Petersburg, 1907), 189.

[76]Merzheevskii, *Sudebnaia ginekologiia*, 207–9. See also V. M. Bekhterev, "Lechenie vnusheniem prevratnykh polovykh vlechenii i onanizma," *Obozrenie psikhiatrii, nevrologii i eksperimental'noi psikhologii*, no. 8 (1898): 594.

[77]Zhbankov, "Polovaia prestupnost'," 86, 88.

[78]The journal was V. M. Bekhterev's *Obozrenie psikhiatrii, nevrologii i eksperimental'noi psikhologii*, which published P. Ia. Rozenbakh, "K kazuistike polovogo izvrashcheniia," no. 9 (1897): 652–56; A. N. Uspenskii, "K kazuistike anomalii polovogo chuvstva," no. 12 (1898): 927–28; Bekhterev, "Lechenie," no. 8 (1898): 587–97; S. Liass, "Izvrashchenie polovogo vlecheniia," no. 6 (1898): 415–16; and V. M. Bekhterev, "O lechenii onanizma vnusheniiami v gipnoze," no. 3 (1899): 186–89.

sexuality was a 1898 psychiatric study published in *Vrach*, Russia's most influential medical journal, but this exception proves the rule of interpretive class bias. Following Krafft-Ebing and Carl Westphal, Fedor Rybakov, a Moscow professor of psychiatry, presented an example of lesbianism as a degenerative disorder, akin to the hysteria, alcoholism, epilepsy, tuberculosis, brain fever, and "nerves" with which the patient's family was allegedly plagued. The woman's parents had been clever enough to send the girl for treatment, and Rybakov had happily managed to convince her that "her attraction for women was a disease." In his view, neither prostitutes nor their occasional lesbian affairs belonged to the same degenerate syndrome; and the perversion that flourished in prisons, army barracks, and brothels, he maintained, was accidental rather than organic.[79]

The Russian medical literature of the period does offer one extensive discussion of female homosexual love, the work of another Tarnovskii: the gynecologist and obstetrician Ippolit Mikhailovich. In 1895, when his study appeared, Ippolit Tarnovskii had had thirty-five years' experience in St. Petersburg maternity clinics, twenty-three as assistant director of the capital's lying-in hospital.[80] Like most of his colleagues, he characterized homosexuality as "abnormal and [even] pathological," but he avoided the kind of moralizing typical of Veniamin Tarnovskii. Indeed, the conclusion of Ippolit's exercise was that same-sex love—for which he used the strictly neutral term *gomoseksual'nost'* (counterposed to its opposite, *geteroseksual'nost'*, both extremely rare in Russian texts)—was not necessarily a pathological sign.[81]

The centerpiece of his book consisted of a triptych of lesbian cases, which he used to demonstrate that women who had sex with each other

---

[79]F. E. Rybakov, "O prevratnykh polovykh oshchushcheniiakh," pt. 1, *Vrach*, no. 22 (1898): 641–42. On the influence of current theory, including the work of V. M. Tarnovskii, see ibid., 640–43, and pt. 2, no. 23 (1898): 664–67 (the title translates the German *conträre Sexualempfindung*). On Westphal's model, see Gudrun Schwarz, "'Viragos' in Male Theory in Nineteenth-Century Germany," in *Women in Culture and Politics: A Century of Change*, ed. Judith Friedlander, Blanche Wiesen Cook, Alice Kessler-Harris, and Carroll Smith-Rosenberg (Bloomington, Ind., 1986), 128–29; and Wettley and Liebbrand, *Von der "Psychopathia sexualis,"* 53–54.

[80]I. M. Tarnovskii, *Izvrashchenie polovogo chuvstva u zhenshchin* (St. Petersburg, 1895). Lesbianism is briefly mentioned in N. A. Obolonskii, *Izvrashchenie polovogo chuvstva* (St. Petersburg, 1898), 17–18, citing Ippolit Tarnovskii. See "Tarnovskii, I. M.," in *Entsiklopedicheskii slovar' Brokgauz-Efron*, 32A:651. Tarnovskii served on the medical board of the Empress Marie charitable department and was promoted to director of the lying-in hospital in 1898, the year before his death. From a landless family of the hereditary nobility, Tarnovskii was married to the daughter of a lieutenant general and had four daughters; see his service record in TsGIA, f. 759, op. 41, d. 2053, ll. 16–40. Given their dates of birth, patronymic, and social background, Veniamin and Ippolit *may* have been brothers. In the late 1860s they both taught in the St. Petersburg midwifery institute: Arkhangel'skii, *V. M. Tarnovskii*, 11.

[81]I. M. Tarnovskii, *Izvrashchenie polovogo chuvstva u zhenshchin*, 2, 117–18, and V. M. Tarnovskii, *Izvrashchenie*, 51.

not only could be "psychologically and neurologically healthy in every respect" but were to be found in every social class.[82] The trio he presented included a peasant, an urban prostitute, and a well-educated woman of independent means. Evidence on the first two emerged from courtroom records; the third woman provided testimony of her own in the gynecologist's private office. Like Parent-Duchâtelet, who had insisted that physical examinations were often inadequate to uncover the truth of women's sexual experience in the absence of personal testimony, Ippolit Tarnovskii prided himself on listening to the women who were the objects of his professional curiosity. This attention, he boasted, enabled him to elicit information that physicians who relied on "objective" data (that is, on the evidence of their own eyes and fingers) were bound to miss. Sexual feelings, he insisted, were a subjective matter.[83] Women were the subjects of their own desire.

Like the younger, more eminent Tarnovskii, Ippolit was fully acquainted with the European literature on homosexuality.[84] He accepted the current distinction between acquired and innate forms of perversion, which in the case of women were thought to correspond, respectively, to "passive" and "active," or feminine and masculine, types. In line with Westphal and Moll, he considered "active" lesbianism an organic anomaly but not necessarily an example of pathology or degeneration.

> There are obviously women who are completely normal in every respect, though endowed by nature with an unusual inclination . . . toward sexual relations with members of their own sex. Entirely incomprehensible, strange, and unnatural to the ordinary person, such sexual perversion is entirely natural, normal, and appropriate for the women themselves. Not only does it do them no harm, but on the contrary it satisfies their physiological needs. Their sexual lives develop in the same way as the sexual lives of normal people, except that the sexual impulse leads in the opposite direction: toward women, not men.[85]

Unlike Veniamin, Ippolit Tarnovskii took every occasion to demonstrate the "normality" of such homosexual inclinations. He asserted (1)

---

[82]I. M. Tarnovskii, *Izvrashchenie polovogo chuvstva u zhenshchin*, 107, 128. For a translation, see Laura Engelstein, "Lesbian Vignettes: A Russian Triptych from the 1890s," *Signs* 15:4 (1990): 813–31.

[83]Parent-Duchâtelet, *De la prostitution*, 1:203; I. M. Tarnovskii, *Izvrashchenie polovogo chuvstva u zhenshchin*, 12–13.

[84]I. M. Tarnovskii, *Izvrashchenie polovogo chuvstva u zhenshchin*, chap. 4, provides a summary of the major texts, including Parent-Duchâtelet, Martineau, Casper, Tardieu, Westphal, Charcot and Magnan, Krafft-Ebing, and Moll.

[85]I. M. Tarnovskii, *Izvrashchenie polovogo chuvstva u zhenshchin*, 63, 106, 155.

that lesbianism was more common than usually supposed; (2) that the active type, which deviated more obviously from the feminine norm, occurred less often than the passive; (3) that when wives left their husbands for their girlfriends' arms, they did so not, as some experts thought, because they had all along harbored pathology in the germ but because conjugal sex left them frustrated, unhappy, and at constant risk of getting pregnant; (4) that if prostitutes often made love to each other, "as everyone knew," it was because their working lives left them equally frustrated and also disgusted with men; (5) that even the most flagrantly masculine types did not differ physiologically from other women (he had never observed an enlarged clitoris among them).[86]

Though he was conversant with recent medical lore that had come to characterize same-sex love, female as well as male, as a morbid physiological urge, Ippolit Tarnovskii reached back sixty years for an alternative model in the depiction of lesbian affection. Very much in the spirit of Parent-Duchâtelet's 1836 portrait, he credited women of homosexual tastes with feelings of "tender and passionate" love (though sometimes laced with cruel jealousy, it was true). Unlike Parent, however, he provided his readers with a full description of this love's genital component: "The active tribade lies on top the passive one, putting her left thigh between the other's thighs. Embracing her with her right hand, she strokes the edge of her labia minora and vaginal opening with her left index finger. Sometimes she introduces her finger so deeply into the vagina that the back of her other three fingers, clasped into a fist, press firmly and rub against the external genitals. At the same time, the woman moves her body as a man would do in the act of intercourse."[87]

Willing to acknowledge the carnality of female same-sex desire, Ippolit Tarnovskii nevertheless invoked the moral distinction earlier drawn by Johann Ludwig Casper between homosexual love and perverted sexual acts. Like Casper, Tarnovskii denied that lesbians engaged in oral sex, which he considered an "indubitable sign of depravity" employed by self-indulgent heterosexual couples to enhance their pleasure and prevent conception and by women who trafficked in commercial sex unmotivated by love. "Cunnilingus," he wrote, "is rarely employed [among lesbian lovers], and [then] only between public women or by professional tribades."[88] Arbitrary as it might seem to excuse one form of nonprocreative sex while deploring another, the condemnation of

[86]Ibid., 155, 128, 121–23, 125–26, 131. For European discussion of the enlarged clitoris among lesbians, see Martineau, Leçons, 39, 61–62, 74–77.

[87]I. M. Tarnovskii, Izvrashchenie polovogo chuvstva u zhenshchin, 133, 123, 134. On lesbian lovers, cf. Parent-Duchâtelet, De la prostitution, 1:159–68.

[88]I. M. Tarnovskii, Izvrashchenie polovogo chuvstva u zhenshchin, 134, 158. Cf. Casper, Handbook, 3:335, 337.

cunnilingus helped Tarnovskii in two ways: by reminding readers of his own moral standards, and by setting up a foil against which the acts that mimicked heterosexual intercourse could be seen as products of legitimate desire.

His evidence that homosexual practices occurred in every social stratum, and thus were more common than his colleagues liked to think, permitted Ippolit Tarnovskii to assert that they were therefore not as abnormal as many people would have it. At one end of the social scale, he emphasized the extent to which female homosexuality had found a place in folk custom, challenging the illusion that peasants were immune to such things. At the other, he portrayed the educated lesbian in his trio—an example of the "mannish," active type—as admirable, physically attractive, and happy with her lot. Unlike passive male homosexuals, who, in his view, represented all the female vices (cowardice, dishonesty, weak will), "active homosexual women . . . were energetic, honest, upright, strong-willed, intelligent, and extraordinarily discrete"[89]—that is to say, like the best of men. Tarnovskii's praise ill conceals a strong measure of personal identification, which lends this woman's story an intensity the other two lack. Even as she preserved the appearance of a respectable woman, here was someone with whom he could chat "man to man," as it were, and press to reveal the secrets of her sexual constitution.

If Tarnovskii could not entirely resist the pervasive assumption that female sexual assertion must be coded as male, he at the same time derided this idea by showing its roots in popular ignorance and myth. Peasant society, according to the evidence of his first case, accommodated sexual deviation by adapting it to the heterosexual scheme. This case described the fate of one Mariia Shashnina. Having managed to provide a number of the women in her village with what they later described as complete sexual satisfaction "in the manner of a man," she was said by her neighbors to speak in a coarse, manly voice, smoke tobacco, and rule the domestic roost. So completely did her partners believe Shashnina's boast of possessing a male sexual part that they testified in court to having felt and seen it protruding from between her vaginal lips and called her a "double-rigger" (dvukhsnastnaia). This rural temptress had incited a recent bride to murder her new husband, arguing that he was less intelligent and less well off than she (Shashnina) herself, and more likely to get his wife pregnant. But medical inspection, Tarnovskii reported with obvious satisfaction, did not uncover the fabled masculine organ, nor indeed any other genital anomaly. The examining physicians had such trouble describing what they thought they

---

[89]I. M. Tarnovskii, *Izvrashchenie polovogo chuvstva u zhenshchin*, 131.

had actually seen, however, that the court looked skeptically at their findings, especially since they conflicted with the firmly held convictions of the folk.[90] Tarnovskii carefully detailed the results of three successive and inconsistent pelvic examinations to prove how futile such exercises were. Using the authority of science to debunk naive prejudice, he simultaneously debunked his scientific colleagues' extravagant claims to produce the objective truth about female sex.

Among the scientific "truths" Tarnovskii rejected was the notion that women who had sex with each other must be mentally or emotionally unbalanced. Under questioning, he stressed, Shashnina conducted herself with complete self-possession. Likewise, the prostitute in his second case, who fell victim to her husband's murderous wrath when he found her in the arms of her beloved, had conducted her affairs with intelligent purposefulness: she ran a small prostitution business out of a private apartment, managed to secure her own passport, and won her husband's trust and cooperation until the brutal end. Neither woman, the gynecologist concluded, could be considered abnormal in any respect except her choice of sexual partner. It is perhaps significant, for someone bent on overturning scientific as well as popular myth, that Tarnovskii should have presented the prostitute as the most unequivocally feminine of his three cases.

Tarnovskii's third case, the attractive young widow who had come to the gynecologist's office, offered no example of upper-class decadence and urban vice but maintained a discrete, respectable existence in the privacy of her apartment and intimate circle of friends. She had, however, begun her sexual career in the hothouse atmosphere of a girls' boarding school, where she found her first willing partner. It was in this sense that the sexually aggressive, masculine types presented a social danger, in Tarnovskii's eyes. Women like these should not be held accountable before the law, he argued, "since they cannot experience sexual desire in any other way." But since they could not be cured, they should be kept out of schools, taught at home like other handicapped children, and prevented from doing any harm. Tarnovskii also believed that lesbianism, though not a crime, should qualify as grounds for divorce in civil actions. Without such legal recourse, the young widow's husband had shot himself in despair after a brief, unconsummated marriage.[91]

Ippolit Tarnovskii's examples rely for their effect on the assumption

[90]Tarnovskii (ibid., 127) cites a medical report on the Sakhalin penal colony by Dr. A. D. Davydov, in *Ezhenedel'nik*, no. 1 (1895), as additional testimony to the existence of a folk belief that homosexual women were physical hermaphrodites.
[91]I. M. Tarnovskii, *Izvrashchenie polovogo chuvstva u zhenshchin*, 157–63.

widely held in his own social class that if anyone indulged in sexual excess or sexual variation, it was not the traditional common folk but the educated elite, morally corrupted by city life and Western values. His demonstration that sex between women found a place in village life, that prostitutes, even if they were lesbian, were not sexual psychopaths, and that women of social refinement shared the sexual proclivities of the other two types was meant not to reveal the depravity latent in the lower orders but rather to challenge the supposed pathology of lesbian sex. It is clear, however, that his ideas were not always consistent: he agreed that homosexuality was abnormal but felt that people with homosexual desires were not; he could explain the existence of mannish lesbians as an organic anomaly even though the ones he had encountered were mentally and physically sound; he admired the assertive, morally upright traits the mannish types displayed yet chastised them for leading other women astray. Though he recognized that even the "feminine" partners might seem content, he nevertheless thought they might profit from being "cured," whereas the masculine sort could not alter their inclinations. In part, these tensions were inherent in the reigning medical model, but in his general tendency to opt for the normalizing interpretation, Tarnovskii shared his colleagues' reluctance to stigmatize female sexual perversion.

Until the revolution of 1905, most Russian physicians had no interest in classifying prostitution as a sexual pathology or in finding yet more deviant sexual practices, such as lesbianism, with which to mark the public woman as biologically different from the private one. Even Veniamin Tarnovskii and Praskov'ia Tarnovskaia, who as prominent exceptions to this rule adopted organic theories of social and sexual pathology, hesitated to apply them with full force to the case of women. Indeed, the same socially oriented and empathetic, if morally severe, tone that had characterized the work of Parent-Duchâtelet continued to echo in Russian medical writings almost a century later.[92] Such an echo could still be heard, for example, in a piece by Boris Bentovin, first published in the populist journal *Russkoe bogatstvo* in 1904, when it was still unusual for a physician to discuss lesbianism in Russian brothels. Bentovin nevertheless reflected the mood of the day by avoiding a moralizing tone and eschewing the organic interpretation. The affection he claimed to have observed among women in public houses, far from being pathological, at first had the "innocent purity" of a schoolgirl crush. It was only "in the context of their corrupting environment

---

[92]For a tribute to the moral virtues of prostitutes which invokes Parent and rejects Tarnovskaia's view of female crime, see Nikolai Zeland, *Zhenskaia prestupnost'* (St. Petersburg, 1899), 19–20.

[that] this original spiritual feeling later [took] on the character of 'lesbianism,'" he said. Judging from love letters between women in the St. Petersburg syphilis wards, Bentovin described these pairs of simple women as "touchingly affectionate couples" (*trogatel'no-laskovye pary*), without suggesting that they were sexually depraved. Ignorance kept them naive, just as poverty led them to corruption: "The illiterate lines radiate such a keen and tender womanly spirit [*chutkaia i nezhnaia zhenskaia dusha*] that one is painfully aggrieved. One senses how much sincere love and passionate self-sacrifice lurks under the filthy exterior. If the material conditions of the writer's life had been different, she might have made a good, honest person, perhaps a good wife and mother."[93]

Why were Russian physicians such as Boris Bentovin reluctant before 1905 to adopt ideologies of social marginalization in labeling and explaining female sexual deviation, loath indeed to focus on sexual deviation as a problem of any social significance at all? In Europe the prostitute as lesbian belonged to a category of social boundary markers that guaranteed the concentration of cultural and political power in class and gender terms. By insisting on the virility—ultimately, the homosexuality—of public women, medical texts reinforced the positional definition of masculinity. By questioning the femininity of women who sold sexual services on the marketplace and thus avoided their domestic and reproductive roles, such texts gave authority to the notion that denizens of the public sphere must by definition be male, that sexual agency could not by definition be female, and that prostitutes, though necessary to maintain the economy of heterosexual desire, must be excluded from the normal gender system.

Rendering prostitutes pathological was part of a larger strategy by which bourgeois men in their professional capacity helped secure their own monopoly on public life. In bourgeois societies, men of the dominant class defined their own place in the civic arena by determining who was to be left out (women; the popular classes) and by marking the excluded increasingly in biological terms (as physically and mentally deficient).[94] This was not the case in Russia, where socially concerned and publicly active professional men such as Veniamin and Ippolit Tarnovskii, Eduard Shperk, and Dmitrii Zhbankov were themselves

---

[93]B. I. Bentovin, "Iz zhizni prostitutok," *Russkoe bogatstvo*, no. 12 (1904), 160–76; rpt. in B. I. Bentovin, *Torguiushchie telom: Ocherki sovremennoi prostitutsii*, 2d ed., rev. (St. Petersburg, 1909), 125–27.

[94]For the symbolic function of such marginalization, see the introduction to Peter Stallybrass and Allon White, *The Politics and Poetics of Transgression* (Ithaca, N.Y., 1986); also Frank Mort, *Dangerous Sexualities: Medico-Moral Politics in England since 1830* (London, 1987).

excluded until 1905 from an effective role in national affairs. In Russia, candidates for political life were more likely to use socially marginal groups to exemplify their own disempowered condition than to affirm a monopoly of power they did not possess. Thus feminism became a hallmark of the early radical intelligentsia (one thinks of Alexander Herzen, Nikolai Chernyshevskii, and Mikhail Mikhailov), and peasants continued to stand throughout the century as icons of injustice and oppression, not just for revolutionary populists but for respectable society at large. Peasant women (even those who moved to the cities and fled traditional patriarchal bonds) were thereby doubly immune to stigma.

Peasant society, moreover, did not recognize the public-private distinction. It did not accommodate notions of individuality appropriate to the educated members of Russia's largely urban elite. How could the "public woman" emerge in the traditional peasant community, where public and family life coincided, where even the public function of men depended on their patriarchal status in the literal, domestic sense? Rural physicians in fact routinely denied that prostitution existed in the countryside. They associated commercial sex with cities, army camps, and factories, places where men and women lived outside their families in the disruptive circumstances of a Westernized or modernizing world.[95]

As the problem of syphilis eloquently demonstrates, professional and philosophical divisions within the medical community generated divergent responses to the state's role in public health and to the optimal direction of Russia's social development. Among specialists in venereal disease, Veniamin Tarnovskii was perhaps the most closely identified with state authority. As an advocate of the police-medical regime, he was one of the least reluctant among his peers to mark the prostitute as socially marginal and organically distinct. This theoretical position bolstered his preference for traditional administrative methods of social discipline but also coincided with a readiness to acknowledge the inroads of modernity into traditional Russian life, to recognize (though still deplore) the transformation of peasants into streetwalkers and factory hands.

It is perhaps one of the contradictions of the Russian political landscape that the same physicians who promoted self-consciously modern political values and opposed traditional methods of state intervention and control resisted the evidence of ongoing social transformation. By and large, it was the community practitioners—such as the prolific and outspoken Dmitrii Zhbankov—who insisted on medicine's independence of state control and on civil rights and freedom of speech as essential to their professional as well as civic well-being; who opposed state-regu-

---

[95]For more on this aspect of the prostitution question, see chap. 5 below.

lated prostitution as a violation of professional standards and individual rights; and who joined the movement for political reform on the eve of 1905. The last to abandon their populist dreams, they clung to the image of the sexually innocent peasant despite evidence to the contrary uncovered in the course of extensive practical experience. In this they resembled the political populists, who, as Richard Wortman has shown, clung to their ideal of peasant life long after they had noticed its divorce from social reality.[96]

In the European context, the lesbian prostitute both challenged and sustained the sex-gender system. On the one hand, her supposedly mannish constitution established a palpable organic distinction between the private (normal) and the public (perverse) woman. On the other, she provided "normal" men with heterosexual satisfaction in a world that released them from the personal responsibility and burdensome solitude demanded by womanly women and domestic life. In fraternal outings to the brothel, friends demonstrated their manhood by sharing the bodies of women who belonged to them all. As Anton Chekhov showed in his 1888 short story "The Nervous Attack" ("Pripadok"), a reluctance or inability to participate in this male rite could be literally unmanning.[97] The protagonist's withdrawal from his companions' sexual adventure causes him to fall into a womanish fit of hysteria, which testifies to his strength of conscience and weakness of sexual resolve. When he sees prostitutes as female—in need of protection, privacy, and care—the easy male solidarity that depends on the impersonality of public congress is destroyed.

Not all Western authorities believed that sexual perversion could be either detected or explained in biological terms, but that belief grew stronger as the nineteenth century progressed. Interest in lesbianism, both as a symbol of the prostitute's pathology and as a pathology in itself, increased during the same period. Most nineteenth-century Russian physicians, by contrast, refused to invoke biological explanations either for prostitution or for sexual deviation and rarely associated the two. Those who did tended to support old-fashioned notions of state custodianship, whereas defenders of civic and personal autonomy rejected the organic theory's custodial implications, along with the administrative police techniques it sustained. But even Veniamin Tarnovskii did not use homosexuality to symbolize the marginal status of his supposedly degenerate female types. Perhaps Russian men were less anxious than their Western fellows about the sexual appetites of respect-

---

[96]Richard Wortman, *The Crisis of Russian Populism* (London, 1967).

[97]A. P. Chekhov, "Pripadok" (1888), in *Polnoe sobranie sochinenii i pisem*, 6:218–41 (Moscow, 1962).

able women because respectability in Russian privileged circles did not in fact replicate the bourgeois ideal; perhaps they were less worried about competing with women for public space because they themselves could not claim it; perhaps they were less fearful of the social and political challenge of a permanent working class than they would be in the wake of revolution.

It was not until after 1905 that the figure of the lesbian prostitute emerged in Russian medical and popular discourse to any noticeable degree or that sexual pathology itself aroused much interest. This new emphasis, it seems to me, reflected the impact of events that had altered the structure of Russian civic life, reshaping the possibilities of political participation for all of educated society, causing the medical profession in particular to readjust its attitudes toward the state, and changing the reality and perception of class difference.[98]

[98]On the medical profession's changed attitude toward state control, see Hutchinson, "'Who Killed Cock Robin?'" and idem, *Politics and Public Health*.

Chapter Five

Morality and the Wooden Spoon: Syphilis, Social Class, and Sexual Behavior

The class and cultural presumptions that shaped the Russian adaptation of Western ideologies of sexual deviance also affected the way in which Russian physicians interpreted the disease most closely associated with the perils of commercial sex: syphilis. Ambiguous in its clinical manifestations and variable in its pattern of transmission, syphilis provided an arena in which the medical profession worked out its internal conflicts and its relation to the state authorities involved in the regulation of public health. The doctors' approach to the problem of how to understand and control the spread of the disease was of course informed by theories generated in the West, which differed considerably from today's scientific knowledge of the subject.[1]

Until the discovery and application of penicillin in the mid–twentieth century, syphilis was a potent symbol of the risks of unregulated sexual contact. Many physicians employed the venereal threat to justify the police supervision of public women; opponents of regulation used it to promote a uniform standard of sexual continence for women and men alike. But Victorian medicine did not consider syphilis an exclusively venereal disease. The widely cited study *Syphilis in the Innocent*, published in 1894 by the American physician L. Duncan Bulkley, showed in exhaustive detail that the origins of syphilis were not always—perhaps not even primarily—sexual.[2] Genital contact was but

---

[1] For a comparison of nineteenth- and twentieth-century knowledge of syphilis, see Laura Engelstein, "Syphilis Historical and Actual: Cultural Geography of a Disease," *Reviews in Infectious Diseases* 8:6 (November–December 1986): 1036–48.

[2] L. Duncan Bulkley, *Syphilis in the Innocent (Syphilis Insontium), Clinically and Historically Considered, with a Plan for the Legal Control of the Disease* (New York, 1894). The concept of innocent syphilis, at least a century old, was widely popularized in the 1890s

one of a multitude of possible means of contagion. The shared cup, the casual hug, the dirty razor blade, the unwashed towel: these were the innocent means by which innocent people in all walks of life disseminated a germ that was not intrinsically sexual. European physicians likewise invoked the threat of nonvenereal contagion to deter the educated from casual contact with the unhygienic poor, as well as to emphasize the unintended potential consequences of an originally sexual transgression.[3]

The admission that sexual dereliction was not the only risk facing the urban population did not therefore diminish the power of syphilis to represent the dangers of sex. Though Bulkley denied that syphilis was inherently venereal, he considered the disease most frequent among people who lived in proximity and who indulged their sexual appetites in a promiscuous way—that is, among the residents of big cities, the beneficiaries of modern civilization. In the congested urban environment, with its surfeit of unwed and sexually hungry males and its abundance of destitute and unsupervised females, sexual contact remained a central cause for concern. Given the belief in nonvenereal transmission, however, there was no reason to infer that once the disease was imported in the wake of an urban spree, it should not spread as indiscriminately in the country as in the town. Bulkley nevertheless maintained that it became less common "in a pretty direct ratio to the suburban or rural character of the people."[4]

Under certain circumstances, however, Bulkley found the prevalence of syphilis related neither to density of population nor to cultural sophistication. Far from being rare in primitive agrarian societies, he claimed, syphilis was often endemic to entire communities, in which it penetrated the fabric of everyday life. "It is difficult for us, amid the civilization of the nineteenth century," the American doctor wrote, with the comfortable arrogance of the world's most up-to-date land, "to fully understand the circumstances and surroundings which led to the rapid and extensive diffusion of syphilis in years gone by, and amid the crude modes of living belonging to earlier times; but we can better understand it if we consider, for a moment, some facts relating to its more recent extension among some of the less civilized portions of the

by Bulkley, who is cited in, e.g., Alfred Fournier, *Traité de la syphilis* (Paris, 1899–1906), 1:134; A. I. Rozenkvist, "K statistike vnepolovogo zarazheniia sifilisom," *Biblioteka vracha*, no. 8 (1898): 532; M. A. Chlenov, "K kazuistike vnepolovogo sifilisa," pt. 2, *Russkii vrach*, no. 30 (1902): 1093–94.

[3]See Alain Corbin, "Le Péril vénérien au début du siècle: Prophylaxie sanitaire et prophylaxie morale," in *L'Haleine des faubourgs: Ville, habitat et santé au XIXe siècle*, ed. Lion Murard and Patrick Zylberman (Fontenay-sous-Bois, 1978), 246, 250, 259.

[4]Bulkley, *Syphilis in the Innocent*, 3–4.

earth, as in certain parts of Russia." Distinguished by "gross ignorance," to use Bulkley's phrase, Russia was a remnant of the past at
civilization's outermost edge. It shared with other marginal lands a ves-
tigial pattern of disease which largely excluded the sexual component
supplied by modern city life.[5]

True, nonvenereal transmission plagued even the most sophisticated
of lands. But the saturation of entire districts, the predominance of
cases among women and children and village craftsmen, was considered
peculiar to backwardness. European doctors attending the 1899 international syphilis congress in Brussels insisted, for example, that their own
peasants did not suffer the same form of the disease as those in Russia.
In its endemic guise, in fact, syphilis served Europeans as an exact index
of cultural deprivation: its prevalence mirrored existing levels of filth,
poverty, malnutrition, and overcrowding.[6] Or was it the other way
around? Was the disease everywhere the same and perceived differences
a mirror of cultural expectation? Was the symbolic position of Russia,
as an icon of primitive asexuality, a figment of the Western imagination, or could its backwardness truly be seen in the structure of its
disease? Can we ascertain in retrospect if there was such a thing as nonvenereal syphilis, endemic or otherwise? In search of answers, the historian must first interrogate those observers native to the culture in which
the past was said to have been frozen in time.

Not only did Russian physicians accept their nation's place on the
outskirts of modernity, but they supplied much of the data on which
Europeans based their views. Indeed, Russia's notoriety as a breeding
ground of syphilitic contagion was due in no small part to the evidence
supplied by Veniamin Tarnovskii. "Syphilis of the innocent," wrote the
professor of dermatology and venereal disease, in a work often cited by
European colleagues, "is the most serious, socially most harmful, and

---

[5]Ibid., 140, 4. Fournier, a leading authority, did not ignore the dangers of nonvenereal
contagion but considered 90 percent of all cases in France to be venereal in origin: *Traité
de la syphilis*, 1:20. "Endemic," nonvenereal syphilis was said to occur in contemporary
Moravia, Sweden, Norway, Estonia, Courland, and Lithuania; see Bulkley, *Syphilis in the
Innocent*, 5–7; and Etienne Lancereaux, *Traité historique et pratique de la syphilis* (Paris, 1866;
Russian ed., 1876), cited from English trans., *A Treatise on Syphilis: Historical and Practical*
(London, 1868–69), 1:29–43, 46. A French medical dissertation claimed that syphilis was
brought to the countryside by peasant men who visited prostitutes in the towns. It cited
the dangers of wetnursing and congenital syphilis but largely ignored other forms of
nonvenereal transmission and did not characterize the problem as endemic, a term it
reserved for syphilis among the "Negroes" of North America: Léon Issaly, *Contribution à
l'étude de la syphilis dans les campagnes* (Paris, 1895), 7–8, 10–11, 13–14, 28; cited in Corbin,
"Le Péril vénérien," 250–51. Nelly Furman kindly found me a copy of Issaly's text.

[6]P. A. Gratsianov, "Bor'ba s sifilisom, kak predmet obshchestvennoi gigieny," in
Vos'moi Pirogovskii s"ezd: Avtoreferaty i polozheniia dokladov po sektsiiam (Moscow, 1902),
6:245; Lancereaux, *Treatise*, 2:261–62.

in Russia the most common form of syphilis."[7] Some Russians defensively reminded their readers that the risk of nonvenereal contagion was universal: "Culturally deprived Russia is not the only victim of a high incidence of nonvenereal syphilitic transmission," wrote Andrei Rozenkvist in 1898. "More enlightened peoples also contract syphilis in the unexpected way." But the inflated sense of national inferiority expressed by a practitioner familiar with rural life was more typical: "Endemic syphilis," he asserted, "is not found anywhere in Europe, [but] constitutes the sad privilege of Russia alone."[8]

Venereal syphilis was widely thought to have arrived in Europe at the end of the fifteenth century with the returning sailors on Columbus's ships and then to have been diffused on an epidemic scale by the sexual activity of soldiers and public women. Despite much expert controversy, even today still unresolved, about the actual origins of syphilis in Europe, the myth of venereal syphilis as the herald of the modern age has survived all learned disputes.[9] Convinced that the disease had been in evidence under different names and in nonvenereal form since ancient times, Philippe Ricord, one of the founders of nineteenth-century syphilology, nevertheless recognized the virulent outbreak of 1493, which marked the transition from communal to individual patterns of contagion, as the start of a new era in medical and popular consciousness, akin to the political watershed marked by the Great Terror of 1793: the "veritable '93 of the pox," he called it.[10]

Russians interpreted their own national experience in light of this firmly entrenched idea. Eduard Shperk had spent ten years in Siberia studying syphilis among the native population.[11] On the basis of this work, which he published as a doctoral dissertation in 1863, Shperk stressed the "enormous influence" of "material life" on "manners and morals," citing in particular the effect of railroads in fostering demo-

----

[7]V. M. Tarnovskii, *Prostitutsiia i abolitsionizm* (St. Petersburg, 1888), 196. The German edition (Hamburg, 1890) is cited by Bulkley, *Syphilis in the Innocent*, 198, 204. Tarnovskii's work was also read by French physicians: see Corbin, "Le Péril vénérien," 252.

[8]Rozenkvist, "K statistike," 510; D. V. Imshenetskii from Chernigov Province, in *Trudy Vysochaishe razreshennogo s"ezda po obsuzhdeniiu mer protiv sifilisa v Rossii* (St. Petersburg, 1897), 2:150.

[9]Claude Quétel, "Syphilis et politiques de santé à l'époque moderne," *Histoire, économie et société* 3:4 (1984): 543. This reference was a gift from Natalie Z. Davis. For discussion of the origins of syphilis in Europe and the myth of the post-Columbus epidemic, see Ellis Herndon Hudson, *Nonvenereal Syphilis: A Sociological and Medical Study of Bejel* (Edinburgh, 1958), 14–25; also Corinne Shear Wood, "Syphilis in Anthropological Perspective," *Social Science and Medicine* 12 (1978): 47–55.

[10]Philippe Ricord, *Lettres sur la syphilis*, 2d ed., rev. (Paris, 1856), 131. For the development of knowledge about syphilis, see Claude Quétel, *Le Mal de Naples: Histoire de la syphilis* (Paris, 1986).

[11]See Edouard Léonard Sperk [Eduard Shperk], "Esquisse biographique," in his *Oeuvres complètes: Syphilis, prostitution, études médicales diverses* (Paris, 1896), 1:xv–xxxviii.

graphic mobility and weakening traditional community bonds. Venereal syphilis, he argued, was least common in agricultural peasant villages, where family ties were strong and land the foundation of wealth. The growth of trade altered the character of the economic system by producing nonlanded property, or capital, and wage labor. People moved about more often and married later in life; the family lost its dominant role; and sex too became commercialized. Women, like other goods and personal services, Shperk observed, became objects of exchange rather than items of permanent value and attributes of social status. "As the railroad enters a given locality," he wrote, "it increases the number of rented apartments, rented carriages . . . and rented women."[12] In his opinion the great fifteenth-century epidemic had signaled not the appearance of a new malady but the transformation of an old one. The historic turning point marked by crusades, invasions, and the growth of cities provided the conditions that favored venereal over nonvenereal transmission, he believed, and left contemporaries with the impression of confronting a totally different kind of disease. The continuing prevalence of endemic syphilis in Russian villages indicated that Russia had not fully entered the modern age.[13]

Other commentators linked the new form and level of contagion in Russia to the encroachment of alien values and social institutions. Russian cities, Grigorii Gertsenshtein explained, encountered syphilis "in the form common to Western European countries," where, in the words of a colleague at the Academy of Military Medicine, "the main role in the spread of syphilis is played almost exclusively by prostitution."[14] Though no one could deny that prostitution existed in Russia, most commentators described it as a foreign import, the product of an urban commercial culture inimical to traditional native ways. "During the reign of Peter the Great," wrote Ivan Priklonskii, "we first entered into commercial relations with Western European states, and it was then that the concept of commercial prostitution also appeared in our society. Prostitution began to spread among the people, significantly lowering the moral level in all social strata."[15] In turning to Europe for technical expertise, Russia had incurred the risks of cultural transforma-

---

[12]Eduard Shperk, "O merakh k prekrashcheniiu rasprostraneniia sifilisa u prostitutok," *Arkhiv sudebnoi meditsiny i obshchestvennoi gigieny*, no. 3, sec. 3 (1869): 68; ellipsis in original.

[13]Sperk, "Signification des syphiloïdes dans la doctrine et dans l'histoire de la syphilis" (trans. of "Znachenie sifiloidov v uchenii i istorii sifilisa," *Voenno-meditsinskii zhurnal*, August–September 1864), in his *Oeuvres complètes*, 1:164–65, 167, 172.

[14]G. M. Gertsenshtein, "K statistike sifilisa v Rossii," pt. 2, *Vrach*, no. 18 (1886): 335; P. I. Gratsianskii, "Nevinnye puti i sposoby zarazheniia i rasprostraneniia sifilisa," *Zhurnal Russkogo obshchestva okhraneniia narodnogo zdraviia*, no. 11 (1892): 817.

[15]I. I. Priklonskii, *Prostitutsiia i ee organizatsiia: Istoricheskii ocherk* (Moscow, 1903), 53.

tion. Indeed, Peter the Great, architect of Russia's calculated modernization, had welcomed the destruction of outmoded traditional ways. Russian physicians, who were products of this very transformation, nevertheless joined in the critique of imported cultural norms based on formal contracts and commercial exchange.

In 1897 the Ministry of Internal Affairs sponsored a congress in St. Petersburg on the prevention of syphilis in Russia. Veniamin Tarnovskii delivered the prestigious opening address. Though he did not share Shperk's social values, Tarnovskii did not differ in his view of the social context of venereal disease. Whereas syphilis had of course existed in Russia since the end of the fifteenth century, he explained, it had become a serious threat to the nation's well-being only because "the intrusion of Western manufacture and commerce had increased the channels of infection a hundredfold, while the ability of the popular masses to defend themselves had not improved since the time of Ivan the Terrible." The danger lay, he warned, in the "swift turn of agrarian Russia toward industrial development, when the external manifestations of Western civilization bore no relation to the cultural and economic level of the popular masses." "Under certain conditions," Tarnovskii reminded his listeners, "civilization elevates and enriches the people both morally and physically; under others it destroys them slowly but surely."[16]

Was the syphilis to which Russia was prone a product of change or a symptom of inertia, a mark of the incorporation of or isolation from the ways of the modern world? Physicians seemed unsure. They could not decide whether the civilization they envied and feared had corrupted innocent bodies, leaving popular virtue intact, or whether health had departed in the wake of innocence itself. Unfortunately for lovers of medical and historical certainty, the scientific knowledge that might have decided the question on an "objective" basis and, in its updated form, helped historians decipher the cultural bias of the past has not become less ambiguous with the passage of time and the progress of that enviable sophistication. Syphilis remains a cultural puzzle to this day.

The Clinical Picture

The modern dermatology text informs us that syphilis is the work of the spirocete *Treponema pallidum*, a microorganism first isolated in 1905

---

[16]V. M. Tarnowskii, speech of January 15, 1897, in *Trudy Vysochaishe razreshennogo s"ezda*, 2:18–19.

which invades the bloodstream and other bodily fluids but is swiftly destroyed by sunlight, dryness, cold, soap and water, or simple exposure outside the living host.[17] Gone is the old notion of a "poison . . . endowed with the possibility . . . of being preserved for an indefinite period."[18] Genital contact, experts now explain, fosters transmission of the syphilis organism by providing the warmth and moisture in which it thrives and the friction that propels it through the vulnerable mucous membranes. The danger once said to be inherent in dirty cups, towels, and toilet seats belongs to the mythic catalogue of the scientific past.[19]

Studies of disease in preindustrial societies present a different picture, however, one that saves Victorian medicine from some of the ridicule it has retrospectively earned. Genital contact is indeed the usual way for syphilis to spread in modern industrial nations, for two reasons: intimate bodily contact in Western urban culture is largely restricted to sexual relations; and penicillin, introduced during World War II, was able for the first time to eliminate the reservoir of contagion waiting to be dispersed through the infinitude of social interactions. In the peasant communities of twentieth-century Bosnia or the Bedouin villages of Syria, however, where houses are crowded, notions of privacy and individuality differ from our own, penicillin is scarce, and folk traditions of shared drinking vessels and communal eating are still unshaken, syphilis takes the form remarked in Russian villages a century ago. Public health observers, using modern diagnostic tools, have confirmed that certain inanimate objects such as cups and pipes may convey infected fluid when passed quickly from mouth to mouth and that entire communities may harbor low levels of infection that inure them to the dangers of contagion by sexual means. They note that victims are not charged by their neighbors with sexual misconduct, because sexual activity is perceived as irrelevant to their case.[20]

Syphilis is a disease that adapts to the cultural environment.[21] Its shifting representation in the medical literature reflects not only the progress of clinical technique or expanded knowledge of the biological world but also the different manifestations of the disease itself. The shortness of our historical memory, combined with cultural self-cen-

[17]Hudson, *Nonvenereal Syphilis*, 6.

[18]Bulkley, *Syphilis in the Innocent*, 21.

[19]"It is now known that syphilis and gonorrhea are almost never communicated in non-sexual ways": Allan M. Brandt, *No Magic Bullet: A Social History of Venereal Disease in the United States since 1880* (Oxford, 1985), 22. See also Theodor Rosebury, *Microbes and Morals: The Strange Story of Venereal Disease* (New York, 1971), 250–51.

[20]E. I. Grin, *Epidemiology and Control of Endemic Syphilis: Report on a Mass-Treatment Campaign in Bosnia* (Geneva, 1953), 20, 30, 32. See also Hudson, *Nonvenereal Syphilis*, 9, 12–15, 80–81, 83; and Engelstein, "Syphilis Historical and Actual," 1042–45.

[21]Wood, "Syphilis in Anthropological Perspective," 47.

teredness, has allowed us to ignore present contrasts and discount past knowledge. Though nineteenth-century physicians often misconceived the timing and actual mechanism of communication and vastly exaggerated the scope of the danger, not all the perils they enumerated were necessarily the inventions of sexual anxiety or designed merely to fuel projects of sexual control.[22]

Imagination did certainly come into play, however, in the attempt to derive the origin of a given case from its clinical symptoms. Syphilis typically passes through three stages: the primary chancre, which may appear anywhere on the body, though it is usually on the genitals, and whose secretions are highly contagious; the secondary lesions, or papules, which favor the moist mucous membranes and are also infectious; finally, after a long interval, the dread tertiary symptoms—the collapsed nose and the nervous and mental deterioration of Victorian medical demonology. Both the chancre and the secondary lesions vanish of their own accord without treatment; they are separated by an asymptomatic period during which bodily fluids teem with the offending organism but the victim feels no distress. After the secondary eruption, the infection becomes less virulent without entirely disappearing, and contagion no longer occurs. In only a minority of cases do the tertiary symptoms ever appear.[23]

Some nineteenth-century doctors believed, and modern authorities agree, that endemic syphilis may avoid the primary stage.[24] The Victorians had already correctly noted that infants may be born with the disease, although they misunderstood why;[25] they were also aware that the infection could pass between a wetnurse and her charge. But more often than not, circumstances are ambiguous and symptoms uncertain.

[22]Lancereaux listed innocent kissing, sleeping two to a bed, breastfeeding, cupping, tattooing, shared linen, kitchen utensils, and pipes as dangers that were acknowledged as early as the sixteenth century: *Treatise*, 2:214, 231–46. See also Quétel, "Syphilis," 545, 549; and R. C. Holcomb, "The Antiquity of Congenital Syphilis," *Bulletin of the History of Medicine* 10: 2 (1941): 164–65. For an exhaustive inventory, see Bulkley, *Syphilis in the Innocent*, 209–40. For a Russian list of culpable everyday objects "too long to count," see Gratsianskii, "Nevinnye puti," 807; also Rozenkvist, "K statistike," 531. Corbin ("Le Péril vénérien," 250) notes the popularity of such catalogues among French syphilologists of the period.
[23]Stewart M. Brooks, *The V.D. Story* (South Brunswick, N.J., 1971), 37–39; Hudson, *Nonvenereal Syphilis*, 69.
[24]Grin, *Epidemiology*, 32. Hudson (*Nonvenereal Syphilis*, 64–65) explains that constant low-level exposure prevents extreme reaction to any single inoculation of the organism, of the kind that produces chancres in previously unexposed adults.
[25]For mistaken views of congenital syphilis, see Fournier, *Traité de la syphilis*, 1:5–6; also Elizabeth Lomax, "Infantile Syphilis as an Example of Nineteenth Century Belief in the Inheritance of Acquired Characteristics," *Journal of the History of Medicine* 34:1 (1979): 23–39.

The European texts to which Russian doctors turned left ample room for dispute. Some experts insisted that a chancre was inevitable at the initial point of contact, though it might be concealed (as on the cervix) or small and evanescent, and might escape the diagnostic eye.[26] Failure to spot one did not confirm innocence; finding one in some innocuous place did not exclude guilt. Oral chancres, for example, invited debate. There were so many things a mouth could do. "Honi soit qui mal y prête!" was the sanctimonious comment of Paul Diday, writing for the respectable, middle-class French audience.[27] But the eminent professor Alfred Fournier did not hesitate to put "genital-oral" contact at the top of his list.[28] Other organs were less versatile. Few texts denied that a chancre on the genitals meant sexual intercourse; on the anus, anal penetration.[29] The latter, wrote the French forensic specialist Ambroise Tardieu, was an almost certain proof of "unnatural acts," and even secondary lesions in that spot were to be accounted for in the same manner.[30] Fournier admitted that oral-anal contact was "much more frequent than one would dare to believe."[31] An American professor of genitourinary diseases at the Columbia University College of Physicians and Surgeons baldly asserted that chancres on the anus, tongue, and mouth often resulted from "unnatural and beastly methods of indulgence between persons of the same and the opposite sex"—from what Bulkley referred to more chastely as *coitus preternaturalis*.[32]

Whether doctors interpreted syphilis as venereal thus depended to a large extent on their notions of what constituted sexual activity; on a willingness to acknowledge the existence of nonstandard sexual practices, or of standard practice by the wrong people, such as children, pious peasants, or virtuous wives. While some authorities concluded from the abundance of nonsexual ways to contract it that syphilis "looked at in its largest sense . . . cannot, in the light of present knowledge, be any longer regarded as essentially a venereal disease,"[33] others were not so sure. The much-respected Etienne Lancereaux for one, remained uncertain that outbreaks, even on the marches of European civilization, could be as innocent as they seemed: crowding was a problem

---

[26]Fournier, *Traité de la syphilis*, 1:22, 25–26, citing Ricord.

[27]Paul Diday, *Le Péril vénérien dans les familles* (Paris, 1881), 319.

[28]Fournier, *Traité de la syphilis*, 1:133.

[29]See, e.g., Lancereaux, *Treatise*, 2:355–56.

[30]Ambroise Tardieu, *Etude médico-légale sur les attentats aux moeurs*, 7th ed. (Paris, 1878), 230, 249.

[31]Fournier, *Traité de la syphilis*, 1:189.

[32]Robert W. Taylor, *A Practical Treatise on Genito-Urinary and Venereal Diseases and Syphilis*, 3d ed., rev. (New York, 1904), 496; Bulkley, *Syphilis in the Innocent*, 240.

[33]Bulkley, *Syphilis in the Innocent*, 109.

even there, he thought, because "density of population . . . does not fail, in general, to cause immorality."[34]

The Physician's Role

Russian physicians indulged in the same diagnostic squabbles as their European colleagues. Numerous articles in the medical press reported "rare cases" of nonvenereal infection, primary chancres of nonsexual origin on a penis or on the vagina of a nine-year-old girl.[35] Discussions of prostitution and its relation to venereal syphilis were even more profuse. But clinical evidence, because of its very ambiguity, played a secondary role in distinguishing the two competing paradigms of syphilitic contagion. Instead of searching for primary lesions, Russian doctors looked to the social and cultural context for epidemiological clues.

This approach was no cultural innovation but drew heavily on the European medical tradition of environmental public health. Indeed, one of its leading advocates in Russia was the Swiss-born physician Friedrich Erismann, who practiced and taught in his adopted country from 1869 to 1896, holding the first chair of hygiene at Moscow University.[36] Viewing disease as a social problem, public health physicians devoted themselves to public—which is to say, administrative or political—solutions. Erismann's civic orientation, frustrated in the end by the stubborn hostility of an increasingly repressive regime, was matched by that of Russians such as Eduard Shperk, already established in a successful career when Erismann first arrived. Born in 1837, Shperk shared with the 1860s intelligentsia a belief in science as the key to social improvement and an ethic of commitment to the common welfare—without, however, endorsing the intelligentsia's more radical social critique. Active in Russia's largest and most influential public health organization, Shperk taught in the women's courses of the Academy of Military Medicine in the 1870s. He was known for his humanitarian treatment of hospital patients, including the registered prostitutes under his care, and later broke with the St. Petersburg syphilis society over the issue of regulation, which he had come more and more to oppose.[37] Though not all men of his generation shared his views

---

[34]Lancereaux, *Treatise*, 2:262.

[35]E. A. Rotman, "K kasuistike vnepolovogo shankra: Vnepolovoi shankr na penis'e," *Russkii zhurnal kozhnykh i venericheskhikh boleznei*, no. 10 (1905):34; A. I. Tulinov, "Pervichnaia sifiliticheskaia iazva vnepolovogo proiskhozhdeniia na polovykh chastiakh devochki 9 let," *Detskaia meditsina*, no. 4 (1899): 191–202.

[36]N. A. Semashko, "Friedrich Erismann: The Dawn of Russian Hygiene and Public Health," *Bulletin of the History of Medicine* 20 (June 1946): 1–9.

[37]On 1887 break, see Sperk, *Oeuvres complètes*, 1:xxvi–xxvii; on abolitionism, see B. I.

(Tarnovskii, who also taught in the women's courses, provides the sharpest contrast), Shperk established a tone that dominated publicly conscious medical circles into the 1890s.

This outlook was congenial to physicians serving the urban poor, but it came to be associated in particular with those in rural practice, who fashioned for themselves a distinctive medicopolitical ethos. Most of the latter were employed by the zemstvos, local elective bodies created during the Great Reforms of the 1860s, which enjoyed relative autonomy from bureaucratic control and offered a rare chance for members of educated society to serve the public good and exercise a modicum of civic initiative. Zemstvo physicians shared the populist commitment to promoting the interests of the common folk and raising their cultural level, while pressing for political changes that would facilitate their own professional enterprise.[38] The zemstvos, not surprisingly, acted as a focus of reform politics in the pre-1905 era.

Given the community context in which they worked, it was natural for zemstvo practitioners to adopt the view of disease as a demographic or social problem, rather than a matter of individual pathology, whether physical or moral. "In private practice," explained Dmitrii Zhbankov, himself a zemstvo physician, "the doctor deals only with individual patients, unrelated one to the other. The task at hand is simple: to cure the patient, without a thought to the future or to surrounding circumstances. With the emergence of social medicine, doctors were obliged to deal with great numbers of patients and also with the healthy population among whom they lived. Doctors thus witnessed with their very own eyes the close links between the sick and the well, the way in which individual cases, as well as mass outbreaks, depended on environmental conditions." Programs aimed at education and prevention were thus the key to better health, Zhbankov averred, citing village syphilis, the classic community scourge, as a case in point.[39]

The use of syphilis to illustrate a question of public policy was not original with Zhbankov or unique to those who shared his critical

Bentovin, *Torguiushchie telom: Ocherki sovremennoi prostitutsii*, 2d ed., rev. (St. Petersburg, 1909), 239.

[38]For the European background, see George Rosen, "The Evolution of Social Medicine," in *Handbook of Medical Sociology*, ed. Howard E. Freeman, Sol Levine, and Leo G. Reeder (Englewood Cliffs, N.J., 1963). For the zemstvo medical philosophy, see S. N. Igumnov, "Zemskaia meditsina i narodnichestvo," in *Trudy odinnadtsatogo Pirogovskogo s"ezda*, ed. P. N. Bulatov (St. Petersburg, 1911), 1:78–79; Nancy Mandelker Frieden, *Russian Physicians in an Era of Reform and Revolution, 1856–1905* (Princeton, N.J., 1981), chap. 4; and Samuel C. Ramer, "The Zemstvo and Public Health," in *The Zemstvo in Russia: An Experiment in Local Self-Government*, ed. Terence Emmons and Wayne S. Vucinich (Cambridge, 1982), 279–82. For its decline after 1905, see John F. Hutchinson, *Politics and Public Health in Revolutionary Russia, 1890–1918* (Baltimore, 1990), chap. 3.

[39]D. N. Zhbankov, *O deiatel'nosti sanitarnykh biuro i obshchestvenno-sanitarnykh uchrezhdenii v zemskoi Rossii* (Moscow, 1910), 2–3.

views. The recurrent disappearance of symptoms, the mysterious process of congenital transmission, and the seeming disconnection between the various stages of the disease, made syphilis the perfect symbolic vehicle for the doctor's assertion of professional authority. It provided the ideal clinical avenue between the private domain (meaning both personal and secret) and the public (meaning both civic and revealed). Veniamin Tarnovskii, for example, treated the question of syphilis control with the rhetoric of national honor and national self-defense. "No external, visible enemy can faze us," he announced to the St. Petersburg syphilis congress. "But we can be frightened by the secret, internal enemy that imperceptibly destroys the people's physical and moral well-being."[40] Syphilis, agreed a staff physician at the Kiev military hospital, "is not a visible enemy one can fight with the obvious weapons; it sneaks up unnoticed and ruthlessly destroys whole families, whole generations."[41] A lecturer at Moscow University, Nikolai Fedchenko, echoed the French expert Toussaint Barthélemy in complaining that syphilis caused less alarm than cholera only because "the harm wrought by syphilis, which is well known to doctors, does not catch the attention of the crowd."[42] In this view the physician was to put his occult knowledge, acquired in the discreet examination of private parts, at the service of public policy.

Putting their skills to public use was nothing unusual for Russian physicians, the vast majority of whom were employed by the state bureaucracy or other public institutions.[43] They expressed opinions and exchanged technical information in the pages of the numerous medical journals that promoted the discussion of social themes. Some 450 of them attended the 1897 congress, including nineteen of the forty-two people (thirty-eight men and four women) on whose observations and interpretations this chapter draws.[44] More than half of the forty-two were at least forty years old in 1897, which meant they belonged to a generation that had spent its youth in the 1860s, the optimistic early years of Alexander II's reign; the rest, their juniors, had trained during the repressive aftermath of his assassination in 1881. Half the group had acquired the advanced degree of doctor of medicine, in addition to the basic medical diploma. The sample is dominated by dermatologists, urologists, and specialists in venereal diseases and includes several pediatricians, but physicians typically held a variety of positions in the

[40]V. M. Tarnovskii, in *Trudy Vysochaishe razreshennogo s"ezda*, 2:21.
[41]V. K. Borovskii, "K voprosu ob istochnikakh zarazheniia sifilisom," *Voenno-meditsinskii zhurnal*, no. 8 (1894): 412.
[42]N. P. Fedchenko, "O zarazhenii sifilisom pri brit'e," *Meditsinskoe obozrenie*, no. 1 (1890): 25.
[43]Frieden, *Russian Physicians*, 210–11.
[44]On the congress attendance, see *Trudy Vysochaishe razreshennogo s"ezda*, 2:259–66.

course of their careers, thus widening the range of their social and clinical experience.[45] Whatever their philosophical and educational differences, all Russian doctors shared an acquaintance with European medicine, and all confronted the workings of a state determined to keep a tight reign on the process of social change and cede none of its authority to autonomous groups, a state on which they depended for their livelihood but which impeded the attainment of their professional goals.

In Europe, questions of public health and public morality were closely intertwined as objects of both state policy and medical expertise. Enemies and partisans of regulated prostitution shared a common concern with lower-class sexual comportment and bolstered their devotion to middle-class decorum with a medical rationale. Despite similar (though not identical) standards of respectability, Russian physicians nevertheless found themselves in a different moral universe. Because in Russia, as in Europe, judgments of sexual propriety were shaped by perceptions of social class and of gender, the peculiarities of the Russian class system produced distinctive moral expectations, which were in turn modified and increasingly confused by the turmoil in which the class system was caught.

In the vast majority of cases with which Russian physicians dealt, traditional mores seemed still intact, and syphilis appeared not as the result of sexual promiscuity, the egotistic search for private pleasure in disregard of the collective norm, but as the result of social promiscuity, a reflection of collective tyranny and the weakness of self. This was a weakness the physicians deplored, for it inspired in them a sense of their own helplessness. Yet they persistently rejected evidence of sexual misbehavior that testified to the crumbling of traditional bonds. Though they were eager for signs of personal autonomy that could be disciplined in nontraditional ways, through self-regulation guided by medical expertise, the sexualization of syphilis was nevertheless a strategy most Russian physicians before 1905 did not willingly embrace.[46]

Syphilis in the Russian Countryside

Medical observers without exception characterized syphilis as endemic to the peasant population and nonvenereal in origin.[47] If prostitu-

---

[45]Career and biographical data are taken from *Rossiiskii meditsinskii spisok* (St. Petersburg/Petrograd, 1890–1916).

[46]The link between individuation and modern strategies for imposing behavioral norms is, of course, a central point of Michel Foucault, *Discipline and Punish: The Birth of the Prison* (New York, 1977).

[47]E.g., M. A. Chistiakov, *Protokoly sektsii sifilidologii na pervom s"ezde russkikh vrachei*

tion was "the main source of infection" in Russian cities, "in the countryside," said Konstantin Shtiurmer of the St. Petersburg medical police, "it plays only a minor role, and sometimes none at all."[48] "Rural syphilis," wrote the zemstvo physician Pavel Govorkov, "is a misfortune that befalls the innocent. The statistics on this subject show that the number of people in the countryside who suffer from venereal forms of the disease, acquired as a result of sexual activity, is negligible."[49] On the basis of reports from towns and villages throughout the empire, the 1897 congress on syphilis concluded that prostitution was a factor in the spread of syphilis only in large and medium-sized cities; smaller towns, like peasant villages, suffered an epidemic of the disease in its nonvenereal form.[50]

The noted syphilologist Grigorii Gertsenshtein was typical in defending the moral integrity of peasant life:

Neither prostitution, nor soldiers, nor unmarried young men from urban factories and manufacturing centers transmit the disease, but rather innocent children and the women of impoverished towns and villages. The disease spreads not through sexual relations but in the course of everyday domestic contacts between healthy and diseased members of single families, neighbors, and casual visitors. The infection is spread even further by the sharing of bowls and spoons, by an innocent child's kiss, but not through dissolute behavior or unregulated prostitutes as is the case in the big cities.[51]

Like most practitioners, Gertsenshtein blamed the prevalence of rural syphilis on poverty, ignorance, and traditional customs. Nonvenereal syphilis, he wrote, "occurs only where the population is extremely ig-

---

1885 g. v S. Peterburge (St. Petersburg, 1886), 41; G. M. Gertsenshtein, "Sifilis v Novgorodskoi gubernii i voprosy o bor'be s nim na VII i IX s"ezdakh zemskikh vrachei 1888–1895 gg.," *Vestnik obshchestvennoi gigieny, sudebnoi i prakticheskoi meditsiny*, no. 4 (1896): 25 (henceforth *Vestnik obshchestvennoi gigieny*); Gratsianskii, "Nevinnye puti," 817–18; O. V. Petersen, "O sifilise i venericheskikh bolezniakh v gorodakh Rossii," in *Trudy Vysochaishe razreshennogo s"ezda*, 1:127; A. I. Rozenkvist, "Redkii sluchai vnepolovogo zarazheniia sifilisom: Iz ambulatorii Miasnitskoi bol'nitsy v Moskve," *Vrach*, no. 9 (1899): 244; N. S. Speranskii, *K statistike sifilisa v sel'skom naselenii Moskovskoi gubernii* (Moscow, 1901), 125; V. M. Tarnovskii, quoted in Chistiakov, *Protokoly*, 6; M. S. Uvarov, "Sifilis sredi sel'skogo naseleniia," in *Trudy Vysochaishe razreshennogo s"ezda*, 1:73–85.
[48]K. L. Shtiurmer, *Sifilis v sanitarnom otnoshenii* (St. Petersburg, 1890), 45.
[49]P. A. Govorkov, "Polovaia zhizn' garnizona," pt. 1, *Vrach*, no. 37 (1896): 1016.
[50]Congress resolution cited by, e.g., A. G. Petrovskii, "Bor'ba s sifilisom v gorodakh," pt. 1, *Izvestiia Moskovskoi gorodskoi dumy*, no. 5, Obshchii otdel (1905): 3.
[51]Gertsenshtein, "Sifilis v Novgorodskoi gubernii," 47.

norant and poor, at a relatively low level of civilization."[52] Not virtue but cultural deprivation was at issue, in this view.

Indeed, physicians called syphilis "the Russian people's everyday disease" (bytovaia bolezn' russkogo naroda).[53] The word bytovaia conveyed the three related implications of this phrase: "ordinary," in the sense of common; "social," because rooted in interpersonal exchange; and "customary," as reflecting a way of life (byt). Under this rubric an article on syphilis in Tambov Province listed its causes as the crowding of children in village huts; the impossibility of observing standards of hygiene under circumstances of grueling poverty; the burden of field work on peasant women, preventing them from supervising their children's health and behavior; and carelessness stemming from general ignorance and unfamiliarity with the specific nature of the disease. "Measures to combat prostitution," the article concluded, "are meaningless as far as the rural population is concerned, since there is no prostitution in the countryside, and it thus has absolutely no bearing on syphilis at all."[54]

Russian peasants thus bore the same relation to the denizens of Moscow and St. Petersburg that Russia as a nation bore to Germany and France. Under endemic conditions, Etienne Lancereaux had written in 1866, "the means of [syphilitic] transmission are, in general, kitchen utensils, linen, or other objects, sometimes actual contact; but rarely, it appears, the act of coition."[55] Russian physicians acknowledged "that instances of transmission not involving sexual intercourse constitute a negligible percent of all syphilitic infection," but the countryside was the exception that proved the rule. "In the absence of sexual opportunity," one such practitioner pointed out, "poverty, overcrowded housing, and the lack of preventive measures sometimes reverse this relationship. In remote Russian villages, for example, the majority of syphilis cases are communicated by other [nonsexual] means—infected crockery, shared beds, and so on. But even these instances serve only to demonstrate how much the possibility of nonvenereal transmission depends on especially favorable circumstances."[56]

Unlike the city dwellers who routinely (though not exclusively) contracted syphilis through moral dereliction, peasants, in the doctors' eyes, bore no responsibility for their ills. If the common folk did them-

[52]G. M. Gerstenshtein, Sifilis v Rossii, (1885), and D. N. Zhbankov, both quoted in K. V. Goncharov, O venericheskikh bolezniakh v S. Peterburge (St. Petersburg, 1910), 5, 23.

[53]V. M. Tarnovskii, quoted in Chistiakov, Protokoly, 6; see also Shtiurmer, Sifilis, 44.

[54]V. I. Nikol'skii, "Neskol'ko zamechanii o krest'ianskom sifilise v Tambovskom uezde," Vrach, no. 41 (1886): 738.

[55]Lancereaux, Treatise, 1:44.

[56]M. Shmelev, "Predokhranitel'nye mery protiv sifilisa," Sbornik sochinenii po sudebnoi meditsiny, no. 2 (1872): 219–20.

selves harm, Gertsenshtein observed, it was not deliberately, through mistaken or ill-considered acts, but unknowingly, through blind adherence to "age-old popular habits."[57] "Without knowing it," wrote Shtiurmer, "the peasant is his own worst enemy. The remarkable carelessness toward his own and others' health, the positive savagery of certain customs cannot be characterized as anything other than ignorance."[58] Among these savage customs doctors counted the practice of spitting in a person's eye to cure sties, the feeding of children directly from the mother's own mouth, and the habit of sucking a baby boy's penis to calm him.[59] Accounts did not suggest that this form of sedation be considered a sexual act, for it fell into the category of backwardness, not moral transgression.

Doctors attributed the peasants' backwardness in part to the dominance of collective norms over personal development. Though they believed that traditional patriarchal society protected its members from venereal infection by restraining sexual expression outside the bounds of family life, physicians considered these traditional institutions responsible for habits of dependence and passivity that encouraged syphilis to spread in the endemic form. "The patriarchal family, with its strict moral rules," wrote Mikhail Uvarov, one of the country's most prominent zemstvo physicians, was no defense against syphilis but in fact encouraged its rapid spread. In the smallest villages, where the entire population "sometimes bears a single family name and relations are close, syphilis occurs with greater intensity than in larger villages."[60] The peasants were sick because they lived cheek by jowl. A disease that "knows no bounds" flourished among people who respected no boundaries: they ate from one bowl, slept in the same beds, welcomed the itinerant beggar or tradesman. The disease thus "moved from one family to the next," observed Vasilii Borovskii of the Kiev military hospital, "sparing neither children nor old people, engulfing neighboring houses and nearby villages."[61] No one act—sexual or nonsexual—could be singled out as the first link in the fatal chain: "The original source of infection has long been forgotten," wrote Gertsenshtein, "the agent that first introduced the poison has perhaps long been

[57]G. M. Gertsenshtein, "Peredvizhnye vrachebnye otriady dlia bor'by s sifilisom," *Vestnik obshchestvennoi gigieny*, no. 10 (1896): 25.

[58]Shtiurmer, *Sifilis*, 45; see also Gertsenshtein, "Sifilis v Novgorodskoi gubernii," 36, 38.

[59]Rozenkvist, "K statistike," 529.

[60]Uvarov, "Sifilis sredi sel'skogo naseleniia," 82–83. Uvarov's career in zemstvo medicine is outlined in N. I. Afanas'ev, *Sovremenniki: Al'bom biografii* (St. Petersburg, 1909, 1910), 2:427–30.

[61]Borovskii, "K voprosu," 413.

dead, but the sad consequences have nonetheless unfolded to their full extent."[62] Observers described the disease as spreading "passively" or "spontaneously," without conscious agency. Often victims did not recognize their own disease, so little did the individual case stand out against the background of general ill health.[63]

Mothers were said to play a central role in the spread of infection. Not only did those who carried the disease in their blood pass it to the unborn and the nursing child, but their unsanitary domestic habits threatened all members of the family, as Dmitrii Zhbankov recounted:

> The doctor will no sooner have finished explaining how to stop the spread of syphilis and is barely out the door, when the mother begins to feed her healthy child with the very same spoon she has just used to feed the sick one, or wipes the drooling saliva from the sick child's mouth with the same rag or kerchief she will then use to wipe her own face or the face of the healthy child. Better yet, a healthy family member will reach for a piece of bread with the hand he has just used to dry the sick child's running mouth.[64]

The recognition that syphilis could proliferate by nonvenereal means, however, did not constitute proof of popular chastity, for in a population already saturated with infection from casual social intercourse, sexual relations might occur without consequences of their own. It was not necessary to argue that sexual activity, licit or illicit, was unknown to the sufferers of endemic syphilis but only that the preexistence of infection made sexual contact irrelevant to the organism's dissemination; as Eduard Shperk had contended in 1864, syphilis was already so widespread in the villages as to render rural prostitution, insofar as it existed at all, irrelevant to its control.[65] There is no implication in the medical literature that sexual indulgence on the part of peasant women might contribute to the spread of disease; indeed, there is no suggestion that these women were sexually active outside the context of family life, where they were thought to perform an exclusively reproductive role, routinely conceiving eight to ten children in the

---

[62]Gertsenshtein, "Peredvizhnye vrachebnye otriady," 26.

[63]Ibid.; idem, "Sifilis v Novgorodskoi gubernii," 47; Fedchenko, "O zarazhenii sifilisom," 19. See also the zemstvo physician D. D. Sandberg, cited in P. I. Messarosh, "K voprosu o rasprostranenii sifilisa v Rossii," *Vestnik obshchestvennoi gigieny*, no. 7 (1896): 50.

[64]D. N. Zhbankov, *Materialy o rasprostranenii sifilisa i venericheskikh zabolevanii v Smolenskoi gubernii* (Smolensk, 1896), quoted in Rozenkvist, "Redkii sluchai," 245.

[65]Sperk, "Signification des syphiloïdes," 122–24.

course of married life.[66] Constant childbearing, in fact, symbolized their immunity to sexual desire.[67] Aleksandr Efimov, studying the effects of syphilis on the "sex life" (*polovaia zhizn'*) of peasant women, did not discuss frequency of intercourse or other modes of sexual pleasure but examined menstruation, fertility, and pregnancy.[68]

To a certain extent these views mirrored observed social reality; one may safely conclude that the Russian village had not yet experienced the kind of demographic and cultural upheaval that had transformed the life of European peasants several decades before.[69] Yet village life was perhaps not so inert as it seemed. There is reason to suppose that extra-marital sexual activity, whatever its actual extent, was obscured by the Russian peasants' high level of nuptiality, their early marriage age, and the late onset of menstruation among women.[70] Few births occurred out of wedlock, even if conception had.[71] Contemporary observers, in any case, seemed reluctant to challenge appearances, as is clear in their response to evidence that violated their cherished beliefs.

The most traditional element in a tradition-bound milieu, peasant women shone as unsullied examples of peasant virtue; they were so depicted even by such men as Dmitrii Zhbankov, who held untradi-

---

[66]This well-known figure is mentioned by D. D. Sandberg in her discussion of syphilis among peasant women in Tambov Province: "Sifilis v derevne," *Vrach*, no. 26 (1894): 741. See also S. A. Olikhov, "K voprosu o plodovitosti krest'ianok Kineshemskogo uezda Kostromskoi gubernii," *Zemskii vrach*, no. 52 (1890): 823.

[67]A striking example of the powerful reluctance to see reproductive sexuality as a form (or consequence) of desire can be found in a text dating from the period when this contradictory belief had begun to weaken. Mariia Pokrovskaia could insist that peasant men "restrained their sexual instincts" in the village and at the same time condemn as "barbaric" the "law of nature" in the village, which decreed that children should be conceived without forethought and then allowed to die in massive numbers: *O polovom vospitanii i samovospitanii* (St. Petersburg, 1913), 9, 37.

[68]Appendix to A. I. Efimov, *Sifilis v russkoi derevne, ego kharakternye cherty i vliianie na sanitarnoe polozhenie naseleniia* (Kazan, 1902). Many physicians showed an interest in the "physiological aspects of peasant women's sex lives"; see the studies cited in V. S. Sergiev, "K ucheniiu o fiziologicheskikh proiavleniiakh polovoi zhizni zhenshchiny-krest'-ianki Kotel'nicheskogo uezda Viatskoi gubernii," in *Trudy antropologicheskogo obshchestva pri Imperatorskoi Voenno-Meditsinskoi Akademii*, 5:175–99 (St. Petersburg, 1901).

[69]See J. Michael Phayer, "Lower-Class Morality: The Case of Bavaria," *Journal of Social History* 8:1 (1974): 79–95; also David L. Ransel, "Problems in Measuring Illegitimacy in Prerevolutionary Russia," *Journal of Social History* 16:2 (1982): 111–27.

[70]For the demography of the traditional peasant village, see B. N. Mironov, "Traditsionnoe demograficheskoe povedenie krest'ian v XIX–nachale XX v.," in *Brachnost', rozhdaemost', smertnost' v Rossii i v SSSR: Sbornik statei*, ed. A. G. Vishnevskii (Moscow, 1977). Studies concluded that the average peasant woman began menstruating between the ages of sixteen and seventeen: Sergiev, "K ucheniiu," 180.

[71]Ransel ("Problems in Measuring Illegitimacy") attributes higher illegitimacy figures in the cities in part to the peasant women's practice of bringing unwanted infants to urban foundling homes. Stricter rules for admission to foundling homes in 1890 produced a drop in illegitimacy statistics, which Ransel thinks concealed an actual rise in illegitimate births during the 1890s.

tional views of women's capacity for cultural advancement. Nor were these notions destroyed by prolonged personal contact with rural life. Himself the illegitimate son of a serf and her master, Zhbankov spent years in zemstvo practice, studying the effect of social mobility on the rural way of life.[72] Examining the moral impact of male seasonal labor in Kostroma Province, he described the lonely wives of workers off in the city as faithful to their absent mates. Far from feeling the pangs of sexual frustration, the women lost every trace of sexual feeling, he claimed, through a combination of hard work and lack of opportunity. Peasant women often failed to menstruate during the heavy work season, Zhbankov explained, because "the organism is so exhausted from hard labor that it has nothing to spare for the sexual functions; this is why during the work season and in the absence of their husbands women experience no sexual need. Moreover, seasonal labor has been practiced in this region for a long time, and the women may have therefore developed a *hereditary weakening of sexual desire*."[73] Zhbankov, like Efimov, equated menstruation—an aspect of female reproduction—with sexuality itself. He was not unusual in viewing both peasant men and women as modest and morally restrained as long as they lived in their villages. Grigorii Gertsenshtein complained, for example, that the peasants' "natural bashfulness" (*estestvennaia stydlivost'*) prevented them from seeking medical aid for sexually marked disease.[74]

When faced with undeniable evidence of illicit sexual activity in the countryside, medical observers redefined it as an urban problem. The St. Petersburg physician Aleksandr Vvedenskii assured the congress on syphilis that "prostitution is completely alien to the patriarchal customs of our peasantry. There is no doubt that it appears in the small towns and villages only when the conditions of rural life have fundamentally altered. With the introduction of railroads, the growth of trade or manufacture, or the quartering of troops, the countryside loses its original character; it acquires the distinguishing features of urban life." Then, Vvedenskii lamented, formerly innocent customs became occasions for sexual debauch; local girls sold themselves for the sake of mere amusement; peasants massed on tobacco plantations and at fisheries lost all moral restraint.[75]

---

[72]See his argument for the admission of women to university education: D. N. Zhbankov, "O dopushchenii zhenshchin v universitet," *Russkii vrach*, no. 6 (1902): 209–13. Zhbankov's background is given in S. I. Mitskevich, *Na grani dvukh epokh: Ot narodnichestva k marksizmu* (Moscow, 1937), 96.

[73]D. N. Zhbankov, *Bab'ia storona: Statistiko-etnograficheskii ocherk* (Kostroma, 1891), 91 (original emphasis).

[74]Gertsenshtein, "Peredvizhnye vrachebnye otriady," 30.

[75]A. A. Vvedenskii, "Prostitutsiia sredi sel'skogo (vne-gorodskogo) naseleniia," in *Trudy Vysochaishe razreshennogo s"ezda*, 1:1–2.

Few joined him in believing that country mores had changed so profoundly for the worse. Most physicians clung to the prevalent notions
of popular decorum, even when promiscuous sexuality was clearly a
factor in spreading disease. Far from emphasizing the risks of sexual
contact, they focused instead on environmental dangers or on the consequences of purely social interaction. Prostitutes who congregated in
local taverns, Mikhail Uvarov told the congress on syphilis, were less
dangerous to health than the dirty tables, unwashed glasses, and stuffy,
unventilated rooms in which customers rubbed shoulders and gave each
other the disease. "Rural taverns have been singled out as breeding
grounds of disorder and vice," Uvarov said, "but given the complete
absence of sanitary controls, their filth and slovenliness are undoubtedly
a much greater source of nonvenereal infection. It is enough to recall
the eternal towels used to wipe dirty glasses which are then refilled with
beer."[76]

Physicians were deeply concerned with the effect of syphilis on rates
of miscarriage, infant mortality, and childhood impairment; they considered the endemic and venereal forms equally dangerous in this regard. Thus Veniamin Tarnovskii blamed hereditary syphilis for physical deformity among peasant children, and a study of Tambov Province
cited it as a major cause of infant death.[77] Modern research has shown
that prenatal transmission does not occur when the disease is endemic,
because women who themselves acquire it before puberty are not contagious by the time they reach childbearing age.[78] Of course, healthy
nursing women were vulnerable to infection from syphilitic infants,
posing a risk for their later pregnancies. But if childhood infection had
been common enough, women would rarely have contracted the disease late enough in life to harm their unborn.

At least one observer declined to share the prevailing sense of alarm:
Dmitrii Zhbankov offered evidence consistent with what modern medicine now knows about the disease in its endemic form. In contrast to
the findings of contemporary experts in the West, his data did not show
that syphilis increased the rate of miscarriages and premature births
among Russian peasant women.[79] Two circumstances, however, may
have sustained other observers in their contrary view: a tendency to
confuse the symptoms of pre- and postnatal infant syphilis, which are

[76]Uvarov, "Sifilis sredi sel'skogo naseleniia," 85.

[77]V. M. Tarnovskii (*Prostitutsiia i abolitsionizm*, 206) asserts that hereditary syphilis had
the same impact among peasants as among the educated classes. See also Sandberg,
"Sifilis v derevne," 741.

[78]Hudson, *Nonvenereal Syphilis*, 150–52.

[79]D. N. Zhbankov, "K voprosu o plodovitosti zamuzhnikh zhenshchin," *Vrach*, no.
13 (1889): 311.

in fact distinct; and the contribution of other diseases to high infant mortality in rural districts where syphilis was also widespread.[80] If in retrospect one could somehow ascertain that congenital syphilis was indeed rampant in the Russian countryside, then one would have to conclude that true endemic syphilis coexisted with a certain frequency of adult acquisition, whether venereal or nonvenereal, and thus that venereal transmission (by definition involving postpubescent women) was at the very least more common in peasant society than the doctors were willing to concede. The coexistence of both modes in the same population would account for the high level of syphilis observed among young peasant children and also for examples of early syphilis found among adults.

At the time, however, not only were the archetypal extremes hard to untangle by direct observation, but the social environment itself abounded in contradiction. City and country did not exist in isolation: peasants and workers went back and forth, carrying disease; the living conditions and sanitary habits of the urban lower classes were in no way distinguishable from those in the villages from which they had come; nonvenereal contagion was not unknown in the town or sexual transmission outside it. All these conditions should have shaken physicians' confidence in the reigning scheme, and some did not hide their doubts.[81] But the many ways in which the scheme did account for the evidence at hand, especially given the technical inadequacies of the time, served to reinforce physicians' attachment to the cultural paradigm upon which it was based.

City versus Country

The doctors' reluctance, given their social expectations, to interpret the rural case as evidence of venereal transmission encouraged them to draw similar conclusions about peasants who had migrated to the cities. Nevertheless, the association of urban life with a different kind of social interaction produced a shift in the resulting medical picture. Because city life was "more individualized," explained Nikolai Fedchenko, the pattern of contagion was perceived in a different way: "Among the peasants nonvenereal contagion envelops entire families and social networks; in the cities even nonvenereal transmission affects individuals. Among those who live in town but still maintain the traditional peasant

---

[80]On the astounding level of infant mortality in Russian peasant villages, see David L. Ransel, *Mothers of Misery: Child Abandonment in Russia* (Princeton, N.J., 1988), 165–66.

[81]M. I. Pokrovskaia, "Mery, preduprezhdaiushchie rasprostraneniia sifilisa," pt. 2, *Russkii vrach*, no. 11 (1903): 413.

way of life, nonvenereal syphilis is as prevalent as in the countryside. Thus, cases of nongenital infection have been observed among artisans, soldiers, factory hands, servants, and others; but *such cases are nevertheless described as individual sicknesses.*"[82]

City and country thus constituted, to the medical profession, two social and epidemiological models. Rural syphilis was a disease of community, not individuality; urban syphilis was the reverse. City life, culture, education—these forces undermined the traditional communal structures, in which sexual behavior was neither a matter of choice nor a means of personal gratification but an obligation, a reproductive rather than an expressive function. But city and country were not simply stages in a cultural progression; they were locations on the map. What was their relationship in the spread of venereal disease? To answer this question, physicians were obliged to confront the issue of class, since population movement was the key both to the transformation of social categories and to the spread of social disease. Gender, in turn, modified class distinctions, producing an epidemiological grid that mirrored the process of social change.

In the village, men and women shared the same status at one pole of the class-gender grid: instruments of reproduction, obedient to traditional collective norms, they lacked personal autonomy and hence sexual agency. At the other end of the scale, educated urban men bore full moral responsibility for their sexual acts. Respectable women of the same class, though shaped in some respects by urban culture, nevertheless in sexual terms fell into the same passive, hence morally irreproachable, category as peasants. As mothers and housekeepers, they figured—along with little children—as archetypal victims of *syphilis insontium*.[83] In common with their male relatives, however, and unlike the lower-class casualties of collective disease, each enjoyed her own personal affliction, her private tragedy.

The sexual-epidemiological classification of male and female workers, by contrast, reflected their transitional place in the class hierarchy. As recent urban immigrants they enjoyed the moral immunity of villagers, yet some earned the luxury of moral censure: the common soldier, the male seasonal worker, and the prostitute were deemed capable, like the educated male, of sexual license and were hence held responsible for the propagation of sexually defined disease. The dual

[82]Fedchenko, "O zarazhenii sifilisom," 19 (emphasis added).
[83]That doctors viewed such women as lacking in personal autonomy is evident in the practice of concealing from them the nature of their disease while confiding the unpalatable news to their husbands—to maintain "professional confidentiality" between physician and husband. For a critique, see Z. Ia. El'tsina, "K voprosu o rasshirenii mer bor'by s sifilisom," pt. 1, *Russkii vrach*, no. 26 (1902): 969–71.

identity of working-class men and women thus provided the key to the puzzle of urban-rural interaction. Their uncertain status made them the objects of considerable diagnostic ingenuity and inconsistency.

Some physicians proceeded from the conviction that all syphilis was sexual in origin to the belief that rural syphilis must owe its start to an urban germ. The Moscow police-medical committee reported in the early 1850s that working-class men and women engaged in promiscuous sexual contacts that facilitated the spread of the disease.[84] Male peasants who left the village for the factory or workshop, physicians complained in the 1890s, might frequent prostitutes or sleep with female co-workers or dormitory cooks.[85] "The [young] workers or soldiers are removed from their families at the moment of greatest sexual development and find themselves in the unfamiliar urban environment," wrote Vasilii Borovskii. "[They] contract syphilis and bring this sad product of 'civilization' home to their native backwoods."[86] What began as venereal syphilis in the urban setting, Eduard Shperk had explained in 1869, evolved under rural conditions into the endemic form: "Los[ing] its capacity to spread in the chancre stage, it encounters the ideal conditions for spreading in the secondary stage."[87] "Take any case of syphilis," asserted Veniamin Tarnovskii in 1897, speaking in the interests of his own favorite cause, "though apparently unconnected to prostitution, and trace it back through the chain of [infected] individuals, sometimes through generations, to the original source. You will always find a prostitute at the end of the line."[88]

Without necessarily contesting its sexual roots, other experts denied that syphilis started in the cities. No causal link between factory work and rural syphilis had been established to the satisfaction of Mikhail Uvarov, though he suspected there might indeed be one.[89] More emphatically, Nikolai Speranskii concluded that "syphilis among the population of Moscow Province ha[d] no constant or direct relationship to the development of manufacture or seasonal work."[90] Grigorii Gertsenshtein, for his part, rejected the hypothesis of urban origins on the grounds that sexual contact could not account for the extensiveness of the disease: "The enormous number of cases involving contagion

---

[84]TsGIA, f. 1297, op. 6, ed. khr. 28 (1850), ll. 1–7; ed. khr. 31 (1852), ll. 2–15. Unfortunately, the file contains no further reports of this kind after the 1850s, so it is impossible to know whether the committee's attitudes changed over the years.

[85]Shtiurmer, Sifilis, 41. Peasant women also worked in the factories (hence the availability of sexual partners) but were not themselves designated as carriers of disease.

[86]Borovskii, "K voprosu," 414.

[87]Shperk, "O merakh k prekrashcheniiu," 69.

[88]V. M. Tarnovskii, in Trudy Vysochaishe razreshennogo s"ezda, 2:10.

[89]Uvarov, "Sifilis sredi sel'skogo naseleniia," 56–58.

[90]Speranskii, K statistike sifilisa, 91.

through shared beds, common dishes, and borrowed underclothing completely engulfs the relatively few cases of venereal transmission." Russia, he claimed, did not show the expected epidemiological patterns: "The army, which usually serves as the principal index of syphilis and venereal disease infection in a given locality, in Russia shows a lower level of syphilis than the civilian population. . . . Thus in Russia the normal direction of syphilis infection, from city to countryside, is reversed; here syphilis moves less often from factories, plants, soldiers, and cities to the villages than the other way around."[91]

One could not, then, maintain at the same time the moral innocence of the village and the iniquity of the town without facing insuperable obstacles to consistent diagnosis. Most prostitutes, even in the big cities, were, as everyone knew, recent peasant immigrants. Upon arriving in town, many worked as household servants before turning to the sexual trade. In fact, servants constituted the largest single group of lower-class women in urban syphilis wards.[92] But was their disease the result of moral corruption or the stigma of innocent backwardness?

Physicians tried to have it both ways. As newcomers to the city, prostitutes might arrive with a burden of contagion acquired in the course of everyday domestic life, which they would then help spread through public sexual contact. Infected perhaps by a neighbor's child, a woman might start her new profession while still in the contagious stage and contaminate her male partners. In fact, however, prostitutes could spread venereal syphilis only if they themselves and their clients were vulnerable to infection by the genital route, and early exposure to infection in the endemic form might have prevented them from passing it on or falling ill in their adopted sexual role. Clients from their own social class would have been protected in a similar way. Yet the fact that many prostitutes did suffer from syphilis in the contagious early phase indicated that peasant life had left many adults untouched by the endemic scourge. When such women returned to their native villages on seasonal holidays, they could easily import a client's disease, which might then begin to circulate by nonvenereal means.

Venereal and nonvenereal, urban and rural patterns of contagion were difficult to disengage, yet most physicians persisted in viewing

[91]Gertsenshtein, "Peredvizhnye vrachebnye otriady," 17, 27. Shtiurmer (Sifilis, 45) agreed that infection moved from village to city. For a modern assertion of the same principle, see Hudson, Nonvenereal Syphilis, 192.

[92]Shtiurmer, Sifilis, 42; Z. Ia. El'tsina, "Sifilis i kozhnye bolezni sredi zhenskogo rabochego naseleniia Peterburga," pt. 1, Vrach, no. 42 (1896): 1178–79; Gratsianskii, "Nevinnye puti," 806; M. M. Kholevinskaia, "Otchet ob osmotrakh prostitutok na Samokatskom smotrovom punkte Nizhegorodskoi iarmarki za 1893 god," Vrach, no. 17 (1894): 487.

moral or sexual habits as a function not of person but of place. The majority eschewed the more consistent, if no less improbable, position favored by Veniamin Tarnovskii, who solved the problem of moral ambiguity, as we have seen, by classifying most prostitutes as an inherently degenerate type. The "predisposition to vice," a genetic trait, provided the organic subsoil for sociological growth, he insisted: "It is a mistake to think that city life acts as the only corrupting influence on the large population of migrant peasant women that pours in from the small towns and villages. No! A certain number of rural girls, no longer virgin and predisposed to vice, arrive in the big city, ready material for the prostitutional class." The inborn prostitute might appear in any social milieu, Tarnovskii insisted: "No matter what her social class, the girl with a predisposition to vice will find the opportunity to fall [from virtue] as soon as her sexual instinct comes to life, and will more or less gradually enter the ranks of active prostitution." That most prostitutes were poor proved not that moral compunction had succumbed to economic need but that the poor had no moral compunction. Only the kind of training provided by family life in the cultivated classes, he asserted, could prevent a degenerate woman from satisfying her reprehensible biological urge.[93]

The essence of the prostitute's pathology, in Tarnovskii's view, was her sexual aggressiveness. Abolitionists, he complained, persisted in seeing these unfortunate women as victims of poverty, circumstance, and male depredation, whereas in fact they were responsible for the seduction and downfall of many an innocent man. "The prostitute continually, repeatedly, daily offers herself for sale," wrote Tarnovskii. "She does not do so at the customer's insistence, but herself seeks him out and inveigles him into the deal." As the victim of her own pathological constitution, in Tarnovskii's depiction, the prostitute was not the mistress of her fate; but insofar as this constitution endowed her with sexual desire, he granted her a subjectivity that other women lacked. He associated the sexual agency she manifested in her trade, however, with city life and male entrepreneurship: "The prostitute brings to her trade all the active, commercial initiative [*aktivnaia, promyslovaia deiatel'nost'*] without which success is unthinkable," the professor wrote, "an initiative that responds to competition, demand, and the requirements of the marketplace."[94] Despite his willingness to believe that the village as well as the town might harbor the potential for sexual delinquency, and despite his untypically critical opinion of rural morality, Tarnovskii ulti-

[93]V. M. Tarnovskii, *Prostitutsiia i abolitsionizm*, 177, 179–81, and in *Trudy Vysochaishe razreshennogo s"ezda*, 2:10–11.
[94]V. M. Tarnovskii, *Prostitutsiia i abolitsionizm*, 133–36, 159, 164, 179.

mately pictured the public woman as citified and masculine. He thus created a special category that explained female sexual activity while allowing him to retain his conventional notions about female sexual passivity.

In that respect he was no different from the majority of his colleagues, who found it easier to think of peasant men than peasant women as disengaged from family or domestic ties. Outside the moral confines of the village, men were more likely to be viewed as free sexual agents. Even so, their habits and their diseases remained those of the countryside. In transition between communal and individual modes, they seemed to suffer the disadvantages of both. A study of syphilis in the Kiev garrison, for example, reported that commissioned and noncommissioned officers frequented prostitutes in equal numbers but that only the socially more humble noncommissioned men—many of whom had risen from the ranks—contracted syphilis in other ways as well. Their crowded, unsanitary living quarters, the study noted, were as dangerous to their health as were the freelance prostitutes in the local bar.[95] Career officers might also engage in sexual promiscuity, but they kept their distance from each other and were more likely to wash. Their diseases were those of individuals, not of the mass.

Another study of military health, equating moral with class distinctions, declared that venereal syphilis was three times as common among officers as among enlisted men because the latter were more likely to be married and faithful to their wives. Fresh recruits arrived with village virtues intact, reported Pavel Govorkov in 1896: "The purity of country ways is obviously so resilient that even under circumstances in our opinion most favorable to infection, the young men are nevertheless the most virtuous in the garrison." First encountering prostitutes after joining the army, they did not immediately yield to temptation but only gradually succumbed. The common soldier, Govorkov asserted, wanted not sex but innocent diversion: frequenting the nearby tavern with its cheap beer, cheap prostitutes, and broken-down billiard tables, the young man was more likely to confine himself to a game and a drink than to bother with the women.[96]

Unlike the midcentury studies that had identified the factory population as a source of contagion transmitted by indiscriminate sexual activity, reports from the turn of the century did not depict the urban poor as sexually licentious.[97] On the contrary, most observers then argued

[95]Borovskii, "K voprosu," 417–18.

[96]Govorkov, "Polovaia zhizn'," pt. 1, 1017; pt. 2, *Vrach*, no. 38 (1896): 1050–51.

[97]TsGIA, f. 1297, op. 6, ed. khr. 28 (1850), ll. 1–7; ed. khr. 31 (1852), ll. 2–15. The earlier reports described factory women as sexually promiscuous, if not professional pros-

that unskilled workers, like the raw recruits fresh to army life, had not yet assimilated the city's cultural values and continued to manifest the rural pattern of disease. "The less educated [*kul'turno*] the population," observed Moscow municipal doctor Aleksandr Petrovskii in 1905, "the worse the conditions in which it lives, the more likely that syphilis will be transmitted by nonsexual means."[98] Pavel Shiriaev, who treated Moscow workers, called the prevalent form "common" (*khodiachii*) or "communal" (*obshchezhitel'nyi*) syphilis. "This *everyday phenomenon*," he wrote in 1902, applying the familiar epithet, "has its *raison d'être* in the socioeconomic conditions of life in big cities, particularly among the laboring population." Like peasant syphilis, it was a sign not of depravity but of propriety and faithfulness to the family role: "This common syphilis is the disease of the average, one might say respectable [*dobroporiadochnyi*], worker," Shiriaev contended, "who still maintains close ties with his village and family home [*rodnaia sem'ia*]."[99]

Physicians did emphasize that working people threatened their betters with contagion. "The syphilis of our servants and workers," Shiriaev cautioned, "stalks us everywhere: in the bosom of our family, in hotels, restaurants, theaters, . . . in factories, plants, and so on."[100] The violation of class boundaries could thus have ominous results. The most obvious danger was in intercourse with prostitutes, but physicians insisted that nonvenereal avenues of transmission posed a more insidious threat. Service personnel, they believed, introduced the organism through everyday contacts of an intimate though nonsexual kind. Among the trades that worried physicians were some whose practitioners needed little skill or knowledge of the city and might have fitted the profile of the recent rural immigrant, as well as others associated with a greater degree of cultural assimilation. These people included sales clerks, cab drivers, tailors, bakers, shoemakers, waiters, house superintendents, seamstresses, laundresses, and especially domestic servants, wetnurses, and barbers. Cooks contaminated the food, nannies hugged the children, housemaids folded the linen, laundresses cleaned the underwear, wetnurses took infants to their breasts, and barbers shaved unsuspecting gentlemen with dirty razors and wiped their

---

titutes. Cf. the assertion in 1902 that factory work protected women against the threat of prostitution: M. I. Pokrovskaia, *O zhertvakh obshchestvennogo temperamenta* (St. Petersburg, 1902), 52.

[98]Petrovskii, "Bor'ba," pt. 2, *Izvestiia Moskovskoi gorodskoi dumy*, no. 7, Obshchii otdel (1905): 3.

[99]P. A. Shiriaev, "Organizatsiia vrachebnoi pomoshchi pri sifilise i venericheskikh bolezniakh sredi rabochego naseleniia v bol'shikh promyshlennykh i torgovykh tsentrakh," *Meditsinskoe obozrenie*, no. 13–14 (1902): 151–52.

[100]Ibid., 152.

cheeks with dirty rags.[101] "The syphilitic barber is no less dangerous to society than the [infected] prostitute," warned Nikolai Fedchenko.[102]

This was not a sexual threat, the doctors argued; nonvenereal syphilis must be seen as a product of social conditions. Morality was not the key, wrote Andrei Rozenkvist of the Moscow women's hospital for venereal disease, rejecting the popular idea "'that only complete chastity can save us from syphilis.' . . . For centuries syphilis has been called the 'disease of sexual indulgence [liubostrastnaia bolezn'],' but contemporary knowledge has rendered this notion obsolete."[103] "The question of syphilis and the questions of sexual life, depravity, and prostitution associated with it have been considered taboo and have never been discussed as widely and openly as they deserve to be," wrote Aleksandr Petrovskii. "This conception of the disease . . . must be be radically altered."[104] Mariia Pokrovskaia, a feminist public health physician and an ardent foe of regulated prostitution, argued in a similar vein that because "syphilis is spread in a variety of ways," it should not "be tied to depravity and considered a shameful, secret disease."[105] Both she and Petrovskii objected to the focus on prostitution, regarding it as a diversion from more serious and more basic health issues.

Indeed, Pokrovskaia questioned whether the working class could reasonably be implicated in sexual transmission at all:

> Since the very same population—which is to say, the so-called common folk [prostoi narod]—predominates in the big cities as well as the towns and villages, the question arises as to why this population should begin to spread infection through sexual channels in the city, when it has previously done so through domestic contacts alone. . . . Do the so-called common folk . . . suddenly change their customs and habits, becoming cleaner, more careful, and circumspect, and thus cease to transmit syphilis the way they do in the countryside? . . . In fact, the peasants who come to work in St. Petersburg live under the same sanitary conditions as in the villages, if not even worse.

[101]On servants, see Z. Ia. El'tsina, "Vybor prislugi," in Pervyi zhenskii kalendar' na 1903 god, ed. P. N. Arian (St. Petersburg, 1903); Gratsianskii, "Nevinnye puti," 802, 806; P. A. Pavlov, "Ob otnoshenii vnepolovogo zarazheniia sifilisom k polovomu mezhdu srednim klassom g. Moskvy," Meditsinskoe obozrenie, no. 1 (1890): 13–14; Shtiurmer, Sifilis, 42. The literature on wetnurses is extensive, for they were employed by both private families and state foundling homes. See, e.g., Z. Ia. El'tsina, "Zhelatel'nye sposoby vskarmlivaniia grudnykh sifiliticheskikh detei," Vrach, no. 4 (1894): 101–3; Shtiurmer, Sifilis, 43; Gratsianskii, "Nevinnye puti," 779–88.
[102]Fedchenko, "O zarazhenii sifilisom," 24.
[103]Rozenkvist, "K statistike," 510, 512.
[104]Petrovskii, "Bor'ba s sifilisom," pt. 2, 23.
[105]M. I. Pokrovskaia, "Mery," pt. 3, Russkii vrach, no. 12 (1903): 455.

They continue to share the same wooden spoons and the same towels, to sleep one on top of the other, and so on. Why should these circumstances promote the spread of syphilis in the countryside and lose their meaning in the big city?[106]

The tendency to interpret similar data in opposite ways, Pokrovskaia argued, reflected both contrasting types of medical care and the investigating physicians' preconceived ideas. The rural zemstvo physician treated his patients in the context of their families and social networks; the grateful victims brought their "friends and relatives," and the doctor then "easily traced the origin of the disease." In the city, however, adult men and women came singly to anonymous hospitals specializing in venereal disease, where physicians treated them out of context, as isolated individuals. The anonymity of urban medical facilities and the doctors' immersion in the professional rather than the social world meant that "the source of infection among the common folk of the big cities escaped calculation." Moreover, the urban physician tended to think in individualized—that is to say, sexual—terms, Pokrovskaia observed: "The majority of medical specialists are convinced that the prostitute is the main cause of syphilis in the cities. This preconceived idea cannot help influencing the data they collect." Because rural physicians understood the communal nature of peasant life, she asserted, they avoided this bias and produced more reliable statistics.[107] In trying to stem the spread of infection, complained the Minsk municipal doctor Petr Gratsianov, echoing her theme, the law itself shared this individualized, sexualized bias. Defining syphilis as a disease passed from person to person in the course of discrete sexual acts, administrators focused their efforts on the control of prostitution. As a result, he said, the police merely penalized the urban lower classes and, in particular, lower-class women, while the real epidemiological problem escaped their grasp.[108]

Pokrovskaia's urban-rural contrast did not, however, account for the interpretations offered by those urban practitioners who downplayed sexual transmission because of the moral stereotypes they associated with particular categories of patients. While some physicians primarily

---

[106]Ibid., pt. 2, *Russkii vrach*, no. 11 (1903): 413.

[107]Ibid., 413–14; pt. 3, 455. Pokrovskaia was not the only one to link the sexualization of syphilis to the tendency to see it as a problem of individual behavior; see also Gratsianov, "Bor'ba s sifilisom," 247. Shperk had earlier suggested that clinicians in urban hospitals fostered the venereal interpretation of syphilis, while provincial physicians drew different conclusions from their rural experience: "Signification des syphiloïdes," 186–87.

[108]Gratsianov, "Bor'ba s sifilisom," 247–50.

affiliated with the Academy of Military Medicine or the medical police tended to focus on the prevalence of venereal syphilis as a problem linked to prostitution and moral disorder,[109] others preferred to interpret the ambiguous symptoms that came their way as nonvenereal. Veniamin Tarnovskii's assistant, Zinaida El'tsina, for example, described the case of a shoemaker's apprentice, a girl only eight or nine years old, who turned up with papular lesions of the secondary stage on her genitals and mouth. El'tsina explained the sores as the result of eating from the same bowl as the other girls and sleeping two to a bed on the same pillows.[110] She did not suggest that the girls had either been subject to unwanted sexual contact or solicited such contact themselves.

Diagnoses of this sort show to what extent scientific knowledge allowed—and still allows—considerable room for nonscientific judgment. Although El'tsina's conclusion was consistent with contemporary knowledge of the disease in its endemic form, a number of authorities warned against exaggerating the danger of casual physical contacts, insisting that sexual relations accounted for the overwhelming majority of such symptoms.[111] El'tsina's reading of the clinical signs as proof of nonvenereal transmission confirmed the presumption of prepubescent asexuality that most of her colleagues shared. In so doing, they were rejecting diagnostic alternatives of which they could not have been unaware. For example, a commission established by the Ministry of Internal Affairs in 1847 to investigate unregistered streetwalkers found "still innocent young girls" between the ages of nine and fourteen infected with syphilis, which it explained as either the nonvenereal result of living among syphilitic prostitutes or as the consequence of "inexcusable naughtiness [shalosti] bordering on prostitution, which they commit with infected men."[112] To the Victorian culture that ignored or repressed precocious awakening, the child prostitute—evidence that little girls were capable of sex—was to childhood what the adult prostitute was to the virtuous peasant wife: a contradiction to be explained away.

[109]For the split between urban and zemstvo doctors at the 1897 congress on syphilis, see A. A. Tsenovskii, Abolitsionizm i bor'ba s sifilisom (Odessa, 1903), 52.

[110]Z. Ia. El'tsina, "Nedostatochnost' nadzora za maloletnimi v artel'nykh masterskikh i neobespechennost' detei sifilitikov bol'nichnymi mestami," Vrach, no. 19 (1900): 579. El'tsina's career is described in Afanas'ev, Sovremenniki 1:97.

[111]Hudson, Nonvenereal Syphilis, 81–82; cf. Diday, Le Péril vénérien, 317.

[112]"O prostitutsii v Rossii," Arkhiv sudebnoi meditsiny i obshchestvennoi gigieny, no. 1, pt. 3 (1869): 104–5. On the presumption of asexuality, see Sander L. Gilman, Difference and Pathology: Stereotypes of Sexuality, Race, and Madness (Ithaca, N.Y., 1985), 41. Also see chap. 8 below. Not all societies at all times have taken for granted the sexual innocence of children. French clergymen of the sixteenth to eighteenth centuries feared that four-year-old peasant children who shared a bed would commit "horrible sins": Jean-Louis Flandrin, ed., Les Amours paysannes: Amour et sexualité dans les campagnes de l'ancienne France, XVIᵉ–XIXᵉ siècle (Paris, 1975), 150.

A presumption of chastity likewise governed the conclusions of the numerous observers who attributed sores on the lips and mouths of artisanal workers to communal eating habits. The workers drank from the same cup, they explained, took puffs of the same cigarette, and greeted each other with a kiss, thus spreading so-called "occupational [*promyslovyi*] syphilis."[113] Obviously eager to refute the other possible explanation, Andrei Rozenkvist was led to the improbable assertion that oral intercourse could not be to blame because "*coitus per os* ('sapphism'), as far as I know, is not practiced in Russia."[114] The risk of contagion for centuries associated with certain trades (in particular glassblowing) is not rejected out of hand by all modern authorities; nevertheless, Rozenkvist's discomfort with sexual attribution is clear. His identification of oral-genital contact, even between men and women, with female homosexuality was typical of the period, but denying its existence was not.[115]

Interpreting various nongenital symptoms as sexual would, of course, have led to the admission that the "common folk" might not in fact confine themselves to the basics of heterosexual intercourse, any more than they confined themselves to producing children with wedded partners. Like prostitution, the perversions separated sexual pleasure from the "natural" process of reproduction. They had presumably resulted from the same excess of civilization that had corrupted Europe but gotten no more than a foothold in the Russian empire, still immersed in the natural, agrarian way of life typified by the peasant class. Such preconceptions made it hard for doctors to admit that alternative sexual practices were widespread in any social milieu. Most chancres on the perineum, anus, and buttocks must be sexual in origin, conceded Rozenkvist, "but not in the sense that all chancres on the anus, for example, have resulted from *coitus per anum* ('sodomy'). No, contagion in the majority of cases results from the overflow of secretions from the sexual organs *post actum coitus.*"[116] Although one can find the same suggestion in Alfred Fournier's influential text,[117] equally respected works such as those by Ambroise Tardieu, Robert W. Taylor, and L. Duncan Bulkley would have confirmed a different view.

---

[113]Rozenkvist, "K statistike," 516; A. I. Pospelov, *O vnepolovom zarazhenii sifilisom sredi liudei chernorabochego klassa g. Moskvy* (St. Petersburg, 1889), 45; Gratsianskii, "Nevinnye puti," 805; Uvarov, "Sifilis sredi sel'skogo naseleniia," 61–67.

[114]Rozenkvist, "K statistike," 528.

[115]On the term "sapphism" applied to male-female oral sex, see Louis Martineau, *La Prostitution clandestine* (Paris, 1885), 90. On contemporary frankness, see, e.g., Robert W. Taylor, "Some Unusual Modes of Infection with Syphilis," *Journal of Cutaneous and Genito-Urinary Diseases* 8:6 (1890): 205: "It is very certain that syphilis is not infrequently contracted [by men] from the mouths of women suffering from buccal lesions, and it is well, for many reasons, that physicians should be aware of the fact."

[116]Rozenkvist, "K statistike," 528–29.

[117]Fournier, *Traité de la syphilis,* 1:189.

The Problem of Control

Despite frequent criticism of data on which the claim was based,[118] few medical reports in the 1890s challenged the perception that syphilis was an increasingly severe public health problem, a perception that Russian physicians shared with—or borrowed from—their colleagues abroad. The anxious note struck by Parent-Duchâtelet in 1836 continued to sound throughout the century and across national boundaries with undiminished alarm: "Of all the contagious diseases that affect the human species and cause society the greatest harm, none is more serious, more dangerous, or more to be feared than syphilis."[119] Because of its alleged demographic impact, the problem was said to have reached the proportions of a "question of state."[120] Mobilizing state resources in the medical war against syphilis conceived as a venereal disease resulted in the system of regulated prostitution in nineteenth-century Europe.[121]

Discussion of syphilis control in the Russian medical press and at professional conferences likewise centered on the state's role in policing public behavior. But in contrast to their European counterparts, Russian physicians before 1905 were reluctant to sexualize public health questions and thereby to define sexuality itself as a medical problem. They tended instead to convert sexual into social issues and to extend their authority in an explicitly political direction. Both the preference for nonvenereal diagnosis and the profession's heated debates on public policy reflect the basic issues separating state and society in the fifteen years preceding the 1905 revolution.

Differences of opinion on the question of regulated prostitution within the medical community often coincided with differences in professional location. Russia's leading advocates of regulation held posts either in the Academy of Military Medicine (Veniamin Tarnovskii, Petr Gratsianskii, Grigorii Gertsenshtein) or on the police-medical committees (Aleksandr Fedorov, Konstantin Shtiurmer). Other supporters served in urban hospitals for venereal disease (Oskar von Petersen, Ivan Priklonskii, Aleksandr Vvedenskii). Critics of the bureaucratic approach, by contrast, often worked, or had worked for significant pe-

---

[118]See remarks by L. F. Ragozin in *Trudy Vysochaishe razreshennogo s"ezda*, 2:2.

[119]A. J. B. Parent-Duchâtelet, *De la prostitution dans la ville de Paris*, 3d ed. (Paris, 1857), 1:603.

[120]Borovskii, "K voprosu," 413. Similar statements were made by Mikhail Kuznetsov, *Prostitutsiia i sifilis v Rossii* (St. Petersburg, 1871), 68; Gratsianskii, "Nevinnye puti," 775; Petrovskii, "Bor'ba s sifilisom," pt. 1, 2.

[121]On the nineteenth-century French system, see Alain Corbin, *Les Filles de noce: Misère sexuelle et prostitution (19ᵉ et 20ᵉ siècles)* (Paris, 1978); and Jill Harsin, *The Policing of Prostitution in Nineteenth-Century Paris* (Princeton, N.J., 1985).

riods, either in zemstvo practice (Pavel Govorkov, Mariia Pokrovskaia, Dmitrii Zhbankov) or for city governments (Aleksandr Petrovskii). Individual careers were not always so easily catalogued, however, and ideological divisions not always so neat: Eduard Shperk, chief physician at St. Petersburg's Kalinkin Hospital from 1870 to 1891, had from the start expressed serious doubts about regulation; Petr Gratsianov, who had been a municipal physician as well as an army doctor, moved from ardent to qualified support of the system; Petr Oboznenko began as a zemstvo doctor, rose to a post in the medical department of the Ministry of Internal Affairs, served on the staff of the Kalinkin Hospital, and evolved from moderate to highly critical endorsement.[122]

Regulation itself was hotly disputed throughout nineteenth-century Europe, but in Russia even its most energetic partisans found themselves in a quandary when it came to prescribing measures against nonvenereal contagion, particularly in the social environment of the countryside and of the newly emerging urban working class. Control over prostitutes had from the first been linked to the surveillance of migratory or criminal members of the lower orders. The original 1843 instructions called for the medical inspection of male factory workers, seasonal laborers, and "lower-class persons [*litsa nizshego klassa*] of both sexes arrested for disturbing the peace."[123] Some physicians continued to endorse such measures in relation to artisans, factory workers, domestic servants, wetnurses, and barbers. However, even the fierce, internationally renowned opponent of abolition Veniamin Tarnovskii conceded at the 1885 congress of Russian physicians that medical surveillance of factory workers was not a practical idea.[124]

Convinced that prostitutes were a pathological element, combining the traits of inborn criminality with those of insanity, Tarnovskii believed that they knowingly spread evil and disease. As an impaired species they did not, in his eyes, deserve the same treatment as normal people: society, he wrote, "cannot allow them the freedom to harm other, healthy citizens, neither in the name of personal freedom, a concept applicable only to the normal person, nor in the name of morality, a concept they are constitutionally unable to grasp." As a consequence of their "habitual depravity," these "morally depraved" and "physically abnormal" women should not "enjoy full personal liberty," he argued,

---

[122]Most career information is taken from *Rossiiskii meditsinskii spisok*.

[123]Quoted in P. A. Gratsianov, "K voprosu o reorganizatsii nadzora za prostitutsiei v Rossii," *Vestnik obshchestvennoi gigieny*, no. 11 (1895): 141.

[124]On artisans, see El'tsina, "Nedostatochnost' nadzora," 587–79; on servants and wetnurses, Pavlov, "Ob otnoshenii," 17; on all groups, Rozenkvist, "K statistike," 533; on barbers (under medical inspection in St. Petersburg since 1883), Fedchenko, "O zarazhenii sifilisom," 25; Tarnovskii, cited in Chistiakov, *Protokoly*, 50–51.

"because such liberty harms society in two ways: on the one hand, by encouraging vice, and on the other, by spreading syphilis among the entire population." Moreover, "free competition delivers prostitution into the power of capital," Tarnovskii warned, "and puts it on a par with the most immoral commercial operations."[125] Unregulated prostitution, like free enterprise, needed the discipline of paternal state control.

Having defined moral weakness as a biological deformity, Tarnovskii was then able to conclude, with a satisfying sense of paradox, that syphilis was not the product of immoral behavior, as everyone believed, but its cause. Syphilis was not the price society paid for the existence of prostitution, but the reverse: prostitution was the price of disease. Generations of "physically and morally crippled offspring," the damaged fruit of syphilitic wombs, Tarnovskii asserted, supplied the male sexual demand and the female sexual inclination, without which prostitution itself would wither away.[126] Abolitionists might think that regulation perpetuated the system of commercial sex; in fact, in Tarnovskii's opinion, regulation was the key to its demise.

Many physicians, even if they accepted the principle of state-administered syphilis control, rejected such arguments.[127] Defending his work on the St. Petersburg police-medical committee, Aleksandr Fedorov described the impulse behind male sexual demand as "natural [and] lawful [zakonnyi]," not abnormal and depraved.[128] Grigorii Gertsenshtein considered regulation a function of modern, not traditional, government. Prostitution, he argued in the spirit of Parent-Duchâtelet, must operate under the same constraints as other businesses: "Everyone admits the government has the right to monitor hygiene conditions in factories and mills, to prevent them from poisoning the soil, the water, and the air. Naturally, it also has the responsibility to counteract the unhealthy consequences of the prostitution trade."[129]

But endemic syphilis was confined neither to outcast and dependent groups nor to identifiable trades. The sacrifice of personal rights advocated by Tarnovskii for the socially undesirable could not be justified for the population at large; Tarnovskii himself did not attempt to do so.[130] Gertsenshtein, for his part, was emphatic: "Can we subject the entire

---

[125]V. M. Tarnovskii, *Prostitutsiia i abolitsionizm*, viii, 136–39, 159, 161, 189, 242–43, and 1897 speech in *Trudy Vysochaishe razreshennogo s"ezda*, 2:10–11.

[126]V. M. Tarnovskii, *Prostitutsiia i abolitsionizm*, 210.

[127]Gertsenshtein, "Sifilis v Novgorodskoi gubernii," 58.

[128]A. I. Fedorov, "Prostitutsiia v S.-Peterburge i vrachebno-politseiskii nadzor za neiu," *Vestnik obshchestvennoi gigieny*, no. 1 (1892): 73.

[129]G. M. G[ertsenshtein], "Prostitutsiia," in *Entsiklopedicheskii slovar' Brokgauz-Efron* (St. Petersburg, 1898), 25A:486.

[130]V. M. Tarnovskii, *Prostitutsiia i abolitsionizm*, 242, and 1897 speech in *Trudy Vysochaishe razreshennogo s"ezda*, 2:19, 21.

civilian population to similar surveillance, with the goal of detecting syphilitics, isolating, and curing them? The answer is self-evident!"[131] The use of force for such purposes would be ineffective as well as unethical, he concluded, citing the words of a Novgorod zemstvo physician: "Measures such as the medical inspection of shepherds, itinerant shoemakers, wool-beaters, and returning seasonal workers, or the comprehensive examination of villagers, are in the first place always immoral, like all compulsion [nasilie], whatever its pretext, and in the second place they merely induce the victims to conceal their illness in every possible way."[132]

Physicians thus reinforced their defense of civil rights with arguments about the medical effectiveness of state-administered strategies. The army, at once a focal point of venereal infection and a laboratory for the exercise of compulsory regimes, served as a case in point for the antiregulationist position. Pavel Govorkov, who had practiced as a zemstvo doctor in Novgorod Province before being assigned to army work, used his military experience to demonstrate the futility of standard procedures even under the optimal conditions of regimental life. Well-policed prostitution in the vicinity of army camps and the compulsory medical inspection of supposedly "dissolute" enlisted men formed the basic elements of official policy, designed, in Govorkov's caustic words, "to reinforce [and] prohibit, with special attention to such-and-such; to seize [and] punish."[133]

Such "coercive" (prinuditel'nye) tactics directed at "external causes" reflected the prevalent view of syphilis as the product of moral dereliction, a view Govorkov did not share: "Can one consider [the average soldier's] two to three acts of sexual intercourse per year licentiousness," he ironically queried, "especially among single men (except if one measures [their behavior] by the standards of rural virtue)?" The uselessness of a strategy based on these misconceptions proved how wrong they were. "Nothing is easier," Govorkov commented, "than to impose special regulations in the garrison." Trained to obedience, soldiers did not question orders; hence, nowhere else "can restrictive measures be applied with such force. If contagion nonetheless persists at its former rate," Govorkov summed up, "then one is fully justified in concluding that the officially listed causes [of syphilis] are either dubious or ineradicable, or, finally, that coercive measures cannot achieve their goal."[134]

Physicians had insisted that syphilis was a "question of state," a mat-

[131]Gertsenshtein, "Peredvizhnye vrachebnye otriady," 28.
[132]A. A. Desiatov, quoted in Gertsenshtein, "Sifilis v Novgorodskoi gubernii," 37.
[133]Govorkov, "Polovaia zhizn'," pt. 2, 1052.
[134]Ibid., 1052–53.

ter of national importance to be tackled by the joint efforts of government and society. But in a period of increasingly strained relations between these two forces, they criticized the character of government intrusion into the business of popular health.[135] For the control of rural and working-class syphilis, coercive bureaucratic measures were worse than useless. In the enduring tradition of the 1860s intelligentsia, physicians saw themselves as reaching out to the people with the power of enlightenment, not compulsion, which they identified with the repressive state.[136] They wished to have nothing in common with the callous official; they hoped to cultivate trust rather than obedience, to instill respect and knowledge instead of fear. "One knows how the peasants react to any kind of inventories or censuses," wrote Aleksandr Efimov, "to anything that smells of bureaucracy, intruding into their dark, frightened world. How many calamities have arisen in the peasants' world from this source! Thus, it is obviously critical when performing medical examinations to eliminate anything bureaucratic, official, and, most important, anything coercive, so as not at the outset to undermine [the peasants'] confidence in what is going on."[137]

Physicians did not hesitate to underscore the political implications of their view. Without freedom of speech the educated classes could not make their contribution to the general welfare; ignorance and ill health would remain the common lot. Delegates to the congress on syphilis deplored the extreme difficulty of obtaining official permission to present educational material to popular audiences. The regime tolerated no free public expression, and therefore the physicians' attack on censorship conveyed a demand not only for professional autonomy but also for political reform. "Our existing law allows us to give lectures, but only on the basis of prepared texts," complained Petr Gratsianov. "I may read, but not express myself in my own words. . . . I am merely asking," he assured his listeners, who included officials from the Ministry of Internal Affairs, "that the physician be granted the right to communicate with the element with which he deals." No doubt the rector of Kazan University and the representative of the Kharkov medical so-

---

[135]On the radicalization of public health circles between 1895 and 1905, see E. I. Lotova, *Russkaia intelligentsia i voprosy obshchestvennoi gigieny: Pervoe gigienicheskoe obshchestvo v Rossii* (Moscow, 1962), 39.

[136]What was needed, Dmitrii Zhbankov later wrote (*O deiatel'nosti*, 3), was "the most active intervention in the people's life, not by means of force but through persuasion and trust." On the intelligentsia tradition and opposition to state intervention, see Nancy M. Frieden, "Physicians in Pre-Revolutionary Russia: Professionals or Servants of the State?" *Bulletin of the History of Medicine* 49:1 (1975): 22–23, 28. On the demise of this tradition after 1905, see Hutchinson, *Politics and Public Health.*

[137]A. I. Efimov, "Sravnitel'naia otsenka raznykh sposobov izucheniia derevenskogo sifilisa," *Vrach*, no. 51 (1900), 1553.

ciety, who endorsed this statement, understood the analogies they had been warned not to draw.[138]

Many critics of regulated prostitution condemned the system as a feudal relic, incompatible with the modern principle of civil rights.[139] Such language echoes also in the rejection of proposals to submit the peasantry to compulsory examinations: "What is striking about such lack of respect toward the peasant folk," wrote Efimov, "is its origin in the old feudal attitude [krepostnicheskii vzgliad], which considers them some sort of lower species [nizshaia poroda] that one doesn't have to treat with kid gloves." Respect for the peasants' privacy and personal dignity must acknowledge their individuality: that is to say, their humanity. Every person, Efimov continued, "has an interior world, which is inaccessible to government authority."[140]

Plans to isolate infectious cases in special colonies were rejected as gross violations of individual rights.[141] The analogy between syphilis and other antisocial debilities such as insanity or criminality, promoted by Tarnovskii and his followers to justify their demands for quarantine, were equally condemned.[142] The entire process of selection, segregation, and penalization, argued Zhbankov, was incompatible with the values of the medical profession, not to speak of social justice; its logical conclusion, he pointed out, would be "the death penalty, which [Ernst] Haeckel, if I am not mistaken, recommended as the suitable fate for criminals, the mentally ill, syphilitics, tuberculars, and others. Then, at least, we would *thoroughly* cure humanity of all its ills, if, that is, even a single person survived the strict enforcement of such measures."[143]

Doctors active in the rural context argued for integration, not segregation, of syphilitic patients as the more effective medical strategy. Similar arguments were made in relation to artisans and factory workers. Pavel Shiriaev, consultant to Moscow's working-class hospital, advocated treatment on an outpatient basis, a technique already suc-

[138]P. A. Gratsianov, A. G. Ge, and A. Kh. Kuznetsov, in *Trudy Vysochaishe razreshennogo s"ezda*, 2:28.

[139]See, e.g., P. E. Oboznenko, *Podnadzornaia prostitutsiia S.-Peterburga* (St. Petersburg, 1896), 39.

[140]Efimov, "Sravnitel'naia otsenka," 1552.

[141]For such a proposal, see Uvarov, "Sifilis sredi sel'skogo naseleniia," 113.

[142]For the equation of the infected prostitute with an armed, homicidal maniac, see V. M. Tarnovskii, *Prostitutsiia i abolitsionizm*, 190—an analogy condemned by Gertsenshtein, "Sifilis v Novgorodskoi gubernii," 56. Julie Vail Brown notes that Russian folk tradition did not welcome incarceration even of the insane; after 1905 the Russian psychiatric profession itself abandoned its earlier defense of incarceration as indispenable to treatment: "The Professionalization of Russian Psychiatry, 1857–1911" (Ph.D. diss., University of Pennsylvania, 1981), 331–33.

[143]D. N. Zhbankov, "O s"ezde pri Meditsinskom Departamente po obsuzhdeniiu meropriiatii protiv sifilisa v Rossii," pt. 2, *Vrach*, no. 30 (1897): 832 (original emphasis).

cessful among the middle class. Incarceration, he noted, was particularly hard on the poor, who could not afford to stop working. And hospitalization only emphasized the outcast status of lower-class syphilitics; it assumed they could not be responsible for their own treatment but must be subjected to administrative constraint in this as in all else. Like Mariia Pokrovskaia, Shiriaev deplored the social consequences of linking syphilis to sexual misconduct. Moral self-righteousness and exaggerated fear of infection, he argued, caused people to deny syphilitics the compassion they reserved for victims of other, equally dangerous, diseases such as tuberculosis.[144] Petr Oboznenko, even though he had trained under Tarnovskii, also recommended that syphilis be treated no differently from other infectious conditions. Even in the big cities, it was no cause for special alarm, he emphasized, but simply the "most commonplace" occurrence. Emergency measures, in any case, were never an appropriate response to medical problems but always a reflection of ignorance and fear.[145]

The caution that sexualizing the problem of syphilis could have repressive social consequences was not new in Russian medicine. Thirty years earlier, Eduard Shperk had attacked police regulation for treating prostitutes as social outcasts who had no claim to personal or legal respect. The task of medicine, he warned, was to guard the public health, not to "punish the sins of individual persons or of society as a whole." If physicians raged against the horrors of "the evil disease" (*durnaia bolezn'*) or against "fallen women," it was to disguise their own shame at dealing with such an "indecent" subject. "Shoemakers should not bake pies," he admonished, and doctors should not meddle in legal and moral affairs.[146] The physicians' meddling in public policy, however, continued to function as a strategy of professional legitimization, inseparable from educated society's larger revolt against bureaucratic despotism. In the fight against syphilis, doctors argued, medical authority must replace official regulations; local and municipal organizations must substitute for the police.[147]

---

[144]Shiriaev, "Organizatsiia vrachebnoi pomoshchi," 153–54.

[145]P. E. Oboznenko, "Po povodu novogo proekta nadzora za prostitutsieiu v Peterburge, vyrabotannogo komissieiu Russkogo sifilidologicheskogo obshchestva," *Vrach*, no. 12 (1899): 349.

[146]Shperk, "O merakh k prekrashcheniiu," 77–81; idem, "Otvet na stat'iu: 'Zhenskii nadzor za prostitutsiei,'" *Arkhiv sudebnoi meditsiny i obshchestvennoi gigieny*, no. 1, sec. 5 (1870): 7.

[147]See Pirogov Society petition in K. I. Shidlovskii, ed., *Svodka khodataistv Pirogovskogo obshchestva vrachei pered pravitel'stvennymi uchrezhdeniiami za 20 let (1883–1903 gg.)* (Moscow, 1904), 19; for debates in city administrations, see Empe, "Neskol'ko slov o sifilise s sanitarnoi tochki zreniia i o polozhenii etogo voprosa v Peterburge," *Russkii vrach*, no. 13 (1903): 498–500.

One step in this direction was to update the existing juridical status of the regulatory system by emphasizing due process and minimizing administrative caprice. Oboznenko, for example, endorsed the majority resolution at the 1899 international syphilis congress in Brussels, calling for regulation to be reconstituted on a proper legal basis that would eliminate the punitive role of the police. In Russia, he said, the bureaucratic state machine (*kantseliarskaia mashina*) had preserved regulation, a remnant of serfdom, in modern times as a rear-guard tactic against social and economic change. The influx of peasants into the cities had created a floating population of temporarily homeless, unemployed poor. Women in this category were seized by the police on suspicion of moral impropriety and induced to regularize their situation by accepting the registered prostitute's yellow card. The authorities penalized a woman, Oboznenko concluded, "for being poor, having no job, and living in a dirty corner rather than a first-class hotel."[148] Regulation thus contributed to the absolutist regime's reliance on class-differentiated justice, designed to reinforce status distinctions and inhibit social change.

The first to criticize the traditional order for its hostility to individual expression and personal responsibility, physicians nevertheless defended the importance of intermediate institutions as a buffer against the state's intrusion into civil society. As inconsistent as most Western liberals, they left the principle of familial (that is, patriarchal) authority intact. "If the husband cannot always look after his wife," wrote Oboznenko, "the father after his daughter, the brother after his sister, then how can police agents penetrate into the most secret corners in millions of lives? And is such penetration desirable? By virtue of the corrupting influence it exercises on the family and on society at large, is it not worse than any syphilis?"[149] Moreover, by confirming the status of women as merchants of sex, regulation contributed to the decline of popular morals, defining prostitutes in relation not to father, brother, or husband but to their own sexual agency.

Indeed, both state policy and dominant medical opinion revealed conventional assumptions about family life and the relative autonomy of women and men. The original program of syphilis control targeted precisely those groups that had escaped the authority of the traditional, patriarchal institutions supposed to keep the dependent orders in line: peasants who had left the village, women who had left the family. Thus

---

[148]P. E. Oboznenko, "Vopros ob uporiadochenii prostitutsii i o bor'be s neiu na dvukh mezhdunarodnykh soveshchaniiakh 1899 goda," *Vrach*, no. 30 (1900): 910; idem, "Po povodu novogo proekta," 348–49; idem, *Podnadzornaia prostitutsiia*, 39.

[149]Oboznenko, "Po povodu novogo proekta," 348.

working-class men and women fell under state tutelage just at the moment when they moved from patriarchal submissiveness to the possibility of a relatively independent existence.

Of the two errant groups—migrant peasants and independent women—it was the latter who incurred the more serious reprisals. The prostitute's clients were, after all, free to come and go without examination. Extreme regulationists such as Tarnovskii and feminists such as Pokrovskaia joined in condemning the gender inequity of which their colleagues were guiltily aware, but they analyzed it differently. In her fight against the double moral standard, Pokrovskaia denounced prostitution itself, along with all schemes to control or improve it. She insisted that women were primarily the victims, not the agents, of venereal contagion.[150] Tarnovskii preached a reverse single standard of sexual debauchery. Certainly both partners in the commercial exchange of sex were "equally debased," he contended. Like the brothel's inmates, its visitors were "morally insensitive, with an innately intensified sexual drive. . . . Just as the prostitutes infect the healthy population, so they are in turn, in the majority of cases, infected by these moral cripples, the true refuse of contemporary society." Medical surveillance, in Tarnovskii's view, was society's way of maintaining its normality, the equivalent of health. "Every psychologically healthy, normally developed person," Tarnovskii was sure, "will freely admit the injustice as well as the futility of limiting medical surveillance in regulated brothels to the prostitutes alone."[151]

That sexually active men were allowed to circulate freely, carrying the infection that was so relentlessly tracked on female sexual parts, bothered the profession enough to provoke a resolution at its 1885 congress in favor of pulling clients into the regulatory net. The St. Petersburg syphilis society agreed, as did some officials in the Medical Department, but no such proposal ever became official policy, and the 1897 congress rejected the idea.[152] Few physicians accepted Tarnovskii's perverse egalitarianism, according to which sexual excess became a gender-neutral standard of social exclusion. The prevailing arguments against male inspection showed regulation for what it was: a system of controlling women, not disease. Those who accepted medical exams, for all their diagnostic inadequacy, as the only way to weed out sick prostitutes considered them too perfunctory to detect infection among men.[153] Clients were unlikely in any case, they declared, to submit to such a humiliating procedure and would defeat the whole system by

[150]Pokrovskaia, "Mery," pt. 1, Russkii vrach, no. 10, (1903): 374.
[151]V. M. Tarnovskii, 1897 speech in Trudy Vysochaishe razreshennogo s"ezda, 2:13.
[152]Chistiakov, Protokoly, 59, and Trudy Vysochaishe razreshennogo s"ezda, 2:39, 136.
[153]Oboznenko, "Po povodu novogo proekta," 348.

turning to nonbrothel prostitutes.[154] To Zhbankov, his colleagues were guilty of the same abuse of power they complained of at government hands: "By excluding from medical surveillance many individuals who spread syphilis by sexual means, whether deliberately or accidentally, the [1897 syphilis] congress is acting inconsistently and unjustly: compulsion ceases to be compulsion only if it *applies equally to all.*"[155]

The exempt individuals, exclusively men, were not the educated alone; the privilege of self-regulation was accorded to all males who were customers of public sex. The same "dirty," disorderly factory hands covered by the 1843 instructions were free from interference at the brothel door. The very propensity for unthinking violence cited to justify police jurisdiction over the lower classes served in this case to demonstrate the common man's capacity to assert his will and as an argument that his will must be respected. "Houses of the lowest caliber," observed Oboznenko, were frequented by "drunken, brutalized hordes, resembling human beings only in outward appearance. . . . [Their] victims . . . arrive at the hospital with bruises, bites, and scratches. . . . Clients at the better houses behave themselves with greater tact, but even among them . . . sobriety is rare, and those who would submit to examination without making a scene are rarer still."[156] If the gentleman was assimilated to the beast, so the beast was raised to the level of the citizen, in this tableau. Sexual agency, denied by definition to the female gender, was the mark of personal autonomy, the prerequisite for civic agency in its widest sense. To the extent that individuality was tolerated at all by the distrustful regime, it was in the person of the urban male. With few exceptions the spokesmen of educated society, for all their critical views, shared this prejudice.

Self-regulation was possible, at least in theory, because sexual contact was a matter of choice. To reinforce the subject's ability to exercise such personal discretion, the 1897 congress on syphilis called for a program of intensified moral vigilance: prohibition of alcohol, dancing, music, and singing in registered brothels; a crackdown on street solicitation and independent prostitutes; education of youth in the spirit of "moral cleanliness," sexual continence, and respect for women.[157] The turn to moral purity reflected a current European trend but failed to address the dominant Russian problem: endemic rural syphilis.[158] In his

---

[154]A. I. Fedorov, "Deiatel'nost' S.-Peterburgskogo vrachebno-politseiskogo komiteta za period 1888–95 gg.," *Vestnik obshchestvennoi gigieny*, no. 11 (1896): 185; idem, "Pozornyi promysl'," ibid., no. 8 (1900): 1183.

[155]Zhbankov, "O s"ezde," pt. 2, 833 (original emphasis).

[156]Oboznenko, "Po povodu novogo proekta," 348.

[157]Zhbankov, "O s"ezde," pt. 2, 831–32.

[158]Corbin, "Le Péril vénérien," 255–57. The psychiatrist P. I. Kovalevskii counseled sexual abstinence and early marriage as defenses against venereal syphilis but could offer

enthusiasm for sexual continence, even Tarnovskii was obliged to admit that it would not prevent infection in the rural population.[159] The congress agreed on the usefulness of compulsory examinations for certain dependent groups—prostitutes, soldiers, prisoners, servants, and foundlings—but found them inappropriate in the countryside. The congress did not, however, oppose the examination of factory workers, a position some participants criticized on the grounds that rural and working-class living conditions were virtually the same.[160]

On matters of public policy the 1897 congress produced a number of contradictory proposals.[161] One recommended the compulsory examination of all lower-class women without exception; another called for the enrollment of syphilitics in special regiments that would be mobilized to fight in time of war. Above all, however, the assembly emphasized popular medical education. "What we need," said Dmitrii Zhbankov, "are not police ordinances, but the development of public self-awareness [obshchestvennoe samosoznanie]." Self-regulation, he believed, was the only effective regulation; but an ignorant, oppressed populace was incapable of conducting itself in an enlightened way. Information was the key to autonomy, which was in turn the key to public health. "When it comes to limiting contagion by means of medical exams, no prohibitions or sanitary inspectors will be of any use," wrote Zhbankov. "Without developing popular self-awareness and without information about the dangers and symptoms of the disease, it will be impossible to change those ordinary daily relations which are precisely the reason for its spread."[162]

Autonomy was desirable not only for the victims of disease but for those struggling against it. The congress insisted that medical programs be removed from police jurisdiction and placed in the hands of local "public institutions" (obshchestvennye uchrezhdeniia).[163] The defense of local initiative, popular participation in administrative affairs, and corporate professional responsibility was not a theme peculiar to physicians alone; it was a dominant note in the growing chorus of political dissatisfaction in educated society at large, which was to coalesce after the century's turn in the organized liberation movement. The evolution of medical attitudes in accord with that of the dominant political mood

---

none against nonvenereal infection: "Sifilitiki, ikh neschast'e i spasenie," *Arkhiv psikhiatrii, neirologii i sudebnoi psikhopatologii,* no. 1 (1897): 64–65.

[159]V. M. Tarnovskii, 1897 speech in *Trudy Vysochaishe razreshennogo s"ezda,* 2:16.

[160]Zhbankov, "O s"ezde," pt. 1, *Vrach,* no. 29 (1897): 803; pt. 2, 833.

[161]On "sharp" disagreements at the congress, see Petrovskii, "Bor'ba s sifilisom," pt. 1, 6.

[162]Zhbankov, "O s"ezde," pt. 2, 832–33.

[163]Ibid., pt. 1, 801.

may be charted, for example, in successive articles by Minsk municipal physician Petr Gratsianov. Writing in 1895, Gratsianov insisted on the need to restrict prostitutes' freedom of movement. Dishonest and deceitful, such women could not be expected to submit voluntarily to medical control; the police were therefore obliged, in the general public interest, to confiscate their passports and limit the "personal liberty" that permitted them the luxury of harming others.[164] By 1904, however, Gratsianov was criticizing the system of police regulation as an abuse of the prostitutes' civil rights. Only in the hands of public rather than bureaucratic institutions, he had decided, would regulation be subject to the scrutiny of public opinion and prostitutes be guaranteed the right of judicial self-defense.[165]

A passionate though nonrevolutionary radical of populist outlook who stood outside party politics, Dmitrii Zhbankov was a leading figure in the Pirogov Society, Russia's most influential and progressive medical organization.[166] Acutely aware of the political dimension of the profession's claims, Zhbankov objected to any violation of personal rights, whether the humiliating and intrusive compulsory medical exam or the degrading references to stigmatized and underprivileged groups. Syphilis, as the congress acknowledged, was no respecter of social class, yet physicians allowed themselves to be guided by the prejudices of the age, Zhbankov complained, thus bringing dishonor to their calling and defeating their own goals. "I am always struck," he wrote, "by this amazing injustice: the propertied classes are allowed by law to demand medical certificates for their servants and nurses, but the employers themselves and their sick children may with impunity infect the nurses! The objection that the propertied are more cultured and restrain themselves from infecting others is hardly well founded, especially in regard to sexual transmission; otherwise husbands would not infect their wives, a thing we often see among the middle and even the upper classes."[167]

Such an admission, that education and culture were not enough to transform the personal habits governing sexuality and health, was of course fatal to the physicians' insistence that enlightenment was the key to disease control. It was not, however, inconsistent with another implied assumption: that even if extending to the populace the autonomous individuality accorded the upper orders might foster self-indul-

---

[164]Gratsianov, "K voprosu," 166–67.

[165]P. A. Gratsianov, "Po povodu proekta novogo 'Polozheniia o S.-Peterburgskom vrachebno-politseiskom komitete,'" *Russkii meditsinskii vestnik*, no. 1 (1904): 5–7.

[166]Mitskevich, *Na grani dvukh epokh*, 96–97. For Zhbankov's political activity in the 1890s, see Frieden, *Russian Physicians*, 171–72, 191.

[167]Zhbankov, "O s"ezde," pt. 1, 802; pt. 2, 833.

Advertised cures for venereal disease. (*a*) "Capsules by Dr. Valfour of Paris—the most radical treatment for gonorrhea." (*b*) "If you still suffer from gonorrhea." (*c*) "You are suffering from syphilis and the thought does not give you a minute's rest." (*d*) "Quick, permanent cure for syphilis—gluten tablets, without a doctor's help!" (*e*) "Full cure for syphilis, Professor Taylor's capsules." *Satirikon*, no. 21 (1908), nos. 6–8 (1909). Helsinki University Slavonic Library.

gence, doing so would permit the common folk to err in a manner that allowed for correction through self-control. What doctors rejected was the objectification—whether social, political, or physical—of the victims of misfortune, an attitude that left them hope neither of sin nor of redemption.

Свистопляска вокругъ „606“.

„Противоядіе
пайдено!“

"The Devil's Sabbath around '606.'" Figures from urban society cavort around a bottle of Erlich's Medicine, called "606," a new discovery that promised to cure syphilis. *Iskry* , no. 29 (1910). Helsinki University Slavonic Library.

After 1905, medical interest in nonvenereal syphilis as a matter of social policy seems to have succumbed to a new focus on clinical strategy. It is perhaps a coincidence, but the leading medical journal, *Vrach* (called *Russkii vrach* after the death of its founding editor in 1901), which between 1890 and 1905 had teemed with articles on the social context of syphilis, shifted thereafter to technical questions of individual diagnosis and treatment—especially following the 1910 development of the arsenical drug Salvarsan. By then, however, increased peasant mobility and the extent of popular engagement in the revolution of 1905 had shaken the physicians' image of the common folk as victims, not agents, of misfortune and disease. Belief in the moral integrity of village life, which had sustained the diagnostic distinction between venereal and nonvenereal syphilis, could not long withstand the evidence of upheaval and the sense of social crisis generated by the revolutionary events.

Medical interest in syphilis seems also to have peaked in Europe around the turn of the century, for reasons that are not entirely clear.[168] What is clear is the distinctive symbolic function of syphilis in Russia, as opposed to continental Europe, during the heyday of rhetorical alarm. Insofar as Russian physicians adopted the European association between syphilis and illicit sex, it was in keeping with a view of their own cities as islands of Westernized culture.[169] These islands were colonized by wanderers from the domestic mainland of communal life, who acquired by virtue of their displacement attributes of autonomy they had never previously possessed. Regulated by biological rhythms and traditional norms, village sexuality had engulfed the traces of desire in the conspicuous cycle of conception, pregnancy, and lactation—the elements of reproduction and also of birth control. But when the newcomers were cast adrift or merely repositioned in the big city, their sexual desire and its consequences leaped more readily into public view. Urban practitioners, worried about the integrity of working-class family life and the risks of prostitution, devised strategies of medical control. At the same time, however, Russian physicians maintained an alternative, desexualized interpretation of syphilis which corresponded to notions of class and gender appropriate to the traditional framework of rural life. And just as they failed in such cases to link disease and moral disarray, so they rejected the role of moral arbiter assumed by their colleagues in the West.

[168]Corbin, "Le Péril vénérien," and Gérard Jacquemet, "Médecine et 'maladies populaires' dans le Paris de la fin du XIXe siècle," in Murard and Zylberman, *L'Haleine des faubourgs*.

[169]Gertsenshtein, "K statistike," 335, speaks of urban syphilis as "taking the form common in Western European countries."

PART TWO

CONFRONTING DISORDER:
THE WIDENED PUBLIC FIELD

Chapter Six

Eros and Revolution: The Problem of Male Desire

The revolution of 1905 was both an organized political event—a deliberate challenge to established principles of public life—and an inchoate episode of violence, social disorganization, and personal release. The popular classes revolted against the social hierarchy (workers protested the factory regime, peasants defied the power of landlords); educated society, and some workers as well, explicitly challenged the tsar's politics of repression. On all levels the cry was for greater autonomy, for freedom of expression, for the right to self-regulation—for an end to tutelage and police control.

Everywhere social order deteriorated: violence, intimidation, disregard for the law were common on both sides of the barricades, as each camp sought to impose its political ideal.[1] Despite their disruptive tactics, however, the revolutionaries saw themselves as champions of the common good, which they strove to embody in formal structures of orderly civic life. Social Democrats in their slogans and retrospective interpretations stressed the power of collective mobilization, the triumphs exacted on the streets, in meetings, on the factory floor: in short, the force of organized public initiative. Liberals, for their part, transformed informal associations and personal influence into political instruments in the service of public goals and later hailed the institutional fruits of the revolutionary process. Across the board, protesters denounced the government's arbitrary exercise of power and proclaimed their dedication to more rational, and therefore more just, political forms. Enemies of the revolution, by contrast, saw in the massive

---

[1]For descriptions of the general level of violence and disorder, see Abraham Ascher, *The Revolution of 1905: Russia in Disarray* (Stanford, Calif., 1988).

popular challenge the same elements of coercion and unreason that their adversaries identified in the ruling regime.

In the years after 1905, when political convictions wavered but social turmoil refused to subside and public life failed to settle into an acceptable routine, members of educated society—intellectuals, social scientists, political commentators, artists—continued to meditate on themes of transgression, disorder, chaos, and desire. These elements had a private as well as a public dimension. By 1906 or 1907 the anxiety they generated had crystallized into a self-proclaimed sexual crisis, a "sexual question" with the same rhetorical status as the more familiar "woman question" and "social question." "For the last ten to fifteen years," wrote the psychiatrist Aleksandr Bernshtein in 1908, "the sexual question has been studied from all possible angles—biological and economic, social and moral, juridical and pathological, hygienic and religious; and conversely, questions of ethics and law, political economy, and philosophy have been viewed from the viewpoint of sexual life." Sex talk, so this physician liked to believe, was not a frivolous indulgence but a matter of exalted social concern.[2]

The period in which the sexual question emerged as a subject of concentrated public debate coincided with the years of widespread political mobilization. In these years as well, themes that had earlier preoccupied professionals in the context of their own disciplinary concerns began to attract attention in settings open to a more general audience. Though many of the components of the sexual question drew on existing discursive models, the intensity of interest and the subject's heightened political charge were peculiar to the postrevolutionary decade.

The roots of the rhetorical crisis reached back into the 1890s. The political assertiveness of educated society—beginning with the liberation movement, radicalized by the mobilization of university youth, and culminating in the alliance with popular forces—spelled the end of an era of elite self-restraint. Toward the turn of the century, as students began openly to defy public authorities and highly placed men engaged in acts of deliberate insubordination, psychiatrists, pedagogues, and physicians analyzed the dangerous consequences of a repressive political regime in metaphors borrowed from the rhetoric of sexual science. They focused on the classroom both as a microcosm of the larger civic world and as a breeding ground for subservience and disaffection. Only a regime of disciplined rather than prohibited expression, they believed, could enlist the potentially disruptive energy of desire—as the current police order could not do—in the service of civic virtue.

---

[2]A. N. Bernshtein, *Voprosy polovoi zhizni v programme semeinogo i shkol'nogo vospitaniia* (Moscow, 1908), 5. For a similar statement, see Pavel Kokhmanskii, *Polovoi vopros: Razbor sovremennykh form polovykh otnoshenii* (Moscow, 1912), intro.

The revolution itself seemed to confirm this prediction. Sympathetic professionals, looking back, depicted the period of political confrontation as one in which libidinous impulses had shattered existing constraints but were effectively subordinated to the cause of constructing a new political order; in the heat of struggle, self-restraint had mastered aimless, volatile emotion. In the view of such observers, however, the retreat to the humdrum routine of legitimate political process from the explosive climax of December 1905 had reversed the process of sublimation: as the collective movement passed, people sought avenues of personal gratification. The erotic hunger that had been both stimulated and suppressed by the intolerant old regime and temporarily diverted by the revolution now emerged in all its force.

The sexually coded literature about 1905 and its unheroic aftermath reveals a deep anxiety not only about the unruly conduct of the lower classes but about the volatility apparent in the writers' own social world. The self-reflexivity of the "sexual question" is apparent in the preoccupation with male sexuality that characterized medical and pedagogical literature from the 1890s and through the revolution's decline. Attention to the development of male sexual control provided at once a metaphor for the maturation of political life and a program for civic intervention. In a kind of pedagogical analogy to Lenin's famous "spontaneity-consciousness" paradigm, rebellious schoolboys stood for the angry populace, prey to "elemental" urges (Lenin's untutored masses); university students, like organized workers, subordinated the power of mature but untested desire to the discipline of ideological direction; adults, finally, were capable of deliberate restraint, the self-denial and internalized control necessary if responsible citizens were to stay the long and ungratifying course of institutionalized political life or activists to lead the party.

The exclusive maleness of the terms in this developmental sequence underscores the revolution's failure to provoke images of disruptive female passion. The sexually colored anxiety about disorder in the schools and about what kind of knowledge stimulates desire and what kind keeps it in check did not concern women. It was aroused by visions not of popular rebellion but of upper-class disarray. The dangers of uncontrolled female sexuality took rhetorical precedence only when the mass movement had lost its momentum and social antagonisms no longer took organized form. Such contrasts indicate that the actual subject of pedagogical discourse was not a universal male sexuality. Rather, educators wished to train the nation's male cultural elite in the self-control necessary for civic responsibility: that selfless public devotion symbolized by the ability to inhabit the streets without succumbing to the lure of sensual gratification.

Prostitution had long represented the danger of pollution from the social depths, of corruption from within the social body. Now, beyond the persistent concern to regulate and sanitize the supply side of this insidious exchange across class boundaries, a newly intensified interest in disciplining and purifying male demand developed. In the 1890s, adult men were the objects of individual case studies, and continence emerged as a moral and social ideal; then classrooms provided the occasion for physical exams, punitive interrogations, and instruction in sexual self-control; finally, as the political movement reached its height, universities served as laboratories for a sociology of sexual behavior. The results of sex surveys published after 1905 constituted a moral profile of the coming generation. As the political arena opened up, sexual discipline changed from a matter of personal comportment and individual therapy to a science of collective life.

In relation both to their inferiors and to themselves, the professionals' anxiety about sexual and social disorder focused on the problem of youthful insubordination, whether in the form of juvenile prostitution or of classroom chaos. The preoccupation with youth operated on two levels at once. In a practical sense, it recognized that young people had played a central role in the radical movement. By prompting concern over the future of national political life, the new generation offered its elders an occasion to exercise their professional expertise in the public interest. Teachers and physicians now looked for ways to influence and discipline rebellious children and young adults that conformed to the values they themselves wished to substitute for the principle of state coercion. On another plane, the revolt of youth stood metaphorically for adult society's own rebellion against the paternalist political regime that kept it too in a state of dependence. In that sense, this revolt provided observers with a point of identification for their own dilemma, rather than a foil against which to exercise their powers of control.

Sexual Prelude

If there was a moment when the sexual question first became a self-consciously defined cultural issue, it was in 1889, when the manuscript of Leo Tolstoy's provocative novella *The Kreutzer Sonata*—which condemned sexual passion even within marriage as corrupt and degrading—began to circulate among the cultivated elite.[3] The authorities did not grant permission for the tale's publication until 1891, when they

---

[3]My discussion relies on Peter Ulf Møller, *Postlude to "The Kreutzer Sonata": Tolstoj and the Debate on Sexual Morality in Russian Literature in the 1890s*, trans. John Kendal (Leiden, 1988).

finally sanctioned its inclusion in the thirteenth volume of the author's collected works, hoping thereby to limit its sale. Until then, the censors also restricted discussion of the work in the press. The text, the government insisted, was both offensive and dangerous, a challenge to respectability and religion in the crudeness of its language and in its assault on the moral basis of family life.

The nature of Tolstoy's sexual preoccupations and the notoriety of this particular work are well known. *The Kreutzer Sonata* soon achieved an extraordinary circulation, and its impact was felt throughout the literary and professional worlds. "In a few weeks," wrote Vasilii Rozanov, "it had stirred up all of Russia."[4] For two years a version of the manuscript passed from hand to hand and from city to city; it was recopied, lithographed, and read aloud to audiences assembled in private apartments. At the first such gathering in the fall of 1889, the text was read by the well-known legal activist and judge Anatolii Koni. A canvas by Grigorii Miasoedov, exhibited in 1893, depicted a group of men and women, ranging from youth to old age and including a panoply of Russia's most eminent cultural figures, listening intently as the story was read by a young man under the light of a shaded lamp in a plush bourgeois drawing room.[5] An epilogue written in 1890, further detailing Tolstoy's views, circulated in the same manner. Tolstoy received dozens of letters from readers and listeners, many of them women, describing the enormous impact of the work. By the time the text appeared in print, it was well known to the educated public, and even those who had missed the opening round could not avoid the heated debates that followed in the major newspapers and journals of the day.

For the first audiences, as the critic Peter Ulf Møller points out, the experience was intensified by the public circumstances in which it occurred and the lively discussions that ensued.[6] The setting of the tale's early reception thus echoed the structure of the story itself. Traveling by train, the narrator enters into a discussion of current sexual mores with the other people in his compartment. These chance companions represent a broad spectrum of social types: a modern woman intellectual and a lawyer, both of whom defend free love and divorce; an old merchant of traditional views; and a respectable though distraught gentleman called Pozdnyshev with a confession to make. This last man recounts to the narrator the story of his courtship, marriage, domestic

---

[4]V. V. Rozanov, *V mire neiasnogo i ne reshennogo* (St. Petersburg, 1904), 52. See also comments in S. F. Sharapov, ed., *Sushchnost' braka* (Moscow, 1901), 1.

[5]The picture is reproduced in Møller, *Postlude*, 98.

[6]Ibid., 96–97.

disillusionment, and finally murder of his wife on the suspicion of adultery. A crime of outraged honor, the court had decided, in setting Pozdnyshev free. Sex, even within marriage, Tolstoy made clear, demeaned both partners. In the grip of lust, the man created a wife who mirrored his own fantasies and desires, while she used her sexual charms in the war for personal domination. Not love but power was the issue at stake.

This tale of private passion, of sexual and domestic intimacy, is retailed to a stranger in a public setting both intimate and anonymous, just as the manuscript was read in worldly company, outside the family circle or the confines of the listener's own easy chair. The story, about the fatal consequences of male desire, is told from the husband's point of view to another man. Listening in, as they did in mixed audiences, real women intruded on this sexually exclusive exchange by writing privately to Tolstoy: some thanked him for bringing intimate subjects to public attention and cited their own domestic unhappiness in confirmation of his views; others objected to his dismal opinion of married love.[7] Once the discussion shifted from drawing rooms and personal letters to the printed page, however, almost all of the speakers were men.[8]

Even before the discussion issued into print, The Kreutzer Sonata, as text and cultural event, had become the touchstone for an unfolding debate on the "sexual question." It elicited literary and philosophical responses (from Anton Chekhov, Vladimir Solov'ev, and Vasilii Rozanov, to name a few) and attack from all sides: from spokesmen of 1860s radicalism, for denigrating the possibility of sexual equality and women's intellectual development and for seeking to solve sexual problems in personal rather than social terms; from the church, for disparaging the Christian sacrament of marriage and the value of procreation; from moralists who promoted premarital chastity but found the idea of total abstinence extreme; and later from modernist writers who celebrated beauty rather than didacticism in art and pleasure rather than virtue in personal life. Nearly twenty years after the manuscript first appeared, the feminist physician Mariia Pokrovskaia remembered the story as a call for male sexual continence, unions based on mutual respect rather than sensual desire, and sex for procreation only.[9]

If Pokrovskaia in 1908 somewhat distorted the story's message by

---

[7]On the letters of female correspondents, see ibid., 115–21. An indignant letter from a happily married woman is cited in V. A. Zhdanov, "Iz pisem k Tolstomu (Po materialam Tolstovskogo arkhiva)," Literaturnoe nasledstvo, no. 37–38 (1939): 384–86.

[8]Møller, Postlude, 117.

[9]M. I. Pokrovskaia, "Kreitserova Sonata," Zhenskii vestnik, no. 9 (1908): 193–96. No doubt because of its late date, this response is not mentioned in Møller.

mistaking its radical antisex position for Tolstoy's earlier, more limited endorsement of premarital restraint and his former celebration of women's procreative role, she did not err in emphasizing the theme of male sexual continence.[10] This preoccupation was not a personal quirk of either the feminist or the writer. It can be traced, at least since the 1860s, in the pedagogical and medical press, where it surfaced in connection with problems of sexual pathology (particularly masturbation), venereal disease, and the twin "evils" of abortion and artificial birth control.

Continence

Contrary to Tolstoy's impression, few Russian doctors endorsed sexual indulgence for either sex.[11] Indeed, they cited the novelist's authority in favor of male self-control,[12] which even Veniamin Tarnovskii, that ardent partisan of state-regulated prostitution, considered the ideal: "The young man who emerges victorious in the battle with his sexual desires not only protects his own health but strengthens his willpower and gains the moral independence necessary to succeed in the struggles of life."[13] To whatever extent moral purists dominated the Russian medical community, certainly those in favor of sexual abstinence made the loudest noise.[14] Their argument already held sway in

[10]See Møller, *Postlude*, 28–29; and John M. Kopper, "Tolstoy and the Narrative of Sex: A Reading of 'Father Sergius,' 'The Devil,' and 'The Kreutzer Sonata,'" in *In the Shade of the Giant: Essays on Tolstoy*, ed. Hugh McLean (Berkeley, Calif., 1989), 162. The 1890 Epilogue to the *Sonata* presented Tolstoy's views in their most explicit and extreme form, but they were not without contradiction. Complete abstinence, he said, was the Christian ideal. Even intercourse for purposes of procreation violated this ideal, especially since there were already more than enough children in the world. Tolstoy nevertheless took a somewhat pragmatic comfort in the thought that the ideal was no more in danger of being realized on this earth than other high-minded aspirations; see his "Posleslovie," in *Polnoe sobranie sochinenii L'va Nikolaevicha Tolstogo*, 10: 107–17 (Moscow, 1913).

[11]Tolstoy assailed regulated prostitution as an incitement to debauch and said that doctors, the "priests of science," sent men to brothels to improve their health: *The Kreutzer Sonata*, in Leo Tolstoy, *The Death of Ivan Ilych and Other Stories* (New York, 1960), 170. He attacked contraception, supposedly promoted by medical men, as close to murder: "Posleslovie," 108.

[12]See, e.g., G. Rokov, "Bol'noi vopros vospitaniia," *Vestnik vospitaniia*, no. 7, pt. 1 (1902), 70; V. M. Bekhterev, *O polovom ozdorovlenii* (St. Petersburg, 1910), 5, 7.

[13]V. M. Tarnovskii, in *Trudy Vysochaishe razreshennogo s"ezda po obsuzhdeniiu mer protiv sifilisa v Rossii* (St. Petersburg, 1897), 2:15.

[14]For an overview of contemporary views in favor of continence, see A. V. Favorskii, "O polovom vozderzhanii," *Russkii zhurnal kozhnykh i venericheskikh boleznei*, no. 3 (1905): 257–61. Popular literature promoting abstinence includes L. A. Zolotarev, *Chto govorit nauka o polovoi potrebnosti: Populiarno-nauchnyi ocherk dlia roditelei, vospitatelei i uchashcheisia molodezhi* (Moscow, 1900); idem, *Gigiena supruzheskoi zhizni* (Moscow,

the 1890s, and they continued their agitation right up to the eve of World War I.[15] Some physicians proposed male continence as a way to stop the spread of venereal syphilis.[16] Others emphasized the unhealthy consequences of nonprocreative sex in all its forms, whether intercourse with prostitutes, masturbation, or the use of birth control. A doctor's pamphlet on the physical consequences of depravity blamed "civilized man" for degrading the sexual act by separating it from reproduction. Contraception, like masturbation, made people antisocial, morose, unhealthy.[17] Sex for mere pleasure was not only immoral but unnatural, insisted Anatolii Sabinin, a liberal Voronezh public health physician and prominent medical editor. Too much, not too little sex was dangerous for both women and men, he said: "Spermatorrhea, impotence, constipation in girls, abnormal menses, neurasthenia, neuroses and psychoses, and many other functional disturbances are the results of masturbation, early sexual activity, sexual excitement—in short, of physical and psychological depravity."[18] By contrast, "moderation, and even complete abstinence from sexual relations, does nothing but good," wrote Aleksandr Virenius, a St. Petersburg school physician and public health advocate, in the journal of the Ministry of Education. "The energy wasted on sexual release is well employed in other functions, especially those involving the brain."[19]

In 1905 a St. Petersburg physician named Liudvig Iakobzon set out

---

1901); P. P. Viktorov, *Gigiena i etika braka v sviazi s voprosom o polovoi zhizni iunoshestva* (Moscow, 1904); *Grekhi molodykh liudei: Nastol'naia kniga*, 3d ed. (Moscow, 1906); S. Zalesskii, *Polovoi vopros s tochki zreniia nauchnoi meditsiny: Gigienicheskii etiud* (Krasnoiarsk, 1909); *Nastol'naia kniga dlia molodykh suprugov s polnym izlozheniem pravil supruzheskoi zhizni* (Moscow, 1909).

[15]K. P. Sangailo, "Polovoi vopros i shkola," *Pedagogicheskii sbornik*, no. 3 (1913): 341; and M. I. Pokrovskaia, *O polovom vospitanii i samovospitanii* (St. Petersburg, 1913). I am grateful to Laurie Bernstein for allowing me to read her notes on Pokrovskaia's text before I was able to see it.

[16]E.g., P. I. Kovalevskii, "Sifilitiki, ikh neschast'e i spasenie," *Arkhiv psikhiatrii, neirologii i sudebnoi psikhopatologii*, no. 1 (1897): 65 (henceforth *Arkhiv psikhiatrii*); M. I. Pokrovskaia, *Vrachebno-politseiskii nadzor za prostitutsiei sposobstvuet vyrozhdeniiu naroda* (St. Petersburg, 1902), 84; V. M. Tarnovskii, *Bor'ba s sifilisom* (1897), cited in A. Kh. Sabinin, *Prostitutsiia: Sifilis i venericheskie bolezni* (St. Petersburg, 1905), 224; L. Ia. Iakobzon, "Kakimi merami sleduet borot'sia s rasprostraneniem venerichesikh boleznei sredi uchashchikhsia," *Russkii vrach*, no. 43 (1903): 1510.

[17]V. S. Iakshevich, *Plody razvrata* (St. Petersburg, 1904), 16; S. V. Filits, "Sovremennaia polovaia zhizn' s meditsinskoi tochki zreniia," *Meditsinskaia beseda*, no. 3 (1900): 78, 80. See also Sabinin, *Prostitutsiia*, 215; and Zolotarev, *Gigiena supruzheskoi zhizni*, 174, 177–78. For an example of popular scare literature on masturbation, see *Grekhi molodykh liudei*.

[18]Sabinin, *Prostitutsiia*, 215, 224.

[19]A. S. Virenius, "Polovoi vopros: Po povodu sochinenii prof. Avg. Forel', 'Polovoi vopros.' St. Petersburg, 1906," *Zhurnal Ministerstva Narodnogo Prosveshcheniia*, May 1907, 82.

to demonstrate that chastity was the scientifically approved solution to the spread of venereal disease. To discover whether his colleagues agreed, he addressed personal letters to 207 Russian and German physicians, of whom only a few (including Veniamin Tarnovskii and the prominent psychiatrist Emil Kraepelin) bothered to reply—all, not surprisingly, in the affirmative.[20] The weak response suggests that Iakobzon may have exaggerated the extent to which his colleagues dwelt on the issue; nevertheless, his survey, which invoked the authority of foreign as well as domestic experts, was later cited as proof that continence was widely approved by medical authorities.[21]

Many Russian physicians combined insistence on male sexual temperance with a defense of women's rights.[22] In that respect, they indeed differed from the patriarchal Tolstoy. The most prolific exponent of the feminist single standard in Russia was the public health physician Mariia Pokrovskaia. In her tireless war against regulated prostitution, Pokrovskaia struck a Tolstoyan note in stressing the responsibility of clients rather than the moral turpitude of whores.[23] But unlike Tolstoy, she considered chastity the necessary condition for women's equality with men, not an attribute of old-fashioned patriarchal relations. Only

[20]L. Ia. Iakobzon, "Polovoe vozderzhanie pered sudom meditsiny," *Russkii vrach*, no. 18 (1905), 589.
[21]Cited in S. A. Ostrogorskii, "K voprosu o polovom sozrevanii (ego fiziologiia, patologiia i gigiena)," in *Trudy pervogo vserossiiskogo s"ezda detskikh vrachei v S.-Peterburge, s 27–31 dekabria 1912 goda*, ed. G. B. Konukhes (St. Petersburg, 1913), 660–61. Foreign authorities were often invoked in support of continence: A. S. Virenius, "Beseda po voprosu o bor'be s polovymi anomaliiami (onanizmom) uchashchikhsia, dlia roditelei i vospitatelei," *Meditsinskaia beseda*, no. 13–14 (1902): 388–89; V. N. Polovtseva, "Polovoi vopros v zhizni rebenka," *Vestnik vospitaniia*, no. 9, pt. 1 (1903): 27; A. G. Trakhtenberg, "Polovoi vopros v sem'e i shkole," *Voprosy pola*, no. 1 (1908): 30; idem, "Anomalii polovogo chuvstva v shkol'nom vozraste i sistema fizicheskogo vospitaniia," *Voprosy pola*, no. 4 (1908): 25–26; V. Ia. Kanel', "Polovoi vopros v zhizni detei," *Vestnik vospitaniia*, no. 4, pt. 1 (1909): 141, 147, 152, 160. Among the authors routinely cited were esp. Seved Ribbing, Hermann Rohleder, and a Professor Heim of the University of Zurich, plus August Forel and Richard von Krafft-Ebing. See S. Ribbing, *Om den sexuela hygienen och några af dess etiska konsequenses* (Lund, 1888) (trans. 1891); Hermann Rohleder, *Die Masturbation: Eine Monographie für Ärzte und Pädagogen* (Berlin, 1899) (trans. 1901). For Ribbing and Rohleder, see *Biographisches Lexikon der hervorragenden Ärzte der letzten fünfzig Jahre (1880–1930)*, ed. I. Fischer, 3d ed. (Munich, 1962), 2:1291, 1313–14. On Heim's "marvelous brochures," see M. I. Miagkov, "Nekotorye zadachi vospitaniia v sviazi s polovoi zhizn'iu chelovecheskogo organizma," in *Trudy pervogo s"ezda ofitserov-vospitatelei kadetskikh korpusov (22–31 dekabria 1908 g.)*, ed. P. V. Petrov (St. Petersburg, 1909), 322; also Zalesskii, *Polovoi vopros*, ii. For a bibliography of Russian and foreign sources, see N. A. Rubakin, *Sredi knig* (Moscow, 1913), 2:353–55, 424–25.
[22]E.g., Rokov, "Bol'noi vopros," 87–89; L. V. Slovtsova, "Polovoe vospitanie detei," in *Trudy pervogo vserossiiskogo zhenskogo s"ezda pri Russkom zhenskom obshchestve v S.-Peterburge 10–16 dekabria 1908 goda* (St. Petersburg, 1909), 675–78; Bekhterev, *O polovom ozdorovlenii*, 10, 17, 18.
[23]See Leo Tolstoy, *Resurrection* (1899); also his expression of sympathy for down-and-out prostitutes: "Tak chto zhe nam delat'" (1886), in *Polnoe sobranie sochinenii*, 17:33–37.

when people refrained from intercourse for purposes of enjoyment, she argued, would "the sexual enslavement of men and women, which . . . prevents the full development of human individuality, finally disappear."[24]

Pokrovskaia agreed with Tolstoy that men must learn sexual restraint, starting at the earliest possible age, for she believed that passion once awakened was hard to control.[25] She differed, however, in her opinion of women's role in the moral revolution. According to Tolstoy, the modern, freethinking, educated woman—depicted in *The Kreutzer Sonata* as smoking cigarettes, wearing a "mannish coat," and defending divorce—betrayed the true calling of her sex. She was part of the same depraved social world in which other women bared their shoulders and prostituted their drawing room charms for the sake of wealthy matches. The only excuse for sex (and that a feeble one), in the novelist's opinion, was the production of children; the ideal marriage embodied a chaste companionship in which the wife devoted herself to domestic affairs.[26] But Pokrovskaia approved of the modern woman who shouldered public concerns. In 1908 she told the feminist congress meeting in St. Petersburg that women must "act in concert to obliterate the conditions that turn young men into beasts": in private by educating their sons, in public by ridding the street of prostitutes and pornography.[27] Where Tolstoy thought a return to traditional moral standards was the only way to save fallen women and saw the exercise of male self-control as the key to this social rebirth,[28] Pokrovskaia argued for practical reforms, implemented with women's help, for the benefit of women: society must provide poor women with work, education, and welfare services to keep them off the streets and out of the brothels. Though she did not hesitate to blame men for the problem of disorderly desire, she could not leave the solution to them. Indeed, her confidence in men's capacity for self-restraint seemed to diminish with the years. In 1902 she had insisted that sex education aimed at young men, not medical inspection, was the key to stopping the spread of venereal disease. By 1908 she had shifted the responsibility to women's shoul-

[24]See M. I. Pokrovskaia, "Prostitutsiia i bespravie zhenshchin," *Zhenskii vestnik*, no. 10 (1907): 225–31; idem, "Kreitserova Sonata," 195. On the need for moral restraint, see her "Sovremennyi erotizm s fiziologicheskoi tochki zreniia," *Voprosy pola*, no. 4 (March 23, 1908): 30–32. Though Pokrovskaia disapproved of sex for pleasure, like many of her medical colleagues, she defended women's right to abortion; see chap. 9 below.

[25]M. I. Pokrovskaia, "Bor'ba s prostitutsiei," *Zhurnal Russkogo obshchestva okhraneniia narodnogo zdraviia*, no. 4 (1900): 425–27.

[26]See Tolstoi, "Posleslovie."

[27]M. I. Pokrovskaia, "Kak zhenshchiny dolzhny borot'sia s prostitutsiei," in *Trudy pervogo vserossiiskogo zhenskogo s"ezda*, 278–79.

[28]Tolstoi, "Tak chto zhe nam delat'?" 33–37.

ders; their emancipation, she argued, depended on their "having the deciding voice in regulating sexual relations."[29]

As time wore on, Pokrovskaia resorted more often to biological arguments in the defense of sexual restraint, enlisting the latest scientific language in the feminist cause. "The single sexual standard," she wrote in 1910, "by restraining the sexual instinct . . . will free our progeny from the hereditary burden imposed by today's sexual licentiousness. Future generations will be physically and psychologically healthier."[30] Her rhetoric in 1912 was stronger still: "Early sexual activity is largely responsible for the fact that the vast majority of our male students are nervous, debilitated, sickly, and suffer from various mental abnormalities. When they mature, they become veritable beasts and moral monstrosities!"[31] Until the war she continued to denounce the "abnormality" of sexual mores, blaming "benighted" sexual attitudes (by which she meant intercourse without the desire to conceive) for the prevalence of abortion and infanticide and for the deplorable number of materially deprived and genetically defective children.[32]

The Private Vices of Men and Boys

Personal behavior, in such arguments, had enormous power to shape the social body, to alter the quality of public life. Responsible mothers and sexually continent men held the key to civic virtue. This conviction led physicians and educators to focus on the moral and physical health of those young men in a position to heed their advice or to suffer their intervention. Early training, they believed, would produce the kind of adults who would hew to the standards of sexual hygiene deemed essential to the national welfare. The scrutiny and schooling of individual male conduct was not intended for the uneducated masses, which had at best taken only the first steps toward enlightened self-regulation under the intelligentsia's careful guidance but in general still endured the joint tutelage of family, community, and police. Rather, the professionals' interest in moral discipline sharpened as educational opportunities expanded, cities grew, and political activity spilled out of institutional

---

[29]Pokrovskaia, *Vrachebno-politseiskii nadzor*, 84–85. In the same period, however, she called for women to lead the fight against regulated prostitution, saying men's "protection" had gotten them nowhere: *O zhertvakh obshchestvennogo temperamenta* (St. Petersburg, 1902), 48. Cf. idem, "Kreitserova Sonata," 195.

[30]M. I. Pokrovskaia, "Edinaia polovaia nravstvennost'," *Zhenskii vestnik*, no. 4 (1910): 92. She believed that healthy sex (intercourse with a loved one for purposes of procreation) would improve the genetic condition of the race: see Pokrovskaia, *O polovom vospitanii*, 30–31.

[31]M. I. Pokrovskaia, "O prostitutsii maloletnikh," *Zhenskii vestnik*, no. 10 (1912): 196.

[32]M. I. Pokrovskaia, "K voprosu ob aborte," *Zhenskii vestnik*, no. 4 (1914): 105.

corridors and into private clubs, lecture halls, and finally the streets. If not a bourgeoisie in the Western sense, this cultural elite was taking itself in hand.

The concern that peaked in the early years of the twentieth century rested on a tradition reaching back into the eighteenth, when books on child rearing and pedagogy were already enjoying remarkable popularity among the small reading public.[33] The two classic texts on sexual abuse and sexual chastity most frequently cited by Russian educators were, respectively, Simon-André Tissot's dissertation on masturbation (1758) and Jean-Jacques Rousseau's *Emile* (1762). The Russian version of Tissot's volume had attained its fifth edition by 1852.[34] By the mid-nineteenth century the reader might also have acquired various domestic texts, such as the anonymous *Handbook for Men Suffering from Weakness of the Genital Organs Caused by Premature and Excessive Sexual Indulgence, Onanism, Extreme Old Age, or the Effects of Illness* and *No More Onanism, Venereal Disease, Pollution, Male Impotence, or Female Infertility,* which imitated Tissot's combination of sexual diagnosis and home cure.[35]

In his pedagogical guise, Rousseau was embraced by the Russian educational establishment. *Emile* first appeared in Russian translation in 1779, reappeared in three more editions between 1800 and 1820, and yet again in 1866, 1896, 1911, and 1912. The subject of numerous articles in the pedagogical press, this particular work had enormous staying power.[36] In 1908 a St. Petersburg teacher-training program offered a course in "moral instruction according to Rousseau's *Emile*"; experts also invoked his name in discussions of curriculum in the military academies.[37] Educators writing on the eve of World War I still cited Rousseau in promoting male sexual continence as a moral ideal.[38]

[33]See Gary Marker, *Publishing, Printing, and the Origins of Intellectual Life in Russia, 1700–1800* (Princeton, N.J.,1985), 207, 210.

[34]See Aleksandr Nikitin's introduction to S. A. Tissot, *Onanizm ili rassuzhdenie o bolezniakh, proiskhodiashchikh ot rukobludiia,* 5th ed. (St. Petersburg, 1852). For Tissot's influence in the West, see E. H. Hare, "Masturbatory Insanity: The History of an Idea," *Journal of Mental Science* 108:452 (1962): 2–3.

[35]*Vspomogatel'naia kniga dlia muzhchin, strazhdushchikh [sic] rasslableniem polovykh organov, proiskhodiashchem ot slishkom rannego ili slishkom chastogo udovletvoreniia fizicheskikh pobuzhdenii liubvi, ili ot onanizma, ili ot preklonnoi starosti, ili vsledstvie boleznei: Sochinenie prakticheskogo vracha* (Moscow, 1857); *Net bolee onanizma, venericheskoi bolezni, poliutsii, muzhskogo bessiliia i zhenskogo besplodiia: Prakticheskie sredstva snova vosstanovliat' i ukrepliat' zdorov'e, rasstroennoe etimi bolezniami,* 2d ed. (Moscow, 1865).

[36]M. I. Demkov, *Istoriia russkoi pedagogii,* pt. 3, *Novaia russkaia pedagogiia (XIX vek)* (Moscow, 1909), 474–76; N. Bakhtin, "Russo i ego pedagogicheskie vozzreniia," *Russkaia shkola,* no. 12 (1912): 116–17, 119–21.

[37]I. S. Simonov, "Shkola i polovoi vopros: Iz dnevnika byvshego ofitsera-vospitatelia," pt. 1, *Pedagogicheskii sbornik,* no. 1 (1908): 25; "Protokol zasedaniia obshchepedagogicheskogo otdela pedagogicheskogo muzeia voenno-uchebnykh zavedenii 8 fevralia 1908 g.," *Pedagogicheskii sbornik,* no. 5 (1908): 432.

[38]Sangailo, "Polovoi vopros," 335, 341. Other references to Rousseau include Bekhterev, *O polovom ozdorovlenii,* 12; and Pokrovskaia, *O polovom vospitanii,* 5–6.

Moral instruction first became of particular interest to the military authorities in the 1870s, when the reforming war minister Dmitrii Miliutin focused on the social uses of education. His plans to improve the army by imposing literacy on enlisted men are well known; the ministry also supported the expansion of women's higher education. Recognizing that the Great Reforms demanded open discussion of public issues, the central directorate of military academies established a pedagogical journal in 1864. A few years later, the journal published an essay on the problem of "onanism" in cadet boarding schools.[39]

The themes articulated in this early article reflected attitudes that were common among physicians in the West and repeated by Russian writers into the next century.[40] Boys were said to acquire the habit either by touching themselves accidentally or through the corrupting influence of friends at school and of maidservants and nurses at home (nannies were blamed, for example, for rubbing the penis of a baby boy to put him to sleep). Masturbation was then encouraged by the classroom environment: lack of exercise, too much sitting still over one's books, tight trousers. The chronic abuser showed characteristic signs suggesting disturbances of the masculine attributes: weakness, pallor, cowardice, lack of concentration, and certain disfigurations of the penis (retracted foreskin, according to this observer; rashes or enlargement, according to others). Consequences of the vice included various mental and physical debilities, from headaches to early death. Its remedies were constant supervision, physical inspection of the boys' private parts, sober public discussion in a scientific light, exercise and games.

On the question of who gave in to the vice, writers typically fell into contradiction. On the one hand, they declared it a disease that afflicted all children, no matter what their social class; on the other, they considered it a product of idleness and luxury, certain to be less common among the hardworking poor.[41] The masturbation literature echoed the rhetoric on prostitution: both practices were said to be ancient in origin but exacerbated by the recent progress of civilization. In the case of prostitution, the urban environment was to blame; in the case of masturbation, the increase in private luxuries—rich food, soft beds, material comforts. In a country where the archetypal laborer, the peasant, represented a supposedly uncorrupted historical past, the imagined class

[39]V. Ditman, "Tainyi porok," pts. 1–3, *Pedagogicheskii sbornik*, nos. 3–5 (1871): 367–75, 551–56, 654–63. On *Pedagogicheskii sbornik*, see Demkov, *Istoriia*, 3:468.

[40]For Tissot's view of onanism as a disease with physical and mental consequences, see Hare, "Masturbatory Insanity." On the nineteenth-century literature, see H. Tristram Engelhardt, Jr., "The Disease of Masturbation: Values and the Concept of Disease," *Bulletin of the History of Medicine* 48 (1974): 234–48. A Russian example is *Net bolee onanizma*, 13 (the vice can be read on the countenance); 42 ("masturbation is the surest if not the most direct road to death").

[41]Ditman, "Tainyi porok," pt. 1, 367, 371.

distribution of vice marked Russia's current stage in the process of social transformation.

Most of the scientific literature designed to interpret and regulate sexual conduct concerned socially privileged men. Like prostitution and sexual promiscuity, which physicians could not acknowledge as features of rural life, masturbation emerged as a problem that plagued the elite, not the masses. Thus the military authorities commissioned Veniamin Tarnovskii to write a textbook on sexual hygiene for use in cadet schools, where the nation's future officers were trained. Written for an educated readership and supposedly appropriate for the boys themselves, *Sexual Maturity: Its Evolution, Deviations, and Diseases* had sold out the entire 1886 edition of 6,000 copies by 1891, when a second edition was issued (just as the published version of *The Kreutzer Sonata* appeared in bookstores). Though it remained a standard citation in the professional press, by 1908 the work was hard to find.[42] By then, the obsessed reader or concerned schoolteacher had a proliferation of expert texts to choose from.

As befitted his speciality, Tarnovskii devoted fully two-thirds of the book to the symptoms and prevention of venereal diseases, with particular emphasis on syphilis as a personal and social affliction. For the rest, he inveighed against early sexual activity and commercial sex, counseled restraint, and denounced masturbation. Women figured in this textbook only as the infected prostitutes of whom young men should beware and as the wives whose health they might endanger. Indeed, Tarnovskii never wrote a book instructing respectable women in sexual matters but discussed female sexuality only in connection with the regulation of prostitution.[43] Another subject neglected in his discussion of "sexual maturity" was male homosexuality, which, like prostitution, he considered a pathological deviation from the norm. His earlier volume devoted entirely to that subject, in contrast to his textbook on sexual hygiene, was designed for the edification of specialists like himself, not for the general reader.[44] Perhaps both of the groups Tarnovskii excluded from his concept of the lay public—homosexual men and respectable women—lacked the maleness requisite for responsible, self-regulated lives. They could profit from expert tutelage but could not be trusted with knowledge they might manipulate on their own.

Psychiatry, not public health, was the discipline that claimed perver-

---

[42]V. M. Tarnovskii, *Polovaia zrelost', ee techenie, otkloneniia i bolezni*, 2d ed. (St. Petersburg, 1891), iii, xv. On rarety, see Simonov, "Shkola i polovoi vopros," pt. 3, *Pedagogicheskii sbornik*, no. 5 (1908): 419.

[43]See V. M. Tarnovskii, *Prostitutsiia i abolitsionizm* (St. Petersburg, 1888).

[44]V. M. Tarnovskii, *Izvrashchenie polovogo chuvstva: Sudebno-psikhiatricheskii ocherk dlia vrachei i iuristov* (St. Petersburg, 1885).

sion for its special domain, producing a literature comprising case histories rather than classroom manuals, mass surveys, and statistical reports. The first articles on male homosexuality appeared in the newly fledged psychiatric press in the late 1890s. In 1897 a junior professor at the Academy of Military Medicine described a case of adolescent homosexual attraction associated with other fetishistic tastes.[45] The following year the eminent Vladimir Bekhterev offered two case histories of older men suffering from homosexual desires. Bekhterev supplied a narrative supposedly written for his professional benefit by one of these patients, detailing the writer's sexual experiences in agonized and extremely explicit terms, including his childhood initiation into genital pleasure by the family dog, who managed to suck on his penis.[46] The journal published two other such clinical reports in the same year, one telling of a man who progressed from masturbation to homosexual relations, the other describing the sexual excitation caused by the subject's exposure to horses, an affliction purported to have been cured by doses of "Coca and Cola."[47] Other curiosities examined during this period included the case of a man who obtained pleasure from female urination and that of a youth who loved only elderly women.[48] All these stories, featuring exotic or almost laughable details, concerned male patients from whom the narrating physicians felt as removed and alien as they did from female subjects. The exoticism of male homosexuality was underlined several years later in a spate of articles on its prevalence among non-Christian men in the Caucasus region.[49]

Leaving the sexual margin to the care of "experts,"[50] educators—and some doctors as well—turned their attention in these years to a familiar

---

[45]P. Ia. Rozenbakh, "K kazuistike polovogo izvrashcheniia," *Obozrenie psikhiatrii, nevrologii i eksperimental'noi psikhologii*, no. 9 (1897): 652–56; henceforth *Obozrenie psikhiatrii*.

[46]V. M. Bekhterev, "Lechenie vnusheniem prevratnykh polovykh vlechenii i onanizma," *Obozrenie psikhiatrii*, no. 8 (1898): 587–97.

[47]S. A. Liass, "Izvrashchenie polovogo vlecheniia," *Obozrenie psikhiatrii*, no. 6 (1898): 415–16; A. N. Uspenskii, "K kazuistike anomalii polovogo chuvstva," *Obozrenie psikhiatrii*, no. 12 (1898): 927–28 (the intriguing reference to "Coca and Cola," in English in the text, is not explained).

[48]S. A. Sukhanov, "K kazuistike seksual'nykh izvrashchenii," *Nevrologicheskii vestnik*, no. 2 (1900): 164–68; L. V. Blumenau, "K patologii polovogo vlecheniia," *Obozrenie psikhiatrii*, no. 1 (1902): 17–22.

[49]Grigorii Iokhved, "Pederastiia, zhizn' i zakon," *Prakticheskii vrach*, no. 33 (1904): 871–73; E. V. Erikson, "O polovom razvrate i neestestvennykh polovykh snosheniiakh v korennom naselenii Kavkaza," *Vestnik obshchestvennoi gigieny, sudebnoi i prakticheskoi meditsiny*, no. 12 (1906): 1868–93 (henceforth *Vestnik obshchestvennoi gigieny*); A. Shvarts, "K voprosu o priznakakh privychnoi passivnoi pederasti (Iz nabliudenii v aziatskoi chasti g. Tashkenta)," *Vestnik obshchestvennoi gigieny*, no. 6 (1906): 816–18.

[50]This professional division is made explicit in N. M—ii [N. K. Mikhailovskii], "Bor'ba s polovoi raspushchennost'iu v shkole," pt. 1, *Russkaia shkola*, no. 7–8 (1907): 27; Simonov, "Shkola i polovoi vopros," pt. 1, 26.

subject with which they identified completely: male heterosexuality. Whereas psychiatrists analyzed the misery of their homosexual patients in clinical terms, on the basis of testimony offered by the afflicted, colleagues who dealt with the sexual norm spoke in their own voices. At first they borrowed the case history format, as though otherwise they could not achieve the distance necessary to reveal such sensitive information. Though the data were limited to their own experience, they nevertheless tried to convert their individual lives into statistical sources, moving from the anecdotal treatment reserved for the exotic toward the quantifiable world of the everyday. Under cover of anonymity and cloaked in the dignified mantle of scientific prose, two physicians published autobiographical accounts of their personal sexual histories in the preeminent medical journal *Vrach*: "On the Question of Sexual Intercourse" (1894) and "Personal Statistics on Sexual Need in Relation to the Study of the Sexual Life of Society" (1898).[51]

There is no way to know how idiosyncratic these two men were, but they offered themselves as models of male genital performance. The author of the first article (whom we shall call Dr. A) was born in 1853. He first had intercourse at age twenty-one and married five years later, in 1879. In 1886, after the birth of his last child, he had begun to record every instance of sexual intercourse with his wife, a practice he continued for the eight years up to the article's publication. The author of the second article (whom we shall call Dr. B) was sixteen years the other's senior. Born in 1837 (and thus Veniamin Tarnovskii's exact contemporary), he first had sex at the age of twelve, became sexually active as a student at eighteen, and married at thirty; six years later, in 1873, he began to record his sexual habits, following the example of several friends. He could thus describe in statistical detail twenty-five continuous years of sexual activity. Despite the difference in their ages, the two men shared a fascination with numbers. Dr. A enumerates the month, day, and hour of sexual intercourse over the eight years in one impressive table. Dr. B charts his frequency of intercourse according to the season, days of the week, phases of the moon, and period of his life. Both insist that scientific reports must replace commercial texts catering to prurient interests; both urge their colleagues to undertake more systematic study of normal sexual life.

Though he asserts that his purpose in revealing his own patterns is to help establish a standard of male sexuality to which other men may refer, Dr. A considers his premarital habit of sexual binges (*izlishestva*)

[51]"K voprosu o polovykh snosheniiakh," pts. 1–2, *Vrach*, nos. 1–2 (1894): 8–12, 38–42; and "Lichnaia statistika polovoi potrebnosti v primenenii k issledovaniiu polovoi zhizni obshchestva," *Vrach*, no. 6 (1898): 154–58.

abnormal (*nenormal'na*). Looking back from the age of forty, he notes
that the level of his desire has gradually declined, and his jealous tendencies have correspondingly weakened. From a peak of eighty-five instances of intercourse a year in 1886, he and his wife had fallen to a
mere forty-four in 1893, or from approximately every four days to
every eight—still in excess of his proclaimed ideal of once every ten to
fifteen days for people between thirty and forty. Sleeping in the same
bed, as the couple had done at first, unfortunately increased their desire
but also established a moral rapport and psychological intimacy of
which he approved. They rarely made love in the daytime but usually
in the late evening; at night, he notes, "mutual intimacy aroused us
both." Beyond this brief mention of his wife's desire, Dr. A describes
their pattern as a reflection entirely of his own passions, schedule, and
age. He deeply regrets what he calls the excessive indulgence of his
early years, from which he claims he now suffers nervous ills, irregular
heartbeat, depressed energy, and loss of appetite and memory.[52]

Life, as he recalls it, has been a constant struggle against desire.
Though he rejects complete abstinence as unhealthy (he cites the example of old maids, who lead "damaged" lives), he counsels chastity before the age of twenty-three, when, he believes, the educated person
should first think of marriage. Men should be held to the same moral
standard as women, to avoid such disasters as the one depicted in *The
Kreutzer Sonata*. He disagrees with certain moralists (clearly a reference
to Tolstoy) who think sex itself is harmful; on the contrary, moderate
indulgence strengthens the marriage bond by retaining the spouses' interest in each other. Physical labor is the key to keeping one's desires in
check. The luxury of idleness produces the hyperactive fantasy and untapped physical resources that fuel a sexual self-indulgence that the
needy and careworn common folk cannot afford: "Physical labor is absolutely necessary for the healthy organism and also an excellent regula-
tor of sexual life." He himself gardens and swims for prophylactic reasons. "Equally important for maintaining a correct sexual life," he adds,
"is public service [*obshchestvennaia deiatel'nost'*] that entails direct contact
with the poor. Such activity acts as an excellent corrective to a shattered
imagination, often misdirected from childhood, which is sometimes
one of the main causes of overly frequent sexual relations." Articulating
a belief common to his generation, he suggests that social commitment
will diminish the privileged man's self-absorbed quest for pleasure.[53]
Exposure to injustice and dedication to the common good can deflect
his appetite for personal gratification.

[52]"K voprosu o polovykh snosheniiakh," pt. 1, 8–10; pt. 2, 38, 42.
[53]Ibid., pt. 1, 12; pt. 2, 41–42. See also Pokrovskaia, *O polovom vospitanii*, 49.

Dr. B, of a slightly earlier generation, takes a more optimistic view. Condemning traditional, patriarchal sexual mores, which have directly affected his own life, he believes that sexual attitudes have improved since his mother's time. Born in 1817, she had had her first child at age fifteen and, after producing five more, died of consumption at twenty-three. His own initiation at age twelve he attributes to the then prevalent (by implication, unfortunate) attitude that children had active sexual desires. As a young man he avoided prostitutes but had many affairs, estimating that he engaged in sexual intercourse a total of 2,000 times between the ages of twelve and thirty-five, a number he does not consider extreme. Nevertheless, he warns that young men must be trained to curb their desires: "The intense pleasure achieved by sexual intercourse is so deceptive and enticing," he writes, "that it often paralyzes good sense and silences reason, making the person an unwilling and obedient slave of his sexual passions."[54]

These two confessions in scientific guise justify their narcissism in social terms but retain their very personal focus. Dr. B's article appeared just as his professional peers were collectively about to join a sustained political challenge to the autocratic regime in the form of the liberation movement. The entire decade 1900–1910 was one in which professional discourse was particularly freighted with social and political implications. Far from extinguishing interest in such private affairs as the sex lives of educated men, the revolution put Dr. B's call for systematic inquiry on the professional agenda. Questions of social order, of the relationship between public and private, reason and passion, surged to the forefront of scientific discourse, intensifying and transforming the existing public interest in sex.

Schoolboy Sex

It may be no accident that the medical and educational journals began to strengthen their focus on the regulation of sexual behavior in the schools after student unrest surfaced in 1899.[55] While some writers ruminated on the state of schoolboys' moral health, other researchers planned systematic inquiries into the sexual habits of university students. The dominant theme, emphasized with even greater insistence after the revolution of 1905, was the need to reconstitute the operation of social discipline: to harness rather than repress the natural forces at

---

[54]"Lichnaia statistika," 154.

[55]For a detailed examination of the origins of the student movement, see Samuel D. Kassow, *Students, Professors, and the State in Tsarist Russia* (Berkeley, Calif., 1989), chap. 3.

work in the social body, to engender responsibility rather than obedience. Schoolboys and college men did not, however, simply represent successive stages of personal maturation; they also symbolized two moments in the development of political consciousness. Broken windowpanes and abandoned desks reflected the spontaneity of libidinous release, the joy of flouting authority and evading discipline. Demonstrations and political meetings, by contrast, demonstrated the triumph of ideology over spontaneity: older students put the eros of defiance to constructive work. If revolution was inspired by passion, it was completed by means of intellectual control.

Pedagogical literature had already identified the civic microcosm in the classroom before 1905. The image was sharply delineated in the writing of the St. Petersburg physician Aleksandr Virenius.[56] His 1901 essay "Sexual Depravity in School-Age Children," which appeared in *Vrach* when he was almost seventy years old, remained the standard citation for at least another decade. A proponent of what he considered enlightened, scientific attitudes toward sex, Virenius warned that constraint alone could not contain the latent force of the schoolboys' repressed desires. His anxious description of the orderly classroom seething with longing and revolt can well be read as political allegory:

Take the young man [who] approximates the ideal of the well-disciplined average student. On the surface everything seems fine, but inside, a battle rages between desires, impulses, and passions on the one hand and the demands of duty and school discipline on the other. The imagination looks for ways to free itself, if only for an hour or moment, from the oppressive mood, the melancholy, regimented [*kazennaia*] life; to experience something vibrant or enticing. . . . Since the school environment fails to satisfy young nature's needs . . . , the field is open for all kinds of enthusiasms, for any temptation or excuse to forget the habitual, loathsome daily round. The students' physical and moral oppression makes them yearn for excitement, for artificial ways to revive their depressed vital forces.[57]

---

[56]Professional information in *Rossiiskii meditsinskii spisok* (St. Petersburg, 1890); for membership in the Russian Society for the Protection of Public Health, see E. I. Lotova, *Russkaia intelligentsia i voprosy obshchestvennoi gigieny: Pervoe gigienicheskoe obshchestvo v Rossii* (Moscow, 1962), 21.

[57]A. S. Virenius, "Polovaia raspushchennost' v shkol'nom vozraste," *Vrach*, no. 41 (1901): 1261–65. Examples of late citation: Slovtsova, "Polovoe vospitanie detei," 675; Bekhterev, *O polovom ozdorovlenii*, 4. Virenius had earlier published the basic ideas of this article in *Mediko-pedagogicheskii vestnik* (1886); they had been popularized by L. A. Zolotarev in his pamphlets on sexual conduct: *Chto govorit nauka*, 56; and *Gigiena supruzheskoi zhizni*, 40.

Virenius faulted the school regime for forcing all boys into a single mold in disregard of individual desires, for wanting young children to act like compliant adults, and for undermining their vitality by needless intrusion into their private affairs. He recommended a less rigorous system, with more free time and less supervision, which would encourage independence, creativity, and "the natural development of body and mind." The political order he envisaged in miniature allowed for the autonomous self-expression of subordinate members, individual difference (*lichnye osobennosti*) rather than bureaucratic (*kazennaia*) uniformity, and freedom from the scrutiny and control of authorities. This liberal ideal would foster "naturalness," he said, allowing children to be childlike and pure, free of artificial stimulants and unawakened to sexual desire, whereas current practices produced moral and physical "depravity."[58]

Depravity not only was a moral problem, in the physician's opinion, but had consequences for public health and the national economy. Elsewhere he listed "infertility, the birth of weak and sickly children, a declining birth rate, rising mortality, and the spread of syphilis" as "the horrible, indeed fatal, consequences of sexual depravity."[59] But if the repressive old order fostered unhealthy habits by constraining desire, modern life produced equally harmful results by allowing desire unfettered rein. An enemy of daily comforts and sensual indulgence of any kind (from good wine to fast cars), Virenius also distrusted the commercial world of public pleasures: not merely brothels, but popular entertainment, bazaars, fairs, sporting events, even cheap art were bad for the health.[60] Dirty books, operettas, cigarettes, and alcohol were "agents that pervert[ed] human nature."[61] Neither the regiment nor the pavement provided the social context for the "natural" regime of Virenius's imagination.

Such anxiety about the perils of modernity recalls similar fears among Western advocates of moral purity. The latter, however, were not obliged to combine their critique of the marketplace and of the commercial mobilization of desire with an assault on premodern forms of discipline, as Virenius was compelled to do. Hostility toward the repressive classroom regime was likewise joined with ambivalence about the corrosive effects of modern culture in the doctor's interpretation of schoolboy masturbation, a vice he depicted as both source and symptom of the weakening of collective ties and the growing power of

---

[58]Virenius, "Polovaia raspushchennost'," 1264–65.

[59]A. S. Virenius, "Zhiznennye soblazny i bor'ba s nimi s tochki zreniia gigieny i pedagogii," pt. 2, *Pedagogicheskii sbornik*, no. 12 (1904): 477.

[60]For an extended tirade against "the temptations of daily life," see ibid., pt. 1, *Pedagogicheskii sbornik*, no. 11 (1904): 417–19.

[61]Virenius, "Beseda," 377.

selfish individualism. The masturbator, Virenius wrote, "easily ruptures the bonds of love and friendship, soon cools to his intimates and becomes a vicious egotist, ready to sacrifice everything and everyone to his own individual self." In detailing the habit's ill effects, Virenius merely repeated the commonplaces of the current medical literature, asserting that natural, healthy sex occurred among adults only for purposes of procreation and that premature sexual activity of any kind depleted the organism of energy it needed for other ends. Healthy adults, unlike those who succumbed to the seductions of the kiosk and café, were those who moderated their appetites. Healthy children had no sexual desire at all. Those who showed signs of excitation or sexual interest should be encouraged to feel repulsion at everything to do with sex, until the time when mature desire would lead them to engage in measured reproductive activity.[62]

To determine what the little subjects of schoolroom autocracy were up to, the doctor himself reached into their schoolboy pants and emerged with very contradictory findings. The truly natural state that he contrasted favorably to the distortions imposed by civilized life was nevertheless quite unpleasant: sweat and secretions gave a boy's penis an acrid smell, Virenius reported, and the residue of farts made the underwear less appetizing still. Genitals were dirty; better to keep one's hands away except for washing, "if not every day, then at least two to three times a week . . . with cold tap water, and occasionally with eau de cologne, spirits, or vodka." The civilized penis was a clean one, the product of scientifically prescribed intervention. By contrast, too much touching (unregulated intervention, reflecting moral and cultural depravity) altered the very face of nature: the organ itself grew in size; the scrotum withered. Good boys had small organs, Virenius announced; the cultured man's ideal penis was small, almost atrophied, in proportion to his restrained sexual urge.[63]

Masturbation appears in this account as the intimate equivalent of the world of commercial delights in that both substituted pleasure for production. However, Virenius blamed the supposed prevalence of the vice as much on the suffocating rigidity of the traditional order as on the

---

[62]Ibid., 380–82, 384, 386–87. Other contemporary articles on the harm of masturbation and the need for sex education include V. M. Bekhterev, "O vneshnykh priznakakh privychnogo onanizma u podrostkov muzhskogo pola," *Obozrenie psikhiatrii*, no. 9 (1902): 658–62; A. A. Prais, "Polovaia zhizn' uchashchikhsia," *Meditsinskaia beseda*, no. 23 (1902): 665–71; Rokov, "Bol'noi vopros"; Polovtseva, "Polovoi vopros."

[63]Virenius, "Beseda," 374, 383–84; idem, "Polovaia raspushchennost'," 1262. The notion that sexual acts distorted the shape of sexual organs was upheld by some forensic experts in the West and shared by some, but not all, Russian physicians. On the enlarged penis, see Tarnovskii, *Polovaia zrelost'*, 43. V. M. Bekhterev, by contrast, doubted that the appearance of the penis was a useful guide to sexual practice, though he admitted that perhaps the head might become enlarged: "O vneshnykh priznakakh," 658–62.

unstable enticements of the new. His 1902 study of masturbation in the lower schools appeared in the journal of the politically progressive Voronezh public health society, which stood at the forefront of social medicine's fight against the repressive aspects of the tsarist regime.[64] His tirade against the evils of solitary pleasure was, in fact, part of his activist posture. Convinced that social attitudes represented biological states, Virenius concluded that the selfish were physically frail and the generous healthy and robust.[65] In exhausting bodily energy, masturbation affected the very way in which society operated; in curbing it, the physician could affect the tenor of public life. Virenius's discussion of masturbation thus included a vision of the good society. The onanist's morbid self was not the ideal individuality he wished to free from the rigid constraints of the current pedagogical regime but an antisocial self-involvement produced by those very same external constraints and encouraged by the open market in promiscuous pleasure. What was needed, Virenius believed, was not grudging obedience to external authority but enlightened self-control based on the latest scientific information, which also provided an antidote to the dubious attractions of disease-ridden streets and mass-produced fantasies.

The doctor's scholarly 1902 article shared its publication date with Leonid Andreev's controversial short story "In the Fog," which likewise demonstrated the perils of ignorance and attracted public attention to the subject of adolescent sexuality. Tormented by the first urgings of sexual desire, the story's young hero collects dirty drawings and sleeps with a prostitute, who leaves him infected with syphilis. Unable to discuss his feelings at home, with no one to confide in and no expert advice, he fears the disease is incurable. Lonely, despairing, filled with guilt and self-loathing, he wanders the sinister, fogbound sidewalks of St. Petersburg until he encounters yet another prostitute, as miserable as he. After a desperate and humiliating brawl, he ends by stabbing her to death and killing himself. Such tragedies, Andreev suggested, were the fruits of petty-bourgeois respectability, which substituted prudishness and silence for the refreshing openness that science and literature—socially sanctioned forms of plain-speaking—could provide.[66]

Questions of Class

Like Andreev's story, almost all medical discussions of sexual mal-

---

[64]Virenius, "Beseda," 373–89. On the journal's politics, see Lotova, *Russkaia intelligentsia*, 36–37.

[65]Virenius, "Zhiznennye soblazny," pt. 1, *Pedagogicheskii sbornik*, no. 11 (1904), 412–13.

[66]L. N. Andreev, "V tumane," *Zhurnal dlia vsekh*, no. 12 (1902): 1411–50.

aise drew conclusions about the cultural environment. Virenius believed that the habit of masturbation transcended class lines but nevertheless explained its causes in social terms. Not everyone bore the same relation to the material and cultural luxuries of modern city life: the behavior of wealthy boys, Virenius argued, reflected the corrupt attitudes of their social milieu; the same practice among poor boys should be understood as the consequence of deprivation, of "unfavorable hygienic conditions," as well as lack of guidance at home and at school. He believed, however, that communities that retained "the strict [patriarchal] customs of the past" maintained standards of sexual chastity that also met the demands of modern hygiene.[67] Thus, although ignorance might sometimes be the origin of vice, culture (or privilege) was not always a guarantee of virtue. The sanitary physician Abraham Prais told the Smolensk medical society in 1902 that education itself was the root of upper-class unhappiness and moral failure (Andreev had depicted his young hero as a student with serious intellectual interests). The common folk, Prais attested, adhered to firm moral standards and avoided sexual excess.[68] One educator went so far as to deny that lower-class youth masturbated at all.[69]

In claiming that the comfortable life doomed privileged boys to sexual self-indulgence, Virenius was not original. His colleagues all extolled the healthy effects of hard work (or, as a substitute, sport),[70] deplored the late age of marriage among the upper classes,[71] and denounced the artificial excitement generated by the leisure and eroticized social customs of the well-to-do and the nouveau riche.[72]Tolstoy

[67]Virenius, "Polovaia raspushchennost'," 1261; idem, "Beseda," 376–77.

[68]Prais, "Polovaia zhizn'," 665–66.

[69]Rokov, "Bol'noi vopros," 66. Similar views on lower-class restraint were expressed by socialist writers in Germany in the same period: R. P. Neuman, "The Sexual Question and Social Democracy in Imperial Germany," *Journal of Social History* 7:3 (1974): 274–75. Official investigations of St. Petersburg workers in the 1860s, by contrast, noted that both men and women practiced masturbation: G. I. Arkhangel'skii, "Zhizn' v Peterburge po statisticheskim dannym," *Arkhiv sudebnoi meditsiny i obshchestvennoi gigieny*, no. 2, sec. 3 (1869): 72; henceforth *Arkhiv sudebnoi meditsiny*.

[70]See, e.g., Filits, "Sovremennaia polovaia zhizn'," 79; Trakhtenberg, "Anomalii," 28–29. Favorskii ("O polovom vozderzhanii," 257) offered the physiological explanation that vigorous activity kept the blood circulating and prevented it from accumulating in the genital organs. Zhbankov thought hard physical labor would syphon off the excess energy and consume the extra time essential to the elite's pursuit of sensual pleasure (particularly in perverse forms): "Izuchenie voprosa o polovoi zhizni uchashchikhsia," pt. 1, *Prakticheskii vrach*, no. 27 (1908): 471–72. The idea of work as the antidote to sexual overexcitement was a popular one; see the journalistic A. I. Matiushenskii, *Polovoi rynok i polovye otnosheniia* (St. Petersburg, 1908), 118; also Il'ia Nakashidze, "Bor'ba s nizshimi instinktami cheloveka: K voprosu o polovom samovospitanii," in *Polovoe vospitanie: Sbornik statei, sostavlennykh uchiteliami, roditeliami i vospitateliami* (Moscow, 1913), 82.

[71]Bekhterev, *O polovom ozdorovlenii*, 17.

[72]Kanel', "Polovoi vopros," 154–56, 162–63; Filits, "Sovremennaia polovaia zhizn'," 79. In a journalistic vein, see V. S. Iakshevich, *Plody razvrata* (St. Petersburg, 1904), 7–9 (on *meshchanstvo*).

had endowed his domestic fable with the same class dimension. The guilty husband in *The Kreutzer Sonata* identifies himself as "a landowner and a graduate of the university, and . . . a marshal of the gentry." His morality is that of his social milieu: "Before my marriage," he tells his listener, "I lived as everyone does, that is, dissolutely; and while living dissolutely I was convinced, like everybody in our class, that I was living as one has to."[73]

The tendency in the professional literature was to depict the sexual habits of the uneducated as natural, and the closer to nature, the more natural. Peasant men might behave in unrefined ways, one educator explained in 1902, and use language too explicit for cultivated circles; peasant women might have more intimate knowledge than upper-class women of the biological facts of life. But in comparison with the artificial and hypocritical attitudes of the educated, he was convinced, the peasantry exhibited a moral purity based on physical strength and well-being, a realistic understanding of the opposite sex, and a relaxed and natural relation to sexual experience which represented an uncorrupted human ideal.[74]

Not everyone accepted this picture. The class hierarchy of virtue in its Tolstoyan version was challenged, for example, by the famous writer and country doctor Anton Chekhov in his 1891 short story "Peasant Women" ("Baby"). Conceived as a reply to *The Kreutzer Sonata*, the tale depicts the same love triangle of wronged husband, unfaithful wife, and the lover who disappears from the picture. This time, however, the principals are peasants, and the sexual infidelity cannot reasonably be construed as a figment of the husband's imagination. It is not the power of desire that destroys human life and happiness, in Chekhov's view, but the weight of conventional norms and patriarchal relations. The wife's mistake, as he tells it, is not the betrayal of her marriage vows but her belief in the legitimacy of true love. It is her lover who soon tires of the affair and urges the woman to return to her husband. A mild-tempered man, the husband is willing to take her back. In contrast to Pozdnyshev's wife, who denies having been unfaithful, the peasant woman boldly insists on the moral value of her transgression. When she resists her husband's and her lover's urging to resume her proper place, the two men resort to the traditional male expedient of violence, beating her soundly in turn. As the husband then repents his violent deed, the lover lectures the woman on her conjugal duty. When the husband suddenly dies, she is accused of his murder, and the lover testifies against her: she "had a mind of her own" (*s*

---

[73]Tolstoy, *Kreutzer Sonata*, 167–68.
[74]Rokov, "Bol'noi vopros," 65–69.

kharakterom byla), he says, and did not love her man. On her way to hard labor, the woman dies of a fever; the lover survives to tell the tale.[75]

Certain structural features distinguish the two stories. The narration reported in Chekhov's version once again constitutes an exchange between men, just as the plot itself involves a contest between males. The ex-lover (who, unlike the husband, is in a position to know "what really occurred") recounts his story to an innkeeper. But the innkeeper's womenfolk also listen in and offer their own dissenting commentary on the tale. They, like the details of the story itself and its title, shift the focus to the subject of female rather than male sexuality. The setting of the interior narration in "Peasant Women" is less claustrophobic, more communal, than the dark train compartment in which Pozdnyshev tells his tale to one other traveler. In that confined and individualistic frame, the upper-class male's narrative voice exercises the sole power of interpretation. By contrast, the peasant male's recital gets interrupted, even challenged, as the women mutter among themselves.

In switching the class venue, Chekhov has altered the meaning of the tale. In the first place, he questions the elite's supposed monopoly on sexual immorality. The common folk, in his view, not only are subject to the same passions but are no less capable than their betters of the cruelest hypocrisy and the most brutal conformity. Some (like the woman in this case) are able to rise above moral convention and defend their personal rights, but these few are swiftly brought to their knees— if not to their graves—by the force of communal censure. Second, the author insists that women are no strangers to true sexual passion. Chekhov's focus on peasant rather than upper-class women may have made it easier for him to argue that men are not so much ensnared by the power of female seduction as women are oppressed by the legal and physical power of men.

Chekhov was not the only one to question the purportedly edifying effects of impoverished living. At least one trained observer of children's sexual habits, the physician and Social Democrat Veniamin Kanel', suggested that the apparent virtues of country life might be the result not of old-style patriarchy or nature's beneficent effects but of material deprivation: hard work and poor nutrition might diminish sexual opportunity and desire.[76] Dmitrii Zhbankov, the zemstvo physician

---

[75]A. P. Chekhov, "Baby" (1891), in *Polnoe sobranie sochinenii i pisem*, 7:340–51 (Moscow, 1977).

[76]Kanel', "Polovoi vopros," 151. On Kanel''s radical politics, see S. I. Mitskevich, *Zapiski vracha-obshchestvennika (1888–1918)*, 2d ed., rev. (Moscow, 1969), 230; Lotova, *Russkaia intelligentsia*, 136; John F. Hutchinson, "'Who Killed Cock Robin?': An Inquiry into the Death of Zemstvo Medicine," in *Health and Society in Revolutionary Russia*, ed. Susan Gross Solomon and John F. Hutchinson (Bloomington, Ind., 1990), 8.

from Kostroma, had noted that when village men left for long periods of work in the cities, their womenfolk were obliged to assume more taxing agricultural labor than was usually demanded of wives. He blamed the increased work load for the women's loss of sexual appetite and interrupted menstruation.[77] The example from Kostroma showed the effects of change rather than tradition, but poverty, not morality, was at work in both cases. Even industrial labor might have the same result. A 1902 article on female "anaphrodesia" explained, for example, that "a woman exhausted by factory work, living under the heavy burden of pauperism, finally becomes impotent [*impotentnaia*], no matter how young she may be." Believing that agricultural tasks were lighter, this author suggested that if the woman returned to "her native [*rodnaia*] circumstances or to housekeeping [*domovodstvo*]," her sexual functioning would return.[78]

After 1905, however, the tendency in the literature was to define the pathology of working-class sexuality in opposite terms: not as suppressed desire but as disorderly, promiscuous family life, precocious childhood sex, and adult sexual indulgence. This critical view of working-class mores as the product of urban experience did not disturb the earlier association between peasants and the virtues of nature. Rather, the problem of sexual indiscipline among the lower classes emerged as part of a sharpened concern with the horrors of city life: a perceived increase in crime and in the virulence of prostitution. Though inhabited largely by the poor and uneducated, just as the countryside was, cities represented modernity, luxury, and cultural sophistication. As the urban lower classes became more of a social problem, they acquired the taint of upper-class vice. But distinctions were still made. Depravity in the personal sense—as a failure of self-control, not a social pathology—remained the province of pedagogues concerned with producing a disciplined elite.

Revolution in the Classroom

For physicians such as Virenius and Pokrovskaia, who strove to protect public health and ensure public virtue by training young boys to early sexual restraint, the revolution only made their task more compel-

---

[77]D. N. Zhbankov, *Bab'ia storona: Statistiko-etnograficheskii ocherk* (Kostroma, 1891), 91. This idea is repeated in "K voprosu o polovykh snosheniiakh," 12. See Barbara Engel, "The Woman's Side: Male Outmigration and the Family Economy in Kostroma Province," in *The World of the Russian Peasant: Post-Emancipation Culture and Society*, ed. Ben Eklof and Stephen P. Frank (Boston, 1990), 71.

[78]V. M. Burlakov, "Ob anafrodizii zhenshchin," *Meditsinskaia beseda*, no. 13–14 (1902): 392.

ling. The school had always been a political instrument, whether in the
hands of education ministers eager to limit social mobility and restrict
access to cultural goods or in the plans of war ministers interested in
military reform or under the auspices of zemstvo representatives who
hoped to improve local welfare or in the dreams of radicals who wished
to inspire revolt. In the professional literature published after 1905 the
schoolroom figured both metaphorically and practically: first, as an analogue of the political order; second, as an arena for intervention by
professionals bent on recasting public habits and preparing themselves
(through their progeny) for a more responsible role in a restructured
political life.

The revolution showed that traditional authorities had lost their grip
on the social energy so tightly contained by the police regime. In 1901
Virenius had warned against the dangerous consequences of repressing
desire by external means, of preferring coercion to self-discipline. The
thirst for excitement he had perceived under the students' surface compliance turned into overt rebellion once controls were relaxed. When
the repressive order collapsed, students of all ages entered the thick of
the conflict. Universities and secondary schools became centers of political ferment; teenagers helped build barricades; schoolchildren broke
windowpanes. In part, youngsters identified the older generation with
the school's oppressive disciplinary regime. In part, they imitated their
politically mobilized elders: pupils in the early grades played at student
and Cossack, formed their own right-wing anti-Semitic bands, sang
revolutionary songs.[79]

As Vasilii Rozanov put it in 1906, "Revolution is almost entirely the
work of youth, in both its poetic and its physical dimension. Revolution can be characterized in two words: youth has arrived."[80] During
the revolution, reported one educator in 1907, the monitors of order
had released their hold or rejected their mandate: "Normal studies were
interrupted; parents had no time for their children, teachers even less;
the authorities lost their heads. . . . In a word, chaos reigned in the
schools." It was then that the frustrated charges took their opportunity
for revenge: "Everything that thrives on boredom, idleness, dullness,
and malice, which had long been contained by the corporal's stick—all
shot to the surface. To everyone's horror, pranksters erupted on all
sides, like criminals released from prison after an earthquake or fire."[81]
The violence unleashed by the disarray of established authority, he concluded, echoing Virenius's warnings of six years before, demonstrated

[79]S. Zolotarev, "Deti revoliutsii," *Russkaia shkola*, no. 3 (1907): 1–23.

[80]V. V. Rozanov, *Kogda nachal'stvo ushlo* (St. Petersburg, 1910), 345.

[81]Zolotarev, "Deti revoliutsii," 6. For the rebellious classroom as the emblem of revolution, see also Rozanov, *Kogda nachal'stvo ushlo*, iii–iv.

Большая забастовка г.г. слушателей приготовительнаго
и перваго классовъ.

Резолюція собранія для классныхъ наставниковъ: „Не хотимъ учиться,
а хотимъ жениться"!

"The Great Strike of Students in the Preparatory and First Classes." The meeting of class
monitors declares: "We don't want to study, we want to get married!" *Strekosa*, no. 35
(1905). Helsinki University Slavonic Library.

the sinister consequences of repression, not the need for more constraint. The schools treated students as inmates, training them for either subservience or revolt, just as the police regime bred both passive subjects and rebellious mobs. Unhappy as he was with the young people's disorderly conduct, the educator sympathized with their complaints and urged responsible parents to shoulder the burden of necessary reform.

The political interpretation of sex provided such commentators with a backhanded way of confronting the erosion of political and social discipline. After 1905 many observers felt that conventional moral values had lost their power to shape private norms, just as political beliefs (old values as well as recent hopes) had ceased to channel public behavior in constructive ways. The problem was to replace the discredited authorities with others employing more effective disciplinary techniques. Professionals with scientific expertise stepped into the breach. Science, claimed the psychiatrist Vladimir Bekhterev in 1910, must provide rational guidance in an era when "neither religious nor moral norms are able to contain sexual desire within the bounds necessary for human well-being."[82] A "materialist and liberal," as one historian has called him, Bekhterev believed that scientific objectivity, not political ideology, was the key to social renovation.[83] He called for legal restriction on the choice of marital partners to ensure the production of healthy offspring and prohibition of marriage between people who did not intend to have children. Marriage, he said, "should be regulated not by the ritual aspects of religion but by basic biological goals."[84]

Training the bright light of science and rational discourse on this crucial subject rooted it firmly in the public domain, subordinated the sexual urge to social purposes (procreation), and wrested it from the influence of irrational dogma and the unrefined impulses of the uneducated servants and prostitutes to whom boys would inevitably turn if not properly instructed.[85] Scientists proposed to substitute their own guidance not only for the harmful myths circulating in the popular mass (an indiscriminate public menace) but also for the inadequate direction now provided by parents (the failure of erstwhile private controls). The older generation, complained many observers, had lost its moral influence; children no longer talked to them at all.[86]

---

[82]Bekhterev, O polovom ozdorovlenii, 2.

[83]Daniel Philip Todes, "From Radicalism to Scientific Convention: Biological Psychology in Russia from Sechenov to Pavlov" (Ph.D. diss., University of Pennsylvania, 1981), 345. On the dissociation of biological materialism from political radicalism, ibid., 355; on the tendency in psychology to see social problems in biological terms, ibid., 421.

[84]Bekhterev, O polovom ozdorovlenii, 18.

[85]Ibid., 5.

[86]Trakhtenberg, "Anomalii," 25–26; Slovtsova, "Polovoe vospitanie detei," 674–75.

Vladimir Bekhterev, professor of psychiatry at the Academy of Military Medicine. *Iskry*, no. 12 (1913). Helsinki University Slavonic Library.

Even people outside the scientific community saw the breakdown of
family communication and the supposed increase in youthful self-indulgence as symptoms of political decline. Intellectuals who now rejected
positivism in the name of absolute spiritual values nevertheless found
themselves speaking in a scientific voice. Writing in the famous *Landmarks* collection of 1909, which attacked the intelligentsia for its political radicalism and philosophical materialism, A. S. Izgoev examined the
sorry state of post-1905 youth. He blamed its pitiful condition in part
on the failure of parents to create a moral atmosphere in which children
would refrain from masturbation, a practice he denounced (mixing reli-

gious and biological terms) as an "evil dangerous to the race."[87] In the
same period a number of educators, particularly in the cadet schools,
urged their colleagues to enlist the help of priests in stemming boyhood
vice.[88] Cadet instructors often sprinkled their discussions of sexual morality with explicitly religious language, which their civilian colleagues
would have shunned.[89] It is a sign of the power of scientific discourse in
educated Russian society, however, that the Orthodox chaplain who
contributed to the exchange on childhood sexuality at the 1908 congress
of cadet corps instructors preferred technical to religious language. This
pedagogical priest called masturbation a disease, not a sin; he debated
the accuracy of statistical data, recommended straightforward scientific
talk rather than moral fables for enlightening young boys, and endorsed
the virtues of physical exercise.[90]

Physicians construed their scientific arguments in increasingly organic terms, emphasizing the physiological consequences of sexual vice,
transmissible across generations, to the ultimate detriment of the entire
social body. "Roused from prolonged hibernation," wrote Liudvig
Iakobzon in the heat of 1905, "Russia needs fresh, hearty, healthy people to do her creative work, not a youth saturated with syphilis and
gonorrhea."[91] Excess sexual expenditure, it was thought, depleted the
nation's energy supply; profligacy undermined public health by spreading venereal disease; masturbation had long-term effects. Abusers suffered deformed personalities and weakened bodies that bred damaged
offspring. "The children of such [masturbating] fathers are prime candidates for the ranks of the degenerate, the psychopaths, the weak, the
good-for-nothing," wrote a cadet school instructor in 1913, sounding
very much like Mariia Pokrovskaia.[92]

Physicians and educators not only believed in the link between social
and sexual problems but sought the solution to both in free discussion
and the application of professional expertise. One quoted John Stuart
Mill as saying that "physical ailments, like social ailments, cannot be
prevented or cured unless one speaks openly about them."[93] In the years
after 1905, these professionals joined with public and government fig-

[87]A. S. Izgoev [Aleksandr Solomonovich Lande], "Ob inteligentnoi molodezhi: Zametki ob ee byte i nastroeniiakh," in Vekhi: Sbornik statei o russkoi intelligentsii, 2d ed.
(Moscow, 1909; rpt. Frankfurt am Main, 1967), 99, 102.
[88]N. M—ii, "Bor'ba," 35; Sangailo, "Polovoi vopros," 339; Simonov, "Shkola i polovoi vopros," pt. 2, Pedagogicheskii sbornik, no. 4 (1908): 283.
[89]See the biblical references in Trudy pervogo s"ezda ofitserov-vospitatelei, 353, 358, 380.
[90]See comments of V. V. Vinogradov in ibid., 368–72.
[91]Iakobzon, "Polovoe vozderzhanie," 594.
[92]Sangailo, "Polovoi vopros," 335.
[93]Trakhtenberg, "Anomalii," 25.

Organization Bureau of the Congress of Pedagogical Psychology. V. M. Bekhterev is seated third from left. Standing are D. A. Dril', at far left, and A. N. Bernshtein, fourth from left. *Iskry* , no. 23 (1909). Helsinki University Slavonic Library.

ures in promoting the cause of sexual instruction. In 1906 the first All-Russian Congress of Educational Psychology, meeting in St. Petersburg, discussed adolescent sexuality. In 1907 the Russian Society for the Protection of Public Health heard a leading educator urge teachers to broach sexual questions directly in class.[94] Virenius addressed audiences of adolescent boys at the technological institute. In 1908, public lectures on sex and the schools were presented at meetings of the Moscow Pedagogical Assembly and before the educational division of the Society for the Diffusion of Technical Knowledge.[95] The cadet corps's pedagogical committee met to determine how best to impart sexual knowledge in the classroom, and the first Congress of Cadet Corps Instructors took up the sexual question at the end of 1908 (the voluable and venerable Virenius figured among the guests).[96] Pediatricians dis-

---

[94]The text of his speech appears in Trakhtenberg, "Polovoi vopros," 27–31.

[95]P. F. Solov'ev, "Polovaia pedagogika," *Vrachebnaia gazeta*, no. 30 (1908): 853.

[96]See Simonov, "Shkola i polovoi vopros," pt. 2, 285; pt. 3, 403–4, 408, 419; and *Trudy pervogo s"ezda ofitserov-vospitatelei*, 293–380.

cussed childhood sexuality at their 1912 conference,[97] and the military administration's pedagogical journal continued to publish articles on the subject as late as 1913.

The purpose of this openness, both among professional colleagues and in relation to the pupils themselves, was not to endorse natural eroticism but to de-eroticize sex. Serious "scientific" discussion was to remove the "veil of enticing secrecy and stimulating eroticism" from the subject of sex.[98] Science was the antidote to morbid subjectivity, wrote the psychiatrist Aleksandr Bernshtein in 1908; sex education was the "bright laboratory" that would disperse shadowy, furtive talk in clubs, dormitories, and classrooms. "The atmosphere of secrecy surrounding the question of sexual life," Bernshtein insisted, "has the fatal consequence of directing attention to the intimate, erotic side of sex, to subjective expectations and experiences." Secrecy was the realm of illicit pleasure; science, the realm of public responsibility, exemplified in procreation.[99]

The ideology of scientific objectivity, the historian Daniel Todes has argued, provided a way for physicians and psychiatrists to exercise civic influence while standing above partisan political interests.[100] Indeed, the profession tolerated a wide range of political opinion. But scientific discourse also provided a way to express political ideas in covert form, just as professional expertise offered opportunities for civic intervention. For example, the juxtaposition of secrecy and enlightenment, of sexually mixed and sexually segregated environments, carried a political message if not an ideologically partisan one. The ominously exclusive male preserve that physicians associated with the morbid distortion of sexual life could easily represent the highest institutions of the tsarist regime, which conducted their business behind closed doors and sustained the uncertainty (and unaccountability) of personal rule. Such government, which failed to embody the essential virtues of public life—accessibility, responsibility, and rational procedure—could not promote the common good, which could be served only by open exposure and dispassionate scientific understanding.

Conversely, the archetypal institution of private life—the family—might foster the very qualities on which successful governance relied. If the ultimate burden of controlling desire rested with individual men, who felt sexual passion most strongly, women might contribute to the

[97]Ostrogorskii, "K voprosu o polovom sozrevanii," and N. E. Rumiantsev, "K voprosu o polovom vospitanii," in *Trudy pervogo vserossiiskogo s"ezda detskikh vrachei*, 657–77.

[98]N. M—ii, "Bor'ba," 33.

[99]Bernshtein, *Voprosy polovoi zhizni*, 7, 9.

[100]Todes, "From Radicalism to Scientific Convention," 343.

production of men with the character needed to keep themselves in check. In the eyes of sober pedagogues worried about the effects of unregulated male passion, women of their own social class did not represent the threat of erotic enticement but offered another kind of prophylactic—along with the dispassion of science—in the fight against male depravity. In this conviction, Mariia Pokrovskaia was not alone.

Most physicians and educators favored coeducation and believed that the sexual instruction of women would encourage their interest in maternity and prepare them for the moral tasks it entailed.[101] The neglect of motherhood on the part of modern young women figured in some popular texts as an analogue to masturbation among young men: both were deviations that substituted selfish desire—pleasure without responsibility to others—for the rigors of social obligation.[102] Women's equality, seen as the necessary condition for increasing their disciplinary influence on public life through private intervention, was not infrequently a plank in the professionals' platform of moral reform.[103] Mothers raised the future generations upon which society's moral and even biological well-being relied; hence, Bekhterev argued, "[women's] equality is at the heart of the sexual question."[104] Professionals thus helped articulate a notion of modern domesticity governed by scientific precepts, similar to the one that had evolved in the West, which would replace the old-style patriarchal family and the paternalistic administrative regime as a primary instrument of social discipline.

Sex Surveys

How ready was the younger generation to shoulder the burdens of public life, with its rapidly expanding opportunities? How well had respectable mothers succeeded in raising morally responsible men? These were the questions that prompted the medical profession's most important representative organization, the Pirogov Society, to sponsor a series of four investigations into the sexual conduct of university students

[101]On the moral role of mothers, see Virenius, "Zhiznennye soblazny," pt. 2, 484–85.
[102]For the pejorative view on "the contemporary young woman," see Iakshevich, *Plody razvrata*, 11–14.
[103]Polovtseva, "Polovoi vopros," 18; E. P., "Opyt osvedomleniia v polovom voprose devochki i mal'chika," *Vestnik vospitaniia*, no. 3, pt. 1 (1908): 118; Trakhtenberg, "Polovoi vopros," 29–31; Rokov, "Bol'noi vopros," 74, 82–83, 87–89; Kanel', "Polovoi vopros," 166; Bekhterev, *O polovom ozdorovlenii*, 14, 18; A. S. Virenius, "Period polovogo razvitiia v antropologicheskom, pedagogicheskom i sotsiologicheskom otnoshenii," *Russkaia shkola*, no. 12 (1902): 123; Slovtsova, "Polovoe vospitanie detei," 677–78; Bernshtein, *Voprosy polovoi zhizni*, 14.
[104]Bekhterev, *O polovom ozdorovlenii*, 18.

in the years leading to and following the revolution of 1905, precisely at the height of the society's own involvement in the political movement.[105] Three of these surveys (in Kharkhov, Moscow, and Iuriev) were conducted between 1902 and 1905; findings were announced at the Pirogov congresses of 1904 and 1907, both highly politicized gatherings.[106] The results of the fourth study (Tomsk, 1907), revealed in 1910, offered a retrospective interpretation of the link between private life and political behavior, between sexual morality and revolutionary mobilization.[107]

These larger issues grew out of the investigators' central preoccupation with the problem of venereal disease and its connection to prostitution. The sex surveys represented the obverse of the numerous public health studies of prostitutes: they were the first attempts to chart the male side of the commercial sexual equation. Respectable women were not at issue in these projects. Physicians still had virtually no systematic knowledge of female sexuality that did not relate to prostitution.[108] Some effort was finally made to include women in the Pirogov Society's overall scheme, but it came to very little. The Tomsk survey extended its research to female students (10 percent of respondents), but their answers were not mentioned in the published report.[109] Also in 1907, the Pirogov congress formed a committee to design a questionnaire for women. Among the committee's female members were four physicians, two schoolteachers, a representative of the Union for Women's Equality, and the popular novelist Anastasiia Verbitskaia. All the men were physicians, including Dmitrii Zhbankov, the well-known advocate of community medicine and women's education and a member of the Pirogov Society's governing board. The questions were ultimately distributed to over 5,000 female students.[110]

The results of the women's survey were never compiled, however. The project was terminated by the police for reasons not indicated in

[105]See Ia. Kh. Falevich, "Itogi tomskoi studencheskoi polovoi perepisi: Doklad, chitannyi 18 fevralia 1910 g. na zasedanii Pirogovskogo studencheskogo meditsinskogo obshchestva pri Tomskom Universitete," pt. 1, *Sibirskaia vrachebnaia gazeta*, no. 17 (1910): 197–98; "Polovaia zhizn' iur'evskogo studenchestva," *Vestnik obshchestvennoi gigieny, sudebnoi i prakticheskoi meditsiny*, no. 7 (1907): 1162–63; N. P. Malygin, "Iz itogov studencheskoi perepisi v Iur'eve (Derpte)," *Zhurnal obshchestva russkikh vrachei v pamiat' N. I. Pirogova*, no. 1 (1907): 20–31.

[106]On the ninth Pirogov congress (1904), see Nancy Mandelker Frieden, *Russian Physicians in an Era of Reform and Revolution, 1856–1905* (Princeton, N.J., 1981), 231–61; on the tenth (1907), see M. A. Chlenov, *Polovaia perepis' moskovskogo studenchestva i ee obshchestvennoe znachenie* (Moscow, 1909), 5, 11.

[107]Falevich, "Itogi," pts. 1–13, *Sibirskaia vrachebnaia gazeta*, nos. 17–29 (1910).

[108]They sporadically deplored their own ignorance: "Lichnaia statistika," 157; Bernshtein, *Voprosy polovoi zhizni*, 13.

[109]Respondents totaled 636, of whom 573 were men: Falevich, "Itogi," pt. 2, 209.

[110]Zhbankov, "Izuchenie voprosa," pt. 1, 474; pt. 3, *Prakticheskii vrach*, no. 29 (1908): 503, 505–6.

the available sources.[111] We do know that not only the police but even
the audience at the 1907 congress had revealed a certain anxiety about
interrogating respectable women. Some delegates worried that asking
questions about sex might destroy the innocence of young female
minds. Zhbankov disagreed: no one of the postrevolutionary student
generation was innocent of sex, he assured them. What had once constituted inside information or private pleasure was now unfortunately
open to public view: in the press, in the stores, on dirty postcards, in
"erotic" texts that no one could avoid. City women—at least those
who could read—no longer had any illusions. The ill effects of such
dubious lore, Zhbankov argued, could be countered only by scientific
knowledge.[112] What he could not admit was that the market in cheap
ideas had broken a taboo that respectable scholars had dared not breach,
but could now profit from doing.

The furtive attempt to peer behind the scenes of female student life
occurred during the depoliticized postrevolutionary lull, but the un-
blinking disclosure of male students' sexual habits accompanied their
emergence onto the street in a political capacity. What did physicians
discover when they turned from inspecting the classroom habits of
young boys to investigate the behavior and attitudes of young men,
asking questions—as befitted subjects capable of rational response—
rather than groping for genital organs? Most of their findings were little
different from those of Virenius and the cadet academy instructors: stories of adolescent self-indulgence, private vice, and, in this case, commercial consolation. Nor were the proposed solutions anything new.
Mikhail Chlenov, who conducted the Moscow survey, concluded in
1907 that only "personal prophylaxis"—that is, sexual restraint—could
solve the twin social ills of prostitution and venereal disease.[113] Iakov
Falevich, who presented the Tomsk findings in 1910, favored male sexual abstinence as the only way out of Russia's moral impasse.[114] As the
term "prophylaxis" suggests, Chlenov did not think simple moral fervor would do the job; science and enlightened pedagogical practice
were the essential underpinnings of the reformed sexual life.

Some respondents and a few doctors did go beyond the liberal pre-
scription. The zemstvo physician Nikolai Malygin, commenting in

[111]Falevich, "Itogi," pt. 1, 198, mentions suppression without further comment. Sheila
Fitzpatrick, "Sex and Revolution: An Examination of Literary and Statistical Data on the
Mores of Soviet Students in the 1920s," *Journal of Modern History* 50:2 (1978): 258, cites a
report from the early 1920s indicating that Zhbankov conducted a sexual survey of
women in 1914, but I have come across no other mention of it.

[112]Zhbankov, "Izuchenie voprosa," pt. 1, 471; pt. 2, *Prakticheskii vrach*, no. 28 (1908):
486, 488.

[113]Chlenov, *Polovaia perepis'*, 97–99.

[114]Falevich, "Itogi," pt. 13, 343.

1907 on the Iuriev study, insisted that the "so-called culture of the bourgeois-capitalist type" could breed only sexual exploitation and misery; it would take democratic socialism and women's liberation to purify public morals and establish genuine public health.[115] And among Tomsk students—most of whom echoed their teachers in favoring co-education and supporting sex instruction for both men and women, more openness about sexual matters, and dialogue between parents and children—several believed that socialism alone could eliminate the underlying causes of prostitution and upper-class depravity.[116]

A twenty-two-year-old medical student in 1910, Falevich seems to have fallen back on the old self-help formula, although it was precisely the nature of the connection between politics and morality that the Tomsk survey, unlike its predecessors, had intended to establish with scientific rigor. One of its stated goals had been to explore "the possible influence of party work on attitudes toward the sexual question and on sexual experience itself." Of the men who responded to the survey, 44 percent belonged to political parties: 80 percent of them to the radical left, 18 percent to the liberal-constitutionalist center, and a tiny fraction to the radical right. Another 13 percent of respondents considered themselves sympathizers: 90 percent on the left, 10 percent in the middle. Half the respondents had taken part in "revolutionary work," but only half of these activists claimed that their involvement had altered their sexual feelings: 80 percent thought the revolution had exercised an "ennobling" effect on their sexual life; another 13 percent described their sexual appetite as having diminished; a mere 2 percent believed it had grown; and only 3 percent said the revolution had drawn them to sexual abnormalities.[117]

Falevich concluded that the revolution had in general curtailed sexual activity by monopolizing young people's energy and time. Chlenov too, citing the perception that venereal disease had diminished in 1905, surmised that political interests took the place of sexual indulgence during moments of intense revolutionary conflict.[118] Tomsk students who boasted of greater sexual restraint attributed their improved conduct to ideological as well as personal experience. Working closely with women had taught men to view them as comrades rather than objects of sexual desire, they reported, an attitude encouraged, so they said, by the Marxist parties and reinforced by the reading of August Bebel and Lily Braun. Sex, the students' radical mentors taught them, could be

[115]Malygin, "Iz itogov," 21–22.
[116]Falevich, "Itogi," pt. 12, 329; pt. 11, 318–19.
[117]Ibid., pt. 2, 209–10.
[118]Chlenov, Polovaia perepis', 74.

justified not as a source of pleasure but only "as a physiological need" (*stol'ku po skol'ku etogo trebuet priroda*). In sum, Falevich decided, "the revolution itself seems to have improved people's attitudes in general and in particular their views on women and on the sexual question."[119]

Such evidence indicated that the vision of schoolboy libido running wild had its limits. Discipline, not disinhibition, characterized university youth, who appeared to have attained heights of sublimation during the revolutionary months. The apparently contrasting behavior of boys and young men provided physicians with biological models for the two aspects of revolutionary mobilization: its angry, destructive component on the one hand and its purposeful dedication on the other. The libidinous energy released in the schoolroom was both presexual and prepolitical; the same impulse liberated in young adults served definite civic purposes. Thus frustrated preteens broke windows and built barricades, while their older siblings joined political parties, attended all-night meetings, and were too tired for love. Once the revolution was over and the civic dream had failed, however, the recently activated libido returned in full force in its original sexual form, more imperious for having been liberated from the discipline and constraint of the old regime. The political crisis past, passion returned from the abstract to the bodily, and the thirst for sensual gratification increased as substitute excitement died down: political and sexual virtue deteriorated apace. The ban on public involvement caused people to turn inward, dwelling on personal and, in particular, sexual affairs.[120]

Echoes of this interpretation could be heard as far afield as Berlin, headquarters of enlightened sexual science. The 1908 edition of Iwan Bloch's influential *Sexual Life of Our Time in Its Relation to Modern Culture*, for example, included the story of a Russian anarchist who claimed to have gratified his masochistic sexual inclinations in the *Angst-Ekstase* of political confrontation during the revolution of 1905.[121] Another German sexologist portrayed the revolution as a check on sexual desire: "Radical circles discouraged the discussion of sexual questions and considered any encouragement of sexual activity to be unrevolutionary," the reporter claimed; in any case, when "personal feelings were absorbed in the struggle for social change, the need for sexual expression was minimal." But after the proletarian movement abated, "the intel-

---

[119]Falevich, "Itogi," pt. 3, 221–22.
[120]Ibid., 222; Zhbankov, "Izuchenie voprosa," pt. 2, 487, 490.
[121]Iwan Bloch, *Das Sexualleben unserer Zeit in seinen Beziehungen zur modernen Kultur*, 6th ed. (Berlin, 1908), app. to chap. 21, 646–68. Bloch notes that the anonymous personal account had been transmitted to Magnus Hirschfeld in Berlin in 1906. Bloch's volume appeared in Russian translation in 1910.

ligentsia abandoned [the struggle] and devoted its energy to sexual relations."[122]

In this curious reversal of the Bolshevik paradigm, according to which the party restrains the formless outbursts of popular feeling, the sober proletariat checks the spontaneous expression of its leaders' passionate desire; set adrift from the workers' noble and constraining cause, students and intellectuals revert to the erotic introversion typical of their class. But alongside the persistent stereotypes of popular virtue, new and less idealized images of lower-class sexuality emerged after 1905, especially in relation to city life. To educated observers the laboring masses represented the subordination of private to collective interests, of personal sensibility to communal norms—a model to which the individualized upper classes could aspire only in fleeting moments of civic self-absorption such as revolution. But at the same time, workers had participated in organized politics, demonstrating a capacity for the kind of rational behavior considered the mark of educated men. Each cultural world, the popular and the elite, thus borrowed for an instant its opposite's defining trait. When the moment of political transcendence had passed, all those left in the city—the site of commerce, individuality, and desire—acquired the attributes of modern life and the sexual volatility it represented.

[122]Werner Daya-Berlin, "Die sexuelle Bewegung in Russland," *Zeitschrift für Sexualwissenschaft*, no. 8 (1908): 494–95.

Chapter Seven

End of Innocence and
Loss of Control

Before 1905 the professional literature linked class and sexuality in the following manner: Steeped in the secular culture of the modern West and trained in the exercise of reason, the educated elite could achieve virtue only by submitting desire to intellectual control. The sexual mores of the privileged classes seemed as much the product of artifice as the other conventions of their sophisticated, largely urban lives. The common folk, by contrast, represented the power of unreason, expressed in superstition, religious faith, and erratic violence. Though unregulated by rational considerations, the peasants' sexual regime nevertheless reflected the orderliness of the natural world. Each paradigm of class sexuality thus harbored its own opposite. The educated might discipline the body's animal instincts, yet they also succumbed to the lure of artificial gratification, indulging in sexual perversion and "excess." The folk might lack the ability to govern their impulses, yet the logic of nature to which they were bound secured them from the abuses bred by "civilization."

The tensions inherent in these oppositions exploded after 1905. The perceived rise in "unmotivated" crime and sexual "perversion" among the lower classes suggested that nature and traditional values could no longer contain the elemental passions of the common folk. Perhaps more unnerving still, the elite felt they had lost the power of rational command over their own impulses. Both the natural and the intellectual order were in danger. Public interest in sexuality and spirituality, already evident in the years preceding 1905, when the elite were mobilizing against the regime, intensified after the revolution, when educated society lost whatever control it had recently exerted over forces it itself had helped to unleash.[1] On both sides of the philosophical barricades, the intelligenstia tried to establish a grip on the irrational. Philosophers strove to rehabilitate the socially constructive power of religious faith, psychiatrists to establish scientific mastery over the psychic impulses stimulated by the license of revolutionary times.

The Revolutionary Unconscious

Despite the claims of activists and social scientists, political ideology and organization had neither entirely harnessed nor completely exhausted the surge of libidinous energy that fueled recent events. Social Democrats could not boast that the angry masses had achieved political self-awareness, that "consciousness" in the form of party leadership had mastered the elemental, or "spontaneous," forces of revolt. Liberals were uncomfortable with the rebellion of their own children. The failure that chagrined the radicals and their allies came as no surprise to conservatives, who did not share either Lenin's confidence in the ultimate power of reason to dominate rage or the liberals' discomfort with conflict.

Vasilii Rozanov, for example, assumed that a movement whose goals were articulated by lawyers and intellectuals, believers in Karl Marx and John Stuart Mill, must nevertheless make its appeal on the basis of inarticulate emotion. Whatever its leaders might wish, revolution was a "natural" impulse, Rozanov insisted, not the consequence of rational designs. "The revolution is not a contest between programs," he wrote, "but a movement of spontaneous forces [*dvizhenie stikhii*], in which obdurate need and the most ethereal imagination, woven into an inscrutable pattern, play no less a role than formal political parties." Even the people who joined such parties were not in fact moved by abstract ideas, he asserted, but acted from "unacknowledged and obscure [*temnye*, also suggesting ignorance and darkness] motives, passions, feelings, and fantasies. Revolution would not be complete, indeed would not even exist, without the powerful impetus of these irrational components."[2]

Irrationality was no defect, in the philosopher's opinion. He welcomed the fact that it rattled the tyrannical shackles of normal (that is, reasonable, measured, conventional) social existence. Revolution

[1]On the emergence of sexuality and religion as joint preoccupations, see N. N. Starokotlitskii, "K voprosu o vozdeistvii polovogo instinkta na religiiu (v sviazi s opisaniem sluchaia religiozno-erotomanicheskogo pomeshatel'stva)," *Zhurnal nevropatologii i psikhiatrii imeni S. S. Korsakova*, no. 2–3 (1911): 259.

[2]V. V. Rozanov, *Kogda nachal'stvo ushlo* (St. Petersburg, 1910), 322–23.

eroded the constraints of civilized everyday life: "Artificiality diminishes, candor grows," along with a "general loosening of 'bonds and
conventions,' of the 'burdens' of civilization."[3] Rozanov had no use for
abstract thought, let alone political dogma, but he gloried in the un-
predictability of violent emotion—anger and desire, alone or in combination—which gave the revolution its visceral power while threatening
its project of orderly mobilization and logically articulated aims.

Rozanov's sympathy for the revolutionary drama surprised liberal
readers, who, in the words of the poet Andrei Belyi, were "used to
endorsing [the] liberation movement in abstract terms: on the basis of
ethical, religious, political, and socioeconomic principles." "Conversely," Belyi went on, "the conditions of everyday life often appear
as conservative elements in relation to the abstract expression of our
goals." Indeed, Rozanov's philosophical conservatism led him to extol
the sensuous immediacy of everyday life. His insistence on "his indifference to the abstract principles of public life [*obshchestvennost'*]" made
him "seem to be a poet of the outmoded past." Many therefore found
the philosopher's reaction to 1905 unexpected, Belyi noted: "Smiling
affectionately just when one awaited the sullen regard of incomprehen-
sion," Rozanov seized the pathos of the revolutionary moment in its
specific psychological details.[4]

Depicting the revolution as the movement of fantasy and libidinous
release, Rozanov did not endorse its political aspirations but hailed precisely what he viewed as the transcendence of politics in the ordinary
sense and greeted what he saw as the authentic life of passion and spontaneous delight. Perhaps it took a conservative to celebrate a dimension
of the radical project its authors were not themselves ready to confront.
Could abstract principles be reconciled with the particularities of personal experience? Should civic engagement be motivated by private desires? Had the revolution not entirely sublimated the erotic or irrational
impulses that Social Democrats feared?

Despite their dedication to rational ideals, not all liberals ignored the
impulsive component of political life. Indeed, the liberals' ambivalence
toward the chaotic nature of the revolution, into which many had only
reluctantly been drawn for lack of political alternatives, emerged in
their retrospective pronouncements.[5] By 1905 most Russian psychiatrists had despaired of the autocratic regime and endorsed the values of

---

[3]Ibid., 338–39.

[4]Andrei Belyi, "Publitsistika: V. Rozanov, *Kogda nachal'stvo ushlo*," *Russkaia mysl'*, no.
11, pt. 2 (1910): 374–76.

[5]For the attitude of liberal professionals toward the revolution and their discomfort
with their radical allies, see the pamphlet by the psychiatrist N. N. Bazhenov, *Psikhologiia
i politika* (Moscow, 1906), esp. 6–9.

the liberation movement.[6] Once the revolutionary climax had been reached, psychiatrists tried to master through science (the voice of reason) the turbulence of disorderly desire. Psychiatry, they felt, was the "political economy" of the affective domain, diagnosis a medium of political interpretation. In the words of Aleksandr Sholomovich, writing in 1907: "At moments when emerging socioeconomic transformations take such graphic political forms, extremely important changes undoubtedly occur in the depths of the popular organism, attracting the attention not only of political economists but of physicians as well. In such periods psychiatry and psychology have two extremely important tasks: (1) to illuminate the depths of the collective psyche; and (2) to explain the influence of current events on the population's psychic health."[7] In analyzing the connection between revolution and mental health along those two lines—contemplating the revolution itself as a manifestation of collective psychology, and attempting to understand the link between social upheaval and individual mental distress—psychiatrists translated the casual language of the day into the language of the clinic. Count Sergei Witte, at least in Leon Trotsky's memory, said the population had "gone mad" in 1905; the Octobrist politician Aleksandr Guchkov called the general strike "a psychosis that has seized our society."[8]

Some psychiatrists believed that healthy people had withstood the shock, that revolution had elicited mental disorder only in fragile souls already inclined to instability.[9] Even those who argued that the recent upheavals had increased the incidence of psychosis among otherwise stable people did not see the revolution as an abnormal event. It was the autocracy that was ultimately to blame if society had temporarily lost

[6]See Julie V. Brown, "Revolution and Psychosis: The Mixing of Science and Politics in Russian Psychiatric Medicine, 1905–13," *Russian Review* 46:3 (1987): 283–302.

[7]A. S. Sholomovich, "K voprosu o dushevnykh zabolevaniiakh, voznikaiushchikh na pochve politicheskikh sobytii," *Russkii vrach*, no. 21 (1907): 715–16.

[8]Leon Trotsky, *My Life* (New York, 1970), 178; A. I. Guchkov in *Russkie vedomosti*, October 15, 1905, quoted in Abraham Ascher, *The Revolution of 1905: Russia in Disarray* (Stanford, Calif., 1988), 215.

[9]F. E. Rybakov, "Dushevnye rasstroistva v sviazi s sovremennymi politicheskimi sobytiiami," pts. 1–2, *Russkii vrach*, nos. 3 (1906): 65–67, 221–22; idem, "Psikhozy v sviazi s poslednimi politicheskimi sobytiiami v Rossii," *Russkii vrach*, no. 20 (1907): 679 (Rybakov had argued the opposite in 1905 but had since changed his mind); N. I. Skliar, "O vliianii tekushchikh politicheskikh sobytii na dushevnye zabolevaniia," *Russkii vrach*, no. 8 (1906): 222–24; idem, "Eshche o vliianii tekushchikh politicheskikh sobytii na dushevnye zabolevaniia," *Russkii vrach*, no. 15 (1906): 448–49; I. S. German, "O psikhicheskom rasstroistve depressivnogo kharaktera, razvivshemsia y bol'nykh na pochve perezhivaemykh politicheskikh sobytii," *Zhurnal nevropatologii i psikhiatrii imeni S. S. Korsakova*, no. 3 (1906): 313–23; A. N. Bernshtein, "Psikhicheskie zabolevaniia zimoi 1905–6 gg. v Moskve," *Sovremennaia psikhiatriia*, no. 4 (1907): 66; M. N. Zhukovskii, *O vliianii obshchestvennykh sobytii na razvitie dushevnykh zabolevanii* (St. Petersburg, 1907), 38.

its mental balance. The government had for so long and so completely monopolized the exercise of power, the Moscow Society of Neuropathologists and Psychiatrists argued in a statement issued in early 1906, that social groups and individual citizens unused to the burdens of civic activity were strained to the breaking point by the tremendous effort needed to dislodge existing forms. Their sufferings and the shock of change, which had "penetrated to the essence of collective existence [*obshchezhitie*]—social relations," had "created in Russia today an especially difficult social-psychological atmosphere, which cannot help disturbing the equilibrium not only of individual persons but of entire social groups."[10]

Psychiatrists most sympathetic to the revolution or most distressed by the continued violence and repression of 1906–7 did not maintain that volatile personalities were drawn to revolutionary events or that political involvement threatened one's powers of self-control. Rather, they saw acute anxiety as an appropriate response to current political circumstances. The second national conference of psychiatrists, meeting in September 1905, had called the regime "deeply abnormal" for threatening personal safety, making unwarranted arrests, using violence against crowds, and ignoring the recent law against corporal punishment.[11] The state's repressive practices had pathogenic effects not because they were suffocatingly orderly but because they entailed unpredictable violence.

According to this line of argument, it was not the unplanned chaos of revolution but arbitrary police power, violating public expectations of protection and consistency, that had deprived people of the conditions necessary for mental health. As Sholomovich wrote in 1907:

> Current political conditions represent a direct threat to the mental health not only of constitutionally unstable people but of completely healthy ones, not only of [revolutionary] activists but also of bystanders and involuntary witnesses. . . . In areas where armed Black Hundred bands constitute the last word in the political wisdom of local satraps, much of the population lives in a state of chronic terror. Psychoses and neuroses spread lavishly, affecting both the weak and

---

[10]"Doklad psikhiatricheskoi komissii obshchestva neiropatologov i psikhiatrov po voprosu o psikhozakh v sviazi s poslednimi politicheskimi sobytiiami," *Russkii vrach*, no. 23 (1906): 709–11. Also F. E. Rybakov, "Dushevnye rasstroistva v sviazi s tekushchimi politicheskimi sobytiiami," *Russkii vrach*, no. 51 (1905): 1593–95.

[11]"Priniataia s"ezdom rezoliutsiia o roli sotsial'nykh i politicheskikh faktorov v etiologii nervno-psikhicheskikh zabolevanii," in *Trudy vtorogo s"ezda otechestvennykh psikhiatrov, proiskhodivshego v g. Kieve s 4-go po 11-oe sentiabria 1905 goda* (Kiev, 1907), 424.

the robust. . . . The pathological seeds of lawlessness and coercion [*bezzakonie i nasilie*] are being sown with a generous hand throughout the country. They have already provided and will continue generously to provide present and future generations of psychiatrists with the terrible material of insanity.[12]

Sanity, Sholomovich implied, depended on the existence of a reliable, rationally governed political order. Insanity erupted when public arbitrators failed at their job. The psychiatrist Vladimir Iakovenko went even further in insisting that revolutionaries represented the triumph of society's healthy instincts over the destructive repressiveness of tsarist rule; serious pathology, he thought, was to be found primarily among the regime's servants and defenders.[13]

Not all his colleagues agreed. Though many physicians retained their radical values after 1905, the profession emerged from the revolutionary crisis politically divided. Even some who had earlier taken part in organizations representing the socially oriented program of zemstvo medicine, such as the Pirogov Society, now rallied to the defense of the new status quo, repudiating the radical stance taken by the society in 1905 and boycotting its 1907 congress.[14] The clinical language that served the radicals also served their opponents. Instead of blaming the regime for breeding psychic distress, the conservatives interpreted the revolution as a mass social pathology, though it took individual psychopathological forms. In early 1906 Salman Iaroshevskii depicted recent (and presumably ongoing) events in terms very close to those used by Vasilii Rozanov to describe the release of inhibitions and controls, except that Iaroshevskii had a less sanguine view of the resulting chaos:

Everything that for decades, indeed centuries, has been ripening in the deepest recesses of the people's collective soul, everything that has agitated the mind and heart, . . . has [now] erupted with uncontrollable force. In this chaos of feelings, thoughts, desires, and aspirations, it is sometimes hard to distinguish the good from the bad, the useful from the harmful, the real from the utopian, the true kernel of culture and progress from atavistic phenomena that sometimes recall the

[12]Sholomovich, "K voprosu," 720. The Black Hundreds were violent counterrevolutionary mobs mobilized by monarchist groups or government officials.

[13]V. I. Iakovenko, "Zdorovye i boleznennye proiavleniia v psikhike sovremennogo russkogo obshchestva," *Zhurnal obshchestva russkikh vrachei v pamiat' N. I. Pirogova*, no. 4 (1907): 270, 279–80, 283–84, 286.

[14]See John F. Hutchinson, "'Who Killed Cock Robin?': An Inquiry into the Death of Zemstvo Medicine," in *Health and Society in Revolutionary Russia*, ed. Susan Gross Solomon and John F. Hutchinson (Bloomington, Ind., 1990). I thank Professor Hutchinson for making this manuscript available to me before its publication.

traits of our remotest savage ancestors. . . . Without doubt we are experiencing, in a certain sense, a psychic epidemic.[15]

Iaroshevskii considered the revolutionary state of mind a form of collective psychopathology, involving extreme mood swings, loss of judgment, intense anxiety, and unbridled aggression. Far from imposing rational controls on primitive emotions, ideology merely covered over the brute reality of conflict that fragile egos could not withstand. He blamed the radical parties, despite their high-minded ideals, for their "ill-considered" or even "criminal" attempts to lure young people into politics: "What irreparable harm, what colossal misfortune," occurs when "children" (deti) are "wrench[ed] from the peaceful conditions essential to their proper growth and development, and thrust . . . into the hideous maw of the voracious animal called 'politics.'"[16] These misguided youngsters were not the naughty schoolboys recalled by alarmed pedagogues. They were victims of disruption, not its agents.

If young people—whom he called "children" in a rhetorical attempt to emphasize their innocence and malleability—lacked the power of reason and self-restraint necessary to protect them against their own thirst for excitement, women too seemed deficient in this fortifying power. The majority of patients Iaroshevskii presented as fragile souls unbalanced by the revolution were women. He did not suggest a reason. Were women more susceptible to neurosis than men? Did their fragility stem from an even lower suitability for the political life into which vulnerable children had been inveigled? Did women represent that latent social pathology brought to the surface when external controls failed? Or was it simply that his practice was predominantly female?

At the very least, the female sexual constitution seemed peculiarly liable to disabling storms. Another psychiatrist, Lidiia Pavlovskaia, used a female patient to illustrate the relation between political excitement, mental imbalance, and the release of sexual desire. The case history, published in 1906, featured a twenty-six-year-old female student who had participated actively in the major events of the preceding year. Present at the confrontation of Bloody Sunday, speaking often at meetings, the student had had several brushes with the police and found herself in a continual state of extreme excitement. In the midst of the general exaltation, she met and fell in love with a man who she soon

[15]S. I. Iaroshevskii, "Materialy k voprosu o massovykh nervnopsikhicheskikh zabolevanii," *Obozrenie psikhiatrii nevrologii i eksperimental'noi psikhologii*, no. 1 (1906): 1; henceforth *Obozrenie psikhiatrii*.
[16]Ibid., 1–2, 8–9.

realized did not love her. Predisposed by family background to nervous distress (alcoholic father, "nervous" mother, "hysterical" sister), the patient reacted by falling ill. By April 1906 she was suffering from headaches, hallucinations, and paranoid delusions and had unsuccessfully attempted suicide by jumping from a bridge.[17]

This was not an idiosyncratic case, in Pavlovskaia's view. Rather, the psychiatrist claimed to see a connection between "passionate involvement [*uvlechenie*, a word often used in a sexual sense] in public affairs and erotic excitement, between intensified public activity and sudden falling in love, analogous to the link between sexual excitement and religious ecstasy."[18] She was sure there was some emotional tie between "public involvement" and "personal love." Perhaps, she speculated, the physical strain of active political life increased people's "sexual excitability," which might then degenerate into "erotic delirium." Perhaps it was simply that people caught up in revolutionary events felt life was short and wanted to take the full measure of its joys and pleasures.[19]

Psychiatrists skeptical of the wisdom or rationality of political engagement may have used the example of excitable women to discredit the entire revolutionary enterprise: what could be more frivolous than a young girl's romantic infatuation, especially when it was reduced to an involuntary expression of organic urges beyond her control? Such arguments and clinical examples nevertheless rested on the same assumption that guided the evidence presented by political radicals: that science offered diagnostic tools and intellectual categories with which to interpret social and political events and upon which to ground programs or policies.

Only the most ruthlessly consistent conservatives challenged the underlying assumptions of the entire debate. The forensic psychiatrist Vladimir Chizh, who had avoided the 1905 Pirogov Society congress out of discomfort with its radical political stand, insisted in 1908 on the impossibility of demonstrating any kind of connection between political freedom and the incidence of nervous disorder. Claims for such a connection could not withstand scientific scrutiny, he wrote, but merely reflected the profession's changing political mood. At the first national psychiatric congress in 1887, Chizh recalled, Ivan Merzheevskii had blamed the impact of peasant emancipation and emerging capitalism for a supposed increase in mental disease. In 1905, by contrast, the second psychiatric congress denounced the "pathological abnormality" of the

---

[17]L. S. Pavlovskaia, "Dva sluchaia dushevnogo zabolevaniia pod vliianiem obshchestvennykh sobytii," *Obozrenie psikhiatrii*, no. 6 (1906): 419.

[18]Ibid., 422.

[19]L. S. Pavlovskaia, "Neskol'ko sluchaev dushevnogo zabolevaniia pod vliianiem obshchestvennykh sobytii," *Obozrenie psikhiatrii*, no. 9 (1907): 557–58.

existing regime. The one had questioned the wisdom of liberalism and modernization, the other had criticized traditional repressive forms in similar clinical terms. In Chizh's view, both were equally arbitrary conclusions.[20]

Chizh was particularly incensed by a speech given at the tenth Pirogov Society congress in April 1907, in which Vladimir Iakovenko had connected certain mental illnesses or personality types to specific political ideologies. Despite the preeminent rationality of the revolutionary enterprise, Iakovenko admitted, not all participants joined the movement for sensible reasons. Social upheaval brought pathological tendencies to the surface, and both sides of the struggle attracted their share of mentally unbalanced partisans. He was convinced, however, that the extent and quality of pathology differed. The side of change drew a certain number of neurasthenics and hysterics into its ranks, he claimed, acquiring thereby "an admixture of extravagance, nervousness, impetuosity, dreaminess, and self-delusion." The conservative side attracted instead the allegiance of the retarded or senile, of epileptics, degenerates, and sexual perverts. These defenders of the established order, he maintained, often displayed "mental weakness . . . , extreme depravity . . . , and finally the kind of unmotivated cruelty characteristic of epileptics and moral idiots. This prolonged and cold-blooded cruelty has nothing in common with the cruelty sometimes perpetrated by the party of change, usually in connection with transient anger, which is quickly superseded by generosity and general forgiveness."[21]

If the revolution erred in the direction of excessive enthusiasm, according to Iakovenko, the regime's crass brutality and calculated violence amounted to "extreme pathology." Though it posed as the custodian of the national welfare, the regime was in fact injurious to the nation's health. Official propaganda represented the recent repressive measures as a cure for the revolutionary disease; in fact, said Iakovenko, they represented serious malpractice. In indicting such policies, Iakovenko inverted the usual medical metaphor endowing physicians, as guardians of the body social, with political responsibility. The profession must exert its authority in political affairs, he asserted, because the regime was acting like an incompetent physician: "What should we

[20]V. F. Chizh, "Znachenie politicheskoi zhizni v etiologii dushevnykh boleznei," pts. 1–2, *Obozrenie psikhiatrii*, nos. 1, 3 (1908): 3–5, 11, 151. For his 1887 speech, see I. P. Merzheevskii, "Ob usloviiakh, blagopriiatstvuiushchikh razvitiiu dushevnykh i nervnykh boleznei v Rossii i o merakh, napravlennykh k ikh umen'sheniiu," in *Trudy pervogo s"ezda otechestvennykh psikhiatrov, proiskhodivshego v Moskve s 5-go po 11-oe ianvaria 1887 g.* (St. Petersburg, 1887), 19–20.

[21]Iakovenko, "Zdorovye i boleznennye," 282.

say of a surgeon who does not request the patient's consent or even ignores his outright objections but nevertheless starts to bleed the anemic, to pull healthy teeth while protecting the rotten ones, to amputate an entire leg when only the little toe is infected, and finally, under the guise of saving the patient's life, cuts off his head? Before he can accomplish such irremediable harm, such a surgeon must of course be put in a psychiatric hospital."[22]

Such metaphorical extravances were anathema to Vladimir Chizh. While agreeing that "abuse of power, degradation, and brutality" (*proizvol, unizhenie, poboi*)—that is to say, the worst aspects of the unreformed old order—were bad for psychic health, Chizh could not accept the idea that the members of the regime's peacekeeping forces could be either "psychopathic or mentally disturbed." They were simply doing their jobs. Political activism on either the right or the left, he argued, differed from public service in attracting the participation of unstable characters.[23] Perhaps Chizh felt free from the taint of partisanship in tarring all forms of unauthorized political activity with the brush of pathology. But despite his objection to the habit of seeking political causes and remedies for psychological problems and psychological explanations for political events, he could not in fact free himself from the terms of the discourse in which his entire profession was engaged.

Nor did he have the last word on the subject. Many of his colleagues took him to task, one journal calling his position "a reminder of the intensely reactionary period we are now compelled to endure."[24] They also rejected Chizh's plea for supposedly neutral scientific discussion. At its third national convention in late 1909, the profession had returned to the political theme in its broadest terms. The renowned professor and ranking bureaucrat Vladimir Bekhterev—no radical himself—sounded very much like a socialist in blaming capitalism, with its ruthless competition and materialist greed, for increasing the incidence of mental distress.[25] The Kishenev psychiatrist and Bolshevik sympathizer Petr Tutyshkin characterized the revolution's demoralized after-

[22]Ibid., 284, 286.

[23]Chizh, "Znachenie," pt. 1, 4, 8–9, 11. On Chizh's response to Iakovenko, see V. P. Osipov, "O politicheskikh ili revoliutsionnykh psikhozakh," *Nevrologicheskii vestnik*, no. 3 (1910), 460, and Brown, "Revolution and Psychosis," 291–95.

[24]From *Sovremennaia psikhiatriia* (1908), quoted in Brown, "Revolution and Psychosis," 295.

[25]V. M. Bekhterev, "Voprosy nervno-psikhicheskogo zdorov'ia v naselenii Rossii," in *Trudy tret'ego s"ezda otechestvennykh psikhiatrov (s 27-go dekabria 1909 g. po 5-oe ianvaria 1910 g.)* (St. Petersburg, 1911), 57, 66. On Bekhterev's career, see N. P. Zagoskin, ed., *Za sto let: Biograficheskii slovar' professorov i prepodavatelei Imperatorskogo Kazanskogo Universiteta (1804–1904) v dvukh chastiakh* (Kazan, 1904), 2:130–31; and *Biographisches Lexikon der hervorragenden Ärzte aller Zeiten und Völker (vor 1880)*, ed. August Hirsch, 3d ed. (Munich, 1962), 2:85–86.

math as psychotic: "The current reactionary mood in Russian society could well be called 'mass moral imbecility,' a step in the direction of moral insanity." Ironically, given his party loyalties, Tutyshkin sounded less revolutionary than reformist in his critique of absolutism, insisting that current ills could be cured only by "the basic political and economic reform of Russian life." Psychic health was not possible under the rule of "lawnessness" (*bezzakonie*) and "abuse of power" (*proizvol*), he said, reiterating a familiar liberal theme; the concepts of "law" (*pravo*) and "lawfulness" (*zakonnost'*) were the only "prophylactics" the profession could recommend.[26] Once again, the language of medicine provided an analysis—both diagnostic and metaphoric—of social ills and a justification for practical action.

The Revolution in Crime

Whether or not psychiatrists diagnosed the revolution as some sort of collective psychosis, they differed as to whether the revolutionary experience had in fact increased the incidence of mental disorder among individual subjects or changed the quality of mental disease. In the case of social rather than psychic pathology, experts did agree that the revolutionary era coincided with a radical upsurge in crime and with the alteration of its character and demographic profile, stamping the decade 1900–1910 as one in which disruptive passions had gotten the upper hand.

Statistical data on convictions, published and analyzed by the Ministry of Justice, offered empirical backing for the impressionistic view that 1905 had been the work of young people in revolt against familial authority and respectable society. The Ministry of Justice statistician Evgenii Tarnovskii reported in 1913 that crime in general had increased by 35 percent between 1900 and 1910 and juvenile crime by a stunning 112 percent, rising most acutely after 1906.[27] Female crime continued to lag well behind, and girls were only slightly more likely to be convicted than adult women: the latter constituted 12 percent of offenders in 1904, while 13.4 percent of all juvenile offenders were female, a con-

---

[26]P. P. Tutyshkin, "Zadachi tekushchego momenta russkoi obshchestvennoi psikhiatrii," in *Trudy tret'ego s"ezda otechestvennykh psikhiatrov*, 736–37. On Tutyshkin, see Brown, "Revolution and Psychosis," 298–99.

[27]E. N. Tarnovskii, "Dvizhenie chisla nesovershennoletnikh (10–17 let), osuzhdennykh v sviazi s obshchim rostom prestupnosti v Rossii za 1901–1910 gg.," *Zhurnal Ministerstva Iustitsii*, no. 10 (1913): 46. For his career, see *Spisok chinam vedomstva Ministerstva Iustitsii 1900 goda* (St. Petersburg, 1900), 66.

stant proportion since 1890.[28] The most startling change in the crime pattern that emerged from Tarnovskii's analysis was the discovery that "despite the many unfavorable aspects of life among the urban lower strata," as he put it, criminality was increasing at a faster rate in rural areas than in the cities, although in absolute terms urban statistics remained higher.[29]

Tarnovskii himself doubted that the urban–rural contrast was in fact as sharp as the judicial records made it seem. The relatively low level of criminal convictions in the countryside, he suggested, may have reflected village demography and the peculiar organization of the justice system. For one thing, many crimes tried in peasant courts never appeared in official statistics; for another, an unusually large proportion of the rural population consisted of women, children, and old people, precisely those groups with the lowest propensity for crime.[30] Rural men, he thus implied, might well be as lawless and violent as the urban groups with which they were often favorably compared.

Indeed, by the end of 1905 the peasantry's reputation for good behavior had begun to lose its shine. The increasing delinquency of young men was partly to blame, for they were the first to experience the social changes affecting the countryside. More likely than their elders to have achieved a certain level of literacy, to have left for the cities and returned, to have fought as soldiers in the war (perhaps in units affected by mutiny), the younger generation showed its dissatisfaction with the traditional way of life by disturbing the peace, sneering at authority, uttering profanity, harassing women and children, torturing animals, and generally making trouble simply for its own sake. Observers viewed the spread of this behavior to the countryside as the triumph of a pathology originally generated in the towns, and in fact, the problem was said to be most intense in commercial or industrial villages.[31]

Having first attracted attention in the cities, such delinquency had already acquired a special name. In common use by 1900, the term "hooliganism" designated the random mischief, malicious violence, and deliberately provocative misbehavior of boys on the street. Whatever the specific deeds it covered, its defining feature was an apparent irra-

[28]E. N. Tarnovskii, "Raspredelenie prestupnosti po professiiam," *Zhurnal Ministerstva Iustitsii*, no. 8 (1907): 64; idem, "Dvizhenie chisla," 87.
[29]E. N. Tarnovskii, "Dvizhenie chisla," 65, 67.
[30]E. N. Tarnovskii, "Raspredelenie prestupnosti," 71–72.
[31]See the attempt to define hooliganism in P. P. Bashilov, "O khuliganstve, kak prestupnom iavlenii, ne predusmotrennom zakonom," *Zhurnal Ministerstva Iustitsii*, no. 2 (1913): 222–27. Also Neil B. Weissman, "Rural Crime in Tsarist Russia: The Question of Hooliganism, 1905–1914," *Slavic Review* 37:2 (1978): 231; and Joan Neuberger, "Stories on the Street: Hooliganism in the St. Petersburg Popular Press," *Slavic Review* 48:2 (1989): 177–94.

tionality: in the words of a contemporary journalist, hooliganism constituted "illegal, malicious assault on the life, health, honor, or property
of another person, unprovoked by the victim and not inspired by the
expectation of personal gain."[32] After 1905, when this particular type of
wrongdoing seemed to have spread to rural areas, it became the subject
of investigation by a special government commission.[33] Hooliganism
and abortion were the two topics that occupied the Russian group of
the International Union of Criminologists at its convention in 1914.[34]

The pairing suggests a parallel between the male crime of random
delinquency and the female crime of reproductive self-determination.
Both were interpreted as products of city life; both represented challenges to the ability of educated men to control their sexual and social
subordinates. But in other ways they were opposites. Young men had
always provided the largest contingent of sexual offenders, but sexual
assault and various forms of public indecency formed only part of the
hooligans' repertoire.[35] It was their provocative irrationality that constituted their menace in contemporaries' eyes, not the extent to which
they indulged in specifically sexual license. Abortion, by contrast, demonstrated an excess of rational control over supposedly natural impulses. Both abortion and hooliganism, however, flouted patriarchal
authority in the interests of self-gratification. Both involved violence.

A release of disruptive passions, particularly those of a sexual nature,
seemed to follow the revolution, in the perception of the passionate and
ever-vigilant Dmitrii Zhbankov. In his 1908 essay "Sexual Bacchanalia
and Sexual Violence: Feast during Times of Plague," written for a medical journal and later revised for a more general audience, Zhbankov
argued that vicious forms of sexual transgression had reached unprecedented proportions during and after 1905.[36] Like war, he observed, revolution not only raised the general level of violence in society but also
stimulated powerful desires, "unloosing passions and leaving behind it a
long trail of brutality and a taste for all manner of violence."[37]

Like hooliganism, such ferocity did not so much satisfy personal appetites or express individual depravity as constitute a social malaise.

---

[32]"Khuliganstvo," *Novoe vremia*, no. 13318 (April 9/22, 1913): 4.

[33]Weissman, "Rural Crime," 228, 230–31.

[34]*Otchet desiatogo obshchego sobraniia Russkoi gruppy mezhdunarodnogo soiuza kriminalistov, 13–16 fevralia 1914 g. v Petrograde* (Petrograd, 1916), 103–212.

[35]On the rate of sexual assault according to the marital status of the attacker, see M. F. Zamengof, "Brak, sem'ia, i prestupnost'," *Zhurnal Ministerstva Iustitsii*, no. 2 (1916): 157.

[36]D. N. Zhbankov, "Polovaia vakkhanaliia i polovye nasiliia: Pir vo vremia chumy," pts. 1–3, *Prakticheskii vrach*, nos. 17–19 (1908): 308–10, 321–23, 340–42. Zhbankov repeated many of these arguments for a more general audience in "Polovaia prestupnost'," *Sovremennyi mir*, no. 7, sec. 2 (1909): 54–91.

[37]Zhbankov, "Polovaia vakkhanaliia," pt. 1, 308.

The erosion of normal inhibitions against aggression, Zhbankov pointed out, could be seen in the high levels of sexual assault that had characterized the recent wave of pogroms: "Women and girls were attacked regardless of age and condition," he reported. "Gangs of men raped women in public, not only on the streets but at home before their families' very eyes; women's breasts and nipples were sliced off, their bellies ripped open and stuffed with garbage; fetuses were torn from pregnant women." Such horrors might be dismissed as the work of benighted mobs encouraged by counterrevolutionary groups or by the local police to vent their anger on vulnerable neighbors—women and Jews. But Zhbankov did not consider sexual assault an incidental consequence of the revolution. Sex had become an arena of violence, in his opinion, not only because the brutality of civil conflict encouraged aggression of every kind but because the movement for political change had called into question the nature of personal relations as well as the organization of public life. "The liberation movement," he wrote, "shook society to its foundations and provoked the reexamination of all human ties. It could not therefore ignore existing forms of sexual relations: marriage, divorce, prostitution, childbirth, free love, and other questions, large and small, aroused the liveliest debate." After 1905, society found itself caught between moral systems: "The old beliefs, the time-honored foundations, the customary forms are cracking up. They have lost their meaning and force. Thanks to the triumph of political reaction, new foundations and renovated forms have not yet emerged, and people have lost their way."[38]

In this moral vacuum, sex had taken center stage not only in popular life but in everyday culture as well. If the streets teemed with violence, literature and the press overflowed with sexual license, pornographic images, and the depiction of sexual perversions, among which Zhbankov catalogued "homosexuality, sodomy, lesbianism, and bestiality." Everywhere one found dirty postcards, dirty films, and dirty books. The energy once devoted to the public cause had turned inward toward "personal life." But satisfaction was hard to come by: "Uncontrolled passions and lusts could not be sated with normal sexual intercourse," Zhbankov declared, "but demanded ever more intense stimulation: cries, moans, violence [were] needed; finally the blood and torment of the unhappy victims of human voraciousness."[39]

Most shocking to this loyal populist, sexual perversion could no longer be considered the privilege of the spoiled elite. Not only children but the common folk had lost their innocence, Zhbankov admitted,

[38]Ibid.
[39]Ibid., 308–10.

bemoaning the fate of the "lowly estates formerly known for the naturalness and simplicity of their [sexual] relations": "The thirst for something unusual, for extra stimulation, has seized people of all social ranks. Not only the sated representatives of the propertied classes, who have always indulged in various sexual perversions, but peasants and workers now also participate in raping women and girls." Still trying to soften the shock to his cherished ideal, Zhbankov contended that rapes committed by lower-class men were less offensive than those committed by their social betters. Educated men, he claimed, acted from cool calculation, whereas less sophisticated rapists only obeyed the natural force of sexual desire.[40] This artificial distinction fitted uncomfortably with Zhbankov's insistence that many rapes were being perpetrated by groups of men as part of mob action. His horrified fascination with the gang rape suggests that it may have represented the moral perversion of precisely that collective popular spirit he so admired and tried valiantly and illogically to defend.

What was the cause of this pervasive sexual disorder? Revolution had stimulated both aggression and desire; political repression had prevented these dangerous impulses from being reintegrated into a constructive social order. The rapists, Zhbankov wrote, "have been nursed on the blood and violence of recent years. They are the product of a troubled society, whose former ideals and principles have crumbled but which has been forbidden to fashion new conditions for correct public life. . . . When civic and political life cease and healthy forward movement stops, violence, perversions, the dark side of human nature rise to the surface. Without healthy public life and civic freedom there is only stagnation, living for the moment, the arbitrary imposition of authority [proizvol], and violence." If the problem was political, so were the solutions, in Zhbankov's view: "an end to the arbitrary imposition of force from above as well as from below; freedom and the widest popular participation in public political activity; the basic transformation of the country's socioeconomic life. Without these changes, the bloodletting, violence, suicide, and sexual bacchanalia will continue."[41]

Syphilis: The Survival of Old Contradictions

If Zhbankov, despite his anguished doubts, still clung to the icon of popular sexual virtue even in these troubled years, he was not the only

---

[40]Zhbankov, "Polovaia prestupnost'," 82, 86, 88; idem, "Polovaia vakkhanaliia," pt. 3, 340–41.

[41]Zhbankov, "Polovaia vakkhanaliia," pt. 3, 342.

physician to do so.[42] The post-1905 medical literature on syphilis also shows the persistence of old paradigms, even as it reveals their subtle erosion in the face of graphic social transformations. The neat division between the sexual transmission associated with city life and the non-venereal contagion of the village still exerted a powerful ideological force, but it could no longer sustain the weight of confusing and contradictory social evidence. To be sure, this tension had always existed; the social and epidemiological taxonomies had never perfectly fitted. After 1905, however, the disjuncture became more obvious in physicians' own accounts, though few were willing to abandon the dichotomy outright.

Many simply modified the scheme to accommodate the complexity of class categories and social situations they could not help observing. In 1904, for example, Nikolai Pismennyi addressed the Society of Factory Physicians on the hygienic conditions in the Serpukhov textile plant to which he was attached. His report, published in 1906, acknowledged that the workers could no longer be considered migrant peasants: "A group of factory workers has by now separated itself completely from the other social classes"; yet he maintained that syphilis still spread in the classic rural pattern. "In the factory population, syphilis is primarily the result of daily routine, not sexual activity," the doctor insisted. "Less than a quarter of all sufferers contract syphilis by sexual means, and few of these get it from extramarital relations. . . . Both adults and children are often infected by fellow apartment dwellers or casual visitors."[43] Pismennyi retained the hallowed association between class and epidemiology to the point of noting social gradations among the workers: "Workers at the textile printing factory live a more urban life than those at the cotton mill," he observed. "The first contract syphilis primarily in ways common to city dwellers or seasonal laborers, whereas the latter suffer the disease mainly in its everyday, or rural form." Therefore, Pismennyi concluded, "the best way to keep the [factory] population healthy is to maintain its ties with the village, to impede the formation of a factory proletariat, while raising the workers' cultural, intellectual, and moral level."[44]

Other physicians attempted to reconcile the conflict between ideal

[42]On the natural, healthy condition of lower-class sexuality, see D. N. Zhbankov, "Izuchenie voprosa o polovoi zhizni uchashchikhsia," pt. 2, *Prakticheskii vrach*, no. 28 (1908): 489; M. A. Chlenov, *Polovaia perepis' moskovskogo studenchestva i ee obshchestvennoe znachenie* (Moscow, 1909), 86–87; V. M. Bekhterev, *O polovom ozdorovlenii* (St. Petersburg, 1910), 17.

[43]N. N. Pismennyi, "K voprosu o sifilise v fabrichnom naselenii," *Vestnik obshchestvennoi gigieny, sudebnoi i prakticheskoi meditsiny*, no. 11 (1906), 1726, 1730–31; henceforth *Vestnik obshchestvennoi gigieny*.

[44]Ibid., 1733–34.

type and clinical experience by making the opposite argument. In 1907, for example, the St. Petersburg syphilologist Mikhail Chistiakov recounted an incident in which a woman factory worker had given birth to a syphilitic infant in her apartment. The woman who assisted her, in the manner of a village midwife, tried to revive the sickly child by breathing into its mouth. As a result, she herself became infected. "The everyday customs of the countryside," Chistiakov concluded, "would seem improbable in the capital. . . . [But this case shows that] transmission of syphilis in the rural manner can occur in the capital as well."[45] Rural habits, in this view, constituted not a protection against the disease, as Pismennyi contended, but a threat to working-class welfare. According to at least one physician, addressing a conference of factory doctors in 1910, typically rural patterns of nonvenereal contagion could still be observed among "unenlightened" (nekul'turnye) workers even in Moscow.[46] As in the years before 1905, arguments for the ubiquity of this pattern often served to deflect attention from the possibility of sexual transgression and thus to sustain the innocence of ideologically protected groups. Admitting to a conference of pediatricians in 1912 that teenage girls in poor families were frequently molested or raped by adult men, Zinaida El'tsina still attributed the majority of cases of childhood syphilis to routine everyday contacts.[47]

Some writers sustained the familiar oppositions even while undermining them with more current observations. In 1907 the newly minted physician Sergei Kolomoitsev of the Kazan military hospital reiterated the old saws about the prevalence of nonvenereal syphilis in the Russian countryside, especially in "the remote, godforsaken corners of our fatherland, carefully guarded against the penetration of culture and enlightenment." Forced to admit that things had begun to change, however, he asserted that the decrease in nonvenereal syphilis among enlisted men was evidence of their improved cultural level and increased self-awareness.[48]

The clichés died hard. A physician at the clinic run by the Moscow

[45]M. A. Chistiakov, "Sluchai derevenskogo sifilisa v stolitse," Prakticheskii vrach, no. 52 (1907): 939.

[46]N. E. Anderson, "Zabolevaemost' sifilisom sredi fabrichnogo, remeslennogo ï torgovo-promyshlennogo muzhskogo naseleniia g. Moskvy po dannym ambulatorii Miasnitskoi bol'nitsy za 1910 g.: Mery bor'by protiv sifilisa," in Trudy vtorogo vserossiiskogo s"ezda fabrichnykh vrachei i predstavitelei fabrichno-zavodskoi promyshlennosti, izdannye pravleniem Moskovskogo obshchestva fabrichnykh vrachei, ed. I. D. Astrakhan, vyp. 2 (Moscow, 1911), 42–43.

[47]Z. Ia. El'tsina, "Priobretennyi sifilis detei, ego etiologiia i bor'ba s nim," in Trudy pervogo vserossiiskogo s"ezda detskikh vrachei v S.-Peterburge, s 27–31 dekabria 1912 goda, ed. G. B. Konukhes (St. Petersburg, 1913), 230–31.

[48]S. V. Kolomoitsev, "K voprosu o vnepolovom sifilise," pt. 2, Prakticheskii vrach, no. 40 (1907): 713.

Artisans' Society declared in 1908 that "cases of nonvenereal syphilitic contagion are rather rare in the cities."[49] In 1910 Vasilii Zhukovskii, professor at Iuriev University, could still describe working-class women as "completely healthy" and morally "irreproachable" migrants from the countryside, who became infected through contact with the capital's men.[50] A 1912 report from Kostroma again insisted that rural syphilis was not 'spread by sexual contact, as it was among the urban classes.[51] In his 1910 doctoral dissertation on syphilis in St. Petersburg, Konstantin Goncharov of the Academy of Military Medicine reiterated that urban and rural patterns of contagion were completely distinct.[52]

Often physicians modified the argument to keep it alive. Semen Listov, a physician of socialist leanings, still maintained in 1910 that professional prostitution did not exist in the countryside because village communities imposed strict moral standards and because they lacked a population of unattached males eager for sexual services. Although he conceded that "depravity" *might* occur "even in villages remote from industrial and commercial centers," he claimed that "it rarely takes the form of prostitution organized as a trade or special occupation." Where the demand for sexual services did arise, as at large rural fairs, prostitution flourished, but Listov was sure the countryside had not adopted the city's crass pursuit of wealth and pleasure. It was in urban households, he reported, that servant girls first encountered the possibility of luxury, first realized the value of money, first learned to dress properly and keep themselves clean. Victims of economic and sexual exploitation, they turned to the street when they could no longer find work or wished to finance a new-found desire for comfort: "They might do so without fear of gossip because in the big cities ordinary people are as imperceptible as a grain of sand in the sea."[53]

Even those physicians who began to challenge the faith in peasant virtue did not always do so with complete consistency. The Saratov physician Sergei Starchenko, for example, criticized the Pirogov Society in 1912 for concentrating its educational efforts on improving peasant hygiene and ignoring the problem of sexual transmission in the spread of syphilis. Perhaps this focus was appropriate in village prac-

---

[49]T. M. Gershun, "K kazuistike vnepolovogo zarazheniia sifilisom: Redkii sluchai pervichnoi skleroznoi iazvy iazyka," *Vrachebnaia gazeta*, no. 24 (1908): 720.

[50]V. P. Zhukovskii, "K voprosu o rannei detskoi smertnosti v S.-Peterburge i o merakh bor'by s neiu," *Vrachebnaia gazeta*, no. 24 (1910): 737.

[51]"Sifilis v Kostromskom uezde za 16 let (1895–1910)," *Vestnik obshchestvennoi gigieny*, no. 4 (1912): 631.

[52]K. V. Goncharov, *O venericheskikh bolezniakh v S. Peterburge* (St. Petersburg, 1910), 24.

[53]S. V. Listov, "Zhenskaia domashniaia prisluga, prostitutsiia i venericheskie bolezni," *Vestnik obshchestvennoi gigieny*, no. 4 (1910): 485–87.

tice, Starchenko conceded, since most peasants indeed fell victim to nonvenereal contagion, but they were not the only ones affected by the disease. A year later he had apparently changed his view of village life: this time, in again protesting physicians' neglect of sexual contagion, he contended that it was rendered more serious still by the fact that non-venereal transmission "hardly play[ed] a major role even in the country-side." Yet even as he called for educating the public in the use of con-doms and disinfectants, he acknowledged the limitations of such a strategy. Having shifted to a more individualized conception of syphilis prevention, he fell once again into a habitual rhetorical posture in insist-ing that "the vast majority of prostitutes come from the gray country-side and are merely rouged and costumed to look urban." Given their still low cultural development, they could hardly be expected to take precautionary measures against disease. In any case, as his colleagues had lamented for decades, personal prophylaxis was of no use in solv-ing the epidemiological problems of the countryside.[54] Starchenko thus ended his argument in contradiction.

The equally contorted reasoning of another physician suggests the recognition by some observers that the stark contrast between city and country was a matter more of terminology than of social reality. In 1906 A. Balov described the ways in which sexual barter had indeed found a place in village life: some peasant women were earning money from sexual encounters "on the side" (*mezhdu delom*), he reported, as a "covert trade" (*podsovnyi promysel*).[55] Some who had worked in provin-cial towns and as laborers on large agricultural estates, where sexual mores were loose, returned to the village with bad habits and syphilitic infections. Other part-time prostitutes, he had observed, were drawn from the ranks of young widows, wives of soldiers away on service, religious sectarians, and landless peasants—in other words, women on the margins of community and family life.

Like other physicians, Balov repeated the commonplace that com-mercial sex was most likely to thrive where cash and unencumbered men abounded: in cities, in areas surrounding factories and markets. "Obviously," he wrote, "there is no demand for prostitution in small remote villages of ten to twelve households, in which all the mature men are married and all the young men are out of the way, either in military service or at seasonal labor." But Balov did not stop there. Some peasant women, he admitted, might sell their sexual services just

[54]S. N. Starchenko, "Populiarizatsiia svedenii o venericheskikh bolezniakh," *Vrachebnaia gazeta*, no. 18 (1912): 713–14; idem, "K voprosu o bor'be s venericheskimi bolezniami," *Vrachebnaia gazeta*, no. 33 (1913): 1138, 1140–41.

[55]A. Balov, "Prostitutsiia v derevne," *Vestnik obshchestvennoi gigieny*, no. 12 (1906): 1865.

"for the fun of it" (isskustva radi). Though rare, such types were not confined to "suburban villages or the neighborhood of factories, plants, estates, and trading centers," he confessed, "but [existed] even in the most remote and isolated villages." Having acknowledged their existence, the physician then denied that they could be part of normal society: "Such women," he announced, "must of course be regarded as mentally disturbed [psikhicheski bol'nye]."[56] Though Balov explained the rarity of prostitution in sturdy village communities as a function of social circumstance rather than moral superiority, he was still unable to imagine ordinary peasant women pursuing transgressive desire for its own sake.

It had always been easier for physicians and other educated observers to imagine the working class and the urban poor in general as vulnerable to the sexual temptations supposedly bred by city life, though the ambiguity of class boundaries had weakened the force of the rural-urban contrast. In order to retain the distinction, for example, Konstantin Goncharov introduced certain modifications into the social map. He denied that the industrial outskirts of large cities qualified as urban space, since the workers who lived there maintained their former village customs. Only in the depths of the urban lower orders, among common laborers and prostitutes, he claimed, had peasant ways been lost and sexual contact become the prime source of venereal disease. Unable to abandon the conventional notion that village women were free from urban vice, Goncharov insisted that "those very same simple women who constitute the overwhelming majority of prostitutes in the cities resort to prostitution very rarely in the countryside."[57]

Despite this stubborn conviction, Goncharov in fact registered one of the major post-1905 shifts in the syphilis paradigm by arguing that the urban poor, in their misery and promiscuous sexual habits, threatened the physical welfare of the well-to-do.[58] If physicians had earlier invoked the fluid boundary between woman worker and streetwalker to deny that prostitutes were sociologically distinct from honest women, they now invoked the close connection between the two types to explain the danger posed by all subordinate women. The housemaid who was once thought to spread syphilis by folding laundry and serving soup, and who therefore symbolized the ubiquitous threat of non-venereal contagion, could now appear also as a link in a chain of sexual transgressions.

[56]Ibid., 1865–68.
[57]Goncharov, O venericheskikh bolezniakh, 24, 84, 97.
[58]Ibid., 91. As we noted in chap. 5, reports by the Moscow police-medical committee from the early 1850s characterized factory workers as sexually promiscuous, but such statements do not appear in the medical press later in the century.

In portraying the health menace constituted by servile women, Goncharov employed language that underscored rather than obscured the sexual dimension of the threat. Female servants are no less ridden with syphilis than professional prostitutes, Goncharov warned, and they "enjoy the most intimate relations with their masters and employers. They penetrate everywhere: into high society mansions and luxurious boudoirs, business offices, nurseries, and bedrooms; and at the same time, into miserable hovels, flophouses, and every possible den of poverty and iniquity."[59] Rather than stigmatize prostitutes as a breed apart, as Veniamin Tarnovskii had done, Goncharov preferred to think of the entire working class as a wellspring of infection, in which venereal and nonvenereal transmission played inextricable roles.

So long as an ideological line was drawn between prostitution as a trade and the more casual or erratic exchange of sexual services, a case could still be made for the difference between city and country. The contrast eroded, however, as more observers admitted that rural women often made money on sex, and others began to recognize a wider range of sexual practices among lower-class urban women. David Gurari, for example, conceded in 1910 that many shopgirls supplemented their earnings and relieved the boredom of their jobs by having affairs with educated men willing to pay for the illusion of disinterested love. Neither poverty nor "naked desire" was to blame for these women's behavior; rather, Gurari explained, "confused sexual attraction, in connection with the thirst for culture aroused by class contrasts, pushes such women into the arms of the first man who promises to take her away from her drab daily existence, if even for a time."[60] Here was the admission that female sexual transgression was not necessarily a sign of self-degradation but perhaps a symptom of cultural aspiration or, at the very least, of a desire to circumvent the power relations of everyday life. It was this provocative element in prostitution that allowed it to symbolize the threat of working-class disorder in the wake of 1905.

Naughty Nellies

Caught between village and town, the urban poor had never enjoyed the moral prejudice displayed in the peasants' favor, as the pre-1905 syphilis literature shows. Labor militance in 1905, combined with the

[59]Ibid., 75, 86.
[60]D. L. Gurari, "K voprosu o neprofessional'noi prostitutsii," *Gigiena i sanitariia*, no. 4 (1910), 287.

general rise in urban crime and the particular salience of juvenile delinquency, deprived the working class of its last vestige of innocence. The
professional focus on the sexual problems of educated males now
merged with a general anxiety about the city as the site of a newly
threatening sexual youth culture, in which the classes mingled to their
mutual harm. "Sex has become a weighty and agonizing problem as a
result of the new complexity in social relations, the turbulent feverish-
ness of public activity, and the development of urban life," wrote the
socialist physician Veniamin Kanel' in a 1909 article on children's sexuality. "The appearance . . . of large groups of people unable to satisfy
their vital needs in proper fashion has made relations between the sexes
abnormal, indeed pathological."[61] City living undermined folk customs
and weakened traditional family ties, releasing not only women but
children too from the chastening bonds of patriarchal authority. The
result was sexual as well as social chaos, the emergence of a new public
space inhabited by female creatures with the independence and energy
of men and by men who sought a way to exercise power, or to claim
power of a new kind, outside the frame of respectable family life.

Curiously, neither the rowdy young man from the slums nor the
educated adult male, who dominated the statistics on hooliganism and
sexual crime, respectively, figured as the symbolic expression of the
urban sexual threat. That role was filled instead by the teenage prostitute.[62] Worried for decades about the problem of commercial sex, professionals and public figures now focused their rhetoric on its juvenile
dimension. Their intensified concern with the precocious sexuality of
teenage and preteen girls from the lower classes paralleled their interest
in the sexual habits of young boys from their own social milieu. In each
case they diagnosed misdirected desire, though pointed in opposite directions. The sexual problems of schoolboys were confined to interior
space: the bedroom, the playroom, the dormitory, the guarded places
that prepared them for entry into the public world when they grew
older. Their vices were those of privacy; solitary self-pleasure was antisocial, a retreat from interpersonal exchange.[63] Though many physicians
recognized that masturbation was not confined to boys, some were

[61]V. Ia. Kanel', "Polovoi vopros v zhizni detei," *Vestnik vospitaniia*, no. 4, pt. 1 (1909):
139–40.

[62]On the high level of conviction for sexual crimes among members of the liberal
professions, see E. N. Tarnovskii, "Raspredelenie prestupnosti," 95–96. On the prostitute
as female analogue of the hooligan, see Neuberger, "Stories on the Street," 185. For
adolescent prostitution in the American context, see Christine Stansell, *City of Women:
Sex and Class in New York, 1789–1860* (New York, 1986), 181–85, 190.

[63] Onanism was "an individualized form of lust, an end in itself rather than a means of
perpetuating the species": S. Zalesskii, *Polovoi vopros s tochki zreniia nauchnoi meditsiny:
Gigienicheskii etiud* (Krasnoiarsk, 1909), 17.

convinced that "pubescent girls" rarely practiced it.[64] The vices of young girls were primarily public: they took what should have been private (female chastity) and exchanged it on the street for a share in the worldly resources available to adult males.

If postrevolutionary observers did not invent the link between sexual and social disorder, they did change the tone in which long-recognized problems were debated and the conceptual terms in which they could be understood. Zemstvo physicians, as we have seen, had stressed hygiene and health education over administrative measures as the key to syphilis control. Opponents of regulation rejected Veniamin Tarnovskii's claim that the system was needed to limit the harmful influence of a dehumanized breed; they condemned it rather for dehumanizing those who came into its clutches. Many who justified the regulatory system did so in the spirit of the French public health expert Alexandre Parent-Duchâtelet as prophylaxis against disease, not against sexual pathology. After 1905, as John F. Hutchinson has argued, distrust of the state's role in public health declined, and faith in technical solutions grew among physicians who had until then relied on the principles of community medicine for practical guidance and political self-definition.[65] At the same time, physicians shed their earlier reluctance to think of prostitution in pathological terms.

The contest between social and biological interpretations of deviance did not disappear. Many professionals, still rejecting the idea that criminals or prostitutes were constitutionally defective, stressed the environmental circumstances responsible for their fate.[66] But the impression was widespread that prostitution itself had taken new, distinctly pathological forms—in particular, that underage girls were to be found on the streets in unheard-of numbers, in states of abjection and degradation formerly unknown.

[64]Denying the practice among girls: Iu. Fridlender, *Nravstvennye epidemii* (St. Petersburg, 1901), 48–49. Admitting it: A. Kh. Sabinin, *Prostitutsiia: Sifilis i venericheskie bolezni* (St. Petersburg, 1905), 258; B. I. Bentovin, "O prostitutsii detei," in *Trudy pervogo vserossiiskogo s"ezda po bor'be s torgom zhenshchinami i ego prichinami, proiskhodivshego v S.-Peterburge s 21 po 25 aprelia 1910 goda* (St. Petersburg, 1912), 2:442; *Net bolee onanizma, venericheskoi bolezni, poliutsii, muzhskogo bessiliia i zhenskogo besplodiia: Prakticheskie sredstva snova vosstanovliat' i ukrepliat' zdorov'e, rasstroennoe etimi bolezniami*, 2d ed. (Moscow, 1865), 41; A. N. Bernshtein, *Voprosy polovoi zhizni v programme semeinogo i shkol'nogo vospitaniia* (Moscow, 1908), 13; *Grekhi molodykh liudei: Nastol'naia kniga*, 3d ed. (Moscow, 1906), 107–8 (lurid description of the consequences for women).

[65]Hutchinson, " 'Who Killed Cock Robin?' "

[66]P. E. Oboznenko, "Obshchestvennaia initsiativa S.-Peterburga v bor'be s prostitutsiei," pt. 2, *Vestnik obshchestvennoi gigieny*, no. 12 (1905): 1864, 1894–95; I. I. Kankarovich at the 1910 prostitution congress, reported in N. P. Danilov, "Pervyi vserossiiskii s"ezd po bor'be s torgom zhenshchinami," *Izvestiia Moskovskoi gorodskoi dumy*, no. 9, Otdel obshchii (1910): 98; Goncharov, *O venericheskikh bolezniakh*, 97, 105.

Juvenile prostitutes. M. K. Mukalov, *Deti ulitsy: Maloletnie prostitutki* (St. Petersburg, 1906). Helsinki University Slavonic Library.

The horrors of prostitution in any form were an old story, as was the use of biological language even in texts depicting prostitution as a social problem. In his widely read 1864 series on the St. Petersburg slums, Vsevolod Krestovskii had described brothels as "the gangrene of our society," "the sores of the social organism." Women experienced in the trade soon lost all human qualities, he said. Worse than depraved, they were bestial, mindless, apathetic, repellent: "faceless creatures, lacking character, independence, will power, or an understanding of life outside the narrow confines of their profession. . . . The turbid atmosphere in which they move seems to them completely natural, normal, and fitting." He did not, however, depict them as originally defective but blamed "the conditions of the [women's] existence, . . . their social situation," and particularly the "ruthless and corrupting slavery" engendered by the system of regulated brothels for "leading to the complete destruction of their individuality and human rights," reducing them to the level of dumb animals.[67]

As for juvenile prostitution, evidence had long been available that young girls offered commercial sex on the streets of the capital city. A midcentury commission on "vagrant women of depraved comportment" had found "girls of nine to fourteen years old" who engaged in "inexcusable naughtiness [shalosti] . . . verging on prostitution."[68] Dostoevsky's 1861 novel *The Insulted and Injured* centered on the attempted sexual exploitation of a little orphan girl called Nelly. In 1869 a reporter for *Otechestvennye zapiski* deplored the sad fate of girls who were deflowered at an early age, sometimes by their own fathers, left home for the streets, and ended in syphilis wards. Their misfortune testified, he said, to the "horrifying character of our morals."[69] The story of a miserable prostitute who sold the sexual services of her own thirteen-year-old daughter served Leo Tolstoy in 1886 as an emblem of the moral depths to which society had fallen.[70]

Such moral melodramas obviously appealed to respectable readers, eager for tales of danger and redemption. The themes mined by Dostoevsky and Tolstoy echoed in the pages of medical reports, newspapers, and popular pamphlets. In 1888 the physician Vladimir Okorokov celebrated the work of Moscow's two-decade-old Mary Magdalene shelter for wayward girls in his pamphlet *Returning Fallen Girls to Honest Labor*, designed to arouse pity for the juvenile victims of a "complex and far-flung system engaged in the capitalist exploitation of depravity."[71] Okorokov pictured the victimized girls as daughters of

[67]V. V. Krestovskii, "Peterburgskie trushchoby," in *Sobranie sochinenii Vsevoloda Vladimirovicha Krestovskogo* (St. Petersburg, 1899), 2:559–61.

[68]"O prostitutsii v Rossii: I. Iz otcheta o deistviiakh v S.-Peterburge komissii dlia razbora brodiachikh zhenshchin razvratnogo povedeniia v techenie 5-ti let, s 27-go aprelia 1847 g. po 27 aprelia 1852 goda," *Arkhiv sudebnoi meditsiny i obshchestvennoi gigieny*, no. 1, pt. 3 (1869): 104; henceforth *Arkhiv sudebnoi meditsiny*. The committee's observation was noted in Mikhail Kuznetsov, *Prostitutsiia i sifilis v Rossii: Istoriko-statisticheskie issledovaniia* (St. Petersburg, 1871), 225.

[69]L. R., "Sovremennye zametki: Uchebno-vospitatel'noe zavedenie dlia nesovershennoletnikh prostitutok," *Otechestvennye zapiski*, no. 9 (1869): 125–58.

[70]L. N. Tolstoi, "Tak chto zhe nam delat'?" (1886), in *Polnoe sobranie sochinenii L'va Nikolaevicha Tolstogo* (Moscow, 1913), 17:36–37.

[71]V. P. Okorokov, *Vozvrashchenie k chestnomu trudu padshikh devushek* (Moscow, 1888), 12 (repeated almost verbatim in *Prodazha devushek v doma razvrata i mery k ee prekrashcheniiu* [Moscow, 1899]). More on the history of charitable homes for girls in B. I. Bentovin, "Spasanie 'padshikh' i khuliganstvo (Iz ocherki sovremennoi prostitutsii)," *Obrazovanie*, no. 11–12, pt. 1 (1905), rpt. in Bentovin, *Torguiushchie telom: Ocherki sovremennoi prostitutsii*, 2d ed., rev. (St. Petersburg, 1909); also D. G. Mordovtsev, *Zhivoi tovar: Postydnaia mezhdunarodnaia torgovlia molodost'iu i krasotoi i mery protiv beznravstvennykh sovratitelei zhenshchin* (Moscow, 1893), 71–91. See also Laurie Annabelle Bernstein,

the working class, both artisanal and industrial, earning miserable wages on their own. He described them not as teenagers who had chosen the street over available employment but as orphans or as children of working parents who were unable to exert a "moral influence" upon them. Apprenticed at an early age, employed in large factories, or sent into domestic service, ill-paid and ill-treated girls easily fell prey to crafty agents who entrapped them with promises of jobs and luxuries. Their seducers, in Okorokov's opinion, were part of the same entrepreneurial system that had rendered them destitute in the first place. "The trade in fallen girls is a veritable branch of industry," he wrote. "Girls are sold to regulated brothels by special dealers," he informed his readers, then added a casual anti-Semitic note: "The men, who are mostly Jewish, are known as 'Maccabees.'"[72]

Aware of theories that described public women as degenerate types, Okorokov admitted that some young victims of the sexual trade might harbor the germ of "psychopathology," which would flourish once they were extracted from a "normal" life of constant labor. It is clear that Okorokov only dimly understood the import of these ideas, however, since they did not shake his faith in the power of salvation. By teaching the girls good work habits, he was sure, the shelters would turn the most reprobate into decent citizens again.[73]

Another antiregulationist diatribe, *Living Merchandise: The Shameful International Trade in Youth and Beauty and Measures against the Immoral Perverters of Women* (1893), which borrowed directly from Okorokov's text, accused "the purveyors of white slaves on the international prostitution market" of forming "Black Hundred bands" that preyed on women and girls as young as eleven.[74] The idea would not have shocked readers of antiprostitution literature. The feminist writer Serafim Shashkov had already declared in 1871 that an "extensive and well-organized system of commercial depravity envelopes all of Europe. It operates through numerous agents, many intrigues, stratagems, subterfuges, and even newspaper advertisements," preying on naive young women.[75] The Brockhaus encyclopedia article on prostitution, written by Grigorii Gertsenshtein in 1898, also testified to the existence of such

---

"Sonia's Daughters: Prostitution and Society in Russia" (Ph.D. diss., University of California, Berkeley, 1987), chap. 5. See also Oboznenko, "Obshchestvennaia initsiativa," pt. 2; Vseslav, "Prizrenie maloletnikh prostitutok v Rossii," *Novoe vremia*, no. 9137 (August 12/25, 1901): 4.

[72]Okorokov, *Vozvrashchenie*, 14.

[73]Ibid., 60–64.

[74]Mordovtsev, *Zhivoi tovar*, 2. See also Okorokov *Vozvrashchenie*, 59; K. Griaznov, *Prostitutsiia, kak obshchestvennyi nedug i mery k ego vrachevaniiu* (Moscow, 1901), 112–14.

[75]S. S. Shashkov, *Istoricheskie sud'by zhenshchiny: Detoubiistvo i prostitutsiia* (St. Petersburg, 1871), 575–78.

an international trade in "living people" and to the market in girls as young as nine.[76]

The prevalence of these themes reflects the fact that Russians had followed the growth of "white slavery" as a public issue in Europe.[77] The opinions and work of the British antiregulationist Josephine Butler were well known, and her colleague, William Coote, visited Russia in 1899 and had an audience with the foreign minister.[78] Russian newspapers reported on international congresses, which Russians regularly attended. Andrei Saburov, a prominent legal expert and State Council member, went to London in 1899, to Paris in 1900, to Amsterdam in 1901, to Frankfurt in 1902, to Zurich in 1904, and to Brussels in 1912.[79] When Vladimir Dmitrievich Nabokov wished to emphasize the importance of strengthening the laws against child molestation, he cited W. T. Stead's 1885 *Pall Mall Gazette* exposé, "The Maiden Tribute of Modern Babylon."[80]

In 1900 Russians formed their own antiregulationist association on the British model, calling it the Society for the Protection of Women.[81] William Coote was an honorary member. Princess Elena Saksen-Al'-tenburgskaia presided, along with Andrei Saburov; Princess Evgeniia Ol'denburgskaia acted as the official sponsor; the ministries of internal affairs, foreign affairs, justice, and trade and industry were represented on its rolls. By 1913 the society had about four hundred members, almost half representing the Jewish community. Over the years, the

---

[76]G. M. G[ertsenshtein], "Prostitutsiia," in *Entsiklopedicheskii slovar' Brokgauz-Efron* (St. Petersburg, 1898), 25A: 484–86.

[77]On the European movement, 1881–1910, see Alain Corbin, *Les Filles de noce: Misère sexuelle et prostitution (19ᵉ et 20ᵉ siècles)* (Paris, 1978), 386, 406–12, 424–27, 431–33. Russian reports appear in, e.g., V. F. Deriuzhinskii, *Politseiskoe pravo: Posobie dlia studentov*, 3d ed. (St. Petersburg, 1911), 380–83; P. E. Oboznenko, "Vopros ob uporiadochenii prostitutsii i o bor'be s neiu na dvukh mezhdunarodnykh soveshchaniiakh 1899 goda," *Vrach*, no. 30 (1900): 912. For the influence of European congress resolutions on Russian legislation, see D. A. Koptev and S. M. Latyshev, eds., *Ugolovnoe ulozhenie (stat'i vvedennye v deistvie)* (St. Petersburg, 1912), 903–4.

[78]A. I. Elistratov, *O prikreplenii zhenshchiny k prostitutsii: Vrachebno-politseiskii nadzor* (Kazan, 1903), 232, 260–62; Edward J. Bristow, *Prostitution and Prejudice: The Jewish Fight against White Slavery, 1870–1939* (Oxford, 1982), 267.

[79]L. G., "Belye rabyni," *Novoe vremia*, no. 9553 (December 8/21, 1902): 2–3; Oboznenko, "Vopros ob uporiadochenii," 913; V. F. Deriuzhinskii, "Pamiati A. A. Saburova," in *Rossiiskoe obshchestvo zashchity zhenshchin v 1915 godu* (Petrograd, 1916), 10–11.

[80]V. D. Nabokov, "Plotskie prestupleniia, po proektu ugolovnogo ulozheniia," *Vestnik prava*, no. 9–10 (1902), rpt. in V. D. Nabokov, *Sbornik statei po ugolovnomu pravu* (St. Petersburg, 1904), 91. Stead's piece was also mentioned in Griaznov, *Prostitutsiia*, 160.

[81]Oboznenko, "Vopros ob uporiadochenii," 915; idem, "Obshchestvennaia initsiativa," pt. 1, *Vestnik obshchestvennoi gigieny*, no. 11 (1905): 1672–75; I. M. Faingar, "Detskaia prostitutsiia," *Vestnik psikhologii, kriminal'noi antropologii i pedologii*, no. 3 (1913): 39–42.

membership included many prominent medical and legal practitioners: Veniamin and Praskov'ia Tarnovskii; the physicians Mariia Pokrovskaia and Zinaida El'tsina; the jurists Sergei Gogel', Vladimir Deriuzhinskii, Anatolii Koni, Vladimir Nabokov, and Arkadii Elistratov; the psychiatrist Vladimir Bekhterev; and the social activist Dmitrii Dril'.[82]

The society's two greatest accomplishments were the enactment of the antiprostitution statutes in 1909, for which it had strenuously lobbied, and the organization of a congress on prostitution, held in St. Petersburg in 1910. The statutes strengthened the criminal penalties against traffickers in commercial sex; they were the only part of the chapter on sex in the 1903 reformed code to be given the force of law.[83] This achievement was the fruit of backstairs influence and pressure in high government quarters.[84] The congress, by contrast, was a broadly based public affair that drew participants from a range of professional, civic, and official domains. Altogether, three hundred people attended, of whom only a small fraction were society members. Various charitable organizations, feminist parties, bureaucratic offices, and educational institutions sent representatives. Only 16 percent of those present (49) were physicians. The strongest single contingent consisted of women. The society itself was divided evenly by sex except for the Jewish section, which was 80 percent female. At the 1910 congress, almost two-thirds of the participants (194) and nearly half the physicians (23) were women, many clearly motivated by feminist concerns. Half of the women (including ten of the female physicians) had also participated in the conference on the woman question sponsored by organized feminists in St. Petersburg two years before.[85]

Mounted by a social organization, the 1910 event reflected the new civic initiatives of the urban social elites: attorneys and physicians played a subsidiary part and respectable women took the public stage to address the issue of public sex. In both respects this congress on prostitution differed from the first meeting at which the question of pros-

---

[82]*Rossiiskoe obshchestvo zashchity zhenshchin v 1913 godu* (Petrograd, 1914), 133–50; TsGIA, f. 1335, op. 1, d. 24 (Alfavitnyi spisok chlenov Rossiiskogo obshchestva zashchity zhenshchin s 1901 goda), ll. 20–69.

[83]See chap. 2 above.

[84]On Saburov's role, see Deriuzhinskii, "Pamiati A. A. Saburova," 7–9. The archival record of the society's involvement is in TsGIA, f. 1405, op. 543, d. 512.

[85]On individual feminists, see Richard Stites, *The Women's Liberation Movement in Russia: Feminism, Nihilism, and Bolshevism, 1860–1930* (Princeton, N.J., 1978), 192, 225–26. For conference attendance, see "Spisok chlenov," in *Trudy pervogo vserossiiskogo s"ezda po bor'be s torgom zhenshchinami,* 1:7–21; "Spisok chlenov pervogo zhenskogo s"ezda," in *Trudy pervogo vserossiiskogo zhenskogo s"ezda pri Russkom zhenskom obshchestve v S.-Peterburge, 10–16 dekabria 1908 goda,* 907–20 (St. Petersburg, 1909); and Danilov, "Pervyi vserossiiskii s"ezd," 94.

titution had been a focus of open discussion. The conference on syphilis held in 1897 had been a strictly professional affair; sponsored by the Russian dermatological association with the approval of the Ministry of Internal Affairs, that gathering had included only physicians and government medical officials. And of the 450 persons present, only two were women: Veniamin Tarnovskii's loyal assistant Zinaida El'tsina and a physician from Poltava.[86]

Since 1897, not only had physicians broadened the context of their social activism from the narrow institutional and professional confines imposed by the pre-1905 regime, but the composition of the profession had evolved as well. Between 1898 and 1916 the number of registered physicians increased by 58 percent; the proportion of women practitioners rose from 3 to 15 percent of the total; and the names of one-third of the earlier group of doctors were absent from the medical directory of 1916.[87] The profile of the 1910 gathering reflected the generational change. A number of leading figures had departed: Eduard Shperk died in 1894, Grigorii Gertsenshtein in 1899, Veniamin Tarnovskii in 1906, and Praskov'ia Tarnovskaia (who had attended the 1908 women's congress) in 1910. Only three of the male physicians present in 1910 had also participated in the 1897 syphilis congress.

The disparity in numbers may deprive a comparison of any statistical meaning, but it is nevertheless striking that the two groups also reveal different occupational contours. Over half of the doctors present in 1897 were associated with zemstvo or municipal medicine; one-third worked for the police, bureaucracy, government administration, or venereal disease hospitals. In 1910, zemstvo and public hygiene specialists were few (6 percent of the 49 physicians present) and government employees only somewhat more numerous (12 percent). Venereal disease hospital staff constituted the largest group (20 percent); the rest worked at a variety of hospitals and in private practice. In short, the prostitution problem had been recontextualized; the dominance of zemstvo medicine had come to an end; and socially and politically active women had entered the public debate in unprecedented numbers.

These developments may help explain the different ways the two gatherings handled the central social issue with which they were concerned. At the first, the question of prostitution entered into the political struggle between physicians and the state, in which the experts asserted their right to pronounce on social policy questions and

---

[86] "Spisok chlenov s"ezda," in *Trudy Vysochaishe razreshennogo s"ezda po obsuzhdeniiu mer protiv sifilisa v Rossii*, 2:259–66 (St. Petersburg, 1897). I thank Ann Toohey at the National Library of Medicine for making this list available.

[87] Cf. *Rossiiskii meditsinskii spisok* (1898 and 1916). The volume for 1910 was not available to me, so I use 1916 as an index of the rate of change.

challenged the regime's administrative control over their own professional affairs. The participants could not agree, however, on the value
of regulation. Against the protests of a large minority, the 1897 conference adopted an ambiguous resolution, insisting on a greater role for
physicians in a system it criticized as inadequate but failed to condemn
outright. The 1910 conference, by contrast, denounced regulation; it
advocated voluntary medical care and laws to suppress prostitution.[88]

The published record does not indicate how the participating physicians voted on a position that certainly affirmed the attitude of organized feminists. Many of the doctors present, especially those who were
also active in feminist concerns, would surely have agreed. But Zinaida
El'tsina, the lone female holdover from 1897, remained true to the
spirit of her late mentor, Veniamin Tarnovskii: "I do not sympathize
with regulation from the ethical point of view," she said in a separate
dissenting opinion, "but I must consider the sanitary interests of the
empire's millions of rural inhabitants, whose welfare is dearer to me
than the fate of the insignificant numbers who enter the ranks of prostitution."[89] Her tone of scientific pragmatism was not shared by the
moral purity majority, which preferred to penalize rather than regulate
illicit sex. Arkadii Elistratov, professor of police law at Moscow University and a participant in the 1910 congress, later drafted legislation to
end regulation and outlaw prostitution, which the Kadet deputy and
physician Andrei Shingarev presented to the State Duma in 1913.[90]

Partisans of the regulatory system, such as Veniamin Tarnovskii, had
seen prostitution as the root of all syphilitic contagion, even the endemic variety common in the countryside. Zemstvo practitioners had
been more likely to downplay the role of sexual transmission and also
to oppose the administrative imposition of medical care. Focusing on
nonvenereal contagion, antiregulationists had argued, first, that because
only sexual activity (not everyday life) could be policed, compulsory
exams and treatment were useless in the countryside; and second, that
the police-run public health regime violated the right of prostitutes to
civil dignity and personal autonomy. The antagonism to regulation in
1910, however, seems to have reflected not the legacy of zemstvo medicine, whose moral authority had by then declined, but rather changing
attitudes toward sexuality. Abolitionists still spoke in the name of civil
rights but at the same time emphasized the dangers inherent in uncon-

[88]D. N. Zhbankov, "O s"ezde pri Meditsinskom Departamente po obsuzhdeniiu
meropriiatii protiv sifilisa v Rossii," pt. 2, *Vrach*, no. 30 (1897): 831; Danilov, "Pervyi
vserossiiskii s"ezd," 114.
[89]"Osoboe mnenie chlena s"ezda Z. Ia. El'tsinoi," in *Trudy pervogo vserossiiskogo s"ezda
po bor'be s torgom zhenshchinami*, 2:604.
[90]"Bor'ba s prostitutsiei," *Zhenskii vestnik*, no. 12 (1913): 265.

trolled sexual desire. Unlike the regulationists, however, they wished to transfer control from agents of the state to individual subjects: by educating men and women in sexual self-restraint, suppressing cultural stimulants to desire, and rendering criminal what had earlier been administratively contained, they hoped to force participants in the sexual trade to take responsibility for their actions.[91]

Perhaps it was intensified concern with individual responsibility that made juvenile prostitution a central issue in both the 1909 statutes and the 1910 congress deliberations, for the line between adulthood and minority divided the legally accountable from the legally immune. The behavior and treatment of children were certainly matters of public concern. Special courts for juvenile offenders had been established in 1897.[92] The authors of the 1903 criminal code, as we have seen, identified the sexual abuse of children as a primary consideration. The laws adopted in 1909, based on the 1903 text, penalized men who profited from the vulnerability of underage girls to entice them into a life of vice, especially if the girls had been transported across national borders. Physicians too had broached the subject. At the 1897 conference on syphilis the issue of juvenile prostitution had arisen as a technical point in the endless debates over regulation, which applied only to adult women. The businesslike exchange of opinion focused on legal distinctions between various definitions of minority, but participants thought the question incidental to the more contested problem of regulation itself.[93] By 1910, however, the tone of discussion had decisively altered, and so-called child prostitutes came to symbolize the entire social problem in its most acute and menacing form.

The Crisis of Proletarian Family Life

The concern with juvenile prostitution was embedded in a larger anxiety, which had already emerged before 1905, about the urban lower-class family. Its expression built on existing rhetorical structures, complete with their own internal contradictions. In an 1886 essay on the "scum" or "plague" (iazvy) of St. Petersburg, Vladimir Mikhnevich had assailed the disintegration of the nation's moral fabric while continuing, in a fashion typical of his generation, to celebrate the unspoiled instincts of the common folk. As society suffered the wrenching transi-

---

[91]The 1897 congress had also evinced an interest in cultural purity, and the Society for the Protection of Women harped on this theme until the war. See chap. 5 and 10.

[92]I. M. Tiutriumov, "O zashchite detstva (zasedanie komiteta 8 ianvaria 1900 g.)," in Mezhdunarodnyi soiuz kriminalistov, russkaia gruppa, 1899–1902 (St. Petersburg, 1902), 183.

[93]Reported in Trudy Vysochaishe razreshennogo s"ezda, 2:94–99.

tion from traditional to modern life, he observed, families were undergoing a crisis that extended "from the patriarchal muzhik's blackened cottage to the magnificent mansions of the highborn, brilliant grandees."

Pervasive as this crisis might be, it did not affect all social classes in the same way. For the common folk, the threat to family integrity came from the attractions of the big city, which lured village youth away from home to a life of impoverished and frustrated bachelorhood. Peasants who "in their primitive state [v primitivnom sostoianii] . . . strongly endorse the principles of family life and are by nature not disposed to depravity," Mikhnevich reported, ". . . los[t] all moral sense and freely violate[d] the seventh commandment" when they came to work in the towns. Since most of these migrants were young men, either unmarried or separated from their families, was it "conceivable that chastity and virtue in sexual relations should flourish?"[94] It was the call of nature in unnatural circumstances that led the peasants astray.

The "depravity" of privileged urban society, by contrast, consisted precisely in the denial of nature, the subordination of "healthy instincts" to "social propriety, status snobbery [soslovnye predrassudki], and material interest." In prostitution, however, the degradation of city life affected women across the social spectrum. Commercial sex lurked everywhere, Mikhnevich warned: in "café-chantants" and "dance classes," in the private cabinets of restaurants and baths. From respectable ladies who financed luxurious tastes to exploited seamstresses oppressed by their jobs, from servant girls seduced by employers, lackeys, and house superintendents to female operatives pressed against men on the factory bench, women traded sexual services for cash and consideration. Yet despite the leveling effect of prostitution, Mikhnevich still considered common-law relationships among the poor less degrading to the female partners than comparable arrangements among the upper classes.[95]

The supposed moral superiority of working-class couples, which Mikhnevich associated with the virtues of honest labor and the habits of moral order learned in the countryside, was not supported by the later literature on the housing question, which denounced the sexual claustrophobia of working-class families and the libidinous disorder to which it purportedly gave rise. In the 1890s, public health physicians and social observers began to emphasize the morally harmful effects of crowding on the development of lower-class children[96] and to distin-

---

[94]Vladimir Mikhnevich, Iazvy Peterburga: Opyt istoriko-statisticheskogo issledovaniia nravstvennosti stolichnogo naseleniia (St. Petersburg, 1886), 352, 356–57.

[95]Ibid., 358–59, 366–67, 417–25.

[96]A. N. Rubel', "Zhilishcha bednogo naseleniia Peterburga," Vestnik obshchestvennoi

guish the congestion of urban households from the similar though allegedly less pernicious crowding that also afflicted peasant lives.[97] By 1905, however, and especially after 1910, the literature was insisting on a specifically sexual connection between the problems of juvenile crime, juvenile prostitution, and the chaotic state of proletarian family relations. By then the "naughtiness" of midcentury runaway girls had become a sign of sexual pathology, produced by the social pathology of working-class life.

The authors of such views were not necessarily hostile to the popular cause: among them were the physician Veniamin Kanel', a Social Democrat, and Anatolii Sabinin, a liberal public health activist from Voronezh.[98] In opposing regulated prostitution, Sabinin argued that extramarital sex should not receive official support because it was "abnormal" (*anormal'noe*) rather than simply immoral. The normality he had in mind did not, however, correspond to the state of affairs in lower-class families. "Thanks to the conditions in which they live," the doctor wrote, "the children of the poor become acquainted with the sexual functions at too early an age. It is fairly common for girls to indulge in unnatural sexual practices when they mature." The habit of masturbation, for example, helped the girls to calm desires aroused by the "benighted family situations" in which they were raised. "Under such circumstances," Sabinin intoned, "it is but a single step to prostitution."[99]

A socialist pamphlet on prostitution published in 1906 described housing conditions in which "old and young are all mixed together in a heap, men and women sleep side by side on the floor. It is not far from such intensely close living to disorderly sexual relations. . . . Children, of course, also live here, imbued with depravity, alcohol, and the at-

---

gigieny, no. 4 (1899): 442–43; M. N. Gernet, "Prestupnost' i zhilishcha bedniakov," *Pravo*, no. 43 (1903): 2396, 2399; I. Verner, *Zhilishche bedneishego naseleniia Moskvy* (Moscow, 1902), 13–14, quoted in P. V. Vsesviatskii, "Prestupnost' i zhilishchnyi vopros v Moskve," *Pravo*, no. 20 (1909): 1266. For a similar theme in midcentury French texts, see Joan W. Scott, "'L'Ouvrière! Mot impie, sordide . . .': Women Workers in the Discourse of French Political Economy, 1840–1860," in Scott, *Gender and the Politics of History* (New York, 1988); for fin-de-siècle Vienna, see Sander L. Gilman, *Difference and Pathology: Stereotypes of Sexuality, Race, and Madness* (Ithaca, N.Y., 1985), 42–43.

[97]B. I. Bentovin, "Istochniki prostitutsii," *Russkoe bogatstvo*, no. 11 (1904), rpt. in Bentovin, *Torguiushchie telom*, 42–43.

[98]Kanel', "Polovoi vopros," 152–53. On Sabinin, see E. I. Lotova, *Russkaia intelligentsiia i voprosy obshchestvennoi gigieny: Pervoe gigienicheskoe obshchestvo v Rossii* (Moscow, 1962), 36–37.

[99]Sabinin, *Prostitutsiia*, 241–47, 258–59. Other experts believed that early sexual experience, often in the form of sexual assault, aroused the sexual instincts and doomed the girls to a life of compulsive depravity: Bentovin, "O prostitutsii detei," 451–54; Faingar, "Detskaia prostitutsiia," 36; V. I. Gal'perin, "Prostitutsiia detei," in *Deti-prestupniki*, ed. M. N. Gernet (Moscow, 1912), 369–71.

mosphere of crude verbal abuse."[100] Numerous commentators mentioned the frequency of incest between young girls and their fathers, uncles, and brothers.[101] A sociological report of 1912 explained how poverty led to moral perdition: "The girl-prostitute almost always spends her childhood in a dank, dark, and tedious corner, ... accustomed from her earliest years to see the most indecent caresses, to hear the most cynical words. She thus falls easily into the maelstrom of depravity [*omut razvrata*]."[102] A similarly ominous picture was drawn by Dmitrii Dril', a prominent social reformer, in his presentation to the 1910 prostitution congress in St. Petersburg.[103]

Political radicals, civic activists, professional sociologists, and moral purity crusaders all manipulated the same rhetoric of sexual crisis to convey the menace of a dangerously unstable social environment. A moralist at the 1910 congress invoked the image of a nine-year-old runaway girl who was taken to a shelter "with a cigarette in her teeth," cursing boldly.[104] A criminologist described "proletarian children" in 1913 as "unwilling witnesses to various family scenes, to broken lives and drunken sexual intoxication." This environment, he said, encouraged "chronic indulgence in onanism," a habit that left the girls with a "pathologically irritable sensibility, which they attempt to gratify by unnatural means (*masturbatio, cunnilingus*)." He described them as degenerate types with "weak wills, pathologically developed imaginations, and a tendency to drink and smoke."[105]

The urban milieu was ultimately to blame, it seemed, in all these accounts. Various observers suggested that rehabilitation homes be established in the countryside, far from the temptations of city life. Working-class families, they contended, could not provide the psychological conditions for normal development still supposedly characteristic of peasant households.[106] If for some professionals that mythic household still figured as a fount of health and moral integrity even as late as 1913, the juvenile prostitute—a product of the city streets—had come to symbolize the extreme degree to which urban sexuality was thought

---

[100]D. Gremiachenskii, *Sovremennyi stroi i prostitutsiia* (Moscow, 1906), 19.

[101]A. Ia. Gurevich, "O zhenskom fabrichnom trude i prostitutsii," in *Trudy pervogo vserossiiskogo s"ezda po bor'be s torgom zhenshchinami*, 1:154–55; N. A. Zakharov, "Prichiny rasprostraneniia prostitutsii nakhodiatsia ne stol' v ekonomicheskikh, skol' v moral'nykh usloviiakh," in ibid., 1:203.

[102]Gal'perin, "Prostitutsiia detei," 376.

[103]D. A. Dril', "O zabroshennosti detstva, kak mogushchestvennaia prichina detskoi prostitutsii," in *Trudy pervogo vserossiiskogo s"ezda po bor'be s torgom zhenshchinami*, 1:84–88.

[104]Zakharov, "Prichiny," 203.

[105]Faingar, "Detskaia prostitutsiia," 30–31, 36.

[106]Oboznenko, "Obshchestvennaia initsiativa," pt. 2, 1897; Faingar, "Detskaia prostitutsiia," 46; E. P. Kalacheva, reported in Danilov, "Pervyi vserossiiskii s"ezd," 99.

to have lost its moorings in the "natural" realm of reproductive activity and to have escaped the discipline of traditional family bonds.

Perversions of Desire

Indeed, the girls were intent on escaping the family and any other institution that tried to confine and control them. Since men legally governed their womenfolk's freedom to move, restlessness and insubordination signified resistance to male authority. In fact, "vagrant women" with no permanent domicile might be apprehended by the police on suspicion of immoral conduct. Movement itself acquired a sexual connotation. What struck observers of sexually rebellious girls was their habit of running away. One sociologist declared that premature sexual activity created a state of continual excitation, sometimes to the point of mental illness, leading to a taste for a "wild, free-ranging life." Taken to charitable homes, the girls fled; restored to their parents, they again disappeared. A nine-year-old girl who escaped a sexually abusive father and a mother who "tied her to the bed and mercilessly beat her" was diagnosed as psychologically and sexually disturbed for having run away.[107] A shelter administrator attributed the desire for escape to "sexual neurasthenia," against which "the most severe domestic punishments, including beating," were useless.[108] The director of the juvenile division of a St. Petersburg shelter told of a habitual runaway with "unbridled instincts" for whom "the street was everything."[109]

For such children, wrote a journalist in 1906, "home in the sense of family does not exist—their family is the street." Their parents, whom he identified as factory workers, petty tradespeople, and unskilled laborers, were too busy or impoverished to give them proper care.[110] "In areas frequented by prostitutes," wrote a physician that same year, "you will now meet whole groups of ten- to twelve-year-old girls who look you straight in the eye, promise astounding pleasures, and disgorge foul words."[111] The children of factory workers, said Dmitrii Dril' in 1910, were doomed to follow the "dead-end path to the 'broad street,' the 'big road,' and a single unavoidable fate: imbibing the hor-

[107]Gal'perin, "Prostitutsiia detei," 373–74, 376.

[108]Zakharov, "Prichiny," 202.

[109]"Obsuzhdenie voprosa o detskoi prostitutsii: Preniia po dokladu B. I. Bentovina," in Trudy pervogo vserossiiskogo s"ezda po bor'be s torgom zhenshchinami, 2:553.

[110]M. K. Mukalov, Deti ulitsy: Maloletnie prostitutki (St. Petersburg, 1906), 9.

[111]B. I. Bentovin, "Tainaia prostitutsiia (Iz ocherkov stolichnoi prostitutsii)," Obrazovanie, no. 10, pt. 1 (1906), rpt. in Bentovin, Torguiushchie telom, 166.

rible moral poison that causes the progressive degeneration of the race, both physical and psychic."[112]

Like the dissolute proletarian family, criminal street children were not new to the literature of social indignation. But sexual precocity had not figured in the earlier picture; now it was at the heart of the discussion. The man who did the most to make it a public as well as a professional issue was Boris Bentovin. Trained as a specialist in venereal disease, Bentovin at the same time was active as a journalist and playwright.[113] He thus crossed the cultural divide between medical science and popular entertainment—precisely the intersection that generated the most intense interest in the question of public sex. Bentovin used the culturally oriented socialist journals, both populist and Marxist, to bring the physicians' social concerns to a more general serious public. During the five years after the revolution of 1905, he made the specific issue of juvenile prostitution his particular concern. In an article first published in 1905 (reissued four years later in a collection of essays titled *Trade in Bodies*—a stock phrase in the antiregulationists' moral arsenal and a sign of Bentovin's sensitivity to the popular appeal of his theme), he traced the link between hooligans and prostitutes. Excluded from the community ("thrown overboard by society," Bentovin wrote), the prostitute suffered the "civic death" (*grazhdanskaia smert'*) of moral stigma. In her loneliness she turned to the adolescent street gangs, which offered a highly structured world of ritual, loyalty, and attention that supplanted the world of kinship to which the girls no longer belonged.[114]

Such groups might provide a social alternative to the respectable working-class community for those who flouted its basic values, but their very existence was beginning to erode the moral cohesion against which they rebelled. In Bentovin's view, the girls and their boyfriends represented the deprofessionalization of vice, its integration into the normal social world; they testified to the deterioration of orderly working-class life and the fraying of protective social boundaries. These youthful partnerships, Bentovin asserted, were not as marginal as those of seasoned prostitutes and professional pimps, who represented "the hopeless, rotten dregs of society, incapable of any creative effort. The

[112]Dril', "O zabroshennosti detstva," 84–85.
[113]For Bentovin's career, see S. A. Vengerov, *Kritiko-biograficheskii slovar' russkikh pisatelei i uchenykh: Istoriko-literaturnyi sbornik*, 6:433–35 (St. Petersburg, 1889–1904).
[114]Bentovin, "Spasanie 'padshikh,'" 232. Other evidence of the alliance between hooligans and teenage prostitutes can be found in the report of a juvenile court officer: M. M. Karmina, "Doklad," in *Trudy pervogo s"ezda deiatelei po voprosam suda dlia maloletnikh: S.-Peterburg, dekabr' 1913 g.* (Petrograd, 1915), 238.

'new men,' by contrast, ha[d] not completely broken their ties with work or else ha[d] recently entered the ranks of labor." They were employed as apprentice printers, machine shop journeymen, factory hands, and store salesmen. Eventually, however, "the drunken life and depraved environment take their toll, and the converts to pimping who still have jobs finally stop working and devote themselves full-time to their new occupation. They are in turn replaced by newcomers who lead the same double life—half worker, half pimp." These men could not be dismissed as the "scum of the capital's proletariat," warned Bentovin. They were its robust youthful members.[115]

Energetic and young they might be, but were they truly healthy? Once reluctant to identify popular deviance as pathological in an organic or clinical sense, physicians were now more willing to apply such distinctions. In the aftermath of violence and disorder, popular and medical works alike emphasized the line between the accepted and the proscribed, the commonplace and the abnormal, as they had not done before. The public woman of perverted sexual tastes, long familiar to professional readers of standard Western forensic texts but excluded from Russian treatments of prostitution, finally appeared on the printed page. She figured, for example, in Aleksandr Kuprin's novel *The Pit*, which began publication in 1909 and was widely read and discussed.[116] The madam in this supposedly realistic depiction of brothel prostitution inflicts her "exacting, repulsive love" on unwilling girls, who at first react "with aversion" but eventually acquire a "revulsion . . . to men so deep that they all, without exception, compensate for it in the Lesbian manner and do not even in the least conceal it."[117]

The same foreign sources to which sober physicians had earlier turned a deaf ear inspired the morally earnest Izrail Kankarovich to assert, in a 1907 pamphlet called *Prostitution and Social Depravity: Toward a History of Contemporary Morals*, that Russian prostitutes offered lesbian services to idle upper-class women who were bored with normal sex. With this claim the doctor invoked the stereotype of elite immorality, while admitting (with reference to such authorities as the esteemed Parent-Duchâtelet and the scabrous French journalist Léo Taxil) that brothel inmates had sex among themselves.[118] To Kankarovich

[115]Bentovin, "Spasanie 'padshikh,'" 227–29.
[116]A. I. Kuprin, "Iama," in *Zemlia sborniki*, bk. 3 (1909); bk. 15 (1914); bk. 16 (1915); rpt. in *Sobranie sochinenii v deviati tomakh*, 6:5–311; notes, 453–61 (Moscow, 1964); published in English as *Yama: The Pit*, trans. Bernard Guilbert Guerney (London, 1930).
[117]Kuprin, *Yama*, 294, 299, 113.
[118]I. I. Kankarovich, *Prostitutsiia i obshchestvennyi razvrat: K istorii nravov nashego vremeni* (St. Petersburg, 1907), 190. Among other scandalous pamphlets is Léo Taxil's pseudonymous account of sexual turpitude in which he described the habits of lesbians: Ga-

such conduct reflected the general moral decline that had followed 1905.

Juvenile prostitution represented in itself the ultimate pathology of commercial sex, precisely because children were supposed to be innocent and dependent, not sexually active free agents. "Essentially abnormal," a criminologist wrote, "child prostitution also assumes monstrous sexual forms."[119] A journalistic exposé of the "sexual marketplace" published in 1908 and intended for a general readership began with a chapter on underage prostitutes and proceeded with descriptions of homosexual and lesbian "perversions." The author recounted the affecting story of an orphaned boy and girl who survived on the docks of a southern seaport by responding to sexual solicitations. The prostitution of young boys he explained as a "vile and unnatural" custom peculiar to Oriental culture and to the Western upper classes. Girl prostitution was more widespread, he claimed, but sex with children was "only the first of the transgressive passions [*potustoronnie strasti*]," of which lesbianism was the most extreme.[120] Other journalists too reported that teenage girls routinely offered "to satisfy perverted sexual demands."[121]

Such accounts usually described the girls' clients as disturbed, abnormal, or suffering from the degeneracy typical of the upper class.[122] Mariia Pokrovskaia, by contrast, saw the problem as symptomatic of the normal economy of sex. She insisted that the social endorsement of prostitution as an acceptable outlet for male sexual desire encouraged ordinary men to pursue all kinds of urges, including a bent for underage partners.[123] Indeed, some observers thought such tastes had lost much of their stigma over the years as the supply of willing girls increased.[124] Physicians and social workers were divided, however, as to how to treat the children themselves: were they victims of invidious social forces or of psychic disease?

---

briel Antoine Jogand-Pagès, *La Corruption fin-de-siècle* (Paris, 1891). Despite its lurid character, this text was cited in the serious medical literature: e.g., Cesare Lombroso and Guglielmo Ferrero, *La Femme criminelle et la prostituée* (Paris, 1896), 401; and (in connection with male sexual practices only) V. M. Tarnovskii, *Izvrashchenie polovogo chuvstva: Sudebno-psikhiatricheskii ocherk dlia vrachei i iuristov* (St. Petersburg, 1885), 33, 49, 70, 80.

[119]Faingar, "Detskaia prostitutsiia," 33.

[120]A. I. Matiushenskii, *Polovoi rynok i polovye otnosheniia* (St. Petersburg, 1908), 5–13, 20, 23, 115–16.

[121]See, e.g., Mukalov, *Deti ulitsy,* 24.

[122]Bentovin, "O prostitutsii detei," 447; Mukalov, *Deti ulitsy,* 7–9; I. I. Kankarovich, in *Trudy pervogo vserossiiskogo s"ezda po bor'be s torgom zhenshchinami,* 1:89.

[123]M. I. Pokrovskaia, "O prostitutsii maloletnikh," *Zhenskii vestnik,* no. 10 (1912): 195–96.

[124]Faingar, "Detskaia prostitutsiia," 32, 36; Gal'perin, "Prostitutsiia detei," 367–68, 373.

In contrast both to the pre-1905 syphilis literature and to the studies of schoolboy sex, discussions of runaway and delinquent children were now likely to admit that masturbation did not exhaust the range of juvenile "perversions." Mikhail Sodman, a physician at the Vologda province zemstvo hospital, described a group of adolescent boys from the local correctional institution whom he had treated for syphilis. One had acquired the disease "on the street," the doctor reported in 1909, and had given it to seven of his mates "through pederasty"; two others had been infected by prostitutes. Anal papules, which earlier physicians might have been tempted to explain as the result of infected vaginal secretions or dirty underwear, this observer did not hesitate to blame on "abnormal sexual intercourse." Sodman called the original culprit "an 'incorrigible' character of the degenerate type, with wicked tendencies, . . . morally rotten to the core." While acknowledging the existence of homosexuality among young men of common origins (Sodman used the explicit term *muzhelozhstvo*), the physician nevertheless divided the perpetrators into "repentant" victims and "degenerate" victimizers.[125] The former were somehow not responsible for their own seduction, and the latter could be dismissed as organically unsound.

Both the frankness and the moral opprobrium expressed in organic terms were typical of the moment in which Sodman wrote. Various speakers at the 1910 congress on prostitution viewed the participation of young girls in prostitution as part of a larger sexual disorder that included autoerotic and homosexual acts. "Within the very walls of the shelter itself," one administrator reported, "[the girls] try to satisfy their morbidly excited sensuality by *masturbatio* or *cunnilingus*," or by trying to seduce their fellow inmates.[126] Serafima Konopleva, director of a St. Petersburg shelter, described a particularly obdurate charge whose sexual advances created such an "atmosphere of infectious corruption" in the home that other girls complained.[127] A female correctional officer attached to the juvenile court noted the "suspicious closeness" that often developed between teenage prostitutes.[128] Even Mariia Pokrovskaia, who emphasized the sexual victimization of innocent children, by 1912 was describing juvenile prostitutes as abnormal.[129]

The question of the girls' underlying pathology was posed most

---

[125]M. M. Sodman, "K voprosu o sifilise v ispravitel'nykh zavedeniiakh dlia maloletnikh," *Russkii vrach*, no. 40 (1909): 1353–54.

[126]Zakharov, "Prichiny," 202–3.

[127]"Obsuzhdenie voprosa o detskoi prostitutsii," 553.

[128]Karmina, "Doklad," 240.

[129]M. I. Pokrovskaia, "Zhenskaia bezzashchitnost'," *Zhenskii vestnik*, no. 2 (1905): 46–47; idem, "O prostitutsii maloletnikh," 194. By 1913 Pokrovskaia was willing to consider prostitution itself a form of sexual perversion; see idem, *O polovom vospitanii i samovospitanii* (St. Petersburg, 1913), 12–13.

sharply in Bentovin's presentation to the congress on prostitution. In 1906 he had described what he saw as unprecedented expansion in the market for juvenile prostitutes.[130] He now enlarged on his earlier claims, evoking Léo Taxil's 1891 cameo of the *petites agenouillées* ("little kneelers" or, more scandalously still, "little praying girls"), who supposedly made the rounds of Parisian cafés selling flowers and offering to perform oral sex.[131] In St. Petersburg, Bentovin claimed, girl prostitutes similarly proffered "all forms of unnatural sexual relations. . . . up to sixteen various forms of ecstasy, except the natural one." This was a sign of the times: "Once, child prostitution might have existed only to serve the needs of sexual gourmands: old men, impotent men, fanciers of sexual perversion. It was hard and expensive to obtain a child's body. . . . Now the trade in child-love is conducted with astonishing openness and speed, without any mask or screen."[132]

The lines between open and secret, public and private, were growing less distinct, Bentovin suggested; the normal and the perverse had moved dangerously close. "Nowadays," he complained, "there are so many perverted sex maniacs [*erotomany*] always looking for ways to satisfy their vicious desires. . . . But what is worse, unperverted men seek to gratify simple 'curiosity,' especially after drinking, when they venture a try at unfamiliar, abnormal sensations. Often this attempt leads the novice to acquire the taste and become a regular fancier of one or another sexual perversion." Bentovin believed that some older prostitutes "inclined to lesbianism" now slept with young recruits, preparing them to accept female as well as male clients, since supposedly respectable ladies of lesbian tastes had recently turned to the juvenile trade.[133]

Dismissing the clients as spoiled matrons satisfied the cliché about upper-class depravity: "Lesbian love is the preserve of the well-to-do classes," wrote a contemporary journalist. "It does not descend to democracy."[134] It was harder to explain the willingness with which children seemed to cooperate in perverse adult games. Describing the enthusiasm with which the girls took to the streets, Bentovin took pains to depict them as the objects of manipulation: "These are still completely unformed children," he explained, "who can be directed at will to perform any kind of vice and who regard various depraved refinements as a form of mischief. In a state of almost constant alcoholic

[130]Bentovin, "Tainaia prostitutsiia," 166.

[131]Jogand-Pagès, *La Corruption fin-de-siècle*, 265.

[132]Bentovin, "O prostitutsii detei," 446, 448, 441. At least one other writer on the subject repeated Bentovin word for word: Faingar, "Detskaia prostitutsiia," 32, 34.

[133]Bentovin, "O prostitutsii detei," 446, 444, 448–49.

[134]Matiushenskii, *Polovoi rynok*, 115–16.

intoxication, they perform the depraved 'tricks' that adults teach them with special bravado and even pride." In a perverse version of "normal childish imitation," immature prostitutes caricatured the public woman's masculine habits: "Girls of eleven to thirteen use such incredibly foul language as to make even grown prostitutes blush. These same girls become proficient in unnatural forms of love from which more than one adult prostitute has fled in loathing. Girl-prostitutes are almost all drunkards, and they drink with panache, with pleasure, for the fun of it and to make a row." Here was a carnival of vice, in which the innocent surpassed the corrupt in depravity.[135]

Most were ordinary children, Bentovin claimed, led astray because they were young. Some had been raped and then forced into prostitution. But certain girls, he asserted, like some adult women, were constitutionally impervious to moral distinctions and actually enjoyed the prostitute's life. "We should save those who want to be saved," Bentovin advised the conference on prostitution, "and protect those who waver on the brink of perdition. But we should not waste material resources or moral energy trying forcibly to wrest from the depths of depravity those who feel comfortable there. . . . Torn from their dens, [such girls] will take the first opportunity to return to the 'gay' life." By contrast, he argued, those who found themselves there by "the force of circumstance" retained "an inborn sense of respectability, instinctively feeling the entire horror of their experience, while their wearied bodies tremble from the horror of their shame." In view of the limited resources at society's disposal, Bentovin concluded, hard choices had to be made, and the potentially more rewarding strategy ought to be favored.[136]

Such unflinching pragmatism reflected the outcome of a somewhat zigzag evolution in Bentovin's views. In 1904 he had rejected Veniamin Tarnovskii's assertion that prostitution was the expression of innate morbid propensities, favoring instead a social explanation for women's taking to the streets.[137] In 1906 he went so far as to blame the demand for underage prostitutes on their clients' "perverted instincts," describing the girls themselves as victims of the mercenary wiles of female procuresses and sometimes even of their own depraved and desperate

---

[135]Bentovin, "O prostitutsii detei," 446, 450. On the Victorians' double image of the juvenile prostitute as both innocent and depraved, see Leonore Davidoff, "Class and Gender in Victorian England: The Diaries of Arthur J. Munby and Hannah Cullwick," *Feminist Studies* 5 (Spring 1979): 93.

[136]Bentovin, "O prostitutsii detei," 442, 449–53, 457; "Obsuzhdenie voprosa o detskoi prostitutsii," 555.

[137]Bentovin, "Istochniki prostitutsii," 38.

parents.[138] In 1905 he had distinguished between girls so deeply deformed by their experience in the trade that they could no longer be
disengaged from it and those who might still be susceptible to moral
suasion or saved from desperation by the practical services of charitable
homes.[139] But not until 1910 did he take the final step of arguing that the
hopeless were organically destined for their fate.

Bentovin's audience objected less to this characterization of the problem than to the solution he proposed. In suggesting to the congress that
philanthropists and social workers abandon their efforts to force young
girls off the streets and into homes for moral rehabilitation, he aroused
a storm of disapproval. His proposal that custodial care be limited to
"normal" girls was rejected by a majority of participants in the discussion, though opinions about the clinical status of the incorrigibles varied. Vadim Slanskii, a hospital staff physician, was even more emphatic
than Bentovin in asserting that "juvenile protitutes are without doubt
mentally ill [dushevno-nervno-bol'nye]" but drew the opposite conclusions: far from disqualifying them from professional attention, their pathology made supervision all the more imperative. "Society places the
mentally ill under special care," he reasoned. "One cannot therefore
object to the compulsory education [prinuditel'noe vospitanie] of juvenile
prostitutes."[140]

Other speakers disputed Slanskii's extreme formulation. Dmitrii
Dril' said that disturbed girls constituted only part of the problem;
most young prostitutes were the "mentally normal" victims of harsh
circumstance. Mariia Blandova, a member of the Russian League for
Women's Rights, agreed that not all underage prostitutes could be considered abnormal but considered it "inhuman" to deny any of them
special care. Serafima Konopleva admitted that many of her intractable
charges were "pathologically vicious," but they did not deserve less
attention for that reason, she argued; rather, they merited a special regime in separate institutions for "morbidly perverted" children, where
they would be isolated from the healthy ones.[141]

In the end, Dril' and Konopleva joined with two others in offering
an alternative to Bentovin's idea. Their resolution, which the assembly
adopted, called for correctional (vospitatel'no-ispravitel'nye) institutions

[138]Bentovin, "Tainaia prostitutsiia," 164–65.
[139]Bentovin, "Spasanie 'padshikh,'" 198. Cf. Krestovskii's distinction ("Peterburgskie
trushchoby") between the "bestialized" veterans and the newcomers who had not yet lost
all human qualities.
[140]"Obsuzhdenie voprosa o detskoi prostitutsii," 550, 556; Danilov, "Pervyi vserossiiskii s"ezd," 108.
[141]"Obsuzhdenie voprosa o detskoi prostitutsii," 551, 553–54.

along with "autonomous social organizations" (*organy obshchestvennogo samoupravleniia*) to share in the "compulsory rehabilitation" of juvenile prostitutes. In short, the problem was one of broad social implications, not narrow technical ones; responsibility must be distributed widely, not concentrated in specialized hands. Even Konopleva seems to have relinquished her suggestion that troubled girls needed expert treatment. Slanskii, who had vociferously insisted on the girls' pathological condition, objected to the amateurish basis of such arrangements: "The treatment, instruction, and supervision of juvenile prostitutes constitute a special responsibility, which demands medical and pedagogical expertise. It cannot remain in the hands of charity." He was alone at this session, however, in concluding that juvenile prostitutes should be confined in psychiatric institutions.[142] Even Bentovin had preferred to neglect rather than discipline the hardest cases.

In the end the meeting decided that no young prostitute should be denied the chance of moral salvation, that all still retained the potential to change. The majority wanted neither to abandon nor to segregate those who posed the greatest social threat. What it did defend was society's right to incarcerate delinquent girls in corrective institutions for purposes of moral reform: this was the meaning of the term "compulsory education." In fact, however, the legal status of such girls was ambiguous. Beginning in 1912, those over ten years of age could be brought to juvenile court and sentenced to a reformatory for engaging in clandestine, or unregistered, prostitution—but there was no statutory basis for such charges, because girls under seventeen were forbidden by law to register with the police. Since they could not comply with the law without violating it, their conviction for avoiding registration was clearly unfair, as some court officers admitted.[143]

The reason for this contradiction, one such officer reported to a gathering of her colleagues in 1913, was the state's reluctance, on the one hand, to protect women adequately from sexual assault (that is, to interfere with male sexual license) and, on the other, to accord young girls adult status in sexual matters (that is, to weaken its custodial authority and hold the girls responsible for their own acts). She pointed out that what the girls themselves enjoyed most about the trade was the illusion that it made them grown up: "As an adolescent, the girl prostitute is charmed by her adult role in the brothel, which she fulfills by virtue of her trade. She is pleased by the long dress and the adult hairdo, . . . by the the rouge and powder."[144]

[142]Ibid., 554.
[143]Karmina, "Doklad," 226.
[144]Ibid., 234.

The choice facing professionals and lawmakers was either to declare the juvenile prostitute abnormal and subject her to a therapeutic regime or to declare prostitution itself illegal and punish both partners to the exchange. The first alternative put authority in the hands of medical experts and consigned the delinquents to the status of wards. The second, in acknowledging the participants' sexual responsibility, exposed them to the risk of punishment but also offered them the possibility of self-defense. As it stood, both adult and juvenile prostitutes remained the objects of administrative, not judicial, sanctions; as juridical minors, they were perpetual objects of custodial intervention. Men were not encroached upon at all. In preferring the criminalization of adult participants in prostitution, feminist advocates nevertheless still hoped to operate through institutions of moral redemption. They had no more interest than the state in extending the autonomy of underage girls. What they wanted was the transfer of custodial authority over sexually delinquent minors from the police (an organ of the administrative state) to society (in the name of charity or community self-determination). Ultimately, they hoped to instill the kind of discipline that would allow the girls to police themselves.

The medical profession, by contrast, had moved toward an acceptance of the first alternative, the recognition of sexual pathology—not, like Veniamin Tarnovskii, to defend continued state regulation (in his case, with the intention to eliminate venereal disease) but to bolster the role of professional authority in social affairs. This last claim entailed the admission that therapeutic or corrective measures imposed from above could not always be avoided, as the zemstvo physicians had once liked to believe.[145] The common people could not always effect their own salvation. The willingness of physicians to include sexual delinquents in distinctive categories identified on scientific grounds corresponded to the decline of the populist ethos and the profession's increasing desire to act as the scientific arbiter of social issues. It expressed their greater eagerness to shoulder the corrective burden previously borne by the administrative arm of the state.[146]

The disagreement over Bentovin's proposal demonstrates, however, that as late as 1910 many Russian physicians and moral reformers still refused to exile prostitutes to the biological netherworld, lost to the benign effects of enlightenment and social action. In the face of this

[145]S. N. Igumnov, "Zemskaia meditsina i narodnichestvo," in *Trudy odinnadtsatogo Pirogovskogo s"ezda*, ed. P. N. Bulatov (St. Petersburg, 1911), 1:82.

[146]Hutchinson, "'Who Killed Cock Robin?'" In 1908 even physicians employed by the police denounced "coercive police measures" in controlling prostitution and called for "moral and educative" methods instead, presumably to be administered by themselves: Goncharov, *O venericheskikh bolezniakh*, 135.

resistance, it is the professionals' growing tendency to stigmatize such women as organically defective and in need of expert care that needs to be explained. Though it was rarely translated into social policy, such language expressed a pervasive anxiety in educated circles as to the actual status of the lower classes, whose patrons and benefactors they yet hoped to remain. In the face of recurrent peasant anger, a seemingly elemental force, one could maintain one's vision of the piously naive and upright folk; centuries of revolt and repression had not altered the basic structure of patriarchal society, with its hierarchy of sex, age, and class deference. But urban revolution had violently challenged the stability of that order; faith in innocence had been shattered. If village dirt and village poverty reflected a natural (if benighted) way of life, city dirt and urban poverty were clearly social products. Likewise, proletarian rage did not share the supposedly naive and spontaneous quality of peasant unrest but responded to the theories and strategies of agitators pursuing calculated goals—or so it seemed.

Even peasants—childlike in their simplicity, submissiveness, and tendency at times to disobey—could no longer be trusted. The prostitute, whom physicians had long considered identical to her sisters in the village, now revealed signs of sexual perversity indicating organic derangement. In this context, children too seemed less benign. After 1905, images of the sexual corruption and victimization of children came to represent, on the one hand, the threat of libidinous disorder posed by classes whose innocence and capacity for self-control (collective or individual) were now in doubt and, on the other, the proven capacity of manipulative elites to elicit the latent forces of violence and desire.

Chapter Eight

Sex and the Anti-Semite:
Vasilii Rozanov's Patriarchal Eroticism

Along with the evils of urban life and the possibility of proletarian aggression, Jews had figured throughout the nineteenth century as prime culprits in the symbolic drama of imagined national peril. They were construed as the opposite of the peasant: alien and cosmopolitan rather than native and rooted in the soil; urban rather than rural; commercial rather than productive; artificial rather than organic; cerebral rather than instinctive; sober rather than drunk; insidious rather than violent. The Jews' symbolic function, however, was profoundly contradictory. They simultaneously intruded an unwelcome particularism into the Christian body of state and society (which in the Russian empire also included a large Islamic segment, conveniently obscured in the stark Jew-Gentile juxtaposition) and managed to convey the threat of supranational universalism. Officially restricted to the margins of the public world, the Jews were thought to retain the archaic patriarchalism threatened by modern life.[1] Yet they also embodied the two mutually antagonistic faces of modernity: capitalist enterprise and the threat of socialist revolution.[2]

Two fantasies current in turn-of-the-century Russia show how these

---

[1] On the Jews' association with private family matters, see Hannah Arendt, *The Origins of Totalitarianism* (New York, 1958), 28; on their resemblance to women by virtue of their civic exclusion, see Jean-Paul Sartre, *Réflexions sur la question juive* (1946; Paris, 1985), 102.

[2] On Jews and capitalism, see Arthur Hertzberg, *The French Enlightenment and the Jews: The Origins of Modern Anti-Semitism* (New York, 1968), 23, 66, 74; Heinz-Dietrich Löwe, *Antisemitismus und reaktionäre Utopie: Russischer Konservatismus im Kampf gegen den Wandel von Staat und Gesellschaft* (Hamburg, 1978), 12, 43, 45. Cesare Lombroso argued that the Jews helped Russia develop the capitalist relations it sorely needed: *L'Antisémitisme* (Paris, 1899), 93. On the combination of capitalism and socialism, see Bernard Lewis, *Semites and Anti-Semites: An Inquiry into Conflict and Prejudice* (New York, 1986), 92.

contradictory anti-Semitic tropes might acquire a sexual dimension. Campaigners for moral purity who denounced the international prostitution trade—white slavery, in the language of the day—claimed that Jewish men from the area of the Russian Pale played an especially active part in transporting innocent non-Jewish girls across national borders to sell them into sexual servitude.[3] The exaggerated image of a worldwide network of Jewish sex traders was persuasive in Russia as well as in Europe because it dovetailed nicely with the anti-Semitic myth of an international conspiracy of commercially rapacious Jews who sapped the autonomy and welfare of vulnerable Christian states. In the white-slave melodrama the Jew was a natural for the villain's role.[4] Representing the pure spirit of commerce, he extracted profit from the mere process of exchange (the objectionable feature of usury, which caused it to be paired with prostitution in the medieval cosmology of despised and useful trades)[5] while reducing intimate relations and personal desire to public and impersonal affairs. White slavery was, in short, the Enlightenment in nightmare form. It was a caricature of universalism, a network of global intercourse, of interchangeable private female parts loosed from the domestic into the public (indeed the international) sphere, transforming the particular (my wife, your daughter) into a public woman accessible to all men.

The second of these sexualized fantasies, the legend of ritual blood murder, associated the Jews not with modern forms of exploitation but with archaic ones. The belief that Jews killed Christian children in order to use their blood for religious purposes got its start in medieval Europe and persisted intermittently into the modern period. It was revived by the tsarist Ministry of Justice in 1911, when it accused the Jewish artisan Mendel Beilis of just such a crime.[6] Though the blood ritual conceit was

---

[3]Edward J. Bristow, *Prostitution and Prejudice: The Jewish Fight against White Slavery, 1870–1939* (Oxford, 1982), 1–3, 81. For Russian charges that Jews played a predominant role in white slavery, see the debate on antiprostitution legislation in Gosudarstvennaia Duma, *Stenograficheskie otchety*, III sozyv, sessiia II, zasedanie 109 (May 8, 1909), 897; and "Sovremennaia negrotorgovlia," *Zemshchina*, no. 1620 (March 24, 1914). On the implication of Jews in the Jack the Ripper panic in 1880s London (including accusations of ritual murder), see Judith R. Walkowitz, "Jack the Ripper and the Myth of Male Violence," *Feminist Studies* 8:3 (1982): 555.

[4]I owe the term "melodrama" in this context to the work of Judith Walkowitz.

[5]See Jacques Le Goff, "Métiers licites et métiers illicites dans l'Occident médiéval," in Le Goff, *Pour un autre Moyen Age: Temps, travail et culture en Occident* (Paris, 1977), 91–107.

[6]A contemporary German study listed 128 trials for ritual murder in Europe and Russia, 1850–1900: F. Frank, *Der Ritual-Mord vor dem Gerichtshof der Wahrheit*, trans. as *Ritual'noe ubiistvo* (Kiev, 1912). Lewis (*Semites and Anti-Semites*, 107) mentions 12 cases in Germany and Austria-Hungary, 1867–1914, all resulting in acquittal. A Russian anti-Semitic newspaper published a table of 199 recorded cases, beginning in ancient times: "Spisok ritual'nykh ubiistv," *Zemshchina*, no. 1369 (June 29, 1913), 5–8. On classic em-

not explicitly sexual, Vasilii Rozanov managed to invest it with homosexual overtones in a series of scurrilous pamphlets he wrote during the
Beilis affair.

These two myths reached heights of symbolic power in Russia during the decade between revolutionary defeat and the outbreak of World
War I. Anti-Semitism, both rhetorical and applied, had intensified during the climactic phase of 1905. In the revolution's wake, as we have
seen, the language of social commentary relied increasingly on emblems
of sexual disorder, and these images in turn were sometimes linked to
the problem of the Jews. Two contemporary writers widely read in
Russia at the time made their names by insisting on this connection: the
Austrian Otto Weininger and the Russian Vasilii Rozanov. Both considered philosophers, they shared an obsession with sex and the Jews
but nevertheless argued their case from opposite points of view. Jean-
Paul Sartre has identified the logic of their positions as the Scylla and
Charybdis of anti-Semitism. The conservative (that is to say, the overt)
anti-Semite, Sartre writes, "wants to destroy [the Jew] as a man, to
leave only the Jew, the pariah, the untouchable; [the democrat] wants to
destroy him as a Jew to preserve only the man, the abstract and universal subject of the rights of man and citizen."[7]

Weininger, the self-styled man of science, expressed a distorted vari-
ant of Enlightenment hostility to the particularity of traditional Jewish
culture.[8] A baptized Jew, Weininger created a sensation in his native
Vienna with a doctoral dissertation published as *Sex and Character*
(1903), whose anti-Jewish spirit he seemed to endorse by committing
suicide shortly after its publication. Translated into Russian in 1909 and
issued in thousands of copies and numerous editions by 1914, the book
was hailed by anti-Semites, denounced by feminists, and read by university students thirsting for new ways to think about sex.[9] Weininger's

---

bodiments of this myth, see R. Po-chia Hsia, *The Myth of Ritual Murder: Jews and Magic in
Reformation Germany* (New Haven, Conn., 1988), and Alan Dundes, ed., *The Blood Libel
Legend: A Casebook in Anti-Semitic Folklore* (Madison, Wis., 1991). On the Beilis case, see
Hans Rogger, *Jewish Policies and Right-Wing Politics in Imperial Russia* (Berkeley, Calif.,
1986), 40–55.

[7]Sartre, *Réflexions*, 68. See also Hertzberg, *French Enlightenment*, 338.

[8]Emanuel Glouberman, "Vasilii Rozanov: The Antisemitism of a Russian Jude-
ophile," *Jewish Social Studies* 38:2 (1976): 120, 133, points out that Jewish advocates of
cultural assimilation often expressed hostility to traditional Jewish ways; see, e.g.,
Lombroso, *L'Antisémitisme*, 18–19, 98. Thanks to Jane Daley Nirenberg for the Glouberman reference.

[9]Otto Weininger, *Geschlecht und Charakter: Eine prinzipielle Untersuchung* (1903;
Munich, 1980). *Knizhnaia letopis'* lists translations by A. M. Belov (Moscow, 1909), 3,000
copies; S. Press (Moscow, 1909), 10,000 copies; V. Likhtenshtedt (St. Petersburg, 1909,
1910), 5,000 and 3,000 copies; and an edition by G. Namiot (St. Petersburg, 1912), 10,000
copies. The catalogue of the Helsinki University Slavonic Library includes two anony-

ponderous academic prose signaled his commitment to the conventions of scientific thought, which he conceived as a masculine bulwark against the uncontrolled libidinous desire represented by women and Jews, both equally incapable of reason. Only males had the power of speech, thought, judgment; only they constituted civil society. Mired in biology, women and Jews represented the impersonal procreative forces that male individuality must subdue, the collective and familial ties civic man must rise above.[10] In Weininger's world of abstract principles (he insisted that he spoke only of types, not real people), the male provided the ordering structure of mind that was to serve as a "straitjacket" to master the seething chaos (*un-Sinn*) of female (non)being (*das Nichts*).[11]

Devoted to community and kin, to sensuous merging rather than analytic distinctions, women and Jews—in Weininger's vision—nevertheless served these timeless tribal interests in modern ways, promoting sexual connection by public and commercial means (*Kuppelei*).[12] Though he was obviously in debt to the current rhetoric of white slavery, in which the globe figured as a vast network of sexual exchange, Weininger condemned procuring not as a threat to domestic integrity and traditional moral values, as abolitionists usually did, but as the triumph of just those sexual impulses proper to the domestic sphere. Indeed, insofar as prostitution diverted sex from its traditional reproductive purposes, Weininger allowed it a certain positive role.[13] Nor did he accept the abolitionists' assumption that women recruited into the trade were the passive victims of male, particularly Jewish male, manipulation. Both Jews and women, in his view, conspired to defeat masculine principles of rational self-restraint, which could triumph only if Judaism and femininity were extinguished.[14]

No less hostile to Jews, Rozanov was a believing (though not obedient) Orthodox Christian, a brilliant and imaginative stylist, and a resolute enemy of abstract thought, rejecting not only the values of the Enlightenment but the constraints of reason itself. Convinced that sexual expression was the source of personal vitality and national energy, he extolled the Jews for their supposed Old Testament fecundity while

---

mous translations (Moscow, 1912; St. Petersburg, 1914) and an edition by A. Gren (Moscow, 1912).

[10]Weininger, *Geschlecht und Charakter*, 249–52, 265–66, 278, 385.

[11]Ibid., 3–4, 343, 398, 384, 246.

[12]Ibid., 351.

[13]Weininger distinguished the mother and the prostitute as opposite ideal types but saw them both as organic female propensities, equally suffused with sensuality, equally indiscriminate in choice of partner for their respective sexual goals (ibid., 281, 284, 288–91). If anything, the prostitute was morally superior because she had freed herself from the procreative urge and rebelled against the passive female ideal (297–98).

[14]Ibid., 418–19 (on Jews), 455 ("the Female as such must disappear").

simultaneously denouncing them as a threat to sexually puritanical Christian culture. Because he praised the Jews for their imagined sexual prowess and patriarchal procreativity, Rozanov was sometimes mistaken (and mistook himself) for a lover of Jews, although his own (philosophically consistent) contempt for intellectual consistency allowed him to abhor them at the same time.[15]

Difficult and ambiguous, his work was sometimes denounced for sexually explicit language but his books did not penetrate far beyond the narrow intellectual circles of which he was part. Yet he cannot be dismissed as an isolated, quirky figure. He was not alone in his intellectual iconoclasm but participated in the cultural and philosophical revival known as the Russian Silver Age and gained the (not always uncritical) admiration of such famous contemporaries as Dmitrii Merezhkovskii, Aleksei Remizov, and Nikolai Berdiaev. He considered himself and was considered Dostoevsky's philosophical heir (anti-Semitism was one of the points they had in common).[16] Often compared with Nietzsche,[17] he was described as "beyond good and evil,"[18] excoriated for lack of principle,[19] and admired for his intellectual daring and eccentricity.[20] Today critics and scholars still respect him as one of modern Russia's most original literary voices.[21] Even his less original, less enlightened views suited the cultural moment: the influential conservative publishing house founded by Aleksei Suvorin paid him a salary, published his books, and, more important, printed his column in its widely read newspaper *Novoe vremia*.[22] The combination of aesthetic courage and

[15]For Rozanov's contradictions, see Glouberman, "Vasilii Rozanov."

[16]D. A. Lutokhin, "Vospominaniia o Rozanove," *Vestnik literatury*, no. 4–5 (1921): 5; E. Gollerbakh, *V. V. Rozanov: Zhizn' i tvorchestvo* (Petrograd, 1922), 58. Lukian, "Rozanovshchina," *Birzhevye vedomosti*, no. 15543 (May 7/20, 1916): 3, accused Rozanov of plagiarism, claiming that his supposed originality was merely rehashed Dostoevsky.

[17]Merezhkovskii, cited in N. K. Mikhailovskii, "Literatura i zhizn': O g. Rozanove, ego velikikh otkrytiiakh, ego makhanal'nosti i filosoficheskoi pornografii," *Russkoe bogatstvo*, no. 8, sec. 2 (1902): 96–97; also in Gollerbakh, *Rozanov*, 4–5. See also Anna Lisa Crone, "Nietzschean, All Too Nietzschean? Rozanov's Anti-Christian Critique," in *Nietzsche in Russia*, ed. Bernice Glatzer Rosenthal (Princeton, N.J., 1986), 95–112. Rozanov himself, in *Opavshie list'ia* (1913, 1915); rpt. in Rozanov, *Izbrannoe*, ed. Evgeniia Zhiglevich (Munich, 1970), 258, denied the resemblance to Nietzsche.

[18]A. A. Smirnov, "O poslednei knige Rozanova: V. Rozanov, *Oboniatel'noe i osiazatel'noe otnoshenie evreev k krovi* (St. Petersburg, 1914)," *Russkaia mysl'*, no. 4, sec. 3 (1914): 45.

[19]Petr Struve, "Na raznye temy: Bol'shoi pisatel' s organicheskim porokom: Neskol'ko slov o V. V. Rozanove," *Russkaia mysl'*, no. 11, sec. 2 (1910): 143–45.

[20]Z. N. Gippius, *Zhivye litsa* (1925; Munich, 1971), 2:9. Nikolai Berdiaev praised Rozanov's insight and intellectual courage in talking about sex: *Novoe religioznoe soznanie i obshchestvennost'* (St. Petersburg, 1907), 158–59.

[21]See Heinrich A. Stammler, intro. to Rozanov, *Izbrannoe*. A recent though disturbing appreciation is A. D. Siniavskii, *"Opavshie list'ia" V. V. Rozanova* (Paris, 1982).

[22]On the newspaper, see David R. Costello, "*Novoe Vremia* and the Conservative Di-

political servility made Rozanov as great a figure of controversy during his lifetime as Weininger briefly became after his death, while his superior talents have earned him a more enduring fame.

Pimps and Patriarchs

If Rozanov was the master of the recontextualized cliché, he had rich material to draw on in making the connection between sexuality and the Jews. From its earliest appearance in English moral reform literature, the term "white slavery" as applied to commercial sex was connected with Jews. In 1839 Michael Ryan wrote of prostitution in London that "the infernal traffic in question is still carried on to a great extent, principally by Jews," whom he referred to as "white-slave dealers."[23] In the 1880s the outcry against prostitution and the abuses to which women were allegedly subject began to focus on procuring as an international trade, and the movement to stop it took on a correspondingly international form, with branches in England, on the continent, and in Russia as well.[24] The Jewish trafficker had by then become an anti-Semitic cliché, especially in central Europe. One German tract, for example, asserted that the Jews were exercising "exclusive" control over the prostitution trade (*Mädchenhandel*), which suited their "international organization," and turning all Germany into a brothel (*Bordellstaat*).[25]

In Russia the perception of the Jews' role in prostitution was part of a larger complex of associations which endowed them with the capacity to evoke the forces of violence and desire latent in the social body. In the minds of anti-Semites, the Jews figured as antagonists of the regime itself, locked with the authorities in a contest for the benefits of popular industry and for influence over the people's state of mind. Depicted as a thirsty bloodsucker who drained the gullible peasants of the fruits of their labor, the Jew was also said to arouse the peasants' blood. To this end he dispensed alcoholic drink, which loosed violent passions;[26] attracted the peasants' hatred by his very presence—a provocation in it-

---

lemma, 1911–1914," *Russian Review* 37:1 (1978): 30–50. On its anti-Semitism, which began in the 1870s, see Effie Ambler, *Russian Journalism and Politics, 1861–1881: The Career of Aleksei S. Suvorin* (Detroit, 1972), 171–74.

[23]Michael Ryan, *Philosophy of Marriage* (London, 1839), 14, is quoted in Bristow, *Prostitution and Prejudice*, 35, as the first use of "white slavery" in this sense.

[24]See Alain Corbin, *Les Filles de noce: Misère sexuelle et prostitution (19e et 20e siècles)* (Paris, 1978), 405–35.

[25]Alexander Berg, *Juden-Bordelle: Enthüllungen aus dunklen Häusern* (1890), 4th ed. (Berlin, n.d.), 11, 33. See Bristow, *Prostitution and Prejudice*, 81, 250.

[26]On the Jews' role in the liquor trade, see I. G. Orshanskii, *Evrei v Rossii: Ocherki ekonomicheskogo i obshchestvennogo byta russkikh evreev* (St. Petersburg, 1877), 94–109.

self—and thus instigated murder and pogroms;[27] aroused class conflict with socialist ideas; and supplied objects of sexual desire. Each of these activities in fact presented a real or symbolic arena for imagined struggle over influence and control: Jewish domination of the liquor trade was replaced in the 1890s by a state liquor monopoly; local authorities either instigated or tolerated pogroms; the police established their own trade unions; and the Ministry of the Interior sponsored regulated prostitution.

Anti-Semites considered sex an instrument in the war for cultural and racial superiority. Not only were the Russian Jews associated with the unleashing of popular passions, but their own desire was a source of suspicion and fear. To reduce their tax burden, Jews were thought to conceal their real numbers from the census takers and at the same time to breed at excessive rates, surpassing even the peasants' prodigious standard.[28] Procreative prowess was not, however, supposed to be their only tool in the demographic struggle. Convinced that the Jews exerted a predominant influence in journalism and in the pharmaceutical trade, anti-Semites posited a sinister mechanism whereby Jews might succeed in reducing the Russian birth rate. Popular magazines carried advertisements for contraception and abortion, complained the anti-Semitic newspaper *Zemshchina* in 1911, and "the press of the capital cities, especially those papers controlled by Jews, often includes information on how to prevent and even interrupt pregnancy." What the Jews wanted, the newspaper alleged, was to stimulate sexual desire while inhibiting its reproductive consequences: "All criminally pornographic works are written and published by Jews. This convincingly demonstrates that the aim of the Jews is not only to pervert the Russian people to the marrow of their bones, both physically and spiritually, but also to diminish their number."[29]

This imagined biological subversion was linked not only to the moral disarray occasioned by the revolution but also to the proliferation

---

[27]In fact, the pogroms of 1881–82 started not in the villages but in the large towns, instigated by workers or recent peasant migrants, but this pattern did not prevent officials from insisting that the Jewish presence in rural areas, rather than the peasants' move to the city, provoked the attacks. See I. Michael Aronson, "Geographical and Socioeconomic Factors in the 1881 Anti-Jewish Pogroms in Russia," *Russian Review* 39:1 (1980): 19, 23, 26–27.

[28]See S. V. Filits, "Sovremennaia polovaia zhizn' s meditsinskoi tochki zreniia," *Meditsinskaia beseda*, no. 3 (1900), 72; and Rogger, *Jewish Policies*, 16, 145. The Jewish growth rate did surpass that of the general population throughout the nineteenth century. Even though (contrary to anti-Semitic belief) Jews married at a later age and bore fewer children than non-Jews, infant mortality (before age nine) was three times as high in the non-Jewish population; Jews had a longer life expectancy (for women three times as long!); Jacques Silber, "Some Demographic Characteristics of the Jewish Population in Russia at the End of the Nineteenth Century," *Jewish Social Studies* 42:3–4 (1980): 269–80.

[29]S. M. K., "Raspad sem'i," *Zemshchina*, no. 602 (March 30, 1911): 3.

of commercial literature that had become available to a relatively unsophisticated audience. As the medical profession was stepping up its campaign to decriminalize abortion, *Zemshchina* commented in the language of medical metaphor:

> Even today, the Slavonic organism of boundless Russia, healthy from time immemorial, has still not regained its strength or fully recovered from the morbid condition induced by the savage nightmare of revolution that engulfed the country in 1905 and 1906, shaking and even sweeping away the age-old religious and moral foundation of national [*narodnaia*] life. At this moment our "friends" the Jews are flooding the Russian book market with hundreds of thousands of copies of all the newest corrupting publications, which advocate abortion and infanticide.[30]

The family, *Zemshchina* concluded, was an essential source of the patriotic sentiment upon which the nation and government relied. Enemies of the family, including the Jews, were therefore enemies of the state, and the government must lead the fight against them.

The Jews' own devotion to family life, however, presented a symbolic dilemma. Even though Jews were defined by official decree as the social and cultural opposite of the peasant population (excluded from the ethnic heartland, from rural areas, and from agricultural pursuits), they managed to stand for the same hallowed principles of social organization that governed peasant life: patriarchy; universal marriage and large families; domestic and collective affairs dominated by religious ritual and traditional, pre-Enlightenment beliefs. Indeed, some of the chestnuts of European anti-Semitism failed to sprout in Russian soil, so struck were observers by the failure of the Jewish community to succumb to modern ways. While Jews in Europe were accused of being an innately criminal race and spreading venereal disease,[31] in Russia even their critics admitted that violent crime and syphilis were relatively rare among them, presumably because they tended to marry at an early age and to maintain strict moral standards.[32] The Jews thus presented a tan-

[30]Ibid.

[31]Berg, *Juden-Bordelle*, 5; and Edouard Drumont, quoted in Stephen Wilson, *Ideology and Experience: Antisemitism in France at the Time of the Dreyfus Affair* (Rutherford, N.J., 1982), 586.

[32]E. V. Erikson, "Nervnye i dushevnye bolezni u evreev," *Nevrologicheskii vestnik*, no. 2 (1913): 222–23, 232, in a respectable psychiatric journal, describes the Jews as a degenerate race marked by distinctive inherited characteristics—distaste for agricultural labor, litigiousness, lying and deceit, cowardice, special gestures—but suggests that the rarity of syphilis might be explained by the fact that Jews were more careful than peasants in the selection of sexual partners and more fastidious. Other medical authorities also recognized

gle of contradictions: patriarchal virtue and commercial vice; the power
of wealth (because many dealt in trade) and the humiliation of need
(because most were dirt poor);[33] closely confined and seemingly ubiquitous. A contributor to the medical journal of the Ministry of Internal
Affairs wrote in the late 1860s: "The Jews are the most slovenly of
nations. Jewish prostitutes are the same: on the surface, silk; underneath, filth in all its splendor."[34] Both respectable and depraved, Jewish
brides were said to earn their dowries by going on the street.[35]

Sociological appearances fed these pervasive stereotypes. Even in the
desperate conditions of the remote Sakhalin penal colony, far from the
Pale, commercial sex was run by Jews, reported Anton Chekhov—not
himself an anti-Semite.[36] Census figures showed that Jewish women
worked as managers of regulated brothels in numbers far out of proportion to their representation in the population as a whole, especially
in Poland and the Russian provinces of the Pale, where they sometimes
had a complete monopoly on the business. This predominance was actually more a function of place than of ethnic predilection, however.
Because both Jews and prostitutes were overwhelmingly urban, the
percentage of Jewish women in registered brothels throughout the empire was higher than their share of the population at large, but it was
well below their proportion of city dwellers inside the Pale.[37]

As brothel managers, Jews were seen by antiregulationists as agents
of the police—subcontractors in the business of sexual control rather

---

that Jews had lower syphilis rates; e.g., O. V. Petersen, "O sifilise i venericheskikh bolezniakh v gorodakh Rossii," in *Obshchie doklady po otdelam po izucheniiu rasprostraneniia
sifilisa i venericheskikh boleznei v Rossii* (St. Petersburg, [1897]), 127. The sociologist E. N.
Anuchin, *Issledovanie o protsente soslannykh v Sibir' v period 1827–1846 godov* (St. Petersburg, 1866), 37, suggested that crime was rare among Jewish women because so few
escaped the confines of family life.

[33]On Jewish poverty, see Löwe, *Antisemitismus*, 33–35; and Rogger, *Jewish Policies*, 17.

[34]N. B—skii, "Ocherk prostitutsii v Peterburge," *Arkhiv sudebnoi meditsiny i obshchestvennoi gigieny*, no. 4, pt. 3 (1868): 73–74.

[35]A. A. Vvedenskii, "Prostitutsiia sredi sel'skogo (vne-gorodskogo) naseleniia," in
Obshchie doklady, 2; Reginald Wright Kauffman, *The House of Bondage* (New York, 1910),
208.

[36]Anton Chekhov, *Ostrov Sakhalin* (1895; Moscow, 1984), 289. Although Chekhov
was closely associated with A. S. Suvorin, he did not share the conservative publisher's
attitude toward the Jews. See Simon Karlinsky, ed., *Anton Chekhov's Life and Thought:
Selected Letters and Commentary* (Berkeley, Calif., 1973), 306–7.

[37]Jews were only 4 percent of the population of the entire Russian empire in 1881 but
as much as 13 percent in the Kingdom of Poland and 12.5 percent in the fifteen Russian
provinces that constituted the Pale: Rogger, *Jewish Policies*, 145. In 1897 Jews constituted
half the urban population of Lithuania and White Russia and just under a third in the
Ukraine: Salo W. Baron, *The Russian Jew under Tsars and Soviets*, 2d ed., rev. (New York,
1976), 68. For figures on brothel managers and prostitutes, see Statistique de l'Empire de
Russie, *La Prostitution d'après l'enquête du 1er (13) août 1889* (St. Petersburg, 1891), 13:16–
17, 21.

than victims of sexual exploitation.[38] When in 1906 the Russian Society for the Protection of Women urged the Ministry of Justice to strengthen the laws against procuring, it noted that "a majority" of white-slave traders were "Russian and Galician Jews."[39] This belief echoed the convictions of anti-Semites at home and abroad[40] and was accepted even by writers of a politically progressive bent. In Aleksandr Kuprin's widely read pre–World War I novel *The Pit*, a purported exposé of the social injustice of prostitution, the most repulsive character is the Jewish procurer and pornography peddler Semen Iakovlevich Gorizont, a dirty-minded, deceitful huckster who panders to the vilest of sexual tastes.[41] Cesare Lombroso complained that anti-Semitism was so pervasive in Russian culture as to have affected the thinking of usually "objective" scientists, with whom the Italian was otherwise in intellectual sympathy. He reported that Veniamin Tarnovskii and Dmitrii Dril', both proponents of criminal anthropology, defended the government's anti-Semitic policies as appropriate to dealing with Russian Jews, whom they considered to be different from—that is, worse than—Jews in other parts of the world.[42]

The Jewish community reacted to such images and accusations on two levels. In 1910 eminent Jews from all over Europe and Russia met in London to consider the problem of Jewish involvement in the prostitution trade.[43] But even earlier, their proletarian brethren had taken to

[38]In Poland and the Pale, Bristow asserts, "the forces of order utilised Jewish lumpen elements to help maintain social control" (*Prostitution and Prejudice*, 86).

[39]Letter from the Russian Society for the Protection of Women to the Minister of Justice, November 12, 1906, signed by Saburov: TsGIA, f. 1405, op. 543, d. 512, l. 1 ob. See also D. A. Koptev and S. M. Latyshev, eds., *Ugolovnoe ulozhenie (stat'i vvedennye v deistvie)* (St. Petersburg, 1912), 904–5. Baron Alexander Günzburg, the society's most prestigious Jewish member, did not think it anti-Semitic; he too believed that Russian Jews were prominent in the traffic in women: *Official Report of the Jewish International Conference on the Suppression of the Traffic in Girls and Women, held on April 5th, 6th and 7th, 1910, in London, convened by the Jewish Association for the Protection of Girls and Women* (London, 1910), 69–70, 53, 30–34, 41. See also Marion Kaplan, "Prostitution, Morality Crusades and Feminism: German-Jewish Feminists and the Campaign against White Slavery," *Women's Studies International Forum* 5:6 (1982): esp. 620, 621n, 622.

[40]E.g., *Novoe vremia* claimed that Odessa was the center of the Russian trade, "where this vile commerce is primarily conducted by Jews": L. G., "Belye rabyni," *Novoe vremia*, no. 9553 (October 8/21, 1902). Though nothing in the record suggests that its motives were anti-Semitic, the government did reinforce the criminal sanctions against trafficking in 1909, despite its refusal to recast other sexual legislation or to abandon regulation; see A. A. Shchipillo, "Sostav prestuplenii, predusmotrennykh zakonom 25 dekabria 1909 goda o merakh k presecheniiu torga zhenshchinami v tseliakh razvrata," *Zhurnal Ministerstva Iustitsii*, no. 10 (1911): 56–102.

[41]Alexander Kuprin, *Yama: The Pit*, trans. Bernard Guilbert Guerney (London, 1930), 131–40. Foreign examples of such caricatures were not hard to find; e.g., Kauffman, *House of Bondage*, 19, partially quoted in Bristow, *Prostitution and Prejudice*, 45.

[42]Lombroso, *L'Antisémitisme*, 28.

[43]See *Official Report of the Jewish International Conference*.

the streets to vindicate the collective honor: in the midst of the revolution of 1905, Jewish workers in Warsaw staged what became known as the "Alphonse pogroms"[44] (which we might call "the Great Cathouse Massacre"). Perceiving the brothelkeepers and pimps (Alphonses) as agents of the police, who were attempting to undermine the Jewish trade union movement, armed factory workers and artisans (all male) raided establishment after establishment. Though the mob seemed to be following the "American method of lynch law," as one newspaper commented, it operated under the sign of virtuous wrath. Nothing of value was taken, but symbols of luxury were destroyed: jewels ground under foot, paper money torn to shreds or "ostentatiously thrown down the drain or into the toilets" where sexual effluvia normally flowed. The men "smashed the furniture and chinaware, ripped the linens to shreds, and destroyed various domestic articles, releasing clouds of feathers from pillows and comforters . . . [until] the streets and sidewalks seemed buried in snow. . . . Pianos, mirrors, and chandeliers were thrown from the windows." Most of the prostitutes escaped, but several madams were assaulted and some procurers killed.[45]

By adopting the quintessentially non-Jewish, even anti-Jewish method of the pogrom against their erring brothers,[46] these Jewish workingmen sought to confirm their own respectability, as though by resorting to violence they might establish the masculine dignity of their race. Needless to say, Jewish Social Democratic leaders did not approve of this gesture,[47] but the events did impress the Russian public. Indeed, the feminist physician Mariia Pokrovskaia applauded the workingmen's outraged response as an exercise in patriarchal duty, a defense of those "nearest and dearest," the sisters and daughters who had fallen victim to "seduction, deceit, or violence." Their attack on the brothels, Pokrovskaia wrote, showed that "the Jewish workingmen wanted publicly to refute the charges of hypocrisy directed against them and dissociate themselves from coreligionists who grow rich by trading in female bodies. Let us note by the way that such charges are well founded. The Jews should take heed and restrain their coreligionists from playing such an extensive part in this disgraceful trade."[48]

Equating poverty with virtue, evil with wealth and exploitation,

[44]Bristow, *Prostitution and Prejudice*, 58–61.
[45]M. I. Pokrovskaia, "Spetsial'naia i lichnaia gigiena: Torgovlia zhenshchinami," *Zhenskii vestnik*, no. 1 (1907): 20–22; also *Bulletin abolitionniste: Organe central de la Fédération abolitionniste internationale* (Geneva), n.s., no. 42 (June 1905), 69–71, citing accounts in Warsaw and Bern newspapers.
[46]The struggle was "fratricidal": *Bulletin abolitionniste*, 71.
[47]Bristow, *Prostitution and Prejudice*, 61.
[48]Pokrovskaia, "Spetsial'naia," 21.

some populists had hailed the 1881 pogroms as an expression of primitive anticapitalism.[49] Pokrovskaia's distinction between the good Jew (poor, powerless, and sexually chaste) and the bad Jew (rich, coercive, and morally corrupt) reflected the same populist dichotomy. It also managed to reconcile the competing stereotypes of Jewish sexual comportment by intruding the element of class. Such attitudes were so widespread among the Russian educated public that no one at the time would have interpreted Pokrovskaia's patronizing tone as offensive to the Jews.[50] She was not, after all, the first Christian moralist to congratulate the Jews on censuring their own propensity for vice.[51]

Weininger's denunciation of Jewish sexual rapacity fell into this category of self-censure. By 1909, when the Russian text of *Sex and Character* finally appeared, the book was already enormously popular in Europe,[52] though it is hard to tell which of Weininger's twin phobias attracted more praise or blame: Paul Julius Möbius, a flagrant misogynist in his own right, waxed indignant at Weininger's calumny of the Jews,[53] while a psychiatrist condemned the young man's hatred of women as the mark of a typically Jewish degenerate type.[54] In Russia the reception was decidedly one-sided: almost all commentators attacked Weininger for his attitude toward women; not one took him to

[49]Nora Levin, *Jewish Socialist Movements, 1871–1917* (London, 1978), 53; and Stephen M. Berk, "The Russian Revolutionary Movement and the Pogroms of 1881–1882," *Soviet Jewish Affairs* 7:2 (1977): 24–26, 28–29. On socialist anti-Semitism, see Lewis, *Semites and Anti-Semites*, 111, 113–14.

[50]Though Pokrovskaia published at least one article ("Iarmochnaia prostitutsiia") in *Novoe vremia*, no. 9130 (August 5/18, 1901), she was probably no more an anti-Semite than Chekhov was.

[51]E.g., Michael Ryan, *Prostitution in London* (London, 1839), 193–96.

[52]See Bram Dijkstra, *Idols of Perversity: Fantasies of Feminine Evil in Fin-de-Siècle Culture* (New York, 1986), 218; Emil Lucka, *Otto Weininger, sein Werk und seine Persönlichkeit* (1905), rev. ed. (Berlin 1921), 9–10, 15–16; David Abrahamsen, *The Mind and Death of a Genius* (New York, 1946), cited in Hans Kohn, *Karl Kraus, Arthur Schnitzler, Otto Weininger: Aus dem jüdischen Wien der Jahrhundertwende* (Tübingen, 1962), 37. For the book's impact on racist thought, see George L. Mosse, *Nationalism and Sexuality: Respectability and Abnormal Sexuality in Modern Europe* (New York, 1985), 145. For misogynist appropriations, "Briefe August Strindbergs," in Weininger, *Geschlecht und Charakter*, 650–51; Niki Wagner, *Geist und Geschlecht: Karl Kraus und die Erotik der Wiener Moderne* (Frankfurt am Main, 1982), 155–56.

[53]Paul Möbius's misogynist tract *Ueber den physiologischen Schwachsinn des Weibes* (Halle, 1900) appeared in Russian in 1909, as did his attack on Weininger's book: *Geschlecht und Unbescheidenheit: Beurteilung des Buches von Otto Weininger "Ueber Geschlecht und Charakter"* (Halle, 1904). Though Möbius knew he had not himself invented misogyny, he thought Weininger had appropriated and distorted his ideas (*Geschlecht und Unbescheidenheit*, 3, 7, 13); he found particularly odious Weininger's equation of mothers and whores and his attack on the Jews (24, 27).

[54]Ferdinand Probst, "Der Fall Otto Weininger: Eine psychiatrische Studie," in *Grenzfragen des Nerven- und Seelenlebens*, vol. 31 (Wiesbaden, 1904), 4, 35–40. Probst and Greta Meisel-Hess's expanded attack on Weininger appeared in Russian translation: *Anti-Veininger: Izlozhenie i kriticheskii razbor knigi "Pol i kharakter"* (St. Petersburg, 1909).

task for his opinion of the Jews. His first translator praised him as a genius, a moralist of the stature of Tolstoy, and considered his portrait of the Jews "brilliant," though "negative."[55] His ideas appealed to avant-garde writers who toyed with notions of bisexuality,[56] but Andrei Belyi dismissed the book as an intellectually worthless piece of popular journalism, a parody of scientific and philosophical discourse, with nothing useful to say on the question of women. Belyi did not mention the question of Jews, and neither did Weininger's feminist critics, who focused instead on the harm he might do to the local struggle for women's rights.[57] More surprising, Weininger did not lack Jewish defenders. Jews were to be found among his Russian translators, and at least one reviewer with a Jewish name found nothing objectionable in the young man's book and commended his adoption of the "remarkable" ideas[58] of Houston Stewart Chamberlain, the British anti-Semite who married Richard Wagner's daughter and became a German citizen.

Unabashed anti-Semites, of course, had a field day. Mikhail Men'-shikov, a notorious columnist for Suvorin's *Novoe vremia*, noted the irony of Weininger's reception with irony of his own: "A cluster of scribbling Jews has gathered around the name of Otto Weininger," he wrote. "As thoroughly Jewish as the Bible, this book is weighty, disturbed, nervous, tragic, and false." The falsity was all on one side, however: "One senses in the book of this immature Jew-boy [*nezrelyi Evreichik*] the traces of sexual psychosis. . . . No one has ever expressed such furious contempt for women, such malice toward feminine charms." Weininger's ideas about the Jews, by contrast, reflected the bitter and accurate fruits of self-knowledge. The author, Men'shikov declared, might very well have killed himself "from despair at being a Jew," one with just enough Aryan sensibility to feel "in his blood and nerves something deeply offensive" that deserved to die.[59] Not an idio-

---

[55]A. M. Belov, introduction to Weininger, *Pol i kharakter* [Russian trans. of *Geschlecht und Charakter*] (Moscow, 1909), xvi–xx.

[56]See, e.g., Temira Pachmuss, *Zinaida Hippius: An Intellectual Profile* (Carbondale, Ill., 1971), 92; Anatolii Kamenskii, *Zhenshchina: Rasskaz, pamiati Otto Veiningera* (St. Petersburg, 1909), about a man who adopted female attire; Bruno Mikhel'son, *Pol i krasota: Beseda o krasote po povodu stat'i Otto Veiningera "Erotika i estetika"* (Moscow, 1909), 8.

[57]Andrei Belyi [Boris Bugaev], "Na perevale: Veininger o pole i kharaktere," *Vesy*, no. 2 (1909): 77–81;M. M. Ianchevskaia, "Zhenshchina u Veiningera," in *Trudy pervogo vserossiiskogo zhenskogo s"ezda pri Russkom zhenskom obshchestve v S.-Peterburge, 10–16 dekabria 1908 goda* (St. Petersburg, 1909), 557–66.

[58]N. A. Rubakin, *Sredi knig* (Moscow, 1913), 2:353, cites as the best edition the Likhtenshtedt translation (St. Petersburg, 1910). On Chamberlain, see I. Ashkinazi, "Otto Veininger," *Obrazovanie*, no. 11, sec. 1 (1908): 123. See also M. O. Men'shikov, "Pis'ma k blizhnim: Evrei o evreiakh," pt. 1, *Novoe vremia*, no. 11815 (February 1/14, 1909): 4.

[59]Men'shikov, "Pis'ma k blizhnim," 4. Another anti-Semitic newspaper, praising Weininger for having "expose[d] his tribe's pettiness, spiritual backwardness, racial insig-

syncratic lapse, Men'shikov's failure to couple racism with misogyny reflected the peculiarly uneven outlook of *Novoe vremia*, which held enlightened views on issues such as women's higher education and their right to independence from male legal control, as well as divorce and inheritance reform, while promoting the crudest expressions of racist chauvinism.

The conservative camp in these years indeed accommodated a wide range of responses to a variety of themes. Not only did *Novoe vremia* continue to publish Rozanov's sexually explicit material, but in equally paradoxical fashion the Suvorin publishing house issued a collection of the essays Rozanov had written during and immediately after 1905, depicting the upheaval in ecstatic terms as the creative outburst of normally repressed desire, the happy revolt of youth against sterile parental restraint.[60] Rozanov soon recanted his enthusiasm (though he never regretted his momentary flirtation with political revolt, since it so happily contradicted the rest of his assertions), and other intellectuals who had earlier sympathized with the goals of 1905 also distanced themselves from the radical tradition that had inspired the revolutionary events. They expressed their new conservatism in an enormously influential and controversial collection of essays called *Landmarks*, which first appeared in 1909 and was hailed by *Novoe vremia* as a welcome assault on the left's moral credentials.[61] Reflecting on the state of Russia's youth, one contributor, A. S. Izgoev, deplored its intellectual arrogance and love of subversive ideas, which he equated with the thirst for sexual excitement. Once fanatically attached to populist theories, university students now read Otto Weininger with similar illicit passion, Izgoev claimed. The change of subject matter, merely a new form of the old libidinous urge, was "hardly a cause to rejoice!"[62] Making the same connection that Rozanov had fleetingly discerned between sexual release and political disorder, Izgoev gave it the opposite moral cast.

Izgoev was not the only one to use Weininger as a symbol of moral decay. His reference to the Austrian philosopher was picked up in 1914 by the right-wing Duma deputy Vladimir Purishkevich to prove that

---

nificance, and lack of talent," explained that Weininger had "committed suicide exclusively because he clearly recognized the impossibility of remaining a Jew by blood, even if he had managed to approach moral perfection": L. Zlotnikov, "Iudei v iskusstve," *Zemshchina*, no. 767 (September 21, 1911): 3.

[60]V. V. Rozanov, *Kogda nachal'stvo ushlo, 1905–1906 gg.* (St. Petersburg, 1910), intro. and "Oslabnuvshii fetish." See chap. 6 above.

[61]Marshall Shatz and Judith Zimmerman, eds., "*Vekhi* (Signposts): A Collection of Articles on the Russian Intelligentsia," *Canadian Slavic Studies* 2:2 (1968): 153.

[62]A. S. Izgoev [Aleksandr Solomonovich Lande], "Ob intelligentnoi molodezhi: Zametki ob ee byte i nastroeniiakh," in *Vekhi: Sbornik statei o russkoi intelligentsii*, 2d ed. (Moscow, 1909; rpt. Frankfurt am Main, 1967), 104.

even leftists (to him the *Landmarks* authors were no different from the rest of the disloyal intelligentsia) admitted that students had become a degenerate lot.[63] This decline, he asserted, was the fault of the Jews, who had engineered the student movement. Elaborating the same anti-Semitic themes that had been current during the revolution,[64] Purishkevich warned that Jews *"recognize no moral principles* and will stop inflaming the student masses only when they have gotten them totally under control and turned them into a dumb herd, fit only to satisfy base Jewish instincts." With educational quotas relaxed, "the slippery Jews instantly weaseled their way [*prolezli*—an anti-Semitic catchword] into all the student organizations . . . [where they promoted] orgies, unruliness, and crime."[65]

Much as he detested Weininger for his Jewishness and sexual preoccupations, Purishkevich was prey to the same obsessions. The most potent instrument of Jewish revolutionary subversion, in the deputy's view, was female sexuality. In permitting women to attend lectures, shortsighted liberals had opened the door not only to the starry-eyed woman student (*idealistka-kursistka*) but to the uneducated and depraved common streetwalker (*ulichnaia publichnaia devka*). Where the revolutionaries had failed with their pamphlets and incendiary speeches, Purishkevich shrilled, "woman, with her animal influence, would succeed!" Some of these "supposed pioneers of women's rights" were really no better than whores, in his opinion, and the revolution itself was a cesspool of Jewish animality: "How many pure young lives have succumbed in the filthy revolutionary wave, enticed by the mirage of Yid female charms!"[66]

The Inverted World of Judeophilia

If Rozanov came to share Purishkevich's distaste for educated women, it was not because they threatened male control with the lure of sexual delight but because they had abandoned their true sexual calling. Unlike Weininger, Rozanov did not fear the destructive power of desire and had no wish to put the disturbing chaos of reality into the

[63]V. M. Purishkevich, *Materialy po voprosu o razlozhenii sovremennogo russkogo universiteta* (St. Petersburg, 1914), 267.

[64]See examples in S. Zolatarev, "Deti revoliutsii," *Russkaia shkola*, no. 3 (1907): 2–3, 20–21.

[65]Purishkevich, *Materialy*, 34–35 (original emphasis). Blaming 1905 on the Jews was a standard anti-Semitic turn; see, e.g., M. O. Men'shikov, "Tragikomicheskoe plemia," *Novoe vremia*, no. 13501 (October 12/25, 1913): 4–5. Rozanov shared the conviction: *Opavshie list'ia*, 409.

[66]Purishkevich, *Materialy*, 35–37.

straitjacket of reason. Quite the contrary: he allowed the messy texture of the subconscious process to surface in the idiosyncracy of his prose.[67] "I either laugh or cry," he boasted. "Do I reason [*razmyshliat'*] *in the strict sense? Never.*"[68] To the masculine constraints of science (*muzhskoi, nauchnyi slog*) Rozanov preferred the contradictory, uninhibited, libidinous delight of disconnected sensuous detail, precisely the concreteness of everyday life, the little pleasures of the armchair, the great pleasures of the conjugal bed.[69] Obsessed by the individual, the particular, the immediate, national, spiritual, and organic, Rozanov scorned every form of abstraction and despised propriety, moral categories, ideological divides, good manners, and discretion.[70] He even regretted the uniformity of print, which erased the uniquely personal quality of handwriting; and he rejected both truth and falsehood in the name of sincerity, a directness that transcended any "moral scruple."[71]

Though he criticized the Orthodox church for its formality and repressive moralism and quarreled with Christian doctrine, Rozanov was deeply religious, hostile to secular intellectual life and to public affairs. Along with other writers and thinkers who sought intellectual direction in religious thought after the turn of the century, Rozanov rejected politics in both the practical and the ideological sense.[72] "Politics must be destroyed," he wrote in 1915, "apoliticism established. [. . .] By confusing all political ideas. . . . By making 'red yellow,' 'white green.'" Very much in the tsarist bureaucratic spirit, Rozanov conceded that "'administration' will remain, 'the course of affairs,'—but only in its

[67]"Every movement of my soul is accompanied by *verbal expression* [*vygovarivanie*], which I then without fail must *write down.* It is an instinct": V. V. Rozanov, *Uedinennoe* (St. Petersburg, 1911); rpt. in Rozanov, *Izbrannoe*, 34. Both his speech and his writing were described as "unplanned, unexpected, even by him," and his thinking as lacking in logical sequence: Lutokhin, "Vospominaniia," 6. Gollerbakh (*Rozanov*, 4–5) called his style "physiological" and his work organic, "unsystematic and hodgepodge, but internally harmonious and unified."

[68]Rozanov, *Opavshie list'ia*, 123 (original emphasis).

[69]Rozanov, *Liudi lunnogo sveta: Metafizika khristianstva*, 2d ed. (St. Petersburg, 1913), 228. A friend said: "He was not a thinker but an artist of thoughts; he could think only in images. His nonlogical mind was the type Weininger described as feminine": Lutokhin, "Vospominaniia," 6. Other comments along these lines include: "The flesh and blood of living images" (Gollerbakh, *Rozanov*, 3); "Psychological, not logical" (A. Volynskii, "'Fetishizm melochei,'" pt. 2, *Birzhevye vedomosti*, no. 15348 [January 27/February 9, 1916]: 2).

[70]Rozanov, *Opavshie list'ia*, 131, 417 (respectability is repulsive); 154 ("I change convictions as I change my gloves").

[71]Rozanov, *Uedinennoe*, 5 and 54: "Carelessness" was the path to a kind of truth that was not a moral category (partly cited in Crone, "Nietzschean," 98). Rozanov's concept of sincerity describes very well the process of free association.

[72]See Jutta Scherrer, *Die Petersburger Religiös-Philosophischen Vereinigungen: Die Entwicklung des religiösen Selbstverständnisses ihrer Intelligencija-Mitglieder (1901–1917)* (Berlin, 1973). Among the intellectuals in Rozanov's circle were Zinaida Gippius, Dmitrii Merezhkovskii, Nikolai Berdiaev, Aleksei Remizov, Andrei Belyi, Fedor Sologub, Viacheslav Ivanov (Lutokhin, "Vospominaniia," 5).

Caricature of Vasilii Rozanov. *Satirikon*, no. 50 (1909). Helsinki University Slavonic Library.

empirical sense: 'here's a *fact*,' 'because it is *necessary*.' . . . Without any transitions to theory and general passion."[73] Rather than oppose political principles on principled grounds, Rozanov practiced a kind of literary terrorism designed to "disorganize" public discourse. His strategy took the form of simultaneously adopting opposite political positions in mu-

[73]Rozanov, *Opavshie list'ia*, 230–31 (original emphasis). Because Rozanov frequently used points of ellipsis, I have enclosed in brackets those marking my own omissions from his texts (but not when I cite other writers).

tually hostile publications (a trick for which he was roundly denounced by those who took their ideas seriously)[74] and of introducing disruptive stylistic maneuvers. Random facts (many wrong), intimate details, metaphorical flights, playful nonsense—all these strategies wreaked havoc with the conventions of serious prose and responsible argument.[75]

Rozanov turned literary mores inside out by celebrating the domestic, not in the bourgeois sense of respectable private life but in the archaic tradition of the *Domostroi*, the fecund, patriarchal hearth.[76] Viktor Shklovskii, the formalist critic, explains how Rozanov translated his admiration into a radical textual strategy: Just as Dostoevsky, another conservative who broke with stylistic convention, adapted aspects of the boulevard novel to his literary ends, so Rozanov made intimacy an expressive device; each writer thus insisted on the power of vulgarity, the force of crude emotion and commonplace experience, to unsettle the complacency of accepted ideas. If Dostoevsky celebrated the awesome power of negation, irrationality, and faith, Rozanov represented the return of the repressed: "Forbidden themes continue to exist outside the literary canon," Shklovskii writes, "just as the erotic anecdote has always existed, just as repressed desires persist in the psyche, occasionally surfacing in dreams." Rozanov's "domesticity," which he paraded in the columns of *Novoe vremia*, was one such forbidden and shocking theme; before Rozanov, Shklovskii maintained, "family life [*semeinost'*], quilted comforters, kitchens and kitchen smells (*without satiric* intent) did not exist in literature."[77] Anti-Semitism was another suppressed theme (though not one Shklovskii mentions): though hardly absent in the broader culture, it was at once invisible for being so widespread and considered in bad taste in intellectual circles (though just how thoroughly it was censured we shall see).

Everything that was taboo, pushed to one side, kept under cover,

---

[74]Struve, "Na raznye temy," 139–41, compares Rozanov's incompatible pronouncements on the subject of the 1905 revolution.

[75]One liberal critic called him Mephistophelean—not like Goethe's "clever, calculating, agile demon," but a "crude and clumsy, vulgar, and unscrupulous [*nechistoploten*]" devil, with a powerful and "absolutely independent" intellect that "changed its aspect like a chameleon, winding among the rocks of thought with the speed of a young snake in the mountains": N. Asheshov, "Pozornaia glubina (V. Rozanov, *Opavshie list'ia, korob vtoroi i poslednii*, Petrograd, 1915)," *Rech'*, no. 224/3247 (August 16/29, 1915): 2.

[76]Renato Poggioli, *Rozanov* (New York, 1962), calls him "bourgeois" or petty bourgeois (39–40) and sees his interest in psychology and family life as a kind of democratization or bourgeoisification of spirituality (80, 88), but it seems to me that Rozanov's self is not the same individual that emerged from the Enlightenment, with both a private and civic aspect, but a universal and all-encompassing ego with uncertain boundaries. See also Siniavskii, "*Opavshie list'ia*," 102, 106.

[77]V. B. Shklovskii, *Rozanov: Iz knigi "Siuzhet, kak iavlenie stilia"* (Petrograd, 1921), 10, 17 (original emphasis).

called by a different name, Rozanov took pains to reveal, to exaggerate, to speak out loud. He rejected the euphemisms of science in favor of crude psychological realism, dismissing Weininger's elaborate scheme of abstract types and his regard for scholarly conventions as an unsuccessful attempt to hide the obvious. "From every one of Weininger's pages can be heard the cry, 'I *love men!*'" Rozanov declaimed. "'Well, fine: *you're a sodomite.*' And with that, one may close the book."[78] Though more outrageous than the younger man in his stylistic and philosophical iconoclasm and impatient with respectable society's sexual constraints, Rozanov was committed to traditional values and remained untroubled by the power hierarchy embedded in family and sexual life. Weininger's tormented struggle with the Kantian moral imperative had no meaning for Rozanov, for whom morality was of no concern.[79]

Weininger had considered himself a champion of downtrodden womanhood, which he aspired to rescue from its humiliating and crippling association with sexuality. Man, he argued, had conjured up Woman by using her as the object of his sexual desire, a process he condemned as deeply immoral. Identified entirely with the sexual function, Woman served only as a means to an end, whether pleasure or reproduction. Yet Weininger could imagine agency only as male: only men, in his scheme, could (and should) bring the mutually debasing subject-object exchange to a halt, by renouncing their own sexual desire (even at the cost of human extinction). Without seeming to realize that in substituting mental for sexual domination he was undermining the logic of his ethical position, Weininger suggested that women should cease to serve as objects of desire (brought to life by infusions of semen) and become the objects of knowledge instead (brought to life by the male mind).[80]

If Weininger wished to de-eroticize the mysteries of women and sex by subjecting them to rational dissection (the avowed purpose of his

[78]Rozanov, *Opavshie list'ia*, 95 (original emphasis).

[79]Ibid., 130; "I don't even know how to spell 'morality,' or who its papa and mama were, or if it had kids, or its address—not the tiniest morsel do I know" (*Uedinennoe*, 46); "I am not yet such a scoundrel as to think about morality" (quoted in Gollerbakh, *Rozanov*, 53). See also A. Selivachev, "Psikhologiia iudofil'stva: V. V. Rozanov," *Russkaia mysl'*, no. 2, sec. 2 (1917): 50–51; Struve, "Na raznye temy," 145; Gippius, *Zhivye litsa*, 2:10.

[80]Weininger, *Geschlecht und Charakter*, 401, 456–58, 129–30, 397. Two typical formulations: "Man's only ethical relation to Woman resides not in sexuality or love but in the attempt to understand her" (451); "Not only through the power of . . . the man's sperm . . . but also . . . through the man's consciousness . . . will Woman be . . . fulfilled, impregnated, and transformed" (376). See also Esther Fischer-Homberger, "Herr und Weib: Zur Geschichte der Beziehung zwischen ordnendem Geist und anderen Impulsen," in *Krankheit Frau: Zur Geschichte der Einbildungen*, 2d ed. (Darmstadt, 1984).

work), Rozanov aimed at the reverse: to infuse intellectual activity with erotic power, which he experienced as the fount of personal and creative mastery. Where Weininger denounced the procreative urge as a threat to selfhood, Rozanov considered it the highest form of individual expression.[81] The genitals disgusted Weininger but sent Rozanov into romantic effusions.[82] Happy with women in traditional domestic roles, Rozanov did not perceive them as unsettling. He was consequently not the least dismayed by what he felt to be his own feminine temperament, which did not interfere, he boasted, with the power of his semen to create real children and page after page of prose.[83] Indeed, Rozanov was proud of all the traits Weininger denounced as female: confusion of boundaries, illogic, self-contradiction, the fragmentation of self, immersion in the concrete and everyday. Nor did he sublimate his homoerotic impulses in literary maneuvers, as he perceived Weininger to have done. Instead, he translated his "mystical admiration" (*misticheskaia vliublennost'*)[84] for Dostoevsky into a real, if displaced, sexual link: never having met the novelist, Rozanov married Dostoevsky's one-time mistress, Apollinaria Suslova, in 1880, the year before the older writer's death.[85]

The disturbing quality of Weininger's book was its aggression: the definition of Jews and women as nothingness thinly disguised the desire to see them disappear. On the eve of his suicide (which followed not long after his conversion to Protestantism, a less absolute form of self-obliteration), Weininger is reported to have said that if he did not kill himself, he might well murder somebody else.[86] Rozanov's hatred of the Jews, though no less violent, was a complex mixture of envy and love. To him the Jew represented not only a threat to Christian culture (as Weininger too believed) but also what Christianity lacked: sexual vitality, unshaken patriarchal principles, and a belief in the sacredness of sexual desire.

Like later commentators, contemporaries struggled to reconcile Ro-

---

[81]Rozanov, *Liudi lunnogo sveta*, 33–34. Siniavskii ("*Opavshie list'ia*," 34) cites Rozanov's description of the face as an outward expression of sexuality.

[82]Weininger, *Geschlecht und Charakter*, 319–20, 339, 458; Rozanov, *Liudi lunnogo sveta*, 40.

[83]Rozanov had two female pseudonyms: Varvarina, after his common-law wife, Varvara (see Lutokhin, "Vospominaniia," 5–6); and Elizaveta Sladkaia, which *Novoe vremia* refused to let him use (Gippius, *Zhivye litsa*, 2:55). Rozanov claimed that his writing was based "not on water or blood but on *human semen*" (quoted in Siniavskii, "*Opavshie list'-ia*," 35).

[84]Gollerbakh, *Rozanov*, 58.

[85]George L. Kline, *Religious and Anti-Religious Thought in Russia* (Chicago, 1968), 55. Suslova soon left Rozanov but never granted him a divorce (she did not die until 1918, only a year before he did). In 1891 Rozanov established a household with a much younger woman, Varvara Rudneva, with whom he had five children. Their inability to marry caused Rozanov to champion divorce reform. See Gippius, *Zhivye litsa*, 2:21, 33, 56–57.

[86]Otto Weininger, *Taschenbuch und Briefe an einen Freund*, ed. Arthur Gerber (Leipzig, 1919), 19.

zanov's professed philo-Semitism with his crude anti-Semitic beliefs.
The philosopher himself had a rather literal-minded view of what it
meant to hate the Jews and contribute to their misfortunes: "Neither
Suvorin nor the paper [Novoe vremia] are 'anti-Semitic' in any way," he
insisted, but "simply tell jokes about the Jews and describe the harm
they obviously cause Russia and the Russians, with their greedy ambition to grasp everything in their own hands."[87] As a political naif,
Rozanov does not rank with the crude Men'shikov, though Suvorin
happily published them both.[88] It did not seem to trouble fellow writers
that vulgar anti-Jewish clichés had peppered his work even as he la-
bored to demonstrate the superiority of Old Testament earthiness to the
cold asceticism ushered in by Christ.[89] The St. Petersburg Religious-
Philosophical Society, of which Rozanov was an active member, expelled him only in 1913, on the grounds that his recent outpourings had
acquired a new, objectively dangerous significance in the context of the
Beilis affair and that his position threatened to politicize the society.[90] In
response, Rozanov published an insulting attack on his former associates Dmitrii Merezhkovskii and Dmitrii Filosofov, as well as the historian Pavel Miliukov, in the newspaper Zemshchina. He claimed that
these intellectual paragons despised the Russian common folk, who
alone recognized the sinister threat embodied in the Jews' ritual murder.
Once having "wallow[ed] in the stinking filth of their 'Pale,'" Rozanov
wrote, the Jews now "sweep through St. Petersburg, slapping the
cheeks of Duma deputies and writers."[91]

Even after his expulsion from the society, however, Rozanov continued to attract the company of respected writers.[92] Many of his peers

[87]V. V. Rozanov, "Iz pripominanii i myslei ob A. S. Suvorine," in Pis'ma A. S. Suvorina k V. V. Rozanovu (St. Petersburg, 1913), 50–51.

[88]Gollerbakh (Rozanov, 88) asserts that Rozanov's anti-Semitism was different from
that of Novoe vremia. After the revolution the Bolsheviks shot Men'shikov, but Gorky
interceded to protect Rozanov, who died of natural causes in 1919 (Gippius, Zhivye litsa,
2:82, 84). Some dissenting voices questioned Rozanov's political innocence. "Widely proclaimed and promoted by Novoe vremia, Rozanov was never apolitical. He was always
political, only political in the vilest way": Lukian, "Ocheredi," Birzhevye vedomosti, no.
15143 (October 12/25, 1915): 3.

[89]E.g., Jews are "unclean, unpleasant, and physiologically repulsive . . . stinking":
V. V. Rozanov, "V svoem uglu: Iudaizm," pts. 10–13, Novyi put', October 1903, 110–11
(journal of the St. Petersburg Religious-Philosophical Society).

[90]Gippius, Zhivye litsa, 2:75–78; Scherrer, Die Petersburger Religiös-Philosophischen Vereinigungen, 196. On Rozanov's role in the society, see George F. Putnam, Russian Alternatives to Marxism: Christian Socialism and Idealistic Liberalism in Twentieth-Century Russia
(Knoxville, Tenn., 1977), 61–64, 79–83, 97–98; his departure was described in different
(and offensive) terms in "Melochi iz zhizni i gazet," Zemshchina, no. 1618 (March 22,
1914).

[91]V. Rozanov, "Nasha 'koshernaia pechat','" Zemshchina, no. 1477 (October 22,
1913): 3. Merezhkovskii and Miliukov joined other intellectuals in a collective attack on
anti-Semitism called Shchit (Petrograd, 1916).

[92]Gollerbakh (Rozanov, 84) names Aleksei Remizov, Kornei Chukovskii, and Mikhail

allowed him the right of self-contradiction and the frank irresponsibility of a child,[93] no doubt in deference to his imaginative and intellectual gifts and his original and unsettling voice, which seemed to stem from that same uncensored naiveté. Admirers who have hailed the fragmentary, reportorial quality of Rozanov's prose simply ignore the scurrilous bits that fluttered into his text along with other scraps of the surrounding culture.[94] The émigré writer Andrei Siniavskii, lecturing in Paris in the 1970s, felt obliged to condemn the offending passages as "trivial, vulgar . . . scandalous and shameful" but maintained that Rozanov had in the end publicly apologized for them.[95] Other critics have denied that Rozanov apologized at all, since he had nothing to be ashamed of: love was really all that was ever in his heart.[96]

Though it is true that Rozanov explicitly deplored the physical abuse of Jews (having also assumed they enjoyed being beaten),[97] often sang their praises, and was blissfully innocent of any intention to provoke, he accepted their position as objects of violence in the archaic world of patriarchal domination, just as he accepted women's traditional lot. As womanish creatures, quintessentially and primitively female, like the Russian peasant woman (the *baba*), the Jews got equally rough treatment: "The Yids are beaten because they are *babas*," Rozanov wrote, "just as Russian muzhiks beat their *wives* [*svoi baby*]. The Yids are not *males*, but *females*."[98] Despite their loud complaints about the discomforts of the Pale, the Jews in fact thrived on "poverty, assault, and mockery."[99]

The sad persistence of anti-Semitic overtones in Russian culture (still

---

Kuzmin among those who visited him during this period and remarks that the supposedly moral grounds for Rozanov's exclusion had "nothing to do with real morality."

[93]Lutokhin, "Vospominaniia," 6. Rozanov said of himself that he was "like a child in its mother's womb, who does not want to be born" (*Uedinennoe*, 39).

[94]Shklovskii, *Rozanov*, 33, 40–41. Andrei Belyi called him the poet of everyday life, who worked with "living particles" of concrete existence: "Publitsistika: V. Rozanov, *Kogda nachal'stvo ushlo*," *Russkaia mysl'*, no. 11, pt. 2 (November 1910): 374–76. For one example, see Lutokhin, "Vospominaniia," 6; for a hostile description of this technique, see the populist critic N. K. Mikhailovskii, "Literatura i zhizn'," 80.

[95]Siniavskii, "*Opavshie list'ia*," 92–93. In *Apokalipsis nashego vremeni* (1917–18; rpt. in Rozanov, *Izbrannoe*), 484, 486, Rozanov said he had been mistaken about the Jews' evil intentions toward the Russians and had maligned them; he denied that he was "an enemy of the Jews," but the same text contains anti-Semitic language. Gollerbakh claims that Rozanov repented of his anti-Jewish words (*Rozanov*, 87–88). See also Glouberman, "Vasilii Rozanov," 122–23.

[96]Lutokhin, "Vospominaniia," 6.

[97]V. V. Rozanov, *Oboniatel'noe i osiazatel'noe otnoshenie evreev k krovi* (St. Petersburg, 1914), 25.

[98]Rozanov, *Opavshie list'ia*, 253 (original emphasis). This argument is repeated in V. V. Rozanov, *V sosedstve Sodoma: Istoki Izrailia* (St. Petersburg, 1914), 11.

[99]V. V. Rozanov, "*Angel Iegovy*" *u evreev: Istoki Izrailia* (St. Petersburg, 1914), 6.

linked to the endorsement of traditional social values) may be exemplified by Siniavskii's remark exonerating Rozanov from charges of ill will toward the Jews, not because it is false in a literal sense but because of Siniavskii's own choice of language. Speaking of Rozanov's attack on Beilis, *The Jews' Olfactory and Tactile Relationship to Blood* (1914), Siniavskii writes:

> For Rozanov the Jews are not simply people like all the rest, a nation like any other, but a physiologically gifted, sexually gifted, and hence religiously gifted folk. I would say that Rozanov experiences the Jews through taste and smell [*na vkus i na zapakh*]. He himself, figuratively speaking, has an "olfactory and tactile" relationship to Jewish blood. He bows before them, envies them, is jealous of them, and thus cannot be neutral or indifferent to them. Rozanov either loves the Jews or actively does not love them. This "nonlove" is the consequence of his heightened attitude, in the last analysis, of his attraction to the "Jews" as a special breed [*poroda*].[100]

Although Siniavskii is not an anti-Semite, he admires Rozanov for his conservative values and describes him in appropriately organic terms, as someone who "left us no system, but the very process of thought"; whose "varied qualities and turns of mind" make him resemble "a many-limbed tree, growing from one seed"; whose opinions fluctuated like the ebb and flow of the tide.[101] For Rozanov to share in the Jews' sensual spirituality, their recognition of the sacred body, of the religious significance of the flesh, is something Siniavskii admires.[102] For Rozanov, Jews represented the unspoiled childhood of religion, the sensuous pagan vitality extinguished by Christianity's chilling hostility to sex, with its self-denying focus on the next world. If Jews had a special relation to the taste and smell of blood—their own or someone else's—it was, in Rozanov's view, to their credit.

Rozanov's late texts, however, make clear that his fascination with the Jews was not a counterweight to his obscurantism but its precondition.[103] Several pamphlets that Suvorin's house also published (*Europe*

---

[100]Siniavskii, "Opavshie list'ia," 94–95.

[101]Ibid., 6–7, 22. On Siniavskii's thought and esp. his attitude toward anti-Semitism, see Joseph Frank, "The Triumph of Abram Tertz," *New York Review of Books* 38:12 (1991): 35–43.

[102]Siniavskii admits, however ("Opavshie list'ia," 290), that some of Rozanov's pronouncements have been cited "out of context" to support fascist ideas.

[103]For a contemporary's view of Rozanov's general ambivalence as psychological honesty, see Gollerbakh, *Rozanov*, 64. For a less benign view of such ambivalence, see Lewis, *Semites and Anti-Semites*, 94–95: "While the philo-Semites in their discussion of the Jews often combine contempt with good will, the anti-Semites frequently display a mixture of respect, or even awe, with their malevolence."

and the Jews, "Jehovah's Angel," In the Company of Sodom)[104] are sometimes omitted from selected bibliographies of Rozanov's works.[105] Yet
his much-admired two-volume collection of aphoristic reflections,
Fallen Leaves (1913, 1915), also contains some rather offensive language.[106]
It includes the passage to which Siniavskii objects in strongest terms:
the depiction of the Jew as a spider, sucking the blood of the harmless
flies (the Russian people) it has caught in its web. Pogroms, Rozanov
explained, were the tormented Christian victims' "convulsive" agony,
the beating of their wings as they tried to escape. He did not excuse
murder, even in self-defense, and believed that the Jews should be
physically protected. But having taken an unexpectedly principled
stance, Rozanov more characteristically then indulged himself in the
play of metaphor, declaring that "one must slash away at the web's
supports, remove it, trample it under foot: free oneself from the spider
and sweep the room of webs."[107]

Victims though they might be, Rozanov wrote, the Jews still managed to dominate the Christian world: "There is power in sex. Sex is
power. The Jews have it; the Christians do not, and therefore the Jews
control them."[108] To wrest control from the Jews, he believed, Christianity must revitalize the reproductive principle. Unlike Weininger,
who saw sexual desire as female, Rozanov did not associate the Jews'
libidinous power with their *bab'ia natura*,[109] however earthy and fecund
actual *baby* might be. Instead, he identified the Jews' sexual potency
with their worship of the phallus and the homoerotic bonding enacted

---

[104]Gippius (*Zhivye litsa*, 2:75) claims that some of Rozanov's anti-Semitic pieces were
too extreme for *Novoe vremia* and appeared instead in *Zemshchina*, the newspaper of the
ultraright Duma deputies. See also Lukian, "Ocheredi," 3. After Suvorin died (1912), his
sons published three Rozanov pamphlets in 1914, with runs of 3,000 copies each: *"Angel
Iegovy" u evreev: Istoki Izrailia; Evropa i evrei;* and *V sosedstve Sodoma: Istoki Izrailia.* (The
last work has nothing to do with homosexuality but concerns the moral qualities of the
effeminate Jewish nation. In other works—e.g., *Liudi lunnogo sveta*—Rozanov called
Christians effete.) For the period of the Beilis trial I found only one article by Rozanov in
Zemshchina: "Nasha 'koshernaia pechat'.'" It was reprinted by the Suvorins in Rozanov,
Oboniatel'noe, 108–16.

[105]E.g., V. V. Rozanov, *Four Faces of Rozanov: Christianity, Sex, Jews, and the Russian
Revolution*, trans. and ed. Spencer E. Roberts (New York, 1978), 17–18; and Stammler,
"Wesensmerkmale," xxxiv–xxxv. They are cited, however, in Heinrich A. Stammler,
Vasilij Vasil'evic Rozanov als Philosoph (Giessen, 1984), 60, 62.

[106]Rozanov, *Opavshie list'ia*. In this same period he published a sometimes overlooked
celebration of Russia's mission in the war: *Voina 1914 goda i russkoe vozrozhdenie* (Petrograd, 1915). It is missing from the two bibliographies cited above but is listed in
Stammler, *Rozanov als Philosoph*.

[107]Rozanov, *Opavshie list'ia*, 301–2. Readers recognized the virulence of this metaphor
and its opposition to his explicitly stated defense of the Jews' physical integrity: Asheshov, "Pozornaia glubina," 2. On Rozanov's contradictory use of metaphor, see Glouberman, "Vasilii Rozanov," 212.

[108]Rozanov, *Opavshie list'ia*, 131.

[109]"The Jews' womanish nature is my *idée fixe*" (ibid., 95).

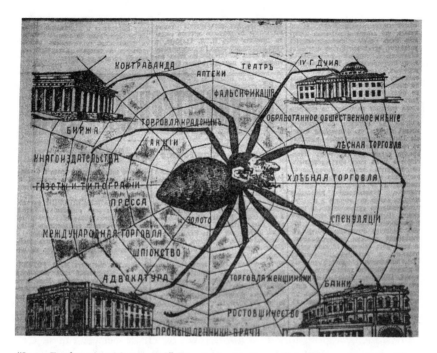

"Is an Explanation Necessary?" Front-page cartoon, *Zemshchina* , no. 1180 (December 6, 1912). The spider, representing the Jew, straddles a web on which the Jews' activities, identities, and places of influence are inscribed: falsification, manipulated public opinion, lumber trade, grain trade, speculation, white-slave trade, usury, commercial medicine, the law, spying, international trade, gold, the press, newspapers and printing, book publishing, the stock exchange, contraband, pharmacies, the theater, and the fourth State Duma. M. E. Saltykov-Shchedrin State Public Library, St. Petersburg.

in their fundamental religious rite, the ceremony that at once marked them off from others and cemented their communal ties: circumcision. By cutting back the foreskin and exposing the "head," the Jews enhanced their sexual potency, their erotic attraction for women, and emphasized their intellectual prowess as well: "Without the external skin, the *exact form* of the head of the membrum virile appears as the *prototype* or *general type* of the structure of the *head* itself." Circumcision made intercourse more pleasurable for both man and woman, Rozanov affirmed, and also more prolonged. He insisted that the Jews' intensified sexual experience explained their "enormous vitality," their "tireless struggle for existence, and their historically tested religiosity."[110]

[110]V. V. Rozanov, *V mire neiasnogo i ne reshennogo*, 2d ed. (St. Petersburg, 1904), 355–58 (original emphasis), describes the improved sexual experience of a man who was circumcised in adulthood.

Much of Rozanov's demonstration of why the Jews might well have murdered Beilis's supposed victim, even if Beilis himself had not in fact done so, centered on his interpretation of the circumcision rite, which (along with kosher animal slaughter) represented the survival of blood sacrifice.[111] In the slaughterhouse, Rozanov claimed, animals were slowly tortured to death with the same expertise as that used to murder Beilis's victim.[112] The simple Russian peasant was justly terrified of such cruel "medieval" practices; if the Jew would only give them up and emerge from his self-imposed cultural isolation (the Pale he had built around himself), the Russians would no longer fear him.[113] As for circumcision, Rozanov explained, it was not the "paper money" of symbolic substitution (the slaughtered lamb instead of the firstborn son) but the "real gold" of literal obedience to divine command, the drawing of human blood. He pictured the ritual circumciser taking the baby boy's mutilated penis in his mouth and imagined him in graphic terms tasting the blood "around his tongue and lips." It would be "hot, sticky, red, arterial—it must be arterial, not dark venous blood, according to the general law and method of all Jewish sacrifices! Only, this is not animal but human blood."[114]

Since Rozanov believed sex to be divine, it is not surprising that he envisioned religious initiation in the form of fellatio between men. Circumcision was the Jewish equivalent of baptism, he explained, the mark of Jehovah's angel descending on each baby boy. Jewish girls gained contact with God only through marriage and procreation and hence were eager brides. Jews were thus "terribly close to each other in a *corporeal* way," he added, "through the ritual bath, circumcision, and marriage. They are essentially all *related* and all a little in love with one another[. . . .] A current of powerful religious electricity flows from 'the American Yids' to St. Petersburg." This homoerotic sexual con-

[111]For a similar attack on circumcision, see Lombroso, *L'Antisémitisme*, 18. Rozanov did not invent the connection between circumcision and ritual murder; anti-Semites often represented Christ's own circumcision as an anticipation of his eventual crucifixion, another bloodletting for which the Jews were to blame: see Stefan Rohrbacher and Michael Schmidt, *Judenbilder: Kulturgeschichte antijüdischer Mythen und antisemitischer Vorurteile* (Reinbek bei Hamburg, 1991), 278.

[112]See "For an end to ritual animal slaughter" and "What I happened to witness," in Rozanov, *Oboniatel'noe*, 73–78, 262–82.

[113]Ibid., 74–77. In *Evropa i evrei* (St. Petersburg, 1914), Rozanov elaborated on the Jews' responsibility for their isolated position, insisting it was the Jews who despised and hated Christians, not the reverse. The Jews' own secretiveness as the source of popular fear was an anti-Semitic cliché; e.g., Kievlianin, "K voprosu ob istinnoi osnove very v ritual'nuiu legendu," *Tserkovno-obshchestvennyi vestnik*, no. 47 (November 28, 1913): 12–13.

[114]Rozanov, *Oboniatel'noe*, 46, 52. Suctioning the blood with the mouth was indeed part of the circumcision ritual among very Orthodox Jews. I owe this information to Eveline Goodman-Thau.

nection (the "sodomitic sting of circumcision") was the secret of the Jews' extreme clannishness, in Rozanov's view. Their bond, he asserted, was "neither human, nor civic, nor social. There are patriots everywhere; many people have intense tribal feelings. But this 'howling in concert from Boston to Odessa' (Dreyfus, Beilis) has parallels only in another world."[115] Their hatred for Christians was as organic as their love for each other, Rozanov explained: "With each 'bit of kosher food,' one could say that the Jew swallows a 'conspiracy' and a 'vow' directed against the Russians; in plain language, he swallows 'the declaration of a pogrom against the Russians,' which is not simply read and discarded but stews in his stomach and is carried by his bloodstream into his flesh and bones, into his blood and brain."[116]

The Jews' rapaciousness, their vampire-like lust for blood, had metamorphosed in civilized societies into a lust for wealth and property, said Rozanov, though occasionally they lost control and went for an innocent Christian boy.[117] The fact that they dominated Russia was demonstrated by the tragic acquittal in the Beilis trial, when the jury, composed largely of Russian peasants, rejected the prosecutor's charge. It was only the goodhearted, soft, and passive Russian character that had enabled the accused to go free.[118] Whereas the Jews derived vitality from an active, passionate sex life, Christians were enfeebled by their sexual inhibitions. Russians in particular exemplified all that was wrong with a sexuality deprived of its original sacred meaning: "as a tribe, in the mass," they suffered from sexual Oblomovitis, Rozanov complained, with reference to Ivan Goncharov's famous literary hero, the pampered nobleman too depressed to get out of bed or woo his beloved. The Russians' numerous children were not robust but "feeble, anemic, untalented, sickly," products of sexual feelings more like a drizzle than a storm. While the Jews had intercourse in the Sabbath holiday spirit, the philosopher claimed, Russians just "threw off their boots," "did the deed," and dropped off to sleep, as though satisfying any other physiological need, "without poetry, without religion, without a single kiss, often without a single word spoken!"[119]

In what way was Rozanov original or extreme in his response to the Beilis affair? Belief in the reality of ritual blood murder was not confined to crackpots or vitriolic rhetoric to the pages of *Novoe vremia* and fringe right-wing journals. Rozanov was not alone, for example, in ac-

[115]Rozanov, *"Angel Iegovy,"* 16–17, 19–24 (original emphasis).
[116]Rozanov, *Evropa i evrei,* 13.
[117]Rozanov, *Oboniatel'noe,* 149–50.
[118]Ibid., 133. On the jury's composition, see Appendix to *Delo Beilisa: Stenograficheskii otchet,* vol. 3 (Kiev, 1913).
[119]Rozanov, *Liudi lunnogo sveta,* 82–84.

cusing the Jews of vampirism.[120] The distinguished linguist Vladimir
Dal', author of the authoritative dictionary of the Russian language,
had been commissioned in 1844 by the minister of internal affairs to
write a book on Jewish ritual murders, in which he explained that the
fanatical Hasidic sect preyed on Christian boys.[121] The professor of psychiatry Ivan Sikorskii, testifying for the prosecution at the Beilis trial,
affirmed that ritual murder had occurred and that only a religious fanatic could have done it.[122] Most voices within the Orthodox community endorsed the blood ritual myth and its relevance to the trial, but
some dissociated themselves from anti-Semitism in general and this belief in particular.[123] The hierarchy did not issue an opinion on the matter, thus implicitly supporting the accusation, yet it appears that no
Orthodox theologian was willing to testify for the prosecution; instead,
a Catholic priest, one Father Pranaitis, was called to the stand.[124] Nevertheless, even among Orthodox figures who deplored anti-Semitism,
some believed that Jewish sects might indeed engage in such ritual acts.[125]

Rozanov was not alone, then, in accepting the blood ritual myth or

[120]See Uranus, *Ubiistvo Iushchinskogo i kabbala*, 2d ed. (St. Petersburg, 1913), 15.

[121]V. I. Dal', *Zapiska o ritual'nykh ubiistvakh: Rozyskanie ob ubienii evreiami khristianskikh mladentsev i upotreblenii krovi ikh* (1844; St. Petersburg, 1913).

[122]*Delo Beilisa*, 2:253–61. Sikorskii's testimony was denounced by his colleagues
Vladimir Bekhterev and Aleksandr Karpinskii, who testified for the defense; by the editors of *Sovremennaia psikhiatriia* ("Khronika," nos. 9–10 [1913]: 754–58, 837–38); and the
St. Petersburg psychiatric society ("V obshchestve psikhiatrov," *Rech'*, no. 294 [October
27/November 9, 1913]: 6).

[123]For defense of the blood murder accusation, see Episkop Aleksii, "Moral' Talmuda," *Pribavleniia k Tserkovnym vedomostiam*, no. 44 (November 2, 1913): 2025–30; T. I.
Butkevich, "O smysle i znachenii krovavykh zhertvoprinoshenii v dokhristianskom mire
i o tak nazyvaemykh ritual'nykh ubiistvakh," pts. 1–4, *Vera i razum*, nos. 21–24 (1913):
281–99, 413–37, 553–608, 723–68; "K kievskomu protsessu," *Tserkovnyi vestnik*, no. 41
(October 10, 1913): 1265–66; " 'Taina krovi' y evreev: Ekspertiza kuratora-ksendza I. E.
Pranaitisa po delu ob ubiistve Andriushi Iushchinskogo," *Missionerskoe obozrenie*, no. 12
(1913): 559–97. The journal for church reform, *Tserkovno-obshchestvennyi vestnik*, denounced the myth and the trial: K. Aggeev, "Tiazhelyi vopros," no. 49 (December 12,
1913): 1–3; V. Pravdin, "Ritual'naia legenda," no. 43 (October 31, 1913): 1–3; Vishnevskii, "Bogoslovskaia ekspertiza," no. 45 (November 14, 1913): 3–5; N. Zaozerskii,
"Tsennost' verdikta prisiazhnykh po delu Beilisa," no. 45 (November 14, 1913): 6–8; and
replies from Kievlianin, "K voprosu," and N. Pisarevskii, "Neskol'ko slov k 'Tiazhelomu
voprosu,' " no. 50 (December 25, 1913): 14–15. The Old Believers sided with Beilis because of their own experience with persecution: "Okolo dela Beilisa," *Tserkov': Staroobriadcheskii tserkovno-obshchestvennyi zhurnal*, no. 42 (October 20, 1913): 1001–3.

[124]Pravdin, "Ritual'naia legenda," 2–3. Rozanov's reply to Orthodox theologians who
rejected ritual murder appeared first in "Vazhnyi istoricheskii vopros," *Novoe vremia*, no.
13485 (September 26/October 9, 1913): 4, then in *Oboniatel'noe*, 42–58. An Orthodox
liberal who blamed Catholicism and the Orthodox hierarchy for anti-Semitism was Vishnevskii, "Bogoslovskaia ekspertiza," 3–4. For the Catholic priest's testimony: " 'Taina
krovi' y evreev," 559–97.

[125]Aggeev, "Tiazhelyi vopros," 1–2, otherwise denouncing anti-Semitism, here
sounds very much like the anti-Semite Butkevich, "O smysle," 767; see also "K
kievskomu protsessu," 1265–66.

in lending it a sexual dimension. Some anti-Semitic texts drew a parallel between the Jews' supposed license to exploit and even kill non-Jews and their authorization to prey on Christian women. The Talmud, supposedly the source of all nefarious Jewish practices, was said to encourage sadism, sodomy, and other violent and depraved sexual acts.[126] Father Pranaitis testified that Jews attributed enormous importance to blood and therefore encouraged their men to deflower Christian virgins, even though they considered intercourse with non-Jewish women no better than sex with an animal. Rozanov himself used this last idea to explain why Jewish authorities did not accept the validity of mixed marriages.[127]

What set Rozanov apart was his recognition of the homosexual undertones of the blood ritual myth, displaced onto his fantasy of circumcision as holy fellatio. Although this interpretation did not accord with his depiction of Jews as paragons of heterosexual vitality, it did resonate with his fixation on semen as the source of sexual power. Flourishing in the life-giving sunlight of ancient piety, the same Jewish men who united in the worship of the phallus, according to him, kept themselves fit in the exercise of conjugal relations, expending their pent-up semen in spurts of energetic passion, to the delight of their fertile wives, who gloried in sensuality for the sake of motherhood and family life.[128] An erotically deficient Christian culture, by contrast, languished in the pale rays of the moon. Lacking semen, it denied the importance of marriage and procreation and had pulverized "the bone-firm principle of 'manliness' [*muzhestvennost'*]," Rozanov complained, allowing "femininity, the 'eternal feminine,'" to "overflow even the souls of men."[129]

Unlike Weininger, who considered gender stereotypes confining,[130] Rozanov deplored the erosion of sexual difference and thus bemoaned what he considered the effeminate nature of Christ. "*Maidenhood*, tenderness, femininity shine through his *masculine* features," he complained. "We worship the Maid in the Man." The true male, by contrast, was heroic, "hard, straight, strong, aggressive, forward-moving, emphatic, masterful"; the true female soft, affectionate, gentle, sweet (though not always good-looking), silent, "compliant, tractable," and "ill-defined," like her soft, receptive, moist, and aromatic genital organs.[131] With typical disregard for the clash of imagery, Rozanov never bothered to explain how the "womanishness" of the Jews differed from

[126]Aleksii, "Moral' Talmuda," 2029.
[127]"'Taina krovi' y evreev," 584, 586; Rozanov, *Evropa i evrei*, 16.
[128]Rozanov, *Liudi lunnogo sveta*, 85–88, 76, 94.
[129]Ibid., 59, 70, 90, 187–88, 200. The title means "People of the Moonlight."
[130]Weininger, *Geschlecht und Charakter*, 69–70.
[131]Rozanov, *Liudi lunnogo sveta*, 189–90, 39–40, 45–46 (original emphasis).

od" of the Christians, though the first were bursting
the latter had none at all. Nor did he reconcile his pride
\eness" with his disapproval of Christ's.

..sier time contrasting traditional with modern women.
..g their procreative mission to receive man's generative seed,[132]
..uern women adopted intellectual pursuits, abandoned their sensual
silence for mannish thoughts and outspoken ideas (in classrooms and
meetings they "fight, curse, read, translate") and avoided marriage.[133]
But for all his celebration of actual childbearing women, Rozanov reserved the true creative power for males. Perhaps it was because generation was phallic that Jewish men could be "womanish" and virile at the
same time. In Rozanov's view, intercourse during pregnancy allowed
the child to be fed by both parents at once. The fetus, he asserted,
"develops to the full only when the husband coddles and caresses the
wife's womb." Mother's milk was a "relatively rational and earthly
substance," he contended, which could not produce a *complete person*
as semen could: "Milk does not grow or develop, but a dewdrop of
semen grows into the complete person and lives in him until he is seventy years old." He denounced the idea that women did not want or
need sexual intercourse during pregnancy as an "impotent, sodomitical"
misconception, since they actually drew strength from sexual contact
with the male.[134]

Both Weininger and Rozanov considered the modern world impotent and feminized, though for opposite reasons: Weininger because it
was suffused with sex and dominated by the Jewish spirit; Rozanov
because it was sexless and Christian.[135] But for all his anger at the sterile
puritanism of New Testament morality and his fury at women who
turned themselves into men, Rozanov, like Weininger, counseled tolerance for people of homosexual inclinations, who sought the same intimacy and affection as everyone else. Believing all forms of desire to be
natural (though he found some particular acts repugnant), Rozanov dismissed scientific notions of sexual pathology as yet another version of
Christian disgust with sex. Norms, he asserted, were merely conven-

[132]"The fate of a girl without children is horrible[. . .]. A girl without children is a
sinner. This is 'Rozanov's law' for all of Russia (except 'lunar people')" (Rozanov,
Opavshie list'ia, 169).

[133]Rozanov, *Liudi lunnogo sveta*, 51–52. In 1904–5 Rozanov had defended women's
education and their participation in public life, but all his opinions of that period he later
contradicted, without in principle repudiating them: see Rozanov, *Kogda nachal'stvo ushlo*,
22, 87–95.

[134]Rozanov, *Liudi lunnogo sveta*, 75–76 (original emphasis).

[135]"The spirit of modernity is a Jewish one . . . this entire modern coitus-culture":
Weininger, *Geschlecht und Charakter*, 441, 443; Rozanov, *Liudi lunnogo sveta*, 197, 200.

tional, part of the culture of abstract ideas he had rejected in favor of a nature that knew no rules.[136]

Rozanov was no more consistent in his attitude toward the pale, "lunar" passion of people who refused to breed, however, than he was toward the Jews. Admitting that unmarried "maiden-men" and "male-maidens" who devoted their untapped domestic energies to social causes were "indispensable to civilization," he claimed at the same time that lack of family feeling deprived them of a stake in the future, making them a threat to the social fabric, an isolated group "at odds with humanity, negating its very roots."[137] In this sense, they resembled the Jews, that self-enclosed body hostile to the surrounding culture, whom Rozanov also accused of designs against humanity itself (symbolized in their plotting of Christ's crucifixion).[138] Ultimately, the principle of homosexuality would spell the end of the human race, he warned, a prospect that Rozanov (unlike Weininger) did not view with equanimity: "Rather than humanity going to its grave, better the sodomites should do so."[139]

Along the road to demographic apocalypse, the same subversive instinct threatened the political order. Homosexuality, Rozanov claimed, accounted for the self-denying spirit of the Russian radical intelligentsia, the ascetic ideal of that "half-*Urning*" Chernyshevskii: "The entire 'storm' [of our 1860s] issued from the beaker of homosexuality."[140] Excluded from the Christian fraternity, the Jews likewise subverted established ways, despite the strength of those family ties that homosexuals lacked. Their domestic values did not seem to keep them from the public enterprises inimical to traditional life—commerce and revolution. Fortunately, Rozanov reflected in 1917, Jewish responsibility for the current crisis would inspire the common folk to launch a counter-revolutionary pogrom, destroying both the new regime and the Jews. It may seem odd that this notion should occur as part of Rozanov's answer to the question "why one should not in fact stage anti-Jewish pogroms," which he posed in his last published work, *The Apocalypse of Our Time* (1917–18), and equally odd that it should accompany his admission that the Jews had never turned "an evil eye" on the Russians, who had made their lives so hard.[141]

[136]Rozanov, *Liudi lunnogo sveta*, 34, 41–42, 103, 110, 195–96, 220–21.
[137]Ibid., 110–11, 198.
[138]Rozanov, *Evropa i evrei*, 38.
[139]Rozanov, *Liudi lunnogo sveta*, 145.
[140]Ibid., 160, 173. The term *Urning* was coined by Karl Heinrich Ulrichs in the 1860s to designate male homosexuals.
[141]Rozanov, *Apokalipsis nashego vremeni*, 480–81, 487. In this text, Rozanov admitted

In the end, Rozanov decided, Jews and Russians were linked by the very power essential to the Jews' vital spirit: their focus on private life, on emotions and physical connection. "Despite the blows ('pogroms') they have endured, the Jews' view of the Russians, of the Russian soul, even of the Russians' intolerable character is respectful and serious," he wrote, because "the Russians' 'swinishness' indeed encompasses one valuable trait—intimacy [*intimnost'*], sincerity [*zadushevnost'*]. The Jews have it too. Precisely this quality is what binds them to the Russians, with the difference that the Russian is a sincere drunk person, while the Jew is a sincere sober one."[142] Both the traditional Jew and the Russian peasant built their communities on the supporting structure of family and male authority; both were excluded (by law) from full access to public life. Rozanov, who detested politics and social concerns—who had, Merezhkovskii complained, "abandon[ed] public concerns for private life"[143]—here celebrated Jews and peasants alike, as equally far removed from the refinements and decorum of the respectable social world.

Violence and Exclusion

Violence pervaded the patriarchal peasant culture on both its domestic and its communal stage: the husband's habit of beating his wife and the village's right to punish transgressors. Violence was considered an attribute of the male and particularly of the anti-Semite: certainly Weininger took care to assert his masculinity by an act of violence (against his Jewish self), and the Jews of Warsaw proved their manhood by staging their own pogrom (also against their fellows). Rozanov's literary posture was not without elements of violence, as well: his habit of forced disclosure—or exhibitionism, as Merezhkovskii called it[144]—which as a textual strategy "outraged" public sensibility and made some readers call him pornographic;[145] his arbitrary disregard of facts and

---

the Jews had never mocked the Russian people for their faults, but in one of his 1914 pamphlets he had accused them of mocking everything European and Christian: *Evropa i evrei*, 19.

[142]Rozanov, *Apokalipsis nashego vremeni*, 484.

[143]D. S. Merezhkovskii, *Bylo i budet: Dnevnik, 1910–14* (Petrograd, 1915), 226.

[144]Ibid., 223. Gippius calls his intimate manner both charming and repulsive: *Zhivye litsa*, 2:17.

[145]The term "philosophical pornography" (*filosoficheskaia pornografiia*) occurs in the subtitle of Mikhailovskii's "Literatura i zhizn'." When Rozanov's book *V mire neiasnogo* came to the attention of Witte and Stolypin, they are said to have denounced him as a "terrible pornographer": Gollerbakh, *Rozanov*, 49.

logic and his dismissal of female individuality.[146] Rozanov clung to the archaic role of the master who dispenses both pleasure and fear within his own despotic domain.

In regard to Weininger's joint obsession with women and Jews, Sigmund Freud commented that both represented the threat of castration: women in a purely symbolic sense, Jews in the more literal form of circumcision.[147] Rozanov was not the only anti-Semite to fixate, with a mixture of horror and fascination, on this rite of circumscribed sexual violence.[148] Without necessarily accepting Freud's terms of analysis, one can see that the link between misogyny and anti-Semitism was more than accidental: marked equally in sexual terms, Jews and women were both excluded from public life.[149] The Jews' marginal status helped define the nation by forming an ethnic and even geographic boundary. Women's silence and domestic subordination established men as masters of the word and civic life.

By deliberately confounding public and private, Rozanov enacted a confusion he believed to be central to Russian life in its ideal (and historic) construction. His prose was the textual equivalent of village society, where intimate relations were general knowledge, the personal and collective merged, and conduct followed implicit norms never formulated in abstract terms. Rozanov therefore made a distinction between the public and the commercial aspects of prostitution, celebrating the first as an expression of woman's natural responsiveness to male desire (her tender selflessness) and bemoaning the latter as a recent perversion.[150] Though he insisted that the national welfare depended on the strength of marriage and family ties and that extramarital sex was "fatal to the subject, to the personality,"[151] he defended the male's turn to public women as a natural if less than aesthetic choice: "What hiker reproaches himself for drinking from a puddle under the horse's hoofs or from a swamp instead of from a glass? No one finds this shameful or immoral."[152]

As is not uncommon with Rozanov's prose, the image pulls against

[146]One critic noted the element of "sadism" in Rozanov's distortion of facts for his own purposes: Volynskii, "'Fetishizm melochei,'" pt. 2, 2.

[147]Cited in Paul-Laurent Assoun, "Der perverse Diskurs über die Weiblichkeit," in *Otto Weininger: Werk und Wirkung*, ed. Jacques Le Rider and Norbert Leser (Vienna, 1984), 182.

[148]See Wilson, *Ideology and Experience*, 588–89. Sikorskii claimed that victims of blood ritual murders were sometimes circumcized before being killed: *Delo Beilisa*, 2:254.

[149]On writers who have acknowledged this link, see Jacques Le Rider, *Le Cas Otto Weininger: Racines de l'antiféminisme et de l'antisémitisme* (Paris, 1982), 192.

[150]Rozanov, *Liudi lunnogo sveta*, 49.

[151]Rozanov, *Opavshie list'ia*, 391, 255.

[152]Rozanov, "V svoem uglu: Iudaizm," pts. 17–20, *Novyi put'*, December 1903, 126.

the apparent argument. Natural it might be, but certainly disgusting. Rozanov in fact had no good words for the commercial aspect of modern prostitution, in which sex functioned like the seamstress's sewing machine as a tool of the trade, and which desecrated the original sacred impulse to share sexual pleasure. He derided prostitutes as masculine and artificial ("ill, painted, drunk, vulgar, aggressive"), as deadened souls, "ruined creatures."[153] The importance of prostitution was a sign that marriage was failing in its task; the proof was that the Jews, happy in sex and marriage, never ran brothels and were rarely to be found inside them.[154] This assertion, so against the anti-Semitic grain, reflected Rozanov's notion that Jewish family life had not been affected by Christian attitudes and modern social conditions. To him the Jew was an icon of intimacy and therefore did not belong in the marketplace. Insofar as he made the market his home, he also served as a negative icon of the perils of public life. Many of Rozanov's anti-Jewish images concerned the Jews' commercial roles, mostly as exploiters of innocent Russians. But then, Rozanov criticized everyone who operated on the civic stage: sharing one's private thoughts and intimate sensations with the anonymous public was characteristic not only of the whore but of the writer, speaker, actor, lawyer, professor, and public man—of himself as a published author.[155]

Since Rozanov gloried in self-display, the comment is as double-edged as his attitude toward public women. But it is typical of his desire to throw accepted distinctions into disarray and to reject respectable moral categories that he should lump lawyers and whores together and keep Jews out of a picture where, in the anti-Semitic paradigm, they belonged. Unlike Tolstoy, who worried about prostitution and the destructive power of sexual passion, and unlike Mariia Pokrovskaia, who preached sexual continence and worried about predatory males, Rozanov was untroubled by female sexual transgression or male desire. He was troubled by the Jews not because they pandered to Christian sexual need but because they kept their genital power to themselves. His anti-Semitism did not stem from fear of sex, and it was tempered by the belief that patriarchy was in the interests of both Christians and Jews.

The tsarist regime's anti-Semitic policies have been interpreted as an attempt to define the nation by constructing cultural oppositions and geographical exclusions. But Rozanov's thinking reveals the strong elements of attraction, even affinity, that underlay (and perhaps provoked) the antagonism Russians felt toward Jews. For Rozanov not only hated

[153]Ibid., 124; idem, *Liudi lunnogo sveta*, 47; idem, *Opavshie list'ia*, 254.
[154]Rozanov, "V svoem uglu," pts. 17–20, 125.
[155]Rozanov, *Uedinennoe*, 12–13.

the Jews, he also admired and even identified with them, boasting of his own "womanish" nature, procreative might, and patriarchal household. And hard as he worked to oppose Old Testament virility to the effeminancy of Christ, contrasting the warmth of sunlight to the chill rays of the moon, the Jews' vital procreativity to the asceticism—even homosexuality—of devout Christians, he sometimes confused his metaphors, calling the Jews sodomitic and their culture feminized. Rozanov enacted similar confusions on the subject of gender, insisting at times on the need for radical distinctions between female and male and at others on the pervasiveness of sexual ambiguity (his own androgynous persona is a case in point). These thematic contradictions expressed themselves on yet another level in the tension between Rozanov's contempt for "bourgeois" convention and his embrace of archaic social forms on the one hand and the daring modernity of his literary practice on the other. Yet underlying the many paradoxes there is one consistent theme—the wholesale rejection of liberalism, with its defense of the coherent, autonomous self, its distinction between the public and private, and its striving for civic equality and rational standards of truth and social value. The sexuality Rozanov celebrated and wished to free from Victorian restraint was not the expression of individual desire or the defining feature of a modern liberated personality but a spiritual force, like religious faith, which underpinned the powerful hierarchies of the traditional community.

Chapter Nine

Abortion and the
New Woman

If the Beilis case can be seen as a desperate attempt on the part of the
tsarist regime to bolster its legitimacy by recourse to archaic prejudices
and retrograde symbolic politics, such strategies did nothing to cope
with the real social and cultural changes afoot in the empire.[1] The trial
merely reinforced the sense of cultural and political opposition between
the progressive members of educated society and the representatives of
a recalcitrant autocracy. For all the willful elusiveness of Rozanov's position on the Jews, his embrace of the anti-Semitic myth and his ardent
defense of traditional patriarchalism belonged to a coherent intellectual
universe antithetical to the one espoused by the liberal professionals
who embraced what they considered to be the values of modernity:
individual rights, equality of the sexes, government based on secular
values, and public policy based on scientific expertise. Indeed, some
lawyers and psychiatrists, in invoking their professional standing to
discredit the case against Beilis and the principles on which it relied,
rejected the government's attempt to mobilize scientific opinion (the
testimony of Professor Sikorskii) on behalf of what they considered to
be pernicious prescientific values.

In an issue that arose as the Beilis drama was unfolding in the national press, these same professionals pursued their strategy of redefining public issues as professional ones, in order to separate the kind of
authority they represented from what they viewed as the outmoded
principles of the autocratic regime. The issue in question was abortion,

[1]On Nicholas II's reliance on regressive symbolism to strengthen the monarchy's public credit, see Richard Wortman, "Moscow and Petersburg: The Problem of Political
Center in Tsarist Russia, 1881–1914," in *Rites of Power: Symbolism, Ritual, and Politics since
the Middle Ages,* ed. Sean Wilentz, 244–71 (Philadelphia, 1985).

and it involved the problem that both Weininger and Rozanov paired with the problem of the Jews: the situation of women in a modernizing world.

The emergence of both educated and working-class women into the public domain had become more and more obvious in the years after 1905, as some of the barriers to education and employment began to crumble and urbanization continued apace.[2] Legal experts and physicians had long debated the formal status of women, and their special needs in connection with childbirth and child care had engaged professional concern and public intervention (in regard to venereal disease and unwanted children, for example);[3] but now women themselves began to occupy the public platform. When the changes in women's roles and in their sexual behavior were publicly debated at a conference on women that convened in St. Petersburg in 1908 and at another on prostitution two years later, women were not merely the subjects of these gatherings but organizers and participants.[4]

It was in this period as well that legal and medical experts refocused their consideration of women's relation to reproduction, from a traditional concern with the problem of infanticide[5] to a more active interest in the question of abortion and artificial birth control. Attention to contraception began to appear in the professional press in the 1890s but remained sporadic before 1905.[6] Only in the years before the war did the issue of abortion crystallize as a special object of professional discussion. As the tenor of that discussion makes clear, the salience of the issue was directly linked to the emergence of the publicly engaged, seemingly more independent, "modern" woman.

[2] Richard Stites, *The Women's Liberation Movement in Russia: Feminism, Nihilism, and Bolshevism, 1860–1930* (Princeton, N.J. 1978), 166–78.

[3] See Samuel C. Ramer, "Childbirth and Culture: Midwifery in the Nineteenth-Century Russian Countryside," and Nancy M. Frieden, "Child Care: Medical Reform in a Traditionalist Culture," both in *The Family in Imperial Russia*, ed. David L. Ransel, 218–59 (Urbana, Ill., 1978); David L. Ransel, *Mothers of Misery: Child Abandonment in Russia* (Princeton, N.J., 1990).

[4] See *Trudy pervogo vserossiiskogo zhenskogo s"ezda pri Russkom zhenskom obshchestve v S.-Peterburge, 10–16 dekabria 1908 goda* (St. Petersburg, 1909); and *Trudy pervogo vserossiiskogo s"ezda po bor'be s torgom zhenshchinami i ego prichinami, proiskhodivshego v S.-Peterburge s 21 po 25 aprelia 1910 goda*, 2 vols. (St. Petersburg, 1911, 1912).

[5] Some early examples include M. G., "O detoubiistve," *Arkhiv sudebnoi meditsiny i obshchestvennoi gigieny*, no. 1, pt. 2 (1868): 21–55; A. A. Zhukovskii, "Detoubiistvo v Poltavskoi gubernii i predotvrashchenie ego," ibid., no. 3, pt. 2 (1870): 1–13. See chap. 3 above.

[6] E.g., A. G. Boriakovskii, "O vrede sredstv, prepiatstvuiushchikh zachatiiu," *Vrach*, no. 32 (1893): 886–87; S. V. Filits, "Sovremennaia polovaia zhizn' s meditsinskoi tochki zreniia," *Meditsinskaia beseda*, no. 3 (1900): 65–80; E. Ia. Katunskii, "K voprosu o prave roditelei na zhizn' ploda," *Meditsinskaia beseda*, no. 7 (1900): 177–84; I. V. Platonov, "Ob"ekt prestupleniia izgnaniia ploda," *Vestnik prava*, no. 7 (1899): 155–67.

The abortion debate was not a back-alley affair. The opening sallies were fired in April 1910 at the eleventh congress of Russia's leading medical association, the Pirogov Society, which since its founding in 1885 had spearheaded the profession's drive for disciplinary autonomy and become the focus of its oppositional political activity during 1905. This round was followed by a discussion at the fourth congress of Russian gynecologists and obstetricians, meeting in St. Petersburg in December 1911; it attracted 320 participants and was chaired by Dmitrii Ott, the internationally renowned director of the Imperial Gynecological Institute and consultant to the medical council of the Ministry of Internal Affairs.[7] Various local medical societies also drew up position papers,[8] and the Pirogov Society again debated the question at its twelfth congress in the summer of 1913.[9] In February 1914 the Russian Group of the International Union of Criminologists discussed the matter at its tenth national convention.[10]

A professional interest in abortion did not distinguish Russian physicians and criminologists from their colleagues abroad. In Germany, England, France, and the United States, abortion and birth control became contested social issues in the late nineteenth century and early twentieth.[11] What is of interest in the Russian discussion is the way

---

[7]Eleventh congress: T. O. Shabad, "Iskusstvennyi vykidysh s printsipial'noi tochki zreniia," in *Trudy odinnadtsatogo Pirogovskogo s"ezda*, ed. P. N. Burlatov (St. Petersburg, 1913), 3:214–17; and "Iskusstvennyi vykidysh s printsipial'noi tochki zreniia," *Sibirskaia vrachebnaia gazeta*, no. 10 (1911): 115–16. Fourth congress: "Protokol zasedaniia chetvertogo s"ezda obshchestva rossiiskikh akusherov i ginekologov," pts. 1–2, *Zhurnal akusherstva i zhenskikh boleznei*, nos. 3–4 (1912): 386–88, 539–48; S. A. Aleksandrov, "Chetvertyi s"ezd rossiiskikh ginekologov i akusherov v S.-Peterburge, 16–19 dekabria 1911 goda," *Prakticheskii vrach*, no. 1 (1912): 15–16; I. P. Mikhailovskii, "Chetvertyi s"ezd obshchestva rossiiskikh ginekologov i akusherov (Spb. 16–19 dekabria 1911 g.)," *Sibirskaia vrachebnaia gazeta*, no. 4 (1912): 44–47. On Ott, see *Rossiiskii meditsinskii spisok* (Petrograd, 1916); and *Biographisches Lexikon der hervorragenden Ärzte der letzten fünfzig Jahre (1880–1903)*, ed. I. Fischer, 3d ed. (Munich, 1962), 2:1158–59.

[8]See, e.g., the report from the Urals by A. V. Linder, in "Protokol zasedaniia chetvertogo s"ezda," pt. 2, 545–46; and from Omsk, "Doklad komissii po bor'be s iskusstvennymi vykidyshami Omskogo meditsinskogo obshchestva dvenadtsatomu Pirogovskomu s"ezdu vrachei," *Obshchestvennyi vrach*, no. 6, sec. 6 (1913): 683–92.

[9]*Dvenadtsatyi Pirogovskii s"ezd. Peterburg, 29 maia–5 iiunia 1913 g.*, vyp. 2 (St. Petersburg, 1913); I. I. Binshtok, "Dvenadtsatyi Pirogovskii s"ezd i uchenie Mal'tusa," *Prakticheskii vrach*, no. 9 (1914): 123–27; E. L. Stoianovskaia, "Otchet: Dvenadtsatyi Pirogovskii s"ezd, otdel akusherstva i zhenskikh boleznei, 2-oe zasedanie 1-go iiunia," *Russkii vrach*, no. 28 (1913): 1010–12.

[10]See *Otchet desiatogo obshchego sobraniia Russkoi gruppy mezhdunarodnogo soiuza kriminalistov, 13–16 fevralia 1914 g. v Petrograde* (Petrograd, 1916), 233–55, 272–333, 354–400; henceforth *Otchet*. An account of this meeting also appeared in "Desiatyi s"ezd Russkoi gruppy mezhdunarodnogo soiuza kriminalistov," *Pravo*, no. 10 (1914): 809–40.

[11]See Angus McLaren, "Abortion in England, 1890–1914," *Victorian Studies* 20:4 (1977): 379–400; idem, "Abortion in France: Women and the Regulation of Family Size, 1800–1914," *French Historical Studies* 10:3 (1978): 461–85; idem, *Birth Control in Nineteenth-*

medical and legal issues were mapped on the local political and cultural terrain. The implications were not narrow; to listen to the Russians argue abortion is to hear an exchange on basic principles of public authority and civic life. Opinions about women's control over their reproductive lives constituted not only a critique of the existing political order but blueprints for a new and better one. Physicians, criminologists, and jurists examined three distinct models of social organization: (1) the state imposes norms of private conduct through the repressive action of the law; (2) society exerts preemptive restraint through the exercise of professional expertise; and (3) individual self-regulation allows private decisions to determine personal choice.

The Status of Abortion in Russian Law

Infanticide and abortion were linked in the legal and forensic medical literature as two forms of murder, both construed in relation to female criminal agency.[12] The 1845 code defined infanticide as an unpremeditated form of murder: the guilty mother was presumed to have acted impulsively, under the pressure of overpowering emotion, in an abnormal physical and mental state occasioned by "shame and fear" or in the commonplace anguish of postpartum distress. By contrast, the code defined abortion as a premeditated act, a crime of choice, not desperation.[13]

Century England (New York, 1978); idem, *Reproductive Rituals: The Perceptions of Fertility in England from the Sixteenth Century to the Nineteenth Century* (New York, 1985). See also Patricia Knight, "Women and Abortion in Victorian and Edwardian England," *History Workshop* 4 (1977): 57–68; R. P. Neuman, "The Sexual Question and Social Democracy in Imperial Germany," *Journal of Social History* 7:3 (1974): 271–86; idem, "Working Class Birth Control in Wilhelmine Germany," *Comparative Studies in Social History* 20 (1978): 408–28; James Woycke, *Birth Control in Germany, 1871–1933* (London, 1988); Carroll Smith-Rosenberg, "The Abortion Movement and the AMA, 1850–1880," in Smith-Rosenberg, *Disorderly Conduct: Visions of Gender in Victorian America* (New York, 1985); and James C. Mohr, *Abortion in America: The Origins and Evolution of National Policy, 1800–1900* (New York, 1978).

[12]For a comparison of existing legislation on abortion in other states, see Gustav Radbruch, "Abtreibung," in *Vergleichende Darstellung des deutschen und ausländischen Strafrechts. Vorarbeiten zur deutschen Strafrechtsreform*, ed. Karl von Birkmeyer et al., Besonderer Teil, 5: 159–83 (Berlin, 1905). Russians were aware of the differences in Western legislation on the subject; see *Ugolovnoe ulozhenie: Proekt redaktsionnoi komissii i ob"iasneniia k nemu* (St. Petersburg, 1897), 6:104–7; henceforth *UU*, vol. 6 (1897).

[13]Arts. 1451 and 1461–62 in N. S. Tagantsev, ed., *Ulozhenie o nakazaniiakh ugolovnykh i ispravitel'nykh 1885 goda*, 11th ed., rev. (St. Petersburg, 1901), 665 (*ne predumyshlennoe*), 670–71 (*umyshlenno*); henceforth *Ulozhenie* (1885). On attitudes toward the two crimes, cf. the denunciation of abortion as an upper-class indulgence with the insistence on the "abnormal," indeed "fatal," circumstances that drove women to infanticide: D. N. Zhbankov, "K voprosu o vykidyshakh," pts. 1–5, *Prakticheskii vrach*, nos. 31–38 (1914): 423–25, 432–34, 442–44, 452–54, 463–67; and idem, "O detoubiistve," *Prakticheskii vrach*, no. 17 (1909): 316.

The abortion statute did, however, recognize a distinction between the woman's role and that of outside parties. Lawmakers argued that it was necessary to retain the concept of abortion as murder because the fetus "still belonged to the human race . . . even though [it was] not yet completely developed,"[14] but they established milder penalties both for people who performed an abortion with the mother's consent and for the mother who performed it herself.[15] Though the motive of shame and fear, said to mitigate the crime of infanticide, was not written into the abortion law, it was considered grounds for leniency in that case as well. The authors of the code explained this indulgence on the grounds that popular opinion did not consider abortion a serious offense: "because the child in the womb does not yet seem alive; because it cannot itself be attached to life or be the object of other people's tenderness and attachment; because many consider (however erroneously) that not wanting it to be born is virtually the same as not wanting children in the first place; and because even the tenderest of parents do not regret an accidental miscarriage the way they do the death of a newborn child."[16]

Criminal agency in both infanticide and self-induced abortion was limited to women in their capacity as mothers, acting independently and privately in relation to their own children, and the penalties were virtually the same.[17] By employing professional aid, however, the mother became part of a wider social network and enhanced her power to control her reproductive life. Yet even when women sought assistance, the full force of the law was directed not against them but against those who helped them. As agents with greater freedom to act and to control their actions, abortionists had more chance of succeeding and less excuse for wanting to

[14]See note to art. 1867–69 justifying the provisions of the 1845 code: *Proekt ulozheniia o nakazaniiakh ugolovnykh i ispravitel'nykh, vnesennyi v 1844 godu v Gosudarstvennyi Sovet, s podrobnym oznacheniem osnovanii kazhdogo iz vnesennykh v sei proekt postanovlenii* (St. Petersburg, 1871), 618; henceforth *Proekt* (1844). The 1845 code (see chap. 1) was not substantially altered until 1917, despite various amendments and a revision in 1885.

[15]The 1845 code punished anyone convicted of performing an unwanted abortion with loss of civil rights and four to six years of hard labor (up to ten if the woman died), plus branding and the lash if the perpetrator belonged to the unprivileged ranks. Abortion performed by the mother or with her consent also entailed loss of civil rights but only exile or resettlement in Siberia for an unspecified period (and the lash in appropriate cases). The 1885 edition (which followed the elimination of corporal punishment from the criminal code) kept the loss of civil rights in all cases and the same term of hard labor when consent was absent but reduced the sentence for the woman or her assistant to up to six years of corrective incarceration, with additional time for highly trained personnel. See arts. 1932–34, *Ulozhenie o nakazaniiakh ugolovnykh i ispravitel'nykh* (St. Petersburg, 1845), 494–95 (henceforth *Ulozhenie* [1845]); arts. 1461–63, *Ulozhenie* (1885), 670–71.

[16]Note to arts. 1867–69, *Proekt* (1844), 618.

[17]Women in both cases lost all civil rights and would spend a minimum of four years in corrective incarceration; the maximum was slightly higher for infanticide (six years as opposed to five). Cf. arts. 1451 and 1462, *Ulozhenie* (1885), 664–65, 671.

do so. The statute especially restricted the right of trained personnel to substitute their own professional judgment for the woman's expressed desire. Though all abortion was illegal, practitioners could not be held additionally responsible, even for the pregnant woman's injury or death, unless they had performed the operation without her consent.[18] Such cases might include destruction of the unborn child as the result of procedures employed during delivery while the mother was unconscious.[19]

Paradoxically, the legal risks were greatest when the operator was actually trained for the job. The lawmakers explained their attempt to control the activities of "doctors, obstetricians, midwives, and pharmacists"— who were able by virtue of their skill to avoid being caught, since they in fact did less physical damage—on the grounds that in practicing abortion, qualified personnel violated professional principles and abused the public trust.[20] Even the proposed revision of the existing criminal code, published in 1903 but never adopted, retained this antiprofessional bias, though it lessened the severity of punishment.[21] This situation goes far toward explaining why the medical profession, obstetricians in particular, rallied to the cause of abortion reform.

Here the Russians differed markedly from their colleagues in the West. In both England and the United States, laws against abortion increased in severity during the nineteenth century, largely at the initiative of physicians seeking to enforce their professional authority at the expense of female autonomy in reproductive affairs.[22] In Russia, by contrast, the desire to enhance their professional standing led physicians to demand either the reduction of existing legal sanctions against abortion or outright decriminalization; they stressed the convergence rather than conflict of women's interests with their own. Such a stance reflected the awkward position of the Russian medical profession. Ill-paid, overworked, dependent on state support and public employment, and suffering from low social esteem,

[18]Senate decision 18 (1904), re Reizner, in *Resheniia ugolovnogo kassatsionnogo departamenta Pravitel'stvuiushchego Senata za 1904 god* (Ekaterinoslav, 1911), 39–40.

[19]Katunskii ("K voprosu," 184) explained that such crude procedures were most often employed by rural practitioners operating under primitive conditions. For a legal critique explaining why explicit consent was not always possible to obtain, see A. A. Ginzburg, "Izgnanie ploda," *Zhurnal Ministerstva Iustitsii*, no. 7 (1912): 54–55.

[20]Arts. 345–46, *Svod zakonov ugolovnykh* (St. Petersburg, 1835), 121; art. 1934, *Ulozhenie* (1845), 495; note to draft arts. 1867–69, *Proekt* (1844), 618; arts. 1462–63, *Ulozhenie* (1885), 671.

[21]UU, vol. 6 (1897), 167; *Ugolovnoe ulozhenie, Vysochaishe utverzhdennoe 22 marta 1903 goda* (St. Petersburg, 1903), 177 (arts. 465–66). These statutes eliminated loss of rights from the penalties imposed.

[22]See McLaren, "Abortion in England," 389–94; idem, *Reproductive Rituals*, 129, 137–44; Knight, "Women and Abortion," 62–63; Smith-Rosenberg, "Abortion Movement," 217–44; and Mohr, *Abortion in America*, 147–57, 160–64. By the end of the century, however, many European physicians had accepted the legitimacy of abortion for medical reasons: McLaren, "Abortion in France," 470–72.

physicians resented state intrusion, repudiated the regime's high-handed administrative methods (including the liberal use of penal sanctions in everyday affairs), and identified with the cause of the powerless, whether peasants or women. Relatively privileged though doctors might be, especially in cultural terms, they maintained a posture of critical caution in relation to the administrative state that deprived them of professional and civil autonomy. The distrust of some physicians was mitigated by the widened political opportunities that emerged after 1905 and by their disillusionment with radical tactics and populist dreams, but it was never entirely overcome.[23]

The Medical Side

Physicians at the 1910 Pirogov congress stressed that abortion, once the privilege of the well-to-do, had reached "epidemic" proportions in all social classes. Often victims of poverty and economic exploitation, sometimes of rape by "hooligans" and violent mobs (*pogromshchiki*), women seeking abortion seemed to be suffering not only the age-old burden of moral opprobrium but also the specific hardships of the post-1905 years.[24] The main speaker at both the 1911 meeting of gynecologists and the 1913 Pirogov congress was the obstetrician Lazar Lichkus, director of a St. Petersburg maternity clinic.[25] On each occasion he was joined at the podium by a legal expert and other obstetricians—in 1911 by Iulius Iakobson, board member of Prince Ol'denburg's children's shelter and senior staff physician at the St. Petersburg Institute of Gynecology and Midwifery, along with the junior professor Liudvig

[23]For a general picture, see Nancy Mandelker Frieden, *Russian Physicians in an Era of Reform and Revolution, 1856–1905* (Princeton, N.J., 1981). For the post-1905 modifications, see John F. Hutchinson, "'Who Killed Cock Robin?': An Inquiry into the Death of Zemstvo Medicine," in *Health and Society in Revolutionary Russia*, ed. Susan Gross Solomon and John F. Hutchinson (Bloomington, Ind. 1990); and idem, *Politics and Public Health in Revolutionary Russia, 1890–1918* (Baltimore, 1990).

[24]Shabad, "Iskusstvennyi vykidysh" (1913), 214–17; and idem, "Iskusstvennyi vykidysh" (1911), 115–16.

[25]For the texts of his talks, see L. G. Lichkus, "Vykidysh s sudebno-meditsinskoi tochki zreniia," *Russkii vrach*, no. 4 (1912): 109–18; and idem, "Iskusstvennyi prestupnyi vykidysh," *Russkii vrach*, no. 39 (1913): 1358–66. For an acrimonious exchange with an opponent of his position, see N. I. Khokhlov, "Po povodu postanovleniia otdela akusherstva 2-go iiunia na dvenadtsatom Pirogovskom s"ezde, 'Ob iskusstvennom prestupnom vykidyshe,'" *Russkii vrach*, no. 29 (1913): 1048–49; L. G. Lichkus, "Vynuzhdennyi otvet na korrespondentsiiu N. I. Khokhlova: Po povodu postanovleniia otdela akusherstva 2-go iiunia na dvenadtsatom Pirogovskom s"ezde, 'Ob iskusstvennom prestupnom vykidyshe,'" *Russkii vrach*, no. 33 (1913): 1181–82; N. I. Khokhlov, "Pis'mo v redaktsiiu," *Russkii vrach*, no. 38 (1913): 1341–42.

Okinchits; and in 1913 by Iakob Vygodskii, from a Jewish hospital in Vilnius.[26]

Nine physicians spoke out at the 1913 session on abortion. Four were women, five came from the provinces, five worked for zemstvos or municipal hospitals, several (like Lichkus himself) had obviously Jewish names, and seven were between thirty and forty years of age and had completed their training within the previous fifteen years. The two main speakers represented a slightly older generation: born in the 1850s, they had completed their studies by 1882. Both attacked existing laws, Lichkus calling for a broader range of legitimate abortion, Vygodskii for outright decriminalization. Of the women, Liubov' Gorovits from St. Petersburg and Kseniia Bronnikova, a senior staff physician at the St. Petersburg Imperial Maternity Clinic, energetically denounced the anti-abortion laws as a violation of women's rights; Nadezhda Bezpalova-Letova, a staff physician at the Rostov municipal hospital, believed that women's equality would solve the abortion problem; and Nadezhda Zemlianitsyna, from Perm Province, insisted on the criminality of abortion and defended the importance of motherhood.[27]

Both the gynecologists' congress and the gynecological section of the Pirogov Society voted to recommend that the government decriminalize abortion.[28] Not all physicians accepted this position, however, and the debate spilled over into the medical press. Most professional commentators agreed that abortion was a problem that physicians and jurists must address because they saw it as dramatically on the rise, in Russia as well as in Europe, despite the existence of repressive laws.[29]

[26]V. L. Iakobson, "Sovremennyi vykidysh s obshchestvennoi i meditsinskoi tochki zreniia," *Zhurnal akusherstva i zhenskikh boleznei*, no. 3 (1912): 305–18; L. L. Okinchits, "Kak borot'sia s prestupnym vykidyshem," *Zhurnal akusherstva i zhenskikh boleznei*, no. 3 (1912): 319–22; Ia. E. Vygodskii, "Iskusstvennyi vykidysh s obshchestvennoi i vrachebnoi tochki zreniia," *Dvenadtsatyi Pirogovskii s"ezd*, 375–77. The legal expert was M. P. Chubinskii; see his "Vopros o vykidyshe v sovremennom prave i zhelatel'naia ego postanovka," *Zhurnal akusherstva i zhenskikh boleznei*, no. 4 (1912): 461–86; and idem, "Istreblenie ploda i problema ego nakazuemosti," *Iuridicheskii vestnik*, no. 2 (1913): 112–35.

[27]For details of the discussion, see *Dvenadtsatyi Pirogovskii s"ezd*, 84–88, 210–13, 375–77; and Stoianovskaia, "Otchet," 1010–12. Biographical information is from *Rossiiskii meditsinskii spisok* (St. Petersburg/Petrograd, 1894, 1916).

[28]N. A. Vigdorchik, "Vrachebnye otkliki: Ereticheskie mysli o prestupnykh vykidyshakh i o preduprezhdenii beremennosti," *Prakticheskii vrach*, no. 15 (1912): 242; V. D. Nabokov, "Desiatyi s"ezd kriminalistov," *Pravo*, no. 9 (1914): 663; Sergei Iablonovskii, "Prava nerozhdennykh," *Russkoe slovo*, no. 129 (June 6/19, 1913): 2.

[29]Criminal statistics on convictions for abortion do not support this impression. In the decade 1895–1904 only 92 persons were convicted of abortion in Russia. Viktor Lindenberg, *Materialy k voprosu o detoubiistve i plodoizgnanii v Vitebskoi gubernii (Po dannym vitebskogo okruzhnogo suda za desiat' let, 1897–1906)* (Iuriev, 1910), 64, cites the figures but insists that the incidence of abortion was sharply on the rise (72).

The Russian incidence was said to have started its climb in the mid-1890s, with the intensification of industrial and urban development, and to have skyrocketed after the 1905 revolution.[30] One physician drew an analogy between 1905 and the French Revolution of 1789, which, she declared, had also produced an upsurge of abortions.[31] Not only was this rise in itself an ominous sign, but it increased the disjuncture between formal legal principles and both medical and judicial practice: physicians found themselves more often engaged in criminal activity, and the courts came up against the force of contrary public opinion and social reality.[32]

Since the law targeted both the pregnant woman and the medical practitioner as perpetrators of the crime, arguments for legal reform took two directions: in defense of women's rights to physical self-determination and of doctors' rights to make technical decisions on professional grounds. Physicians were unanimous in wishing to substitute their trained judgment for the blanket repression of the law, which, they believed, succeeded only in creating a truly criminal underground of incompetent and unlicensed operators, thus increasing the dangerous effects of abortion. Many jurists likewise called for legalization in all or almost all cases, in the name of individual liberty, juridical secularism, and social welfare. Only a few voices spoke for women's absolute right to make personal choices—and many of these speakers were women.[33] Both professional communities thus rejected the first of the three models of social organization (state control through legal repression), preferred the second (public self-regulation through expert intervention), and hesitated before the third (individual self-determination for women).

Physicians who demanded reform of the existing law (model one, modified to accommodate the principle embodied in model two), as

[30]Iakobson, "Sovremennyi vykidysh," 310–12; O. P. Pirozhkova, "K voprosu o vykidyshe," Zhurnal akusherstva i zhenskikh boleznei, no. 4 (1912): 520, 522; Vigdorchik, "Vrachebnye otkliki," 243; "Doklad komissii," 683; Lichkus, "Iskusstvennyi prestupnyi vykidysh," 1359.

[31]O. P. Pirozhkova, in Aleksandrov, "Chertvertyi s"ezd," 16.

[32]On the impossibility of enforcing a law that conflicted with social practice, see G. Ia. Zak, "Umershchvlenie ploda i ugolovnoe pravo," pts. 1–2, Pravo, nos. 46–47 (1910): 2751–52, 2840; also Dr. Brodskii, in Shabad, "Iskusstvennyi vykidysh" (1913), 216; and congress resolution in Aleksandrov, "Chetvertyi s"ezd," 16.

[33]E.g., the physicians K. N. Bronnikova and L. M. Gorovits at the twelfth Pirogov congress: see Dvenadtsatyi Pirogovskii s"ezd, 212–13, and a hostile account in Khokhlov, "Po povodu postanovleniia," 1048; Pirozhkova, "K voprosu," 523; M. I. Pokrovskaia, "K voprosu ob aborte," Zhenskii vestnik, no. 4 (1914): 102–5. On the legal side, see F. A. Vol'kenshtein, a St. Petersburg attorney, and another participant in the 1914 criminology congress named Oks, in Otchet, 327, 380–81; and M. L. Oleinik, "Prestupnyi abort v doktrine i zakonodatel'stve," in Trudy kruzhka ugolovnogo prava pri Spb. Universitete, ed. M. M. Isaev (St. Petersburg, 1913), 138.

well as those who wished to remove abortion from the criminal code altogether, sanctioned the right of properly qualified medical personnel to perform abortions for medically appropriate reasons. The second group, in rejecting the statutory limitations, did not necessarily define the formal mechanisms through which such authority would be exercised. Most relied on moral assurances. "Let the physician's high calling be society's best guarantee that such operations will not be undertaken lightly," urged the Omsk Medical Society, "for indeed society and the state entrust him with the life and health of all their members."[34]

Few of the Russian physicians who engaged in these debates maintained that abortion was under all circumstances morally and medically objectionable.[35] Disagreement arose rather over two points: whether abortion should be decriminalized across the board, so that both women and doctors would be free to make decisions without fear of reprisal; or whether the law should merely establish a range of cases, to be identified by trained practitioners, in which abortion would be permitted. Those who accepted the second alternative went on to differ over criteria; even therapeutic abortion based strictly on medical considerations (the health and welfare of mother or child) left room for disagreement. More controversial was the question of so-called social abortion: whether physicians were ethically justified in terminating a normal pregnancy that the healthy mother was unwilling to bring to term because of her life circumstances.

Feminist principles sometimes, but not always, induced physicians to support the third model. Mariia Pokrovskaia denounced the abortion laws as an unwarranted restriction on female autonomy and called for full decriminalization on the grounds that only women were in a position to judge the legitimacy of their own needs. There was no absolute standard, she declared, against which to weigh individual claims. Invoking "the example most often cited in defense of anti-abortion laws, the high-society lady devoted to pleasure . . . who does not want children," Pokrovskaia denied that such a woman's motives were any less valid than those of an overburdened peasant wife. "Since the society lady risks her life in resorting to abortion, she obviously considers the birth of a child worse than death itself. Her arguments may seem pitiful

[34]"Doklad komissii," 690.
[35]The most categorical denunciations of abortion I have found are K. V. Goncharov, *O venericheskikh bolezniakh v S. Peterburge* (St. Petersburg, 1910), 118; and A. D. Kalinkovitskii, "Eshche ob iskusstvennom vykidyshe (V zashchitu amfibiopodobnogo zarodysha)," *Vrachebnaia gazeta*, no. 43 (1913): 1533–36, both of whom deplored all but the strictest medically indicated abortions. Goncharov, who had served briefly as a zemstvo physician, eventually joined the faculty of the Academy of Military Medicine. Kalinkovitsiii was a Jewish zemstvo physician in Poltava Province. See biographical note to Goncharov's book, 152–53, and *Rossiiskii meditsinskii spisok* (Petrograd, 1916).

to us, but to her they have enormous meaning." Pokrovskaia did not approve of abortion, which she viewed as a consequence of society's "abnormal" indulgence in sexual pleasure for its own sake, but she defended rationally controlled reproduction as essential to "racial hygiene"—the production of socially and physically healthier children.[36] The concept of voluntary motherhood (*soznatel'noe materinstvo*), by which women regulated pregnancy in the interests of more effective mothering and greater personal independence, belonged to the rhetorical arsenal of Russian as well as Western feminists of the time.[37]

Dmitrii Zhbankov, though he shared Pokrovskaia's commitment to women's rights and women's education, her hostility to regulated prostitution, and her antipathy to the idea that sex was meant for anything but procreation, did not agree that women had the right to make their own reproductive choices. To him the upper-class lady unwilling to bear a child represented all that was corrupt, self-indulgent, and self-destructive about what he called the post-1905 "abortion era." Women could not achieve equal rights, he warned in a 1914 article, unless they rejected the emphasis of contemporary culture on artificial (that is, non-reproductive) sexual gratification and returned to their "natural" function—motherhood, not sex—eschewing birth control as well as abortion. Moreover, Zhbankov—virtually alone among professionals engaged in these debates—cited the opposition of the Orthodox church to abortion in defense of his own highly moralistic views.[38]

Like Zhbankov, many physicians interpreted the increased abortion rate since the 1890s in class and political terms. Most attributed the post-1905 rise to a combination of cultural and economic crises: on the one hand, the urban working class was finding it harder to make ends meet and adopting supposedly upper-class habits of family regulation for economic reasons; on the other hand, the so-called moral decline that followed the collapse of revolutionary hopes had increased the pub-

[36]Pokrovskaia, "K voprosu ob aborte," 103, 105.

[37]Rosalind Pollack Petchesky, *Abortion and Woman's Choice: The State, Sexuality, and Reproductive Freedom* (New York, 1984), 41. English feminists promoted voluntary motherhood, but few defended either contraception or abortion: McLaren, *Birth Control*, 197–99; Judith R. Walkowitz, "Male Vice and Feminist Virtue: Feminism and the Politics of Prostitution in Nineteenth-Century Britain," *History Workshop Journal* 13 (1982): 92; Mohr, *Abortion in America*, 111–12. Russian examples defending both include E. Zinov'eva, "V zashchitu prav rozhdennykh: Pis'mo v redaktsiiu," *Sovremennyi mir*, no. 8 (1913): 250, 252–56; and Sof'ia Zarechnaia, "Neomal'tuzianstvo i zhenskii vopros," *Zhenskoe delo*, no. 27–28 (August 10, 1910): 10–12 (unlike Pokrovskaia, both of these writers identified themselves as socialists). On the ambivalence of European socialists, see Neuman, "Sexual Question"; idem, "Working-Class Birth Control"; and Angus McLaren, "Sex and Socialism: The Opposition of the French Left to Birth Control in the Nineteenth Century," *Journal of the History of Ideas* 37 (1976): 475–92.

[38]Zhbankov, "K voprosu," pt. 1, 425; pt. 2, 432–33; pt. 3, 444; pt. 4, 452 (citing Archbishop Stefan of Kursk; see "Arkhiepiskop Stefan ob aborte," *Russkoe slovo*, no. 64 [March 18/31, 1914]: 5); pt. 5, 467.

lic's interest in sexual expression and lessened regard for traditional sexual ties.[39] Abortion itself was said to be a product of capitalist relations, the movement to decriminalize it a foreign idea.[40] The most advanced nations (France and the United States were always cited) and the most cultivated classes were said to practice abortion at the highest rates.[41]

Infanticide was considered the more likely resort of peasant women, and criminal data seemed to confirm the contrast between the backward countryside and the sophisticated town—the one subject to the tyranny of nature and the desperation born of ignorance and need, the other substituting culture for nature and rational control for the elemental expression of emotion.[42] Though some physicians and criminologists admitted that abortion was not peculiar to any specific social group, their rhetoric betrayed the power of the stereotype. According to them, aborting mothers included, in declining order of symbolic prominence, "the high-society lady accustomed to luxury who need not worry about her daily bread, the prostitute seeking a living on the streets, as well as the housemaid and the woman factory worker."[43] One physician hostile to decriminalization insisted that "real" proletarians were interested in the right to motherhood, not to abortion.[44]

The same rhetoric characterized contemporary medical discussions of

[39]Ia. Kh. Falevich, "Itogi tomskoi studencheskoi polovoi perepisi: Doklad, chitannyi 18 fevralia 1910 g. na zasedanii Pirogovskogo studencheskogo meditsinskogo obshchestva pri Tomskom Universitete," pt. 12, *Sibirskaia vrachebnaia gazeta*, no. 28 (1910): 330; Iakobson, "Sovremennyi vykidysh," 310; "Doklad komissii," 684; V. A. Brodskii, "Iskusstvennyi vykidysh s meditsinskoi i obshchestvenno-ekonomicheskoi tochek zreniia," pts. 1, 3, *Vrachebnaia gazeta*, nos. 18, 20 (1913): 660, 711–12; Pirozhkova, "K voprosu," 521–22; N. A. Vigdorchik, "Detskaia smertnost' sredi peterburgskikh rabochikh," *Obshchestvennyi vrach*, no. 2 (1914): 247.

[40]See Binshtok, "Dvenadtsatyi Pirogovskii s"ezd," 124; Vigdorchik, "Vrachebnye otkliki," 242–43 (ironic reference to belief in foreign origins); L. Rutenberg, "K voprosu o sovremennom vzgliade na dopustimost' sotsial'nogo vykidysha," pt. 2, *Vrachebnaia gazeta*, no. 31 (1916): 489; S. Ia. Elpat'evskii, "Samoistreblenie chelovechestva: Po povodu s"ezda kriminalistov v Peterburge," *Russkoe bogatstvo*, no. 4 (1914): 267.

[41]On France and the United States, see Lichkus, "Iskusstvennyi prestupnyi vykidysh," 1359–60; Zhbankov, "K voprosu," pt. 2, 433–34; Gei in *Otchet*, 328; and Lindenberg, *Materialy*, 61. On abortion among the educated, see, e.g., Oleinik, "Prestupnyi abort," 118.

[42]On this sociological contrast, see Lichkus, "Iskusstvennyi prestupnyi vykidysh," 1358; Vigdorchik, "Detskaia smertnost'," 245; Oleinik, "Prestupnyi abort," 112, 118; and Chubinskii, "Vopros o vykidyshe," 470. For rural and urban data, see M. N. Gernet, *Detoubiistvo: Sotsiologicheskoe i sravnitel'no-iuridicheskoe issledovanie* (Moscow, 1911), 143. On the calculated nature of contraception and abortion, see Elpat'evskii, "Samoistreblenie chelovechestva," 269. Europeans made the same urban-rural comparison: see Knight, "Women and Abortion," 58; and McLaren, "Abortion in France," 475.

[43]Zak, "Umershchvlenie," pt. 1, 2751. Among others who believed that abortion occurred among all social classes are Iakobson, "Sovremennyi vykidysh," 312; Oleinik, "Prestupnyi abort," 110; Dr. Rappoport, in Shabad, "Iskusstvennyi vykidysh" (1913), 215; and Elpat'evskii, "Samoistreblenie chelovechestva," 266–67.

[44]I. I. Binshtok, "Eshche o nakazuemosti aborta," *Prakticheskii vrach*, no. 15 (1914): 215.

birth control. Abortion was sometimes castigated by opponents of contraception as its "most loathsome form."[45] Contraception too was a foreign idea, a product of capitalism, more prevalent among the educated
classes than among common folk, who were too busy with work to
think much about sex in the first place.[46] On the positive side, it was
regarded as a means for women to gain access to "freedom and all the
benefits of science, culture, and civic life [obshchestvennaia zhizn']."[47] As
working people gained cultural skills, they also used more rational
means of regulating their reproductive lives. "The only people who still
produce children without restraint," wrote Sergei Elpat'evskii in 1914,
"are those at the very bottom level of society [nizy obshchestva]—in particular the Russian peasants. That is," he corrected himself, "they used
to." He admitted that contraceptive devices had become available in
provincial pharmacies, where they sold briskly.[48]

Natan Vigdorchik, a public health physician of St. Petersburg, offered similar observations on the behavior of urban workers in 1914:
"The more educated members of the proletariat [bolee kul'turnye sloi]
more often use abortion than the less educated. This is indeed a universal phenomenon, that all types of artificial birth control travel the rungs
of the social ladder from top to bottom." In switching from abortion to
contraception, the Russian working class was only following the example of its more sophisticated comrades abroad. Working women had a
particular interest in limiting family size: "No sooner does the first ray
of consciousness penetrate this dark life than women begin to look for
some way to lighten the double load imposed on them by nature and
society."[49]

These were the views of a physician who accepted the fact that
Russia was changing. The anticapitalist bias of many of his colleagues,
by contrast, cut two ways. The morally objectionable aspect of abortion could be attributed to the materialist individualism of capitalist society and of the spoiled rich (urban middle-class women indulged vanity and the desire for pleasure, and doctors turned a profit from the
women's selfish desires); its morally excusable side could be explained
as a response to the suffering engendered by that same society (working-class women were forced to abort to protect the welfare of their
other children, unwed working girls to protect their honor against hyp-

[45]Filits, "Sovremennaia polovaia zhizn'," 73.
[46]Boriakovskii, "O vrede sredstv," 887; "Po voprosu o neomal'tuzianstve," Sibirskaia
vrachebnaia gazeta, no. 38 (1908), 411; Falevich, "Itogi," pts. 6, 7, 11, ibid., nos. 22, 23, 27
(1910): 270, 271, 318; and Filits, "Sovremennaia polovaia zhizn'," 65, 79.
[47]"Po voprosu o neomal'tuzianstve," 411.
[48]Elpat'evskii, "Samoistreblenie chelovechestva," 269–70.
[49]Vigdorchik, "Detskaia smertnost'," 247–48, 250.

Advertisements for contraceptives (clockwise from top left): "Women! 3,000 well-known doctors and professors not only recommend our patented hygienic method of preventing conception but use it themselves"; "Preservatives: The golden safety device"; "Completely Free: Price list of hygienic rubber articles, indispensable for both men and women, single or married"; "Women: If you are worried about your health, use the preparation tested over the years by Professor Braun. It is the only comfortable and completely reliable scientific method for preventing pregnancy"; "Reflux: the remarkable invention for female hygiene. . . . If personal circumstances or your physician require you to prevent pregnancy, use no other methods—all harmful and unpleasant—but only the Reflux apparatus, which is safe, convenient, hygienic, and easily available. Success guaranteed!" *Satirikon*, 1908, 1909, 1913. Helsinki University Slavonic Library.

ocritical bourgeois contempt and to keep their jobs). In a way, this argument should have undermined the Russian abortion debate: if birth rates were falling in most of Europe, to the point where France thought itself to be dying out, the same could not be said for less-developed Russia, where fertility was still holding its own.[50] "It is true," admitted one staunch enemy of abortion (whose spread he attributed to the baneful effects of feminist agitation), "that Russia's population continues to increase at a very high rate, despite our fantastic mortality level. No matter how far the abortion epidemic may spread, Russia will not soon find itself in the position of France." He nevertheless opposed the practice on the grounds that it destroyed the physical and psychic health of individual women.[51]

The Russian anxiety over abortion reflected cultural more than demographic transformation. Not so long ago, commented Natan Vigdorchik ironically, "who in Russia would have dared come to the defense of these 'disgusting sores of capitalism'?" But now "Russian physicians, those same Russian physicians of whom it is always said that they are steeped in the highest idealism, that they hold high the banner of social responsibility [obshchestvennost'], . . . speak openly before all of Russia . . . advocating (how dreadful!) the decriminalization of abortion and artificial contraceptive means!" Could the sharp rise in abortions be explained only by the role of "special underground doctors, unscrupulous has-beens who sell their expertise for criminal ends? Nothing of the sort. [Abortions are performed by] the most ordinary physicians, your good friends and comrades, people you admire."[52]

Vigdorchik's contention that abortion was endemic and in fact accepted by most (male) members of society, who did not think their doctors criminal for performing abortions on sisters, mothers, and wives,[53] was reiterated by the Omsk Medical Society in its report to the 1913 Pirogov Society congress. The report emphasized that women in particular accepted the termination of pregnancy in its early stages as an everyday occurrence, without feelings of remorse. Vigorously endorsing women's right to seek abortion for "social" reasons (illegitimacy, rape, large families, need to keep working), the Omsk society insisted both on the right of the medical profession to corporate autonomy and on women's right to voluntary motherhood.[54] The doctors thus saw

[50]Lichkus, "Iskusstvennyi prestupnyi vykidysh," 1359.
[51]Binshtok, "Eshche o nakazuemosti," 213–15.
[52]Vigdorchik, "Vrachebnye otkliki," 242–43. Vigdorchik's remarks were warmly endorsed by the Siberian *Medical Gazette*: "V zashchitu aborta i preduprezhdeniia beremennosti," *Sibirskaia vrachebnaia gazeta*, no. 17 (1912): 209–10.
[53]Vigdorchik, "Vrachebnye otkliki," 244.
[54]"Doklad komissii," 684, 687–88, 690.

their own desire for self-governance reflected in the situation of their female patients.

In addition to decriminalization, the Omsk report called for extensive social measures to support mothers, families, and children and for reform of the existing laws on marriage, divorce, and illegitimacy.[55] Still, not all physicians who endorsed such practical prescriptions and the accompanying social critique also accepted the demand for full decriminalization. Lazar Lichkus, for example, believed that the courts should continue to prosecute irregular abortionists even as the law allowed physicians to take the necessary medical steps. "Doctors," he asserted, "stand closer than anyone else to the population and can better understand its needs."[56] (It could well have been countered that untrained abortionists, especially in the countryside, where the job was done by trusted peasant women, were closer to the people than any physician could be, although in fact popular attitudes toward abortion—invoked, for example, by the framers of the 1845 criminal code—were difficult to determine.)[57] Despite such objections, the organizations representing the profession as a whole took the more radical stand. Most important, gynecologists associated with the influential Pirogov Society believed they best served the public welfare by opposing the abortion law outright.[58]

The Legal Side

Meeting in St. Petersburg in February 1914, the Russian Group of the International Union of Criminologists—an association of lawyers, legal scholars, and specialists in crime of varied political views—echoed the Pirogov Society's stand on the decriminalization of abortion.[59] By that time there was general agreement in the legal community that reform was overdue. Even the Ministry of Justice favored a drastic reduction in the penalties imposed, including an end to the loss of civil rights entailed in all cases, although it continued to advocate special chastise-

[55]Ibid., 691.

[56]Lichkus, "Iskusstvennyi prestupnyi vykidysh," 1366.

[57]One ethnographic study maintained on the one hand that peasants denounced abortion as ungodly, and on the other that they employed folk abortive practices that usually did not work, leaving infanticide as the only option: T. Popov, *Russkaia narodno-bytovaia meditsina: Po materialam etnograficheskogo biuro kn. V. N. Tenisheva* (St. Petersburg, 1903), 326–29.

[58]Resolution cited in V. D. Nabokov, "Desiatyi s"ezd," 663, and Iablonovskii, "Prava nerozhdennykh," 2.

[59]On the politics of this group, see S. S. Ostroumov, *Prestupnost' i ee prichiny v dorevoliutsionnoi Rossii*, 3d ed. (Moscow, 1980), 197.

ment for physicians and midwives.[60] But the government did not speak in one voice on this question. In 1912 the highly placed, politically conservative professor of gynecology Georgii Rein, also a member of the State Council, had headed a special commission appointed by the minister of internal affairs to review public hygiene legislation. The Pirogov Society was not represented on this commission, which set out to replace the services of zemstvo and community physicians with a centrally administered medical regime that would subordinate the profession to the state bureaucracy. The outbreak of war prevented its plans from being adopted, but Rein's intentions were already well known by 1913. As part of its recommendations, the commission defended existing criminal penalties for abortion.[61] Precisely because reform seemed to be in jeopardy and because neither the criminal statutes of 1903 nor the proposal of the Ministry of Justice had resolved all the relevant questions, the criminologists felt their deliberations to be of immediate practical import.[62]

The 1914 sessions opened with two keynote addresses: the sociologist Mikhail Gernet urged decriminalization; Evgenii Kulisher, a St. Petersburg attorney who served on the editorial board of the group's journal, called for retention of the law in modified form.[63] Fourteen people spoke on each side of the question—thirteen jurists (including one woman) and one physician (also a woman) for decriminalization, ten jurists (including one woman) and four physicians (all male) against.[64] The female physician was the same Liubov' Gorovits who had argued passionately at the Pirogov Society's 1913 congress for women's right to make their own reproductive decisions, and she once again gave a ringing defense of the dignity and moral value of freely chosen motherhood.[65] Without changing his earlier support for broadening the legal

[60]"Khronika," *Pravo*, no. 16 (April 21, 1913): 1024.

[61]Rein and the commission are discussed in John F. Hutchinson, "Politics and Medical Professionalization in Russia after 1905" (unpublished manuscript), 12–30; and idem, "'Who Killed Cock Robin?'" 19–20. (I thank Professor Hutchinson for making his manuscript available to me.) The commission's proposal slightly reduced the weight of the penalty for self-induced abortion but left the other penalties unchanged and thus significantly more onerous than those provided in the 1903 code: Pokrovskaia, "K voprosu ob aborte," 102.

[62]On the inadequacy of the 1903 code, see Ginzburg, "Izgnanie ploda," 53.

[63]Biographical information on Gernet is available in the introduction to M. N. Gernet, *Izbrannye proizvedenia*, ed. M. M. Babaev (Moscow, 1974), 8–37. For his talk, see idem, "Istreblenie ploda s ugolovno-sotsiologicheskoi tochki zreniia," *Vestnik prava*, no. 3 (1914): 233–38.

[64]Beginning in 1906 women were allowed to study law but prevented from practicing: Linda Harriet Edmondson, *Feminism in Russia, 1900–1917* (Stanford, Calif., 1984), 147–48.

[65]The text of Gorovits's remarks is in *Otchet*, 291–97; for her tone and audience response, see A. N. Trainin, "Na s"ezde kriminalistov: Fakty i vpechatleniia," *Russkoe bo-*

grounds for abortion, Lazar Lichkus found himself in the company of the distinguished obstetrician Dmitrii Ott in defending Kulisher's more cautious position. Six professors of law weighed in on Kulisher's side, against three on Gernet's. The eminent State Council member and criminal law expert Nikolai Tagantsev spoke against decriminalization, and the group's executive committee split 6 to 2 in favor of reform, not legalization. Nevertheless, the congress as a whole voted 38 to 20, with 3 abstentions, to eliminate abortion from the criminal code.[66]

The proceedings themselves were passionate in tone and the reaction was more emotional still. "The February congress has become the center of public attention," wrote Aron Trainin, one of Gernet's supporters. "In the daily press and the thick journals, at lectures and fashionable debates, the congress's basic resolutions are discussed with unflagging passion."[67] The vote to decriminalize abortion had "made a terrible impression on wide reaches of the public," a physician commented.[68] Gernet himself counted seventeen published responses by physicians and legal experts, thirteen in his favor. But defeated opponents of the adopted resolution called it irresponsible and demagogic; the right-wing press denounced it as a machination of the Jews.[69]

Beyond its engagement with the particulars of the abortion question, the debate represented a split within progressive educated society. On one side stood an alliance of liberal and socially conservative reformers who viewed the law as a repository of absolute standards apart from any

---

gatstvo, no. 4 (1914): 261. A fuller exposition is in L. M. Gorovits, "K voprosu o nakazuemosti aborta," *Sovremennik*, no. 5 (March 1914): 36–44.

[66]*Otchet*, 400; G. N. Shtil'man, "Abort na s"ezde kriminalistov," *Pravo*, no. 10 (1914): 777. Trainin, "Na s"ezde," 256–62, is an intelligent summary by a Gernet supporter. Proceedings of the debate appear in *Otchet*, 233–55, 272–333, 354–400, but not all voters are identified. For biographical information on the speakers, see "Imennoi ukazatel'," in *Dvenadtsatyi Pirogovskii s"ezd*, xv–xviii; *Rossiiskii meditsinskii spisok*; "Spisok chlenov Russkoi gruppy mezhdunarodnogo soiuza kriminalistov k 4 ianvaria 1909 goda," in *Russkaia gruppa mezhdunarodnogo soiuza kriminalistov: Obshchee sobranie gruppy v Moskve, 4–7 ianvaria 1909 goda* (St. Petersburg, 1909), vii–xx; "Spisok chlenov Russkoi gruppy mezhdunarodnogo soiuza kriminalistov k 21 aprelia 1910 goda," in *Russkaia gruppa mezhdunarodnogo soiuza kriminalistov: Obshchee sobranie gruppy v Moskve, 21–23 aprelia 1910 goda* (St. Petersburg, 1911), viii–xxiv; "Spisok chlenov Russkoi gruppy mezhdunarodnogo soiuza kriminalistov na 1 ianvaria 1914 goda," *Otchet*, 34–48 (also in *Zhurnal ugolovnogo prava i protsessa*, no. 4 [1913]: 136–44); "Spisok deistvitel'nykh chlenov iuridicheskogo obshchestva pri Imperatorskom S.-Peterburgskom Universitete so vremeni uchrezhdeniia obshchestva," in *Iuridicheskoe obshchestvo pri Imperatorskom S.-Peterburgskom Universitete za dvadtsat' piat' let (1877–1902)* (St. Petersburg, 1902), 115–25; and *Ves' Peterburg na 1913 god: Adresnaia i spravochnaia kniga g. S.-Peterburga* (St. Petersburg, 1913).

[67]Trainin, "Na s"ezde," 248.

[68]Elpat'evskii, "Samoistreblenie chelovechestva," 262.

[69]M. N. Gernet, "K voprosu o nakazuemosti plodoizgnaniia: Otvet moim kritikam," *Vestnik prava*, no. 16 (1914): 489. Objections by the chairman, V. D. Nabokov, appear in *Otchet*, 395, and in V. D. Nabokov, "Desiatyi s"ezd," 656–57.

particular views, whether overtly religious, political, or social. They argued that as a form of murder, abortion occupied a legitimate place in the secular statutes. On the other side stood a mixture of left-leaning liberals and reform socialists who challenged the possibility of legislating such absolute distinctions and made social policy their ultimate concern.[70] No outright revolutionary or ardent reactionary entered the lists, and the range of practical alternatives was not wide: no one wanted to increase existing penalties or justified abortion as a positive good; all favored better welfare services, increased medical intervention, and attention to female exploitation and women's rights. Conflict in the legal profession arose over the role of the law and the state in regulating individual conduct, setting ethical norms, and ensuring social justice— that is, over the same political and philosophical issues with which the physicians were engaged.[71]

At the criminology conference, members of the more radical camp attacked the use of repression (model one) and invoked the right of personal self-determination (model three). Under the first rubric they challenged the very basis of existing law: they denied that abortion was murder, since the fetus was incapable of independent life. An anti-abortion physician who invoked the pathetic "squeak" of a four-month-old fetus being expelled from the womb as an argument in defense of the law aroused cold skepticism from opponents, who insisted that the supposedly criminal act did not have a legally recognizable object. Any other claim smacked of religion or fantasy, not science, they said.[72] To condemn abortion in absolute moral terms as a form of murder should logically prevent the law's defenders from endorsing even the exceptional operation as medically or socially valid. As for criminal agency (the other necessary component of a crime), opponents of the law questioned the presumption of malicious intent. One speaker even presented abortion as a laudable moral choice reflecting the pressure of a responsible guilty conscience. Another wondered why the statute on infanticide seemed more sensitive than the abortion law to women's legitimate motivations.

Assailing the law from the social rather than the juridical perspective, some speakers on the radical side considered the statute unfair to the poor. With medically supervised abortion illegal, poor women were more likely than wealthy ones with personal connections to resort to

<hr />

[70]Representing the two sides were the liberal V. D. Nabokov, who became a member of the Provisional Government but left the country after the Bolshevik Revolution, and the socialist Mikhail Gernet, who stayed on to become a respected Soviet legal sociologist. On Nabokov, see N. I. Afanas'ev, *Sovremenniki* (St. Petersburg, 1909), 1:187; on Gernet, see n. 63 above.

[71]Elpat'evskii, "Samoistreblenie chelovechestva," 274; Trainin, "Na s"ezde," 258–61.

[72]*Otchet*, 284–85, 315–16.

underground practitioners, incurring greater physical as well as legal risks. The decriminalizers further insisted that moral values (let alone religious ones) had no place in modern legislation, because moral norms were not in fact universal, as their opponents claimed, but varied from culture to culture and class to class. Not every nation considered abortion murder, and the attitude of the "people" or the "public," even in one's own land, was notoriously hard to pin down. That individuals had the absolute right to control their private affairs (in the case of women, to make their own reproductive choices) meant that the state had no business regulating intimate conduct or subordinating personal behavior to national policy goals. Political motives for encouraging population growth had no place in a modern rule-of-law state (the entire congress shared this objection). In any case, they were sure that decriminalization would have no negative demographic consequences, since the instinct for motherhood was stronger than any law.[73]

Speakers on the other side of the debate, defending the retention of a modified anti-abortion law, adopted some of the positions articulated by opponents of the law: that the existing statute was unfair; that the state should offer positive inducements to motherhood in the form of social services; and that strictly medical indications were not the only valid grounds for abortion to be sanctioned by law. However, they disagreed with the decriminalizers on three points. First, all maintained that abortion was in fact murder, at least in the later stages of pregnancy, and that the law must defend the sanctity of human life. To allow exceptions, they argued, was not inconsistent with the legal recognition of mitigating circumstances; for example, it was not always a crime to kill a grown person. Second, they all attributed a moral function to criminal law, though in varying degrees. However secular, some said, the law still embodied ethical values recognized by society at large. Others went further and insisted that laws must actively encourage virtuous behavior, both condemning murder and shaping the limits of permissible sexual conduct. A few gravely warned that legal abortion would lead to general debauch (in both sexual and social terms) and the collapse of all moral standards, though even these few refrained from speaking in religious terms.[74] Third, these speakers tended to think that motherhood was in need of legal underpinning. Women were "changing before our very eyes," said one, and motherhood itself was in danger of extinction.[75] Paradoxically, the side that defended women's autonomy in reproductive affairs imputed to women a natural inclination

[73]These same arguments also appear in Oleinik, "Prestupnyi abort," 140–44, 148–49.
[74]For this conservative position, see Platonov, "Ob"ekt prestupleniia," 163–64. But even Platonov (176) did not advocate stricter penalties.
[75]M. M. Isaev, in Otchet, 282.

toward motherhood; those who worried about the social consequences of such autonomy saw motherhood as a social construct rather than a natural urge and hence as something in need of public endorsement.

Even before the vote the eminent liberal jurist Vladimir Dmitrievich Nabokov, who had presided over the session, charged Gernet and his followers with behavior unbecoming a professional gathering.[76] He feared that a resolution in favor of decriminalization would diminish the group's prestige; too radical to affect policy decisions, it would remain without issue (*besplodna*—an odd term to use in the context of abortion—*izgnanie ploda*).[77] It was not that Nabokov was afraid of taking controversial stands on sexual issues: he had earlier faulted the 1903 criminal code for penalizing homosexual relations between consenting adults on the grounds that no one's rights or welfare were adversely affected. In the case of abortion, however, he maintained that individual interests (those of the fetus) were indeed at stake, since the decriminalizers had failed to confront the problem of third-trimester operations.[78]

Not everyone who advocated the legal protection of individual rights, however, believed that it necessitated an anti-abortion statute. Gernet insisted that the issue was one neither of morality nor of legal principle but of class politics. His opponents, he charged, in their concern that the woman of comfortable means might indulge her vanity at the expense of maternal obligation, had lost sight of the real victim of the abortion laws—the poor working woman.[79] This was not, strictly speaking, true. Those who argued for keeping the abortion law on the books did not necessarily deny its unfairness, nor were they attempting to combat the rising incidence of the crime by making the punishment worse. Rather, as in the matter of infanticide, they argued for the importance of mitigating circumstances, for attention to the particulars of each case.[80] Despite their refusal to take the final step and emancipate women from the legal tutelage of others over their bodies, one need not doubt the sincerity of their professed interest in the fate of the poor, who were traditionally the objects of compassion and social concern, as contemporary attitudes toward infanticide made clear.

Defenders of the law were concerned about the threatening implications of the third model: a weakening of the intelligentsia's moral and professional authority over the lower classes, and the acceptance of edu-

---

[76]See Nabokov's closing remarks in ibid., 395.

[77]V. D. Nabokov, "Desiatyi s"ezd," 666–67.

[78]*Otchet*, 397.

[79]Gernet, in ibid., 387; idem, "K voprosu o nakazuemosti," 491.

[80]The case for individualized application of the law was made with particular emphasis by the eminent legal scholar M. P. Chubinskii, *Otchet*, 376–77; see also idem, "Vopros o vykidyshe," 475.

cated women as free agents and civic equals. Nabokov, for example, favored equal rights for women before the law but stopped short of endorsing women's suffrage.[81] Class politics may have been at issue in the abortion debate, but not in the sense Gernet had in mind. The anti-decriminalization rhetoric suggests that abortion represented a problem not of poor women in the traditional sense but of women who had greater resources at their command. Peasant wives, as everyone knew, continued to bear too many children and dispose of unwanted ones after they were born. It was the urban working woman and women of the debaters' own social milieu who ended their pregnancies before anyone was the wiser. Abortion was the product of choice, not of desperation; it symbolized female autonomy and rendered women inaccessible to public control. It became a mass phenomenon when poor women shifted in large numbers from traditional agricultural roles to jobs in cities, losing the innocence they seemed to have enjoyed as village wives and daughters subject to male discipline and abuse. Meanwhile, educated women were questioning their domestic obligations, some even wanting a role in public life.[82] One university student, when asked why he and his wife used contraceptive devices, said that pregnancy "removes women from political and intellectual life and prevents them from being members of society."[83] The abortion debate emphasized the dangerous consequences of women's liberation, however imperfectly achieved.

Who spoke most vociferously for reproductive freedom at the professional meetings? "Precisely the women doctors," noted a male colleague. "At both the Pirogov and the criminologists' congresses, the women who took part were the most cultivated [*naibolee intelligentnye*], the most select of women, and, one must assume, of a mature age. There is no reason to believe they were wild fanatics or emotionally perverted types; they without doubt constituted a female intelligentsia, the defenders, to a certain degree, of women's interests." They expressed the "individualism" of women recently freed from complete dependence on men by the increase in civic and economic opportunity.[84]

A 1913 short story, "Nothing Sacred," by Ol'ga Runova, supplied at least one anti-abortion physician with a literary portrait of women who aspired to sexual autonomy.[85] Assembled in the gynecological clinic of

[81]Edmondson, *Feminism in Russia*, 50, 66–67.

[82]The connection between an increase in abortion, the improved position of women in society, the movement of peasant women to the cities, and the principles of women's rights is made explicitly in Elpat'evskii, "Samoistreblenie chelovechestva," 270–71.

[83]Falevich, "Itogi," pt. 12, 330.

[84]Elpat'evskii, "Samoistreblenie chelovechestva," 265, 267, 271.

[85]O. P. Runova, "Bez zaveta: Rasskaz," *Sovremennyi mir*, no. 10 (1913): 29–69. The story is cited in Binshtok, "Dvenadtsatyi Pirogovskii s"ezd," 123. Ol'ga Pavlovna Ru-

На всероссійскомъ женскомъ митингѣ.

Мужчина. –Простите меня, милостивыя государыни мои! Больше никогда уже мужчиною не буду!

"At the National Women's Meeting." The man groveling before the delegates implores them: "Please forgive me, most respected ladies! I promise not to be a man anymore!" *Strekosa* , no. 22 (1905). Helsinki University Slavonic Library.

the fictitious Dr. Rasmussen, Runova's female characters reinforced the dominant iconography of the woman who wished to control her repro-ductive life: eager for pleasure, excitement, fashionable clothes, and sexual attention, they all had jobs to keep as well as husbands and lovers to entice. They used contraceptives and regarded abortion as

nova (b. 1864), one of the women writers in this period whose fiction described young women's attempts to lead professionally and sexually autonomous lives, regularly con-tributed to the liberal and progressive journals; her stories were collected in 1916 under the title *Lunnyi svet* (Moonlight). I thank Charlotte Rosenthal for this information.

свадьба равно-
правныхъ

Для мужчинъ

СЦЕНКа
въ отдѣльномъ кабинетѣ
ресторана. Равноправіе у туалетныхъ
кіосковъ.

Consequences of women's equality. Clockwise from top: the wedding of equals; equality in the toilets; scene in a private room in a restaurant. *Strekosa*, no. 20 (1905). Helsinki University Slavonic Library.

something all women accepted, even peasant wives and respectable mothers worried about pregnant teenage daughters. Physicians, Runova realized, played a central role in women's lives by regulating their sexuality. Thus the women in her story both fear and idolize the powerful figure who holds their sexual fate in his hands. Disabused of the political illusions he entertained in 1905, Dr. Rasmussen punishes his patients for denigrating motherhood—woman's true calling, in his view—by performing hysterectomies without their consent.

Physicians could narrow or expand the range of women's choices but ultimately could not control them. "Before becoming lawyers, judges, or professors," Rasmussen admonishes, "[women] should fight for the right to be mothers."[86] The young heroine of "Nothing Sacred," the wife of a provincial railroad engineer, is briefly impressed by the doctor's traditionalist fervor but cannot renounce the lover whose attentions relieve the boredom of her marriage. Rasmussen believes she should return to her stolid husband and the tedium of provincial life and resist the allure of the big city, which she observes from the clinic window: "There the street seethed with life. The marquees of the movie houses sparkled; the trams purred as they hurried by; . . . loud and triumphant sounded the brief, sonorous tunes of the automobiles, bearing the happy and gay to unknown destinations."[87] In this world of commercial delights, artificial lighting, and hectic, unfocused movement, seekers of sexual pleasure often avoided reproductive consequences, husbands lacked authority, and even prestigious physicians found their advice ignored. The real problem with which abortion confronted physicians, as Zhbankov in his anxiety recognized, was the moral and political crisis—coded in sexual terms—of educated men faced with workers and women who were challenging established patterns of cultural and civic control in an increasingly less predictable social and cultural environment.

[86]Runova, "Bez zaveta," 65.
[87]Ibid., 47.

Chapter Ten

From Avant-Garde to Boulevard:
Literary Sex

The city was a dangerous place. As the site of commerce and exchange, it was governed by desire, driven by psychological and material need and selfish interests. Like the marketplace, the city challenged established social hierarchies with new and transient relations of wealth. A Jew might achieve power and influence there, not despite his exclusion from the traditional order but thanks to that very marginality. A woman might find there an alternative to the domestic regime. In the realm of cultural representation, the market posed a similar threat, one also connected to the urban environment. On the boulevard strangers rubbed shoulders; prostitutes strolled; shops, cafés, and dance halls offered pleasures for a price. The boulevard gave its name to a new kind of culture. "Boulevard" fiction was the literary analogue of cheap amusements and commercial sex (which occurred literally on the sidewalk as well as in public establishments): a counterfeit of the real thing for those who had neither the material nor intellectual resources to obtain and appreciate the original.

Educated Russians feared the effect of this cultural displacement because they held culture in high regard not only in absolute terms but as a powerful social instrument. They had no doubt that ideas and representations exerted a dynamic influence on public attitudes, political values, and even the contours of private life. Throughout the eighteenth and nineteenth centuries elites shaped their own social conventions and interpreted their personal experiences according to literary models and the teachings of literary and social critics.[1] Nor was the impact of print

[1]See Iurii M. Lotman, "The Poetics of Everyday Behavior in Eighteenth-Century Russian Culture" and "The Decembrist in Daily Life (Everyday Behavior as a Historical-

culture merely reflexive: the democrats of the 1860s, the populists of the 1870s, and many "third element" professionals dedicated themselves to raising the intellectual level of the common folk, to "enlightening" the people. As Jeffrey Brooks has so convincingly shown, the turn-of-the-century proliferation of commercial publishing directed at a popular audience challenged the elite's monopoly on determining the value and distribution of images and ideas.[2]

The boulevard had, of course, an obvious sexual dimension. It provided a location for sexual practices unrelated to family life and free from community control: homosexual encounters, prostitution, public display, adolescent exploration—a network of impersonal relations at once intimate and anonymous. As the scope of city life expanded, the printed word came to enjoy the luxury of relaxed censorship rules and found a wider popular audience. The number of daily papers, periodicals, and tabloids multiplied, as did the range of their circulation.[3] The purposes of the print media were also in the process of change. "Along with freedom of the press, however limited it may be," one critic wrote in 1908, "the bourgeois revolution [of 1905] has introduced certain other traits of the European press—the boulevard and advertisement [bul'var i reklama]."[4]

Besides news, social commentary, and literary texts, magazines and newspapers provided a means of commercial and personal communication, a clearinghouse for the exchange of fantasy and the market in desire. The back pages advertised products: cameras, corsets, skin creams, cocoa, candies, and contraceptive devices; remedies for baldness, twisted spines, and small bosoms; cures for sexual troubles (onanism, impotence, neurasthenia, and venereal disease). They listed the services of divorce lawyers, midwives, wetnurses, and shelters where pregnant women might deliver their babies with discretion; they published the pleas of desperate mothers wanting to give their newborns away. Widowers used the columns to describe the perfect second wife (under thirty, "without a past"); governesses and nannies requested positions with good families; families sought cooks and other household servants. Some notices clearly proffered sexual pleasures: how to send

cal-Psychological Category)," both in *The Semiotics of Russian Cultural History*, ed. Alexander Nakhimovsky and Alice Nakhimovsky, 67–149 (Ithaca, N.Y., 1984); also Irina Paperno, *Chernyshevsky and the Age of Realism* (Stanford, Calif., 1988).

[2]See Jeffrey Brooks, *When Russia Learned to Read: Literacy and Popular Literature, 1861–1917* (Princeton, N.J., 1985).

[3]Caspar Ferenczi, "Freedom of the Press under the Old Regime, 1905–1914," in *Civil Rights in Imperial Russia*, ed. Olga Crisp and Linda Edmondson (Oxford, 1989), 198, 205–6.

[4]L. Gerasimov, "Nasha literatura i pressa posle revoliutsii," *Obrazovanie*, no. 2, sec. 3 (1908): 9.

ВСѢМИ ЛЮБИМЪ.

Шлю свою фотограф. карточку въ знакъ своего спасенія отъ онанизма, которымъ я страдалъ 18 лѣтъ. Много я потратилъ силъ, никогда не былъ веселъ, все что-то меня тяготило, а сейчасъ у меня появилась веселость. Я былъ стыдливъ съ женщинами, былъ недоволенъ собой, а теперь принимаю участіе въ какомъ угодно обществѣ и всѣми любимъ. Сто разъ спасибо за вашъ цѣлебный препаратъ «Біолъ-Ласлей». Т. М. Н.

Если Вы страдаете общею и половой слабостью, головными болями, безсонницею, малокровіемъ, **онанизмомъ** и его послѣдствіями, **робостью, слабой** памятью, послѣдствіями венерич. бол., если Вы нервны, раздражительны, переутомлены — спросите въ аптекѣ коробку „Біола". но только настоящаго «Біола-Ласлей». Вы получите блестящіе результаты. Это извѣстное средство противъ неврастеніи, которое совершенно безвредно. За справками можно обращаться въ С.-Петерб., отд. 7 части почтов. ящ. № 371.   33222

"Loved by Everyone."

"I am sending my photograph as proof that I have been saved from onanism, from which I have suffered since the age of eighteen. I lacked energy, was never cheerful, always depressed. Now I am happy. I was shy with women, dissatisfied with myself, but now I can participate in any kind of association and am loved by everyone. A hundred thanks for your healing preparation Biol-Lasley." T. M. N.

"If you suffer from general and sexual weakness, headaches, insomnia, anemia, **Onanism** and its consequences, **Timidity, Weak** memory, the results of venereal disease; if you are nervous, irritable, exhausted—ask your pharmacist for a bottle of Biol, but only the authentic Biol-Lasley. The results will be stunning. This well-known cure for neurasthenia is completely harmless. For information write St. Petersburg, section 7, P.O. Box no. 371." *Solntse Rossii*, no. 28 (1913). Helsinki University Slavonic Library.

for photographs of naked women direct from France; where to buy dirty postcards in St. Petersburg. "Models" boasting "attractive bodies" offered to pose nude for a fee. "Beautiful young gentlewomen" announced their willingness to serve as companions to "solid gentlemen." Such messages appeared not only in popular magazines but in the respectably conservative *Novoe vremia*, with one of the largest circulations in the business.[5] And despite their hostility to "Jewish pornography," even the anti-Semitic papers carried promises of cure for "sexual neurasthenia."[6]

So ubiquitous were such notices that they became the objects of satire. An ironically erotic magazine called *The Female Model* amused its readers with sophomoric spoofs, beginning with a good-humored ref-

[5]A random check of issues of *Novoe vremia* between 1906 and 1913 revealed numerous examples. Similar advertisements appeared in *Niva*, *Satirikon*, and *Solntse Rossii*. For mention of respectable provincial papers carrying notices for indecent publications and of Russian publishing firms that catered to such tastes, see A. B. Petrishchev, "Iz oblasti shchekotlivykh voprosov," *Russkoe bogatstvo*, no. 9 (1907): 95. For complaints about *Novoe vremia*, see A. V. Peshekhonov, "Na ocherednuiu temy: 'Sanintsy' i 'Sanin,'" pt. 1, *Russkoe bogatstvo*, no. 5, sec. 2 (1908): 118. On *Novoe vremia*'s growth and influence, see Ferenczi, "Freedom of the Press," 205, 207, 210.

[6]E.g., *Zemshchina*, no. 561 (February 15, 1911): 4.

ПОДУМАЙТЕ
О ПЕЧАЛЬНЫХЪ
послѣдствіяхъ половой неврастеніи.

Въ настоящее время такъ много
разныхъ средствъ предлагается пуб-
ликѣ благодаря явно рекламнаго
способа распространенія, что никто
не въ состояніи оцѣнить' дѣйстви-
тельнаго дѣйствія того или другого
средства, дабы больнымъ оказать
помощь. Мы даемъ всѣмъ мужчи-
намъ, читателямъ этой газеты, совершенно безплатно популярную
книгу Д-ра Мед. Генриха Шредера
„О половой неврастеніи и ея лече-
ніи" съ приложеніемъ пробной дозы
„ФОСФЕРРИНА", гарантирован-
наго и безвреднаго средства. За
успѣхъ гарантируемъ. С.Петер-
бургъ, Д-ръ Мед. Генрихъ Шре-
деръ Отдѣленіе Б. 60 Фонтанка 127.
Для почтовыхъ расходовъ просятъ
приложить 2 семи коп. марки.

a

Advertised cures for "sexual neurasthenia" (impotence). (*a*) "Think of the Sad Consequences of Sexual Neurasthenia." Readers who accept a trial dose of Phospherrina (guaranteed harmless) are offered a free copy of Dr. Henrich Schrader's book, *Sexual Neurasthenia and Its Cure*. (*b*) "The Unhappiness of Men": men suffering from impotence are urged to try Dr. Bink's Amus. (*c*) "Sexual Hunger—Thousands of men, young and old, suffer from sexual exhaustion without knowing it. Thousands (so to speak) are chained hand and foot to suffering, grief, fear, fatigue, and complete despair. They are so nervous and weak that they will soon become entirely impotent. The ensuing despair will often drive them to suicide or insanity." Send to Moscow for a copy of Dr. William Manken's book. The two leg irons are labeled, "Weak and nervous" and "Artificial stimulants." *Satirikon*, no. 6 (1909), no. 13 (1911); *Solntse Rossii*, no. 8 (1913). Helsinki University Slavonic Library.

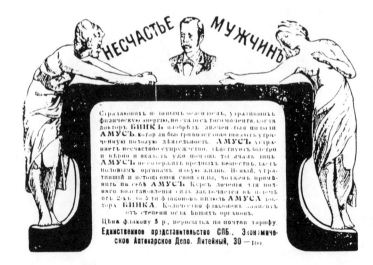

b

erence to Anastasiia Verbitskaia's popular novel of sexual exploration, *The Keys to Happiness.*

Have lost the Keys to Happiness. Please obtain at least a master key. The editor, The Keys to Happiness.

Seeking female secretaries for evening jobs. Trade in artistic post-cards. Mr. Photo.

Seeking lessons in amorous adventures. Poor student.

Lost my head. 200 rubles reward.[7]

[7]*Naturshchitsa*, no. 1 (1914): 4.

Advertised cures for impotence and masturbation. (*a*) Dr. Glaise's Stimulol, a French cure for sexual neurasthenia. (*b*) "MEN suffering from onanism, the effects of venereal disease, depressed, hopeless, discontent with themselves and their fate," are urged to send for Dr. Hamond's new book, *Sexual Impotence and Its Cure*, available at the Moscow bookstore Power of Knowledge. *Niva*, no. 43 (1913) and *Satirikon*, no. 50 (1909). Helsinki University Slavonic Library.

"How to Develop a More Beautiful Bust!" *Niva*, nos. 42 and 49 (1913). Helsinki University Slavonic Library.

Advertisements for pornographic photographs. (*a*) "Men only" are urged to send to a Warsaw address for photographs of naked Parisian beauties. (*b*) An album of 920 photographs for "male connoisseurs" is obtainable from Lodz. Photographs of French and Spanish beauties can be ordered from Spain. *Satirikon*, no. 40 (1909) and no. 46 (1912). Helsinki University Slavonic Library.

a 1452 г.   *b* 1908 г.

Гуттенбергъ. — Я чувствую, что мои многолѣтние труды будутъ служить на пользу человѣчества! Я увѣренъ, что мое изобрѣтение украситъ

Тѣнь Гуттенберга. — Чортъ возьми! О, если бы я зналъ...

Gutenberg and his ghost. (*a*) Gutenberg, 1452: "I believe that my years of labor will benefit humanity! I am certain that my invention will beautify life and ennoble pure art!" (*b*) Gutenberg's ghost, 1908: "The devil take it! If only I had known . . ." He is reading ads reproduced from actual papers: "How to protect oneself from venereal disease"; "Syphilis and its consequences"; "Model offers her services"; "Preservatives" for men and women; and so on. *Satirikon*, no. 34 (1908). Helsinki University Slavonic Library.

The editors of this magazine, designed to titillate its readers at the border of respectability, clearly recognized the function of such advertisements as incitements to the expression of personal desire in ritualized forms of public exchange. The Society for the Protection of Women took a more sinister view: it denounced the genre for encouraging prostitution and general depravity and called upon the Ministry of Justice to suppress it.[8]

[8]Letter to the society from the St. Petersburg House of Charity, May 23, 1914, forwarded to the minister of justice May 31, 1914: TsGIA, f. 1405, op. 543, d. 512, ll. 424–26.

Eroticism found a place not only in daily papers and popular magazines but in respected literature as well. Since the appearance of *The Kreutzer Sonata* in 1889, serious fiction too had expanded the framework of acceptable themes to include not only explorations of the moral and social dimensions of sexual experience but its sensual quality. Beginning in the 1890s, under the influence of French symbolism and fin-de-siècle "decadence," avant-garde Russian poets and prose writers had combined stylistic experimentation with provocative excursions into the realm of deliberate "bad taste," including the explicitly sexual.[9] Such literary innovators rejected the nineteenth-century principles of socially committed realism, with its focus on the public tableau, and turned instead to miniatures of personal introspection. But the thematic shift did not simply coincide with the boundaries of style; writers who perpetuated the realistic tradition also chose erotic themes, using them as vehicles for social commentary. The new interest in individual subjectivity affected even those who clung to more conventional prose forms.

Though the symbolists rejected "naive realism"[10] and the realists decried the "decadents" as "impudent and shameful,"[11] the public perceived both the avant-garde and the neorealists as representatives of a "new" generation, practitioners of what critics called the "young literature."[12] Furthermore, the prevalence of erotic themes in the writing of both schools tended to obscure their philosophical and aesthetic differences and to create confusion between the domain of high-minded cultural production and the disreputable commercial practices of the boulevard. Thus in 1911 the popular journal *Vestnik znaniia*, geared to a relatively unsophisticated urban readership, denounced the neorealists Leonid Andreev, Aleksandr Kuprin, and Mikhail Artsybashev as "erotomaniacs" and "untalented degenerates,"[13] despite the fact that most of

[9]As Vasilii Rozanov remarked in 1904: "Symbolism is drawn to the erotic. The god as old as mother nature, which was driven from the practical poetry of the 1850s–1870s, has returned to the sphere where it has belonged since antiquity. But in a disfigured, strange, and shamelessly naked form": *Dekadenty* (St. Petersburg, 1904), 5.

[10]Andrei Bely, "Symbolism and Contemporary Russian Art," in *The Russian Symbolists: An Anthology of Critical and Theoretical Writings*, ed. Ronald E. Peterson (Ann Arbor, Mich., 1986), 101; trans. of Belyi, "Simvolism i sovremennoe russkoe iskusstvo," *Vesy*, no. 10 (1908): 38–48.

[11]Maxim Gorky, quoted by Nicholas Luker in his *Anthology of Russian Neo-Realism: The "Znanie" School of Maxim Gorky* (Ann Arbor, Mich., 1982), 20, and by John E. Bowlt in "Through the Glass Darkly: Images of Decadence in Early Twentieth-Century Russian Art," *Journal of Contemporary History* 17 (1982): 95. Bowlt provides a good overview of the period's literary and pictorial decadence.

[12]Peshekhonov, "Na ocherednye temy," pt. 1, 104.

[13]Jeffrey Brooks, "Popular Philistinism and the Course of Russian Modernism," in *Literature and History: Theoretical Problems and Russian Case Studies*, ed. Gary Saul Morson (Stanford, Calif., 1986), 101.

their work had first appeared in the respectable liberal or left-leaning press or in serious journals of popular enlightenment.[14] These were, moreover, the same journals that published some of the works that critics denounced as vulgar products of the boulevard.[15]

The boulevard—not confined to private salons or philosophical gatherings, as the avant-garde and the neorealists were—did not target a select audience. By definition indiscriminate, it violated important cultural distinctions. Its productions confused boundaries of genre, mixing high and low in both thematic and stylistic terms. The boulevard also confused boundaries of class by affording pleasure to humble readers for whom pleasure was itself a cultural (and status) aspiration. Finally, the boulevard had the potential to confuse gender categories as well: it provided women writers with a way of making money, garnering public attention, and putting the subject of women at the center of cultural production—in short, fashioning a public self-representation at once transgressive and respectable.

In all its dimensions the boulevard called attention to questions of pleasure and desire; its methods of production, dissemination, and consumption mirrored the message that the imagination was no longer a socially restricted domain. The avant-garde, by contrast, defined itself in terms of the limited nature of its appeal. Both the avant-garde and the neorealists, however, shared the boulevard's obsession with subjectivity and desire, and it was perhaps because of the resulting ambiguity of cultural distinction that critics and other guardians of cultural high-mindedness struggled to denounce the boulevard in particularly vehement terms. Far from constituting a frivolous diversion, the politics of sexual representation formed the core of a debate over the politics of culture in the years after 1905.

The Cultural Construction of Sex

The same physicians and educators who maintained that they could mold civic virtue by training young boys in sexual self-control took a serious view of the impact of culture on the outlook of the nation's

---

[14]E.g., M. P. Artsybashev, "Uzhas: Rasskaz," *Obrazovanie*, no. 3, sec. 1 (1905): 1–20; L. N. Andreev, "V tumane," *Zhurnal dlia vsekh*, no. 12 (1902): 1411–50. Artsybashev's *Sanin* was originally published in *Sovremennyi mir*, nos. 1–4 (1907). Kuprin's collected works were issued by the journal *Niva* in a special supplement in 1912. On *Niva*, see Brooks, *When Russia Learned to Read*, 111–13.

[15]E.g., Verbitskaia's work was called "boulevard" literature in K. I. Chukovskii, *Kniga o sovremennykh pisateliakh* (St. Petersburg, 1914), 8. On her publishing record, see V. Kranikhfel'd, "Literaturnye otkliki: O novykh liudiakh A. Verbitskoi," *Sovremennyi mir*, no. 8, sec. 2 (1910): 69.

АСТРОНОМЪ.

Голосъ изъ публики (въ ужасъ). — Неужели, и та, которая потолще—называется звѣздой?!!
Другой голосъ. — Помилуйте! Не только звѣздой, но цѣлымъ созвѣздіемъ...
— Какимъ же?
— Большой медвѣдицей.

"In the *Café-Chantant*." The legend reads: "Voice from the audience, in horror: 'Is it true that the fat one is called a star?' Another voice: 'If you please, not just a star but an entire constellation.' 'Which one?' 'Ursa Major.'" *Satirikon*, no. 41 (1909). Helsinki University Slavonic Library.

youth. No one believed more strongly in the power of representation to shape moral attitudes, influence behavior, and affect public health in the literal, physical sense. As the pediatrician Izrail Kankarovich reminded his colleagues at the 1910 congress on prostitution, reading had powerful mimetic effects: "Boys who read Jules Verne dream passionately of travel and sometimes even run away from home. Crime novels create criminals. Literature in the genre of Sherlock Holmes stimulates the reader's taste for sleuthing. And erotic art indisputably arouses the sexual instincts and makes people depraved."[16] His linking of Jules Verne, Sherlock Holmes, and erotic representation reflected the contemporary assumption that all "cheap" entertainment provoked desires and fantasies inimical to the social good.[17] One lecturer who spoke to crowded rooms on the "sexual question" insisted that the current interest in sexuality (to which he owed his own popularity) was as ephemeral as the rage for the detective heroes Nat Pinkerton and Sherlock Holmes; they were all subjects that appealed to what he called the readers' most primitive instincts.[18]

Pedagogues interested in strengthening their students' moral fiber were quick to distinguish between harmful and hygienic works. They believed that certain scientific texts had a salutary influence and that some literary works (usually by authors already ranked as classic) could keep the young from going astray. In the fiction department, Tolstoy's *Kreutzer Sonata* and "Family Happiness," Chekhov's "Nervous Attack," Ivan Turgenev's *First Love* and *Rudin*, and Nekrasov's "Russian Women" were considered prophylactic.[19] The works of many other writers, by contrast, were said to produce the same deleterious effects on young men as exposure to the photographs of naked women, displays in store windows, and newspaper advertisements—all features of the cultural boulevard.[20] The texts singled out as symbols of corruption

---

[16]I. I. Kankarovich, "O prichinakh prostitutsii," *Trudy pervogo vserossiiskogo s"ezda po bor'be s torgom zhenshchinami i ego prichinami* (St. Petersburg, 1911), 1:190.

[17]On the harm caused by "commercial" literature on sex, see A. G. Trakhtenberg, "Polovoi vopros v sem'e i shkole," *Voprosy pola*, no. 1 (1908): 27.

[18]Petr Pil'skii, *Problema pola, polovye avtory i polovoi geroi* (St. Petersburg, 1909), 19, 130. On Pil'skii as lecturer, see Peshekhonov, "Na ocherednye temy," pt. 1, 107. On Pinkerton in Russia, see Brooks, *When Russia Learned to Read*, 42–46. The liberal critic Kornei Chukovskii (*Kniga*, 13–14, 16–18) denounced Pinkerton and erotic romances together, as did a reviewer for a liberal newspaper: L. Kozlovskii, "Teper' ili nikogda," *Russkie vedomosti*, January 11, 1911, cited in V. Dadonov, *A. Verbitskaia i ee romany "Kliuchi schast'ia" i "Dukh vremeni": Kriticheskii ocherk* (Moscow, 1911), 72–74.

[19]For scientific texts, see N. M—ii [N. K. Mikhailovskii], "Bor'ba s polovoi raspushchennost'iu v shkole," *Russkaia shkola*, no. 9 (1907): 51; and I. S. Simonov, "Shkola i polovoi vopros: Iz dnevnika byvshego ofitsera-vospitatelia," pt. 3, *Pedagogicheskii sbornik*, no. 5 (1908): 410, 413, 420; K. P. Sangailo, "Polovoi vopros i shkola," *Pedagogicheskii sbornik*, no. 3 (1913): 328.

[20]Among those regularly mentioned were Artsybashev, A. P. Kamenskii, Verbitskaia, Andreev, M. A. Kuzmin, and L. D. Zinov'eva-Annibal. See Sangailo, "Polovoi vopros,"

were not themselves products of the boulevard, however, and for all
their bravado in speaking of sex, anguish was in fact their major theme.
Two widely discussed short stories by Leonid Andreev, published in
1902 ("In the Fog" and "The Abyss"), deal with venereal disease, suicide, prostitution, and gang rape. Mikhail Artsybashev's ominous 1905
tale "The Horror" offered an equally chilling fantasy of multiple rape,
and his 1907 novel *Sanin*, which pretended to exalt the joys of sensual
love, included three suicides. Aleksandr Kuprin's sober account of prostitution in his novel *The Pit*, which appeared serially between 1909 and
1915, was no less grim.

These works offered images of menacing darkness, of yawning pits,
of threatening, engulfing, un(en)lightened hollows where women lost
their sanity (even their lives) and men their self-control and good conscience. The avant-garde texts that aroused critical opprobrium by their
explicit treatment of sex attempted a loftier, more celebratory tone.
Fedor Sologub's satirical touch helped relieve the erotic claustrophobia
of *Petty Demon* (1907), and his fantastical utopia of sexual freedom, *The
Created Legend* (1914), escaped the melodramatic heavy-handedness
that pervaded his colleagues' attempts to explore the libidinous or social
extremes of sex. Mikhail Kuzmin managed to maintain a posture of
aesthetic high-mindedness while preaching the nobility of male homosexual love in his 1907 novella *Wings*, and Lidiia Zinov'eva-Annibal
produced a lyrical hymn to lesbian love, *Thirty-three Abominations* (also
1907), which nevertheless ended in the lovers' estrangement and the
heroine's suicide.[21]

Most of these representatives of the neorealist establishment and the
literary avant-garde borrowed turns from the boulevard repertoire, indulging in melodrama, sensationalism, and shallow psychologizing to
convey the immediacy of sex. In this way they contributed to the erosion of cultural boundaries effected by the commercialization of the
text. Nor was the boulevard itself distinguished by a carefree attitude
toward sexual questions. The works that frankly appealed to boulevard
tastes were no less ambivalent about the possibility of representing sexual satisfaction than those that pretended to a more exalted cultural po-

---

328, 331 (also on photos and store windows); Simonov, "Shkola," pt. 1, *Pedagogicheskii
sbornik*, no. 1 (1908): 17, 20.

[21]Fedor Sologub, "Melkii bes," *Voprosy zhizni*, nos. 6–11 (1905), rpt. separately (St.
Petersburg, 1907), trans. as *The Petty Demon* (Ann Arbor, Mich., 1983); idem, *Tvorimaia
legenda*, serialized 1907–13, rpt. separately (St. Petersburg, 1914); M. A. Kuzmin, "Kryl'ia,"
Vesy, 11 (1906), rpt. separately (Moscow, 1907), trans. as *Wings: Prose and Poetry* (Ann
Arbor, Mich., 1972); L. D. Zinov'eva-Annibal, *Tridtsat'-tri uroda: Povest'* (St. Petersburg,
1907), trans. as "Thirty-three Abominations," *Russian Literature Triquarterly*, no. 9 (Spring
1974): 94–116.

sition. Evdokia Nagrodskaia's *Wrath of Dionysus* (1910), for example, depicted a love triangle of two men and a liberated woman who fails to reconcile creative ambition and personal happiness. The most notorious of them all, Anastasiia Verbitskaia's phenomenally popular *Keys to Happiness* (1910–13),[22] which was thought to celebrate the sexual freedom of the period's new women, featured three suicides, including that of the heroine, who kills herself over the inadequacy of love and the emotional limits of success.

It would seem, then, that most of the writers decried as pornographic had themselves largely undermined their subject's appeal, but they nevertheless aroused a storm of high-minded indignation and culled an avid readership, especially among the young. Physicians who interrogated university students about their sexual habits examined their reading habits as well. Mikhail Chlenov reported in 1909 that a third of survey respondents among the Moscow student body admitted to reading what they called pornography; another two-thirds read popular medical texts. Many cited *The Kreutzer Sonata* and Andreev's "Abyss" and "In the Fog" as favorite literary works and praised Chekhov's "Nervous Attack" as morally uplifting.[23] Iakov Falevich, reporting on a survey in Tomsk, said that almost all respondents admitted reading literary works on sexual themes, including *The Kreutzer Sonata*, "Abyss" and "Fog," and *Sanin*. Asked to describe the moral effect of such reading, most called it "ennobling," though some complained that it was "corrupting" or "titillating."[24] Women students queried in 1912 presented a somewhat different literary profile. They showed less interest in Artsybashev than the men and surprisingly little enthusiasm for Verbitskaia, at the very moment when *The Keys to Happiness* was selling thousands of copies. The author who seemed to appeal equally to students of both sexes was Andreev, the favorite of over half the women who replied.[25]

But questionnaires are crude instruments for measuring aesthetic and moral complexities. Falevich also found that responses to literary texts varied widely and that the moral line between quality and vulgarity was hard to draw. Though most of his male respondents claimed that *The Kreutzer Sonata* encouraged them in the direction of sexual continence,

[22]E. A. Nagrodskaia, *Gnev Dionisa* (St. Petersburg, 1910); A. A. Verbitskaia, *Kliuchi schast'ia* (Moscow, 1910–13). On Verbitskaia's readership, see Chukovskii, *Kniga*, 15.

[23]M. A. Chlenov, *Polovaia perepis' moskovskogo studenchestva i ee obshchestvennoe znachenie* (Moscow, 1909), 49–52.

[24]Ia. Kh. Falevich, "Itogi tomskoi studencheskoi polovoi perepisi: Doklad, chitannyi 18 fevralia 1910 g. na zasedanii Pirogovskogo studencheskogo meditsinskogo obshchestva pri Tomskom Universitete," pt. 6, *Sibirskaia vrachebnaia gazeta*, no. 22 (1910): 258.

[25]*Slushatel'nitsy S.-Peterburgskikh vysshikh zhenskikh (Bestuzhevskikh) kursov* (St. Petersburg, 1912), 122.

one said that it caused him "to regard women as toys," another that it made him fear women. The students disagreed even about Andreev, describing the effects of his work as variously "ennobling, . . . stupefying and burdensome, . . . [or] corrupting, in the sense of arousing sexual desire while reading and of encouraging disrespect for women." One respondent contrasted Andreev favorably with Artsybashev: "'Abyss' and 'Fog' do not increase one's sexual desire," the young man wrote, "but illuminate the psychology of desire and encourage one to think seriously about one's sexual impulses. By contrast, Artsybashev's *Sanin* does not approach sexuality with the same self-conscious attitude imbued with deep psychological insight"; rather, it "leaves the soul coated with a filthy deposit of sensuality."[26]

As the interest in Andreev reveals, his tales of sexual distress and violence put him at the center of the controversy over the literary representation of sex in the years before 1905. The two short stories of 1902 aroused a storm no less fierce than the one that had surrounded Tolstoy in the 1890s, and they brought Andreev instant fame. Newspapers were inundated with letters from readers; literary and social circles buzzed with talk; Andreev was the butt of envy and satire.[27] Attacks came from all quarters. A conservative critic denounced Andreev as a "scribbler-erotomaniac" for whom the cure "was not criticism but the clinic."[28] Zinaida Gippius, of the avant-garde modernist camp, reproached him for cultivating morbid attitudes. Countess Sofiia Tolstoy, who had helped her husband get permission to publish *The Kreutzer Sonata* in 1891, used the pages of *Novoe vremia* to denounce Andreev's work as devoid of higher values: he "revels in the baseness of depraved human life," she complained, "and his love for depravity infects the immature, morally undiscriminating . . . reading public and youth."[29] In private, Tolstoy himself apparently seconded his wife's disgust. Yet Andreev's defenders included Anton Chekhov, Maxim Gorky, and a host of unknown admirers who congratulated the author on bringing the moral agonies of sexual life to the public's attention.[30]

---

[26]Falevich, "Itogi," pt. 6, 258–59.

[27]James B. Woodward, *Leonid Andreyev: A Study* (Oxford, 1969), 71. See also Pil'skii, *Problema*, 25, 97.

[28]N. E. Burenin, "Kriticheskie ocherki," *Novoe vremia*, no. 9666 (January 31/February 13, 1903): 2. Burenin was himself berated by left-wing critics for pandering to vulgar appetites; see G. S. Novopolin, *Pornograficheskii element v russkoi literature* (St. Petersburg, 1909), 45–46.

[29]Sofiia Tolstaia, "Pis'mo v redaktsiiu," *Novoe vremia*, no. 9673 (February 7/20, 1903): 4.

[30]Woodward, *Andreyev*, 72–75. For positive reviews, see E. Koltonovskaia, "Problema pola i ee osveshchenie u neorealistov (Vedekind i Artsybashev)," *Obrazovanie*, no. 1, sec. 2 (1908): 116–19; A. B[ogdanovich], "Kriticheskie zametki," *Mir bozhii*, no. 1, sec. 2 (1903): 1–14.

By focusing on the issue of venereal disease and the problem of adolescent adjustment to the adult sexual world, Andreev's "Fog" raised the personal dilemma of his unhappy protagonist to the level of social analysis; it exposed the negative consequences not of individual indulgence but of society's underlying values. The story of a troubled student who contracts syphilis from a prostitute and, after much tortured self-examination, kills yet another prostitute and finally himself in shame and hopelessness, the tale obviously echoed the current concern with moral education, particularly of boys. It also reflected the preoccupation with venereal disease and with the threat posed by prostitution to the welfare of respectable families. Its rendition of a father-son talk in which the warmhearted, awkward, yet conventional old man helplessly confronts his boy's silent and lonely despair must have wrung the hearts of parents and children alike.[31] It signaled the widening gulf between the generations of educated men, not only in their view of public goals but in their experience of private and family relations as well—in their very sense of self.

"The Sexual Question" Old and New

Observers did not claim that sexual exploration was entirely novel in either Russian literature or Russian life. Well before symbolists such as Valerii Briusov and Fedor Sologub brought erotic themes to poetry and prose in the 1890s, sexual motifs could be found in high as well as low literature. One critic cited the popularity of George Sand in the 1840s and the interest in titillating themes supposedly typical of the "stagnant" 1880s as evidence that sex had long been a preoccupation of Russian letters.[32] The law against literary obscenity was both weak and difficult to apply, and explicit sexual material often managed to elude the censor.[33] The turn-of-the-century focus on erotic themes had its roots, some commentators believed, in the extreme individualism encouraged by the ideas of Friedrich Nietzsche, whose popularity in Russia beginning in the 1870s had purportedly bred "contempt for the conclusions

---

[31]On the pathos (but not pathology) of parent-child misunderstanding, see Bogdanovich, "Kriticheskie zametki," 11–14.

[32]Novopolin, *Pornograficheskii element*, opening chapters and 44, 230.

[33]Petrishchev, "Iz oblasti," 96–97; Peshekhonov, "Na ocherednye temy," pt. 1, 121. For problems in applying the statute on obscenity, see G. Sl—, "Nravstvennoe i beznravstvennoe v proizvedeniiakh iskusstva," *Iuridicheskii vestnik*, no. 8 (1889): 536–37. The text of at least one illicit production, printed in only 100 copies, survives in Harvard University's Houghton Library: *Russkii erot ne dlia dam* (n.p., 1879); I thank Richard Wortman for making a copy available to me.

of social science and for all public tasks, along with an exaggerated sense of the importance of the self."[34]

The paradox of exaggerated individualism, this preoccupation with the self, was that it was not a private affair. The zemstvo statistician and economist Aleksei Peshekhonov, an editor of *Russkoe bogatstvo*, noted in 1908 that the focus on intimacy had become a public spectacle. As crowds gathered to hear talks on the new literary productions, he reported, "sexual excesses and perversions ha[d] suddenly become the center of public attention." Audiences rushed to hear the works of Kuzmin, Sologub, and Artsybashev read aloud, even if they booed or tittered in embarrassment.[35] One lecturer on *Sanin* described the thousands who forced their way into packed halls in Moscow and other cities, insisting on their rights as "*Sanin* fans" (*Sanintsy*), until policemen were obliged to stand guard and contain the overflow crowds.[36]

Such scenes testified that the literary preoccupation with sexual themes bore some relation to the contours of social existence, whether as model or mirror or both. Indeed, the new sexual attitudes took actual as well as literary forms. But public knowledge of current sexual practices was filtered through the highly tendentious screen of journalistic reporting, itself affected by the prurient temper of the times. Journalists were convinced, for example, that young people, in addition to attending lectures, had formed clubs for the discussion of sexual subjects—and perhaps in some cases for sexual experimentation. Minsk supposedly was home to the "Free Love League," Moscow to the "Union of Beer and Freedom," and Kazan, Kiev, and Orel to similar associations. The liberal newspaper *Birzhevye vedomosti* described a club in Kiev whose members were said to practice "free love." But when the governor investigated rumors of yet another club in Ekaterinoslav, he found it did not exist, and the Minsk newspaper report proved to have been a fabrication as well.[37]

But fantastical reports are not necessarily devoid of real meaning. Critics insisted that the current preoccupation with sex, if not new to

---

[34]F. Makovskii, "Chto takoe russkoe dekadentstvo," *Obrazovanie*, no. 9, sec. 1 (1905): 127, 129–30. A later critic noted that neorealists as well as neoromantics cultivated an extreme form of "anarchistic individualism derived from Nietzsche" (Koltonovskaia, "Problema pola," 114). Scholars today characterize the period's dominant literary and philosophical tendency as a defense of individualism: Christopher Read, *Religion, Revolution and the Russian Intelligentsia, 1900–1912* (London, 1979), 15–37; Bernice Glatzer Rosenthal, *Dmitri Sergeevich Merezhkovsky and the Silver Age: The Development of a Revolutionary Mentality* (The Hague, 1975), 84.
[35]Peshekhonov, "Na ocherednye temy," pt. 1, 104, 106–7, 109–12. For his editorial role, see L. G. Berezhnaia, "Zhurnal 'Russkoe bogatstvo' v 1905–1913 gg.," in *Iz istorii russkoi zhurnalistiki nachala XX veka*, ed. B. I. Esin (Moscow, 1984), 60, 64.
[36]Pil'skii, *Problema*, 102–3, 107–9.
[37]Peshekhonov, "Na ocherednye temy," pt. 1, 104, 113, 116–17.

Russian letters, reflected new social processes at work. "We are faced not with the accidental *boutades* of literary daredevils," wrote the respected reviewer Arkadii Gornfel'd in *Russkoe bogatstvo*, "but with a social mood."[38] To socialists, this mood reflected the class consequences of modernization and the supposedly bourgeois character of 1905. Thus a writer for *Obrazovanie* asserted in 1905 that the antisocial "psychosis" of the period was "the property of the wealthiest classes, primarily of the rich urban bourgeoisie. In this milieu the themes of 'blissfully perverted pleasures' [*blazhenno-izvrashchennye naslazhdeniia*] and the peculiar creed of 'anything goes' evoke a spiritual response and inspire imitation. By contrast, among the politically conscious laboring masses and the laboring intelligentsia, these themes arouse the most passionate protest and indignation."[39]

Like the physicians, these critics insisted that morality was a class-bound affair. They went on to translate moral judgments into aesthetic terms: for them profit and bad taste went hand in hand. Not democracy but "idle philistine curiosity" had emerged from the revolution, a contributor to *Obrazovanie* complained. "The printed word has become a matter of profit, the exploitation of sensation a profitable enterprise." Like other left-wing critics, however, this one saved his cruelest words for "the boulevard press," which he called "the legitimate child of the nervous, cowardly, cruel bourgeois epoch, which is at the same time sentimental, unprincipled, and eager for piquant impressions."[40]

Iurii Steklov, a Social Democrat, said the same thing in more emphatic terms. In the tradition of Marxist polemic, his prose drips with sarcasm, swells with hyperbole, and staggers under the burden of political cliché, much as the texts of pornographers groan with the repetition of stock phrases designed to arouse a predictable response in a ready audience. "In today's bourgeois society sexual relations have reached extremes of perversion," wrote Steklov in 1908. Like ruling classes before them, he affirmed, the Russian bourgeoisie had emerged from the 1905 revolution with an appetite for "spiritual masturbation" and crude eroticism instead of "natural" sex, whereas the proletariat adhered to a "morality of collegial collaboration [*sobornoe sodruzhestvo*] and collective labor . . . which protects them from infection by the general illness of this neurasthenic time."[41]

---

[38]A. G. Gornfel'd, "Eroticheskaia belletristika," in Gornfel'd, *Knigi i liudi: Literaturnye besedy* (St. Petersburg, 1908), 28. For Gornfel'd, see Berezhnaia, "Zhurnal 'Russkoe bogatstvo,'" 75–76.
[39]Makovskii, "Chto takoe," 126, 132–33.
[40]Gerasimov, "Nasha literatura," 9–10.
[41]Iu. M. Steklov, "Sotsial'no-politicheskie usloviia literaturnogo raspada," in Steklov, *Literaturnyi raspad: Kriticheskii sbornik* (St. Petersburg, 1908), 1:28–30.

Popular virtue, upper-class decadence. *Satira*, no. 1 (1906). Helsinki University Slavonic Library.

If Steklov accorded the bourgeoisie a progressive role at certain points in history, he could not say the same of its lowly cousin the "petty bourgeoisie," the most despicable category in the Marxist lexicon. He derided the decadent writers for thinking themselves free spirits, rebels against bourgeois convention, harbingers of modernity. In fact, he declared, they belonged only to the "petty-bourgeois swamp [*meshchanskoe boloto*]. . . . Under the pretext of fighting the hypocrisy of the ruling class, they provide us with an apology for a class even more reactionary and morally abject . . . , [composed of] drunken feudal lords [and] guards officers. . . . The love they preach reeks unbearably of the stables and the cheap brothel." The decadents, Steklov concluded with rhetorical panache, had "established their citizenship rights in Russian literature with pornography savage enough to make monkeys blush."[42]

Though Steklov's tone was extreme, his terminology was widespread. The moderate critic Vladimir Kranikhfel'd, writing in the socialist journal *Sovremennyi mir*, complained of Fedor Sologub's "petty-

[42]Ibid., 28, 52. Like his Marxist colleague, Novopolin also considered the new literature to be the work of reaction: *Pornograficheskii element*, 104.

The ruling class feasts and whores while the people (in the corner drawings) suffer and toil. *Satira*, no. 1 (1906). V. I. Lenin State Library, Moscow.

bourgeois [*meshchanskii*] individualism," suffused with "sexual excess." The same Kranikhfel'd, however, warned against the kind of rhetoric in which Steklov indulged: "Reactionaries are always the ones who charge literature with pornography, for every social reaction is accompanied by hypocrisy. . . . Under the pretext of combating indecency it is so easy and—more important—so respectable to tear the heart out of literature and stealthily, without attracting attention, render it lifeless and pale."[43]

In fact, the charge of pornography was thrown at the new sexually explicit literature from both sides of the political divide. Unlike Steklov or Kranikhfel'd—who, despite their differences, identified the left with moral high-mindedness and the bourgeois or even traditional right with moral and cultural turpitude—the liberal religious philosopher Prince Evgenii Trubetskoi blamed radical politics for encouraging "the vulgar pornographic morality" typical of the day's fiction. The noble spirit of the revolution had been defeated not by the political reaction, he de-

[43]Vladimir Kranikhfel'd, "Literaturnye otkliki," *Sovremennyi mir*, no. 5, sec. 2 (1907): 135; idem, "Literaturnye otkliki," ibid., no. 6, sec. 2 (1909): 107.

clared in 1908, but by Marxism's philosophical bankruptcy. "Social Democracy has inculcated people with the cult of material well-being as
the only and unconditional value," he wrote. "Now from Social Democracy's bowels has appeared . . . a new social type, which . . . has
revolted against Social Democracy itself. . . . The revolution, which
could not rise above the level of 'sensations,' now drowns in pornography." To regain its courage, according to Trubetskoi, the "liberation
movement" would have to be "reborn" on a foundation of absolute
religious values.[44]

In these discussions the social analysis clearly served ideologically
partisan ends. It was possible, however, to provide an interpretation of
the current fascination with sexual themes which was social without
being ideological. Afanasii Petrishchev, of *Russkoe bogatstvo*, did so by
telling the story of his own generation's emergence from the small-
town world of petty-bourgeois patriarchy into the light of intellectual
sophistication.[45] His tale illuminates the experience of the men (though
not necessarily the women) who created the cultural debate.

Petrishchev invoked this experience to illustrate the long-term shifts
underlying the current crisis. The son of a provincial family from the
modest urban classes (the much-maligned *meshchanstvo*), Petrishchev had
spent his adolescence in revolt against the traditional sexual values and
abusive sexual practices of his elders. Far from representing the mores
of a decadent and outmoded aristocratic culture, as Steklov imagined it
did, this "petty-bourgeois swamp" could barely be distinguished from
the peasant village, in Petrishchev's telling. The material proof of the
bride's virginity was held up to public inspection at family weddings;
couples covered the holy images when they made love. In principle,
men and women kept a chaste distance; in fact, as the younger generation observed, sexual relations did not conform to these strict injunctions. "The lower-middle-class hut was a crowded place," Petrishchev
remembered. "One would have been happy to hide a lot that went on
inside it, but there was simply no room." Feasts began in traditional
decorum, men and women apart, but ended in adultery; children witnessed fathers beating their wives for refusing to have sex; they noted
that patriarchs who denounced prostitution and disapproved of incest
visited public houses and slept with their daughters-in-law.[46]

It was fathers against sons, Petrishchev recollected, as the boys read
Nikolai Chernyshevskii and Dmitrii Pisarev, thanks to whom "the

---

[44]Evgenii Trubetskoi, "Konets revoliutsii v sovremennom romane: Po povodu 'Sanina' Artsybasheva," *Moskovskii ezhenedel'nik*, no. 17 (1908): 10, 14–15.
[45]Petrishchev, "Iz oblasti."
[46]Ibid., 113–16.

questions of sexual morality, as they touched on our experience, were not at all theoretical questions." Scrambling to find copies of Chernyshevskii's *What Is to Be Done?* in provincial libraries, the youngsters marshaled the ideas of the 1860s in opposition to their parents' way of life. The ideas were powerful because they illuminated a troubling experience. "Our main guide in these situations," Petrishchev wrote, "was the logic of contradictions and the immediate feeling of repulsion for the unsavory arrangements of our fathers' lives." Following the literary models supplied by Chernyshevskii and later Tolstoy, the men of Petrishchev's generation formulated a sexual ethic even stricter than their fathers', though they based it on individual rights rather than religious authority.[47]

In fact, Petrishchev admitted ruefully, the results of this confrontation were ironic: the fathers' old ways changed under the weight of circumstances, not under the pressure of the sons' ideas, yet his contemporaries found their own lives following traditional patterns. "With a few happy exceptions, we ended up in essentially the same circumstances as our fathers, . . . burdened with the same oppressive feeling of dissatisfaction as they had been, with the same unhealthy results—beginning with the desire for venal bodies that sustains prostitution and ending with sadism." Certain improvements had occurred, he admitted, particularly in the realm of women's rights and sexual tolerance, but often the men of his generation failed to honor their high-minded ideals in domestic life. In the end, he considered sexual modernity a mixed blessing. He knew of young couples, unable to afford many children, who took steps to prevent conception which left them "more inconsolable and unhappier" than their religious parents had been.[48]

It was against this background that the traumas of war and revolution had intervened. The Russo-Japanese War had been particularly demoralizing, Petrishchev believed, because it had been fought against Asian peoples. The Russians had treated them with the kind of brutality they would not have inflicted on European foes, because they considered the "yellow-skinned" (*zheltorozhie*) Japanese (he put the disparaging phrase in quotations) not quite human. There was a direct connection, Petrishchev insisted, between disregard for the dignity of other races and the vicious abuse of Russian women by the troops sent to pacify domestic unrest. Not only did racial and sexual brutality echo each other, but the conditions of colonial warfare, he believed, also encouraged sexual license among the military elite, whose bad example then undermined the morals of the metropolis: "The Manchurian de-

[47]Ibid., 113–14, 116–18.
[48]Ibid., 120–22.

bauch [*razgul*] is directly connected to today's Russian hangover [*rossiiskoe pokhmel'e*]." Officers and nurses engaged in shameless orgies before the disgusted eyes of the enlisted men, and officers indulged in "Oriental same-sex (male) 'love,'" as Petrishchev phrased it. "This gift of the 'East,'" which not uncommonly plagued the armed forces, had now penetrated to the center of Russia, he complained.[49]

Though he refrained from polemical invective, Petrishchev too saw the sexual problem as a political one. After the revolution, he claimed, Russians had lost respect for established authority, to which they now submitted out of fear alone. Making the sexual analogy explicit, he likened his compatriots' situation to that of a wife who endures her husband's unwanted caresses because she cannot escape his power: just as such a woman drank continually to dull her senses, so Russians now rushed to *cafés-chantants*, gambling dens, and brothels to drown their moral suffering in pornography. For the wife the drug obliterated sex; for his contemporaries sex was the drug. But if sexual indulgence was an escape from the serious business of civic life, the sexual question was nevertheless a topic worthy of discussion. "The heightened demand for literature dealing with sexual questions," Petrishchev conceded, "can be reduced neither to the thirst for novelty nor to the influence of colonial wars nor even to 'a feast in times of plague'"—the Pushkin title likewise evoked by physician Dmitrii Zhbankov to characterize the excesses of postrevolutionary society. The revolution, Petrishchev observed, had changed "the conditions of life which produce morality in general and sexual morality in particular." A revolution had occurred in daily life (*bytovaia revoliutsiia*) as well as in politics. As with all revolutions, the violent clash of forces—or, in this case, the flagrant expression of sexual desire—represented the culmination of a deep-rooted process of change.[50]

The upheaval of revolution, the violent destruction of families during war and civil conflict, the desire of the educated classes to reconstruct their intimate lives—all these effects of 1905 had only deepened an existing social dilemma. "With rare exceptions," Petrishchev asserted, the problem of sex "torments our entire generation; it is the malaise of everyone accustomed to the heights of European culture." It was therefore also a legitimate public concern, and Petrishchev regretted that the theme had been cheapened by the literary avant-garde, which delighted in scandalous effect. In his opinion, the earnest consid-

---

[49]Ibid., 98–99.

[50]Ibid., 101–3, 112–13. Pavel Kokhmanskii, *Polovoi vopros: Razbor sovremennykh form polovykh otnoshenii* (Moscow, 1912), xx, also noted the analogy between the political revolution and the revolution in everyday life.

eration of sexual issues pioneered by Chernyshevskii and Tolstoy, which had shaped the ideas of his youth, had degenerated into the unpleasant provocations of such writers as Fedor Sologub, Mikhail Kuzmin, and Lidiia Zinov'eva-Annibal.[51] For all its sexual enlightenment, Petrishchev's generation had reached its cultural limit.

Erotic Individualism

If one literary work "expressed the current moment"[52] by conveying the eroticized tone of the post-1905 years, it was Mikhail Artsybashev's novel *Sanin*, the story of a young man without social ideals who spends his time in aggressive pursuit of sensual pleasure, defying respectable conventions and violating the rules of honor for both sexes. Begun in 1902 and finally published in 1907, Artsybashev's tale aroused fanatic partisanship and fervid denunciation.[53] It seemed to have captured the spirit of radical individualism which observers believed to have seized society after 1905. The novel was said to assert "the individual person's right to life and pleasure, demanding that people always be true to themselves and put no limit on their desires."[54] Referring in 1909 to Prime Minister Petr Stolypin's recent reforms, designed to encourage personal enterprise among the peasantry, the critic Vladimir Kranikhfel'd called the novel's eponymous hero "the natural and necessary product of our epoch, which has placed its wager on the 'strong' [Stolypin's motto], on 'people with bold initiative.'" Partisans of capitalism believed that Russia needed the kind of people who had built bourgeois society in the West—creative, energetic, and selfish. "'More individualism!'" Kranikhfel'd ironically declaimed. "In the wager on the strong, this is the first and necessary condition."[55]

But if in some eyes the literary figure of Sanin stood for bourgeois self-interest and naked materialism, to other readers he seemed to challenge bourgeois values by unmasking the hypocrisy of respectable society. Artsybashev's story occurs in that same provincial setting in which Petrishchev spent his youth and learned to repudiate his parents' conventional ways. It depicts a stifling social world in which young men are aimless and bored, the political cause no longer gives meaning to

[51]Petrishchev, "Iz oblasti," 123, 105, 108, 126.

[52]Vladimir Kranikhfel'd, "Literaturnye otkliki: Stavka na sil'nykh," *Sovremennyi mir*, no. 5, sec. 2 (1909): 75; Novopolin, *Pornograficheskii element*, 118.

[53]Michael Petrovich Artzibashev, *Sanine*, trans. Percy Pinkerton (New York, 1926). See V. L'vov-Rogachevskii, "M. Artsybashev," *Sovremennyi mir*, no. 11, sec. 2 (1909): 37.

[54]Koltonovskaia, "Problema pola," 125.

[55]Kranikhfel'd, "Literaturnye otkliki," 75–76.

Caricature of Mikhail Artsybashev. "His motto," an accompanying legend tells us, "is 'In corpore sano, mens sana,' but having given Sanin a 'corpus sano,' he neglected the 'mens sana,' forgetting that ancient statues were not beautiful because they were headless." *Satirikon*, no. 14 (1908). Helsinki University Slavonic Library.

life, and young girls are caught between the relaxation of patriarchal discipline and the persistence of patriarchal attitudes.

For such a crudely polemical book, its message was strangely ambiguous, however. Neither the celebration of selfhood nor the challenge to respectable norms entirely convinced the novel's readers; some, at least, noted the way Artsybashev undermined the apparently positive message of the book—that joy was supposed to be found in the embrace of pleasure, the life of spontaneous impulse and physical sensation. Sanin's final escape from the provincial town as he moves "onward to meet the rising sun"—jumping from a moving train into an open field, from civilization to nature—is rather a product of boredom and loneliness than of devil-may-care sensuality or a talent for fun. Despite its vaunted hedonism, moreover, the book imposes its own scale of sexual values. While exalting the purity of natural desire, Artsybashev condemns sexual conquest for the sake of domination, along with language and actions degrading to women. Indeed, in the view of one contemporary educator, the novel could more convincingly be read as a brief for women's sexual emancipation and equal social standing than as a vindication of the unimpeded sexual appetite of men.[56]

However unconvincing its final effect, *Sanin* was certainly intended to exalt the purity of natural impulses, untainted by social convention or ambition, as the programmatic declarations of the protagonist continually make clear. Peasants in the novel are said to countenance the sexual relations of unmarried young people as part of normal everyday life. In the soothing lull of a warm summer night, with no desire for commitment or sense of remorse, Sanin enjoys a momentary connection with another man's sexually frustrated sweetheart. Indeed, his special role in the narrative is to convince young women who have succumbed to desire that their impulses have improved rather than degraded them. Ultimately, a sense of comradeship develops between the figures who embrace their own sexual transgression. Admitting that defiance of convention is socially more damaging for women than for men, Artsybashev nevertheless contends that the characters' sensual complicity establishes an erotics of parallel status (endowed with overtly incestuous overtones in Sanin's relation to his sister), which contrasts with the conventional erotics of dominance and subordination.

So irritating was *Sanin* to contemporary critics that even its moral earnestness, in the guise of programmatic immorality, did not always qualify as a redeeming feature. The liberal critic Kornei Chukovskii,

---

[56]See A. N. Ostrogorskii, "Pedagogicheskie ekskursii v oblast' literatury ('Sanin' Artsybasheva: K voprosu o besedakh po polovomu voprosu)," *Russkaia shkola*, no. 3, sec. 1 (1908): 1–22.

writing in the Kadet newspaper *Rech'*, noted sardonically that the
work's argumentative tone not only impeded its aesthetic success but
even inhibited its possible erotic appeal:

> However much [Artsybashev] may describe female bosoms and trem-
> bling men, however wild and terrible the words he uses—none of
> this will entice or frighten the reader, because the argument always
> stands in the way. Critics are angry with the novel for its many naked
> bodies and lubricious images. Good God, if only this were so! To . . .
> describe a living, breathing body is indeed a grand and difficult ambi-
> tion for the artist. To chastise Artsybashev for having succeeded . . .
> is to pay him a flattering compliment he does not deserve.[57]

The charge of pornography missed the peculiarly Russian quality of this
sexually explicit—or at least highly suggestive—text. "Russian pornography," Chukovskii wrote, "is not plain pornography such as the
French or Germans produce, but pornography with ideas. Artsybashev
does not simply describe Sanin's lascivious actions but calls on everyone
to follow his example: 'People should enjoy love without fear or inhibi-
tion,' he says. The word 'should' is a vestige of old intelligentsia habits,
a remnant of the old moral code that is disappearing before our eyes."
The work's perverse moralism thus did not save it in Chukovskii's
eyes.[58] Tolstoy, for his part, simply ignored the book's message, dismissing the novel as an expression of "the basest animal impulses,"
lacking any human emotion or ideas. *Sanin* also offended Maxim
Gorky, as a profanation of his own realistic aesthetic.[59]

Behind these various expressions of dissatisfaction lurks the charge
that literature was being somehow misused or cheapened, sexual expression being represented in debased form. Indeed, as Neia Zorkaia, a
scholar of Russian popular culture, interestingly argues, *Sanin* was popular and irritating for the same reason: it translated some of the techniques of rarefied modernism into the accessible idiom of the boulevard; that is, of literature intended for the unsophisticated, vulgar, or
merely provincial audience.[60] Contemporaries, as we have seen, readily

---

[57]K. I. Chukovskii, "Geometricheskii roman," *Rech'*, no. 123 (May 17/June 9, 1907):
2.

[58]Quoted in N. M. Zorkaia, *Na rubezhe stoletii: U istokov massovogo iskusstva v Rossii,
1900–1910 godov* (Moscow, 1976), 147. The vulgar, sterotyped language also provoked
the irony of another liberal critic, Ariadna Tyrkova-Williams, who complained that Artsybashev had merely replaced the old fetish of sacred love with a new idolatry of the
body: A. V., "Krasnyi petukh," *Rech'*, no. 111 (May 13/26, 1907): 2.

[59]Zorkaia, *Na rubezhe*, 149; *Russkaia literatura kontsa XIX–nachala XX v.: 1901–1907*
(Moscow, 1971), 531–32.

[60]Zorkaia, *Na rubezhe*, 151, 154.

translated literary into sociological values, but their categories or claims were not always consistent. Marxists then and now equate commercial culture with what they call "petty-bourgeois philistinism"—the inability to distinguish true aesthetic value. At the same time, they castigate the avant-garde, that bastion of artistic high-mindedness, for "petty-bourgeois decadence"—selfish individualism, a narcissistic regard for aesthetic experience at the expense of social concerns.

Inconsistency is not the only reason, however, for historians to beware of these claims. There is the matter of evidence as well. How do we know who constituted the "boulevard" public, with its love of the cheap and superficial, and who made up the "popular intelligentsia," with its naive moralism and conventional tastes? The little available concrete information suggests that neorealism—both Gorky's sober brand and Artsybashev's supposedly cut-rate version—found readers among the same schoolteachers and white-collar employees who read the serious journals.[61] In fact, the present state of scholarship does not allow us to draw a sharp sociological distinction between the "boulevard," the "popular intelligentsia," and the readers of more sophisticated fare. It is perhaps better, then, to abandon these categories, which in fact serve as markers of cultural value systems rather than an objective social map, and turn to the formal distinctions that separated—or united—the overlapping domains of cultural production.[62]

Such an approach was not unknown to contemporaries. Chukovskii, for example, tried to define in formal terms what elements distinguished the boulevard from supposedly first-class texts. Zorkaia summarizes these characteristics as formulaic monotony, the use of stereotypes, and a reliance on stylistic clichés. In *Sanin*, she points out, the female characters are virtually identical; the same adjectives proliferate endlessly. If Artsybashev supposedly indulged in selfish individualism, Zorkaia comments dryly, his programmatic stance was at odds with his style, which bore all the marks of the dull uniformity imposed by the mass market.[63]

Though formal distinctions helped establish boundaries of genre and hierarchies of taste, thematic connections joined the aesthetically dispa-

---

[61]Brooks finds that a quarter of readers of journals catering to the "popular intelligentsia" were schoolteachers. The same is true for *Sovremennyi mir*, with its predominantly social-democratic orientation. Its 1910 subscriber list suggests that its readership included a lower proportion of clerks, managers, and state employees and a higher share of medical professionals than that of the less prestigious journals, but the social significance of this difference is not clear. See Brooks, "Philistinism," 94 (on *Vestnik znaniia* and *Novyi zhurnal*); and "Sostav podpishchikov," *Sovremennyi mir*, no. 12, pt. 2 (1910): 156.

[62]For the focus on cultural practices rather than sociological categories, see Roger Chartier, "Texts, Printing, Readings," in *The New Cultural History*, ed. Lynn Hunt (Berkeley, Calif., 1989), 169.

[63]Zorkaia, *Na rubezhe*, 151, 160–61, 164.

rate examples of a common sexual preoccupation. It was surely not the market that motivated the poet Mikhail Kuzmin to compose *Wings*, his archly mannered novella about male homosexual love, which first appeared in the avant-garde literary journal *Vesy* in 1906 and provoked as many objections as those that greeted *Sanin*. Kuzmin and Artsybashev belonged to different artistic circles, however: Artsybashev adhered to the conventions of neorealism; Kuzmin was at the center of literary innovation, part of the intertwined society of poets and philosophers who perpetrated the revolt against the dominant canons of socially conscious realism. Kuzmin frequented the Sunday gatherings at the home of Fedor Sologub, became a regular at Viacheslav Ivanov's Wednesdays, and eventually joined the Ivanov household as a permanent guest. He was close to leading figures in the World of Art group, many of whom were also homosexual.[64]

Published as a separate book in 1907 and reprinted in 1908, *Wings* enjoyed a significant *succès de scandale*. Though it reached a more restricted public than *Sanin*, its fame extended beyond the few readers of the modernist press.[65] Despite their stylistic differences, however, the two works reflect a common cultural moment. A homosexual companion piece to Artsybashev's novel, Kuzmin's text also promotes the cult of happy physicality. Kuzmin's account of a young man's discovery and acceptance of his desire for other men ends on a note that is echoed almost literally by the final line of *Sanin*. Having grown the "wings" necessary to rise above the banality of conventional and repressed heterosexual life, Vania "thr[ows] open the window onto a street flooded with sunlight," anticipating (in point of publication date) Sanin's ultimate leap toward the rising sun. And just as Sanin's free-love ethic is counterposed to the suffocating strictures of the Judeo-Christian tradition, the sunshine welcomed by Kuzmin's hero represents the pagan force of "Zeus-Dionysus-Helios," although it shines on pavement and not on grass. The pavement, however, is Italian: it is in Italy—land of antique glory and real sunshine—rather than in philistine Russia that the boy finds his way.[66]

The difference between the two authors in their approach to sex, however, reflects the realists' understanding of representation as mimesis and the experimentalists' defiance of verisimilitude. It is not rustic

[64]On Kuzmin, see John E. Malmstad, "Mixail Kuzmin: A Chronicle of His Life and Times," in M. A. Kuzmin, *Sobranie stikhov*, ed. Malmstad and Vladimir Markov, 3:7–319 (Munich, 1977), esp. 88–89, 93–95, 119–20. On the World of Art, see John E. Bowlt, *The Silver Age: Russian Art of the Early Twentieth Century and the "World of Art" Group* (Newtonville, Mass., 1979).
[65]Malmstad, "Mixail Kuzmin," 99.
[66]Kuzmin, *Wings*, 109.

М.КУЗМИНЪ.КРЫЛЬЯ
К-ВО СКОРПІОНЪ

Book cover by N. P. Feofilaktov for M. A. Kuzmin, *Kryl'ia* (Moscow, 1907), reproduced in N. Feofilaktov, *66 risunkov* (Moscow, 1909). Department of Graphics, M. E. Saltykov-Shchedrin State Public Library, St. Petersburg.

nature that Kuzmin celebrates but the refinements of high culture, the aesthetic tastes that liberate his company of men from everyday life, including the foolishness and sexual intrusions of women. In addition to Vania, the naive Russian youth, the novel features Larion Dmitrievich Stroop, a wealthy, cosmopolitan half-Englishman with whom Vania eventually departs for a life of travel and male companionship. Kuzmin's characters are not muscular and robust, like the virile Sanin, but grow long curly hair and admire their own pouting lips in the glass. They are not disillusioned with politics and do not bemoan the meaningless of life but devote themselves wholeheartedly to the pursuit of aesthetic pleasure. Theirs is an artificial paradise. Unlike Artsybashev, who defends the sexual impulse as part of the natural order, Kuzmin rejects the very distinction between the natural and the unnatural.

Yet in the depiction of sexual activity, neither author manages to emancipate himself from the era's conventional assumptions. Two seduction scenes in *Sanin* leave no doubt that action has occurred, but the message is conveyed in entirely hackneyed terms. *Wings* contains less overtly erotic language. At two points in the story the reader is led to understand that sexual relations are in question; both are oblique, and both involve lower-class men. In the first, Stroop is caught in the company of his valet, whose pants are suggestively unbuckled. In the second, Vania overhears a boy explaining how he earns money by engaging in "hanky panky" with gentlemen in the public bath.[67] For all its aesthetic radicalism and provocative stance, *Wings* follows the line of the reigning clichés, evoking the sexual sophistication (usually branded degeneracy) of the wealthy upper classes and the willingness of popular riffraff to service their needs. The bathhouse passage might have been lifted from Vladislav Merzheevskii's 1878 textbook on forensic medicine, which depicts the same scene and makes the same class distinctions.[68]

Artsybashev and Kuzmin were thus alike in the frankness with which they represented the transgressive passions: heterosexual intercourse outside marriage in one case, sexual relations between men in the other. They also shared a fashionable "pagan" antagonism to Christian asceticism. But there the resemblance ends. Sanin was a kind of Nietzschean superhero, standing above the common herd, spurning both emotion and intellectual ideals: a crude materialist with strong muscles and bold appetites. Though *Wings* had its moments of ironic realism (such as the bathhouse story and various domestic scenes), it trod a more elevated path strewn with classical and literary references and appeals to aesthetic sensibility.

[67]Ibid., 38–39.
[68]See chap. 4 above.

Kuzmin's high-mindedness did not, however, prevent the socially oriented cultural journals from repudiating his work as both politically retrograde and morally depraved. One critic decried the current fashion of insolent sexual posturing as a ploy to arouse "agitation and noise."[69] Another invoked the familiar term of abuse "petty-bourgeois individualism" to characterize Kuzmin's "repulsive novel." Despite the book's classical pretensions, the latter charged, it led no further than the "common Russian bath," where "'good gentlemen' . . . corrupt simple peasants . . . for a pittance." It was the ordinary bathhouse—a notorious haunt of degenerate trash—that Kuzmin had attempted to represent as "the bright kingdom of freedom."[70]

The theme of upper-class degeneracy, transmuted by Kuzmin into the romance of cultural sophistication, struck a chord among readers of a socialist cast, who saw no irony in the cliché. Primed by newspaper reports to believe that wealthy men in St. Petersburg formed a secret society that indulged in cross-dressing and other unnatural tastes, these critics denounced the novel as a reflection of the sorry state of Russian high society.[71] The homosexuality once confined to the Caucasus and to "aristocratic circles," one complained, had for the first time made a brazen appearance on the literary page, where its practitioners were depicted not as the "parasites" they truly were but as "new people, who open everyone's eyes—in Kuzmin's words—to new worlds of beauty."[72]

Kuzmin had his critics among the literary avant-garde as well. Zinaida Gippius, Dmitrii Merezhkovskii (her husband), and the poet Andrei Belyi closed ranks against Viacheslav Ivanov, with whom Kuzmin was closely associated.[73] Though Belyi praised Kuzmin's literary gifts, he felt uncomfortable with the subject of *Wings*. Reducing the work's romantic aura to the level of crude sensationalism, as though it were a boulevard novel, Belyi brusquely satirized the story's main events as a series of admittedly well-crafted but "nauseating" scenes.[74] Gippius accepted the novel's theme but objected to the "tendentiousness" with which it was presented and to the "pathological exhibitionism" it revealed.[75]

---

[69]N. A—vich, review of M. Kuzmin, "Seti," *Obrazovanie*, no. 5, sec. 3 (1908): 107.

[70]Kranikhfel'd, "Literaturnye otkliki" (1907), 133.

[71]Account from *Birzhevye vedomosti*, cited in Peshekhonov, "Na ocherednye temy," pt. 1, 128.

[72]Novopolin, *Pornograficheskii element*, 155–57.

[73]Malmstad, "Mixail Kuzmin," 123.

[74]Andrei Belyi, review of M. Kuzmin, *Kryl'ia*, *Pereval*, no. 6 (1907): 50–51.

[75]Anton Krainyi [Z. N. Gippius], "Bratskaia mogila," *Vesy*, no. 7 (1907): 60–64. The editors of *Vesy* defended *Wings* as part of the renovation of the cultural values to which Gippius pretended to aspire, not a symptom of their desecration. Gippius's intolerance may seem strange for a woman who spent considerable time among openly homosexual

If Kuzmin could not get along with Gippius and Merezhkovskii, he found domestic happiness with Viacheslav Ivanov and his wife, Lidiia Zinov'eva-Annibal, who sponsored a competing artistic salon. In the whimsical, bohemian atmosphere of a drawing room strewn with exotic fabrics and suffused with incense and candlelight, where the hostess often appeared in Greek costume and a playful irreverence reigned, Kuzmin found a welcoming home, which he eventually joined on a permanent basis. Certainly his explicit treatment of homosexual love found no harsh critics here, for it was Zinov'eva-Annibal who produced the female analogue of *Wings*, a short sketch called *Thirty-three Abominations*, which appeared in 1907 in elegant miniature format, dedicated to the author's husband.[76]

Even more mannered and self-consciously "aesthetic" than *Wings*, *Thirty-three Abominations* describes the relationship between an actress named Vera and another young woman, who informs the reader in no uncertain terms of the physical nature of their love: "She kissed my eyes and lips and breasts and caressed my body. . . . Life and death abide in the drunken juice of the rosy fruit of her fresh lips, the sacred phial of my insane love."[77] Only the archly stylized language separates this prose from the routine clichés of contemporary popular novels. Unlike them, however, this story unfolds in a complete social vacuum, on the same ethereal plane where Kuzmin's heroes enact their romantic idyll. Having interrupted her friend's wedding, Vera smothers the failed bride in cloying, jealous, and usually tearful affection, while the abandoned fiancé shoots himself (like the neglected husband in Ippolit Tarnovskii's case study). The women live together in some unnamed city, rarely leaving their room, devoted to the mutual cult of their own beauty. But beauty fades, and female society out of the public eye proves insufficient. Vera decides to let her friend pose for a studio of male artists—the "thirty-three"—so that they may immortalize her appearance. At first resistant, the friend soon welcomes the men's attention as a con-

men; who was on intimate though not sexual terms with Dmitrii Filosofov, Sergei Diaghilev's lover; and who wrote most of her poems in a male persona and was depicted by the artist Lev Bakst in male attire. Simon Karlinsky believes that she found sexual intercourse distasteful, though she described love, in some highly eroticized though unspecific form, as a divine impulse. Merezhkovskii considered androgyny in a positive light in his own writings, though he seems to have imagined it as spiritual rather than corporeal. See Simon Karlinsky, "Introduction: Who Was Zinaida Gippius?" in Vladimir Zlobin, *A Difficult Soul: Zinaida Gippius*, ed. Karlinsky (Berkeley, Calif., 1980), 5–9; and Rosenthal, *Merezhkovsky*, 82–83, 94–95, 108–10, 130.

[76]On her life and exotic habits, see Temira Pachmuss, ed., *Women Writers in Russian Modernism: An Anthology* (Urbana, Ill., 1978), 191–200.

[77]Zinov'eva-Annibal, *Tridtsat'-tri uroda*, 9, 58. I have modified slightly the translations in "Thirty-three Abominations," 94, 109.

firmation of her womanhood; she has become the object of legitimate sexual desire. When she decides to expand her newfound sense of self and personal freedom by accompanying one of the artists abroad, Vera commits suicide.

Though the style is highly decorative and the events remote from real life, the story does investigate the complicated social position of newly independent women in the urban world of social opportunity and cultural constraint. Though the young woman is said to gain independence by going out into the world, in fact the artist's model occupied a position similar to that of the prostitute, making her private attributes available to the public for a fee. Newspaper advertisements proffering the services of studio models often hinted that sexual favors might be involved as well, and a magazine featuring sexually suggestive material was called *The Model*.[78] Though the exchange of money is not mentioned in *Thirty-three Abominations*, the actress promotes her friend's public exposure as a pimp or madam would sponsor a prostitute's affairs.

It was not the troubled story of women's finding a place in the world, however, that disturbed male reviewers; it was simply the openness with which Zinov'eva-Annibal described the physicality of lesbian love. *Thirty-three Abominations* provoked an even less tolerant response from Andrei Belyi, for example, than Kuzmin's novel. "Let's be frank," Belyi wrote, after paragraphs mocking the author's mannered style:

> we are fed up with depictions of unnatural love. Not because they offend our moral sense or because they are unworthy of attention but because these complex riddles and contradictions of human existence have unfortunately been considered by people who lack intellectual integrity, mystical, psychological, or aesthetic, without [which] . . . the subject is nothing more than a passing fashion. But all fashions quickly lose their appeal. Yesterday it was individualism and mystical anarchism [the philosophy associated with Ivanov, Zinov'eva-Annibal's husband], today "Eros." What will tomorrow bring?[79]

If sexual explicitness of any kind aroused discomfort among literary critics, the subject of deviance upset them even more. One self-pro-

---

[78]*Naturshchitsa*, no. 1 (1914). The content of this magazine is lighthearted and satirical, but it was among those castigated as pornographic for their photographs of nude women (in what today would be considered relatively chaste poses) and cartoons, humorous poems, and sketches with erotic themes. See also *Flirt*, 1906–8. William Richardson, *"Zolotoe runo" and Russian Modernism, 1905–1910* (Ann Arbor, Mich., 1986) mentions *The Model* as a pornographic magazine (183).

[79]Andrei Belyi, review of L. Zinov'eva-Annibal, *Tridtsat'-tri uroda: Povest'*, *Pereval*, no. 5 (1907): 53.

claimed defender of progressive cultural values left no doubt that
Zinov'eva-Annibal's "paean to sexual depravity" indicated deep psychopathology on the author's part. Such fantasies bore no relation to
social reality, this Social Democrat declared, but testified rather to the
present "reactionary moment . . . of exhaustion and the desire to forget." Like the majority of physicians in the years before 1905, such
commentators could not imagine that lesbianism, which they associated
with the refinements of Western civilization, might be found in their
own land, which lacked the "culture of comfort" and the "epidemic
neurasthenia, with its attendant sexual perversions," characteristic of
Europe.[80]

While it is thus true that the world of Russian modernism—in both
its neorealist and symbolist camps—accommodated some exploration
of the sexual spectrum, it is clear that not even the most culturally
radical cream of the creative intelligentsia, not to speak of its culturally
stodgy left wing, embraced the representation of sexual deviance as an
artistically appropriate liberty. Those thinkers most at home with the
subject of love and sexuality belonged rather to the intellectual circles
interested in reviving a religious and spiritual consciousness.[81] Vladimir
Solov'ev, Dmitrii Merezhkovskii, and Viacheslav Ivanov, for example,
all wrote about the spiritual meaning of sex, using only the most abstract, philosophical terms.[82] (In 1912 an observer sympathetic to the
democratic left noted the "curious contradiction" that conservatives
were hailing free love, while feminists and Social Democrats were busy
defending women's honor.)[83]

The open-mindedness of the philosophers was limited by the other-
worldly quality of their ruminations. The highly illiberal Vasilii
Rozanov was the only one of their company who thought that talk of
physical sexuality should break the barrier of respectability which separated the discourse of science and the law (professional tongues enjoying a special immunity to speak the unspeakable) from that of the newspaper page and the popular novel. He wanted to write not only about
the spiritual meaning of sex (to which he attributed enormous, indeed
central, importance) but also about the meaning and texture of bodily

[80]Novopolin, *Pornograficheskii element*, 163, 165, 169–70.

[81]See Nicolas Zernov, *The Russian Religious Renaissance of the Twentieth Century* (London, 1963).

[82]Rosenthal, *Merezhkovsky*, 82–83, 106, 108; Read, *Religion and Revolution*, 29. A psychiatrist who remarked on the tendency to link religiosity and eroticism suggested that
religious passion might draw its energy from the sexual instinct: N. N. Starokotlitskii,
"K voprosu o vozdeistvii polovogo instinkta na religiiu (V sviazi s opisaniem sluchaia
religiozno-erotomanicheskogo pomeshatel'stva)," *Zhurnal nevropatologii i psikhiatrii imeni
S. S. Korsakova*, no. 2–3 (1911), 263, 299.

[83]Kokhmanskii, *Polovoi vopros*, 5.

acts themselves: "the joining of one's genital organ with the comple-mentary genital organ of a person of the opposite sex."[84] This daring was part of Rozanov's provocative indifference to aesthetic as well as social convention. While enjoying the rarefied discussions of the artistic and philosophical salon, Rozanov deliberately violated accepted bound-aries of genre and hierarchies of style in his own work. He boasted of loving Sherlock Holmes and Nat Pinkerton, those heroes of popular thrillers despised by high-minded critics. He used vulgar language and called things by their everyday names. Most of his work appeared in the form of newspaper columns, addressed to a wide though highly educated audience.

Willing to broach any sexual topic, Rozanov devoted an entire trea-tise to the question of homosexuality, a subject he approached with characteristic ambivalence but unabashed directness.[85] One may appreci-ate the distance traveled between the 1890s and the post-1905 years by contrasting Rozanov's text with Tolstoy's *Resurrection* (1899), which Rosanov invoked to offset his own intellectual audacity. The story of a common woman's sacrifice to the tyranny of male lust, *Resurrection* was Tolstoy's answer to the question of public sex, just as *The Kreutzer Sonata*, written ten years earlier, had been his answer to the dilemma of domestic desire. In the later work the novelist pursued a favorite and far from original theme, the class contours of sexual morality, contrasting the moral corruption of the ruling elite (typified by male homosexuality in high places) with the moral purity of the poor (typified by the peas-ant woman turned prostitute).[86]

Though Rozanov did not accept the social topography of perversion as a mark of class privilege (one of his most detailed case studies con-cerns a simple peasant lad who cannot resist his wayward inclinations),[87] he did share with Tolstoy a contempt for positivist theories of moral development. In *Resurrection* the novelist derided the fashionable doc-trines of medicine and the law—"heredity and congenital criminality, Lombroso and Tarde, evolution and the struggle for existence, hypno-tism and hypnotic suggestion, Charcot and decadence"[88]—for suggest-ing that prostitutes were abnormal; he considered them martyrs to up-per-class male depredation. Rozanov too denounced Western scientists for their complicity in repression, not of the powerless but of the

[84]V. V. Rozanov, *Liudi lunnogo sveta: Metafizika khristianstva* (1911), 2d ed. (St. Pe-tersburg, 1913), 29.

[85]Ibid.

[86]L. N. Tolstoy, *Resurrection*, trans. Rosemary Edmonds (New York, 1966), 357 (on homosexuality), 391 (on the depravity of upper-class women).

[87]Rozanov, *Liudi lunnogo sveta*, xxx.

[88]Tolstoy, *Resurrection*, 104.

powers of desire: for rendering as sickness and disease (Krafft-Ebing's "sexual pathology") the joyful naturalness of sex.[89] Like those scientists, Rozanov felt, Tolstoy displayed an essentially Christian distrust of the flesh, a tendency to condemn the erotic impulse.

To demonstrate the way cultural inhibitions operated to obscure recognition of sexual desire, even on the part of so refined a psychologist as Tolstoy, Rozanov chose a passage from *Resurrection* in which Tolstoy described the feelings of affection between the imprisoned prostitute and a woman revolutionary who shared her confinement. The two were drawn to each other, Tolstoy wrote, by "the loathing they both felt for sexual love. One hated it because she knew all its horrors, while the other, having never experienced it, regarded it as something incomprehensible and at the same time repugnant and offensive to human dignity."[90] The sexually innocent character—an educated woman who neglected her appearance, who was indifferent to men and devoted to political causes—in fact embodied the portrait of a "female Sodomite," Rozanov insisted, which Tolstoy had drawn "without the . . . faintest suspicion of what exactly he drew."[91]

Rozanov's desire and ability to label what Tolstoy had merely implied reflected not only the peculiarities of the philosopher's own spiritual quest and the limitations of the novelist's personal vision but also the changed social and intellectual environments in which they wrote. Rozanov himself had described such dedicated female types as repressed lesbians tormented by sexual frustration. Unable to obtain the female objects of their unacknowledged desire, they turned to intellectual pursuits and drowned their perfervid emotions in the reading of Karl Marx, he declared.[92] For both Rozanov and Tolstoy, despite their differences, the unfeminine, nonprocreative, but politically active woman represented the moral peril of public existence. They were not alone in their fascination and their fear.

Like a Man

In the years before the war, Russian readers bought thousands of copies of novels that not only questioned conventional standards of sexual morality but examined the troubled place of women in modern society. The most popular of these books were written by women. Some works belonged to the "boulevard"; others managed to win literary

[89]Rozanov, *Liudi lunnogo sveta*, 34, 41–42.
[90]Tolstoy, *Resurrection*, 473.
[91]Rozanov, *Liudi lunnogo sveta*, 106.
[92]Ibid., 53.

approval as well as popularity. They all showed an affinity with the newspaper *feuilleton* and the theatrical melodrama. Stylistically more tame than the productions of avant-garde writers, they were sometimes more daring in their approach to social issues.

The theme of sexual transgression as it was articulated in those literary works denounced as pornographic did not necessarily cast doubt on the boundaries of gender identity. While vindicating women's capacity for sexual pleasure, Artsybashev, for example, reinforced the principle of gender polarity. If he depicted the old-style predatory male as lacking in virility, he exalted the new, sexually defiant male for restoring manhood to its true heroic stature by arousing women's equally avid sexual desire. Artsybashev's female characters are no less feminine for acting on their sexual impulses, even for initiating sexual play; their bodies are portrayed as soft, yielding, and sensuous against the male's firm, muscular physique. Despite its controversial theme, Zinov'eva-Annibal's fictional world likewise leaves ideas about femininity largely intact. Although the narrator of *Thirty-three Abominations* at one point describes herself as "a child: half-boy, half-girl,"[93] the author depicts the pair not as mannish but as beautiful women attracted to each other's bodily charms. If there is anything masculine about Vera, it is her mastery of representation: her ability to appear onstage as an actress and withstand the public eye without losing her sense of self, and her power to impart identity by "seeing" another person. She has the artist's touch—the creative agency usually associated with men—and the license to display herself in public. In the end, however, the women who try to exist in a world of their own meet a conventional fate, either ceasing to be women (the model, when not visible to men) or ceasing altogether (Vera's suicide).

Zinov'eva-Annibal, in company with Artsybashev, may have anguished over the relationship between women and men, but the question had little interest for Kuzmin. His female characters are either suffocatingly maternal, sexually predatory, or frivolous. The men are either rough-and-tumble lower-class boys or delicately constructed (if not exactly effeminate) educated men; they live in a narcissistic world to which women are incidental. Nor do they thrill to the ambiguity of their sexual position. The eroticized oppositions are cultural, not sexual: middle-class vulgarity and refined classicism; things Russian and things foreign; cultivated employer and crude employee. *Wings* in fact does little to shake the established social and cultural hierarchy.

The blurring of gender boundaries was, by contrast, a major preoccupation of Gippius, Merezhkovskii, and their philosophical circle.

[93]Zinov'eva-Annibal, *Tridtsat'-tri uroda*, 48–49; "Thirty-three Abominations," 106.

Rozanov, for example, embraced the fluidity of gender traits. Interest in the notion of gender ambiguity was stimulated by the Russian translation of Otto Weininger's *Sex and Character*, which argued that all people incorporate a mixture of male and female characteristics.[94] The neorealist Anatolii Kamenskii embodied this idea in a short story called "The Woman" (1909), dedicated to Weininger's memory. The tale concerns a young man, on business in St. Petersburg, who takes it into his head to adopt female garb. Venturing out in public, he successfully fools the men on the street and attracts the attention of a young lawyer, who pursues and tries to seduce him. The man in disguise can understand his admirer's motives and strategies (he is in fact enamored of his own temporarily female self) and at the same time experience the lure of sexual enticement he imagines a woman would feel. In the end, pleading mysterious circumstances, he avoids further entanglement and retreats into the anonymity of the city, which has made both the experiment and the flirtation possible.[95]

The story's success in engaging the reader's belief testified to the new possibilities that city life in fact offered to educated Russians. The same anonymous spaces that allowed the young man in his normal guise to pursue women also allowed women to explore the public world.[96] The fictional man-woman maintains her respectability even though she is alone in public and apparently unattached. If she manages to seem genuinely female despite her frankness in conversation and remarkable cultivation, it is because—as she explains to the lawyer, who remarks on the breadth of her knowledge and the modernity of her point of view—she has attended the advanced courses for women. Kamenskii's fictional hero thus can counterfeit femininity because the physical and intellectual latitude allowed modern young women permit his masculine attributes to enhance, not contradict, his artificial persona's claims: she can be outspoken, unconventional, and yet respectable. She can discuss music and literature and also sigh languidly at the mention of love.

The male body underneath the fashionable female attire represented the mixed message of the "new woman." This type, as Aleksandra Kollontai described her in 1913, was independent and single; she earned her own living in a variety of ways, "walking the streets with a busi-

---

[94]See Olga Matich, "Androgyny and the Russian Religious Renaissance," in *Western Philosophical Systems in Russian Literature: A Collection of Critical Studies*, ed. Anthony M. Mlikotin (Los Angeles, 1979), 165–75.

[95]Anatolii Kamenskii, *Zhenshchina: Rasskaz pamiati Otto Veiningera* (St. Petersburg, 1909).

[96]Judith R. Walkowitz brilliantly explores the theme of public space and sexual expression in *City of Dreadful Delights: Narratives of Sexual Danger in Late Victorian London* (Chicago, 1992).

nesslike, masculine tread" in search of work. As she walked the pavement with manly enterprise, her services for sale, the new woman's
situation resembled that of the prostitute. But it was precisely in redefining the public woman that the crux of her identity lay. One way in
which women were making a place for themselves in the public world
was by taking control of the instruments of cultural expression. Zinaida
Gippius and Anastasiia Verbitskaia shaped their own lives on this
model. Zinov'eva-Annibal formed her character Vera in this mold. The
central figure in Evdokiia Nagrodskaia's popular novel of 1910, *The
Wrath of Dionysus*, provides another example. Mania, the heroine of
Verbitskaia's *Keys to Happiness*, was perhaps the most notorious of the
fictional embodiments of the woman artist as gender rebel. In Kollontai's opinion, such literary representations showed that women were no
longer "the *object* of tragedies centered on men but had become the
subjects of their own, independent tragedies."[97]

In novels that violated the boundaries of literary propriety, flirting
with high seriousness while playing to a commercial audience, artistic
performance offered both the woman writer and her heroine a way to
go public respectably. Evdokiia Nagrodskaia came from a family of
active cultural figures: her mother had been a famous salon hostess, her
father a journalist. She herself appeared on stage in the 1880s, before
marrying a Transportation Ministry official. *The Wrath of Dionysus*, her
first novel, appeared in ten editions between 1910 and 1916. Literary
critics as different as the émigré Gleb Struve and the Soviet scholar Neia
Zorkaia have considered it to be a typical product of "the boulevard."[98]
Contemporary commentators, however, recognized a talent that raised
her above the average literary hack. Mikhail Kuzmin, a friend, praised
the daring and delicacy with which she handled the novel's controversial subject.[99]

It was the same subject that he had tackled. In comparison with either *Wings* or *Abominations*, however, *Dionysus* is subdued in style
and sexual theme. It is the story of Tania, a painter, who finds herself
caught between two men: Il'ia, an emotionally stolid character for
whom she feels tenderness but little passion; and Eduard Stark, half

[97]A. M. Kollontai, "Novaia zhenshchina," *Sovremennyi mir*, no. 9 (1913): 153–54, 185
(original emphasis).

[98]Nagrodskaia, *Gnev Dionisa*. See Margaret Dalton, "A Russian Best-Seller of the
Early Twentieth Century: Evdokiya Apollonovna Nagrodskaya's *The Wrath of Dionysus*,"
in *Studies in Honor of Vsevolod Setchkarev*, ed. Julian Connolly and Sonia Ketchian (Columbus, O., 1987), 102–3; Zorkaia, *Na rubezhe*, 159. On editions and print runs, see Brooks,
When Russia Learned to Read, 160.

[99]M. Kuzmin, "Zametki o russkoi belletristike," *Apollon*, no. 9, pt. 2 (1910): 33–35; B.
G., review of E. Nagrodskaia, *Gnev Dionisa*, *Istoricheskii vestnik* 125 (1911): 1160–63.
Dalton, "Russian Best-Seller," 104–5, cites a positive notice in *Sovremennyi mir*.

English (like Kuzmin's Stroop) and half Russian Jew (compounding foreignness with the heightened sensuality of the anti-Semitic stereotype), to whom she is drawn with burning sensual desire. Stark is a name, thinks Tania (no doubt expressing the author's self-irony), "right out of a boulevard novel." But the attraction overcomes her cultural disdain. In Rome, where she is busy in her studio finishing a canvas she calls *The Wrath of Dionysus*, she has an affair with Stark, who poses for the painting's main figure. Here is the inversion of the situation in *Abominations*, where the woman is invested with femininity when she sits as the object of men's regard; in *Dionysus* the male as the object of a woman's regard is unmanned. It is position, not genitalia, that marks gender. With Tania and Stark, the gender roles are reversed: she is resilient, clearheaded, and intent on her career; he is delicate, vain, emotional, even hysterical. Though many people are cross-gendered in this way, Nagrodskaia explains, few enact or even acknowledge their homosexual inclinations. Most, like Tania and Stark, manage to find opposite numbers with whom to form inverted but heterosexual pairs.

Fortunate as Tania may be in stumbling on this arrangement, which spares her the agony of a stigmatized sexual identity, she is finally defeated by the problem of being a woman who does not wish to act like one but respects conventional categories. At first she resists Stark's possessive claims as a threat to her personal and creative autonomy. Returning to Russia, she decides to marry the placid Il'ia, whose gentle affections do not interfere with her work. Stark insists, however, on keeping the child she has borne him, and Tania is obliged to shuttle between St. Petersburg and Paris. When Il'ia eventually dies, Tania remains in Paris; finished with her career as an artist, she marries Stark and devotes herself to her son. However manly she may have been, Tania is still a woman, and therefore cannot happily combine work and passion, the public and the intimate. The retreat to domestic love, when she becomes one man's wife and another man's mother, takes her out of the studio and out of the public eye. This conclusion, so at odds with the work's thematic boldness, displeased Aleksandra Kollontai, who thought it conflicted with Tania's portrayal as a "new woman."[100]

Like the artist's studio and the stage, the boulevard and the salon both permitted women to engage in public self-display and self-assertion. Literary depictions of these social alternatives to conventional family life provided ordinary women with the stuff of dreams. Anastasiia Verbitskaia built a successful literary career on the mixing of fantasy and social realism. By 1913, when the sixth and final part of *The Keys to Happiness* appeared in print, Verbitskaia had published at

[100]Kollontai, "Novaia zhenshchina," 163.

Anastasiia Verbitskaia, in a portrait reproduced in her autobiography, *Moemu chitateliu* (Moscow, 1908). Helsinki University Slavonic Library.

least twenty other works of fiction or drama (many in multiple editions and large print runs), including her novelization of the 1905 revolution, *Spirit of the Times*.[101] She had also sponsored the translation of some twenty texts by foreign authors, with such titles as *The Woman Who Dared, The Women Students, Free Love, The Secretary, High School Girls, The Third Sex, The Right to Love, The Woman Who Loved Nietzsche, A Fashionable Marriage, Socialism Without Politics,* and *The Woman Artist.* She also published her memoirs.[102]

The autobiography, called *To My Reader,* constructed a real life in the form of continuous performance, which by its example summoned young people to search for individual self-fulfillment in artistic careers. It was dedicated to those "who see in the affirmation and growth of the self [*lichnost'*] the dawn of a distant, new, and brilliant life."[103] Like

---

[101]A. A. Verbitskaia, *Dukh vremeni* (Moscow, 1907).

[102]The list of Verbitskaia's publications appears at the back of her novel *Igo liubvi* (Moscow, 1914), complete with print runs, editions, and prices. See also Brooks, *When Russia Learned to Read,* 153–60; Temira Pachmuss, "Women Writers in Russian Decadence," *Journal of Contemporary History* 17 (1982), 125–26; *Russkaia literatura,* 442, 477.

[103]A. A. Verbitskaia, *Moemu chitateliu: Avtobiograficheskie ocherki s dvumia portretami* (Moscow, 1908), 3–4.

Nagrodskaia, Verbitskaia came from a theatrical family: both her grandmother and mother, she claimed, had acted on the stage, though in the humble provincial environment of Voronezh, not the high society in which Nagrodskaia's parents moved. Verbitskaia presented her genealogy as one of female cultural rebellion: the grandmother shocked the community by playing before lower-class audiences; the mother held enlightened sexual ideas and in her declining years still thrilled to the occasion of the 1905 general strike and the October Manifesto.[104] In Verbitskaia's tale, both autobiographical and fictional, truth, beauty, and female independence stood on the side of social justice and public commitment.

It was a tale with a popular following. At least one critic feared that Verbitskaia might surpass Tolstoy in the affections of common readers. Who such readers were, the critic did not know, but they could not be confined, he admitted, to those who savored "the gutter press" (*ulich-naia gazeta*), since Verbitskaia prided herself on the distinguished publications in which her work appeared—the premier journals of the progressive intelligentsia. Though denigrated as a boulevard writer, she considered herself above the "idle and uncultured crowd"—the poet, rather, of the new people.[105] This pretension was received with scorn by the liberal critic Kornei Chukovskii; he compared her with the anonymous authors of such potboilers as *The Secret Knife* and *The Cashier's Three Mistresses*, who were honest enough to admit their commercial aims. Mme Verbitskaia, Chukovskii noted with indignation, had the temerity to join "her boulevard name with the names of Mikhailovskii, . . . Gorky, Andreev, Sologub, Kuprin, and Valerii Briusov."[106]

The claim to aesthetic legitimacy on the part of writers with a broad popular audience seemed to have the same irritating effect on critics as the intrusion of explicit sexual themes into works of literary quality or purported moral seriousness. Was it perhaps the stimulation of desire—in one case the representation of physical satisfaction, in the other the proferring of seductive literary goods—that made the cultural establishment nervous? Certainly the heat with which Chukovskii responded to Verbitskaia's long-winded romance needs explaining. From the elevated vantage point of serious literature, he denounced her products as vulgar trash, her readers as shabby and disreputable. "Boulevard literature," he sneered, "has its traditions and unwritten aesthetics. What kind of novel would it be in which nostrils did not tremble!" Eyes were

[104]Ibid., 23–35, 56.

[105]Kranikhfel'd, "Literaturnye otkliki: O novykh liudiakh," 68–71.

[106]Chukovskii, *Kniga*, 11–14. This review, "Intelligentnyi Pinkerton," first appeared in *Rech'*, no. 51 (1910). Brooks (*When Russia Learned to Read*, 154) writes: "What distinguished Verbitskaia's work from other types of Russian popular fiction was her presumption that her writing was the equal of serious literature."

always glittering, faces were constantly ablaze. "These were not people," Chukovskii chortled, "but rockets." For all his contempt, he admitted ruefully that Verbitskaia's productions sold well—over half a million copies in five years—and public libraries reported that demand for her work outstripped the call for the classics. To Chukovskii's intense annoyance, Verbitskaia viewed the extent of her popularity as a measure of her talent, "because 'figures speak for themselves,' as she says in her shopkeeper jargon."[107]

Who in fact were her readers? Chukovskii could only imagine such "cultural barbarians" as hairdressers and shopkeepers. "We have prostitutes and lackeys, of course, in sufficient number," he added, invoking the most servile inhabitants of the street and turning the boulevard metaphor into a physical location, "but do prostitutes and lackeys read that much nowadays?" He scorned Verbitskaia's claim that her public consisted of "the democratic element": students, workers, artisans, seamstresses, and shop clerks. He doubted that her vulgar nonsense could appeal to Russia's high-minded youth, though he was ready to credit the workers' susceptibility to a "combination of Rocambole and Darwin, Pinkerton and Marx." Despite the left's fervid denunciation of "rotten petty-bourgeois civilization," in the absence of a truly "proletarian art" workers simply "gobbled up" Nat Pinkerton, before devouring Verbitskaia's fare.[108]

Most other critics, regardless of political hue, equally despised the boulevard, but a few recognized the value of Verbitskaia's books. Not all found her democratic sympathies laughable. One such champion accused Chukovskii and his liberal friends of defending "the class privileges of the rich" in attacking her work for the illusory pleasures and secondhand culture it provided her humble readers: "To find access to the ideas and feelings of the simple person; to give him [or more likely her] a few minutes of pure aesthetic pleasure after the day's work—this means, according to the bourgeois critic, to trade in false diamonds." A liberal reviewer "with the aristocrat's contempt for everything democratic and accessible," this advocate wrote, had called Verbitskaia's novel "a real department store" where the reader found cut-rate versions of "the social question, the theory of heredity, Darwin, Nietzsche, Oscar Wilde, and the history of art." But the department store and the popular novel should be seen as opportunities, not travesties: the best the working person could afford and certainly better than nothing.[109]

Critical disapproval certainly seemed to have no effect on the popu-

---

[107]Chukovskii, *Kniga*, 7–8, 14–15.
[108]Ibid., 11, 15–20.
[109]Dadonov, *Verbitskaia*, 71, 75–78 (a hostile review in the liberal newspaper *Russkie vedomosti*).

larity of Verbitskaia's work. "Let Chukovskii make fun of Mme Verbitskaia," wrote Mariia Pokrovskaia in her feminist journal, *Zhenskii vestnik*, "but he has no influence with the public, which continues to read *The Keys to Happiness*."[110] The class dimension, Pokrovskaia pointed out, was but one element in the book's appeal; the women's issue was in fact more central, though Verbitskaia's critics had paid it scant mind. This was not only a story that allowed shopgirls to imagine the pleasures enjoyed by their social superiors; it was also a tale that allowed women to imagine themselves with the sexual prerogatives of men. The fictional Mania took more than one lover and refused to subordinate her autonomy to a single partner, insisting that artistic expression was central to her life. She therefore could be thought to have acted "like a man," as Pokrovskaia phrased it in the title of her review, picking up the novel's explicit theme.

Verbitskaia considered herself a feminist and evoked a feminist response. Men who criticized the novel for immorality were guilty of the greatest hypocrisy, Pokrovskaia felt. After all, Mania made love only to men she knew and cared for, whereas the man in the street bought the favors of numberless unknown prostitutes whom he thoroughly despised. Young people read Verbitskaia not for her political radicalism, in Pokrovskaia's opinion, but in a search for moral guidance: "Today's youth is not satisfied with the morality of their fathers. . . . They reject the double standard as unfair but cannot actualize their ideal. In this situation, they can only allow women the same freedom in love that men enjoy." Critics feared the novel would damage women's morals, but nothing, in Pokrovskaia's view, could match the corrupting effect of organized prostitution. "The sexual frivolity of women," she concluded, "will never reach the heights attained by the sexual depravity of men. In this respect, women will never be 'like men.'"[111]

The Keys to Happiness

Verbitskaia's novel, in six volumes totaling almost 1,400 pages, was indeed a department store of current ideas and preoccupations, and for that reason it is a gold mine for the historian.[112] It tells the story of Mania El'tsova and her road from adolescent hysteria to suicide seven years later, at the age of twenty-two. Born in 1890, Mania is a true

---

[110]M. I. Pokrovskaia, "'Kak muzhchina,'" *Zhenskii vestnik*, no. 12 (1910): 266–68.
[111]Ibid., 268.
[112]See Zorkaia, *Na rubezhe*, 165–79. Volume and page numbers of Verbitskaia, *Kliuchi schast'ia*, are indicated in the text. Because she frequently uses points of ellipsis, I have enclosed my own in brackets.

child of the fin de siècle. In her brief lifetime she achieves fame and
wealth through artistic talent, but the desire for autonomy and professional success struggles with the search for love and sexual expression.
Her prospects are overshadowed from the start by the specter of biological doom in the person of her once glamorous and sexually adven-
turous mother (Mania is the product of an illicit affair), now confined
to her room in a state of wretched psychosis. In boarding school Mania
shows a natural gift for the dance and a propensity for hysterical outbursts but rejects the sexual advances of a classmate (the obligatory lesbian boarding-school scene).

Her character thus defined as passionate but not morbidly sexual,
Mania embarks on her socially significant and heavily symbolic adven-
tures. In Moscow for the winter of 1905, she and her school friend
Sonia help build barricades during the armed uprising. On a visit to the
provincial estate of Sonia's family the following summer, Mania finds
the conservative landowners traumatized by the recent peasant unrest.
There she also encounters a group of revolutionaries (many of them
Jews) being sheltered by the local sugar magnate, one Baron Mark Steinbach, also a Jew, who is sympathetic to the revolutionary movement.[113]
Among the furtive radicals Mania meets one with whom she discusses
the sexual morality of *Sanin*. He explains to her that women should
give themselves to physical pleasure without becoming the slaves of
love, that personal autonomy and sexual satisfaction without emotional
involvement constitute "the keys to happiness." Having imparted his
wisdom, the sage is removed from the scene by an accidental but heroic
death.

Two years later, his graveside serves as the place where Mania meets
Baron Steinbach, who falls madly in love with the passionate young
woman. Mania is both attracted and repelled by his Jewish appearance;
his "rapacious" smile reminds her of a bird of prey. But his background
gives them something in common—morbid heredity. "I am the scion
of a degenerate family," he announces. "The Jews are a dying race"
(1:153–55). Sensuality, he says in self-deprecation, is his only talent: a
typically Jewish trait. Since he is married (to a woman he claims not to
love), he cannot make her his wife, but Mania is not interested in respectability or commitment. "I am a horrible egotist," she tells the
smitten man. "I hate obligations and cannot stand compulsion" (1:169).
Soon to depart for Europe, Steinbach takes Mania to his house, where,
amid luxurious surroundings, she dances to his accompaniment on a
mournful Jewish instrument. "Better than the famous Duncan," the

[113]The Jewish sugar magnate was a stock figure in the anti-Semitic imagination; see
"Zemskaia zhizn': V iudeiskoi kabale," *Zemshchina*, no. 1347 (June 6, 1913): 4.

Baron reflects (1:183). Wrapped in nothing but an exotic shawl, she feels "like a goddess" (1:185). Then "she felt his kisses on her feet. She saw his new, terrible face. [. . .] Everything pressed tighter and tighter, taking her breath away. . . . And then she fell into the abyss that stared at her from Mark's eyes. . . . The secret of life was suddenly revealed to her" (1:186). End of volume 1.

Volume 2 opens with the appearance of Steinbach's rival, the landowner Nikolai Nelidov, who has returned to rebuild his estate, destroyed by the peasants during 1905. Indeed, his finances are in such bad shape that Steinbach now owns part of the Nelidov lands (and holds the mortgage on Sonia's family property as well). An expert horseman, Nelidov is arrogant, right-wing, anti-Semitic, and old-fashioned about women. His favorite newspaper is *Novoe vremia*. He particularly detests England, with its "workers' strikes, parliamentary conflicts, thousands of meetings, feminist congresses, and the Salvation Army" (2:32). The minute he sees Mania, he is struck by her intense expression, her greedy eyes. "The dark whirlwind arose from the bottom of his soul. A secret and terrible whirlwind that took his breath away, quickened his heartbeat, and inflamed his blood with unbearable desire" (2:6). Mania, also enthralled, expresses her erotic feeling in spontaneous dance: "With a cry of unconscious pleasure, she spins about in a kind of bacchanalian intoxication [. . .] ecstasy in her eyes" (2:21). But when Nelidov hears that Mania has been involved with Steinbach, he lashes out at the "Yid," who enslaves those he helps and impoverishes those he exploits: "His millions are the sweat and blood of our people" (2:30).

If *Keys* provided vicarious culture and political caricatures, it also offered vicarious sex—more explicit and even more crudely depicted than in *Sanin*. In fact, Mania's first sexual encounter with Nelidov comes close to rape. In his carriage, Nelidov "kissed her silently, greedily, rapaciously, like a wild beast. Rudely and painfully, somehow primitively he caressed her shoulders, breast, knees, with one sweep of his blind and powerful desire destroying everything that had separated them yesterday and still now." Mania "is shaken and depressed. [. . .] She had not expected this. She had not wanted this. [. . .] He has seized her like his prey, taken her in his arms. [. . .] Like a slave she submits to his will" (2:36).

Thus Mania flirts with the idea of conventional submission in relation to Nelidov, who embodies the traditional patriarchal order with its male sexual prerogative and contradictory moral code. Nelidov expects a wife to be sexually pure but, having seduced Mania, feels honor-bound to marry her—though he knows, having read Tolstoy, that marriage based on passion will not work. He would rather find an uneducated, physically healthy woman who would bear him healthy children. Although Mania has been unfaithful, Steinbach pledges to love her no matter what she does. While she and Nelidov struggle for domination, he will be her faithful "wife": the Jew, ever serviceable and self-abnegating, but devious and ultimately in control.

While Mania wavers between the two men, family and friends debate the question of degeneracy, providing readers with an instant summary of contemporary issues and perhaps reminding them of Chekhov's "Duel" or Tolstoy's *Resurrection*, in which such theories also appeared in the mouths of fictional characters. Nelidov says socialism will promote decadence by protecting the weak; only the ruthless struggle for existence will eliminate imperfect specimens and ensure a healthy race. Steinbach supports socialism and defends the so-called degenerate as those who produce art and beauty. "Normality," he says, is to be found only among the hopelessly petty bourgeois (2:123). Clearly on the side of art and beauty, Verbitskaia thus also lines up behind socialism and the Jews, against reactionary aristocrats and scientific theories of biological inferiority.

Mania's unconventional behavior in the face of narrow-minded ideas dooms her to a life of social marginality, in which she is sustained by the figure who has turned marginality to advantage: the wealthy Jew. Steinbach does not desert her even when he learns that she is bearing Nelidov's child: Mania is the modern woman with the courage to "love like a man," he says. "You have reached the highest wisdom," he tells her, "which consists in being yourself" (2:154). The "new woman" does not love individual men, does not become the slave of passion, but loves only love itself (2:159).

Steinbach has discovered Mania's dark family secret of morbid heredity but pledges to keep it from Nelidov, in the spirit of fair competititition. "How self-assured you are," exclaims Mania, "just like a Jew!" (2:163). Nelidov, however, learns the secret on his own but nevertheless finally proposes that Mania become his wife—on the condition that they abstain from sex and that she obey his mother and provide for his daily happiness. She angrily rejects him. When she proudly admits to having slept with Steinbach because she loves him too, Nelidov abandons his ambivalent suit. As a result, Mania tries to kill herself by eating a jar of poisoned jam. After a heart-stopping night-long vigil, during which Steinbach kneels at her bedside, she finally pulls through. End of volume 2.

Volume 3 is the story of Mania's sojourn in Europe, the kind of imagined voyage that supposedly captivated provincial seamstresses.[114]

[114]Dadonov, *Verbitskaia*, 75.

Mania travels in the company of Steinbach and a German nanny who performs all domestic and maternal services for Mania's child (born without evident hereditary taint). Endless excursions through museums, antique ruins, and moldy palaces, as Steinbach instructs and Mania gushes about European civilization, provoked the critics' indignation. In comparison with the output of the neorealists or even of Nagrodskaia's defter hand, not to speak of the contemporary prose masters, these passages do seem trite and repetitious. To the modern reader's relief, however, the endless pages are frequently interrupted by news from home. Nelidov has returned to his estate, where in fine aristocratic fashion he has raped a maidservant and made her his mistress. Eventually he marries her off to a local peasant and himself weds a woman of good family, trying to drown his longing for Mania in the healthy reproduction of the family line. Meanwhile, Steinbach takes Mania to study theater in Paris, so she can climb "the high tower" where the keys of happiness await her (3:193). Convinced that she has lost Nelidov forever, Mania has also ceased to love Steinbach but accepts his friendship and support.

Despite being a Jew, Steinbach has achieved social respectability, and he can now transmit it to her through art. Though the most prominent patrons of the arts in this period were indeed wealthy merchants and industrialists, they were not Jews. A railroad and steel magnate, Savva Mamontov, financed the World of Art and Sergei Diaghilev's ballet company; Nikolai Riabushinskii, of the great Moscow banking family, underwrote the modernist journal *Zolotoe runo*. The fictional Steinbach is thus associated with the noble civic undertakings of actual men. Like them, he devotes his wealth to selfless causes, contributing to Russia's cultural glory; like them, he translates the dubious coin of material wealth into the pure currency of high culture.

In creating Steinbach's role, however, Verbitskaia manipulates the anti-Semitic conventions she explicitly deplores. The figure of the Jewish baron is replete with the classic contradictions. On the one hand, Steinbach owes his wealth to the exploitation both of Russia's workers and peasants (he owns a factory and an estate) and of the economic weakness of Russia's natural ruling class, the impoverished nobility. On the other hand, he is a socialist, busy financing the subversive activities of fanatic revolutionaries determined to undermine the old order, and unlike the rapacious Jew of anti-Semitic myth, he is self-sacrificing and disinterested to an extreme degree. Yet Verbitskaia's obviously enlightened construction of the heroic and culturally refined Jew is undermined by countervailing associations. Working behind the scenes, providing money, connections, and patient support, Steinbach becomes Mania's artistic pimp. He sponsors her career as a public woman, encouraging her to appear virtually naked on the stage and offer sensual pleasure to the mass of anonymous men who pay to witness her performance.

When the baron puts himself at the service of art and at the disposition of his domineering quasi-mistress (she uses and abuses him without the slightest remorse), he demonstrates how the Jews disguise their power as subservience and undermine the natural relation of the sexes, in which the male dominates both actually and symbolically. Effeminate and sensual (like the half-Jew Eduard Stark in *The Wrath of Dionysus*), he is as submissive and manipulative as a woman, encouraging Mania to overcome her female instincts in the struggle for public glory. Where the virile man asserts his mastery, Steinbach feels impotent before the force of female desire: "As always, he feels himself weak-willed and impassive under the hypnosis of the other's stubborn will. Under the pressure of the other's passionate desire, he loses himself and gives in" (5:121–22). Renouncing his own sexual claims, he tolerates Mania's romantic and sexual involvements with other men, so long as they do not interfere with her work. He is both self-effacing and all-controlling, and without him she would not in fact have risen to the top. He is the slave to whom she owes everything. "While Mania studies and daydreams," Verbitskaia tells us, "Steinbach works for her future with the determination and agility typical of his race" (4:63). Steinbach is a partisan of the "new woman," who sacrifices private pleasures for public performance, overcoming both emotion and desire—the most dangerous obstacles to women's independence. "Passion," wrote Aleksandra Kollontai, "holds women in the most terrible captivity."[115]

Mania's rise to celebrity and her struggle against the "feminine" impulse to subordinate ambition to love are the subjects of the novel's fourth volume. "You men have taken the keys to happiness," Mania says. "Long ago you easily solved the problem we women are still struggling with. But we will finally obtain them! [. . .] Putting love in second place—that is the key to happiness" (4:18–19). Mania finds personal freedom only in the dance studio, where she manifests her "natural" talent for physical expression. In fact, the sensuality of dance replaces the need for sexual fulfillment: "The feverish work and intense creativity seemed to have killed the woman in her. Her desires were dormant" (6:102). In renunciation, she fulfills Steinbach's dream: "The woman dominated by her feelings; the girl created for love—tender and dreamy, jealous and passionate," the Baron tells her, "is the person I decided to liberate from the yoke of love. [. . . from] this awful instinct pushing you women to slavery and self-oblivion" (4:166–67).

---

[115]Kollontai, "Novaia zhenshchina," 163.

In setting the scene for Mania's struggles and eventual triumph, Verbitskaia offers her readers a short course in current artistic fashion. Isadora Duncan is all the rage: "Duncan's appearance on the stage has spread psychological contagion among young girls everywhere. [. . .] They all want to go barefoot" (4:66). Like Duncan, Mania becomes an international ballet star, the "Russian barefoot Marion" draped in elegant clothing, crushed by eager admirers, dancing in the latest "modernist" works. In a piece called *Bacchanales* she "stands half-naked, like a Bacchante, with a tiger skin on her shoulders, adorned by a cluster of grapes" (4:132–33). On another occasion "she wears a light, semitransparent tunic. Her hair is gathered in a Greek knot. Her feet are bare" (5:48). She dances "slowly, with closed eyes, moving like a lunatic across the stage" (4:133). "Une vraie artiste," a critic mumbles (4:136).

Back home in the naive Russian provinces, Mania's friends follow her progress from afar. Examining her photographs in the illustrated magazines, they are shocked by her daring attire, but the most sophisticated of them explains that she is wearing "an antique costume, like Duncan" (4:73). The men are titillated. While pornography for the rural petty nobility clearly came in the form of foreign magazines, Russian newspapers reported on the foreign triumphs of Russian art. *Novoe vremia* includes a *feuilleton* about the Russian ballet in Paris and London, from which Mania's friends learn of her renown. In Russia as in Europe, artistic success depends on the resources of commercial culture, on publicity and the press: "Advertisement is everything!" Mania's dance teacher explains (4:65). Though Mania's talent may be natural, her recognition depends on Steinbach's efforts to mobilize his journalist friends and on his ability to underwrite her experimental art. Thanks to his exertions and to her incandescent personality, she has achieved success in the brutal world of artistic and commercial competition.

But Mania has a Russian conscience: the social question and the woman question torment her. The last two volumes recount her struggle to renounce the glamour and luxury associated with artistic success and her final defeat at the hands of love. The sophisticated productions in which Mania takes part appeal to a privileged and narcissistic public that revels in sensuality and opulence: "Everywhere [in the audience] one saw bare shoulders, bright toilettes, sable and ermine, fresh flowers, artificial pearls, and real diamonds" (5:32). Onstage Mania too shows off her person, appearing "in a blue-gray semitransparent tunic, barefoot, with bare dark-skinned arms" (5:29). Hoping to reconcile the pleasures of being seen with the pricks of social conscience, and having discovered that the common folk are deprived in cultural as well as material terms, Mania develops an interest in the idea of workers' theater. She is convinced that high culture is not inaccessible to the masses. Despite its appeal to the wealthy and idle, Verbitskaia insists, modernism is radical because it violates respectable taboos, bringing cheers from penniless fans in the balcony and boos from the orchestra seats. It is the voice of the young and impoverished—a daring, not degenerate, voice.

In this iconoclastic spirit Steinbach has founded an experimental theater in St. Petersburg, which offers a verse play by a popular writer known as Harold. As the impersonation of modernism, Harold has sacrificed politics and even true love on the altar of "aesthetics." The passionate Mania, who supposedly cares deeply about the social cause, sets out to conquer him. In their duel for survival and domination, the conflict over values and the battle between the sexes overlap. Mania plays the aggressive role: "I love you," she tells him, "the way a man loves a girl who is cold, inaccessible, and ready to give herself to another man. I love you cruelly, avidly, egotistically." "That's not love," he objects. "You are right," she agrees. "It is passion" (5:126). "You don't have a female soul," he objects (5:103), afraid of her elemental power, of the force of desire that threatens his autonomy. And indeed "his personality was dissolving; he was losing his *ego* [*ia*] under the hypnosis of another's will" (5:151). No less susceptible than Steinbach, Harold too, it turns out, is a Jew. His real name, Mania discovers to her disgust and disappointment, is Borukh Isaakovich Mendel'. She prefers "Boris," but Harold on this issue stands firm: "'Yes, I'm a Yid,' he says coldly. 'I have never pretended to be Russian. I chose a pseudonym because it sounds good and protects me from the curiosity of the crowd. But in private life I am no imposter" (5:153). Though self-absorbed and indifferent to social questions, Harold is not a moral coward.

He is no match, however, for Mania's appetite. Having tracked him to his room with unladylike boldness, she gets what she wants. "What they experienced that night," Verbitskaia tells her readers, "was so beautiful, so full and lofty, that words seem paltry and superfluous. For the first time, their souls as well as their bodies came together. And that moment was as ominous and sacred as lightning flashing above the earth. It was the moment when people become gods" (5:165). Having imposed her will and satisfied her desire, however, Mania abandons Harold. Steinbach is not surprised: "As you are now," he tells her, "you don't need a master, you need a slave. [. . .] You don't have a female soul, Mania. You were not created for the yoke like the average woman. You need power and freedom. You demand adoration and submission" (5:175). Unlike Harold, who struggles against Mania's domination, Steinbach is content to serve: he offers to be "friend, brother, [. . .] impresario, or majordomo" (5:178). "In our love," he tells Mania, "you are the man, I am the woman. You command, I obey" (5:179–80).

The reader is by now eager for a resolution to the struggle between Mania's competing appetites for private (sexual) and public (artistic) gratification. Nevertheless, her loyal public, no doubt torn between reluctance to see her renounce her "femininity" and reluctance to see her destroyed altogether, amply forewarned of Mania's impending doom and vaguely aware that morbid heredity has been stalking both central characters since the story's first page, was obliged to endure a number of edifying digressions and subplots (Mania's flirtation with her male costar, Steinbach's attachment to a lovelorn girl who eventually dies alone in a Dostoevskian garret, the romantic dramas of various revolutionary activists). It is only in volume 6 that recurrent allusions to the political euphoria of 1905, intermittent dialogues between political radicals and aesthetic rebels, and an interwoven saga of terrorist intrigue come together with Mania's professional crisis.

The novel's action, which begins on the eve of 1905, when Mania is almost fifteen, has occupied seven years. By 1912 or 1913, when the fictional narrative coincides with calendar time, Verbitskaia was suggesting that the intelligentsia's postrevolutionary disillusionment had been superseded by a renewed sense of political hope and a new commitment to social causes. Back on his estate Steinbach has established a people's theater where local workers are studying art. Mania decides she is fed up with her elite public. "I am surrounded by courtesans, [. . .] snobs, gamblers, cardsharps, and thieves," she writes her friend Sonia. "Also aristocrats and bankers surrounded by wives, mistresses, and daughters; and Americans, whose vulgar philistinism [meshchanstvo] and naked souls I cannot bear" (6:172). She will "renounce wealth, fame, and life's blessings [. . .] to follow the Dream" (6:174). But then Mania changes her mind: "Without art and creativity I cannot live and do not want to." She will become a "people's artist [narodnaia artistka]" (6:201–2).

Her fame secure, Mania finally marries Steinbach, whose wife has died, and they return to the baron's country estate, where her revealing dress—the latest Paris fashion—shocks her friends: "All the lines and shapes of her body were visible through the thin fabric, as if she were undressed" (6:214). Nelidov, who has meanwhile fathered a son and barely escaped death from a terrorist's bullet, has never ceased to long for his early love, feeding his desire with newspaper clippings that display the naked body he never saw in all their passionate lovemaking. Now, no sooner do the two erstwhile lovers cross paths than they are overcome once again by desire: "He presses his lips to hers. Shuddering and pale, in horror and ecstasy she once more lacks the will to resist him. As if the earth were giving way under her feet, the somber voice of mighty instinct calls to her from the mysterious abyss. The dark waves rise and extinguish consciousness. [. . .] The soul craves enslavement. Freedom no longer exists. All is over" (6:253).

In the end, then, the aristocrat prevails over the Jew. "How pale the dark sensual pleasure of those minutes with Mark now seemed. And even paler appeared the cold joy of her love for Harold, a strange, cerebral feeling. Nelidov has been and remains the master. He alone arouses the submissive instinct in her rebellious soul. While she has dominated the others, he dominates her. [. . .] With them she was an independent personality; with him a woman" (6:260). Patriarchy and femininity—"that most traitorous instinct" (6:259)—are triumphant. Even having renounced Steinbach (she leaves him without a second thought), Mania can see no way for her to make a life with Nelidov. With femininity in the ascendant, "she envies every peasant woman who sleeps next to her beloved husband and obediently accepts the coarse caresses that are rightfully hers" (6:258). But she and Nelidov are as incompatible as ever. Even now he cannot accept her past relations with other men, and their love is once again doomed.

Verbitskaia concludes her saga of female self-assertion on a note of radical self-destruction. In despair and anguish the two part, and Nelidov shoots himself in the head. Finding the body, Mania takes his gun and sits down to write Steinbach a farewell letter: "I do not reproach you for anything, dear Mark. [. . .] I am obliged to you alone for all the lofty joys of creativity and success. But the woman in me is stronger than the artist. I am powerless before love. It was not for nothing that I feared it! *He* is dead and I am following him. I cannot remain behind" (6:284).

Epilogue: Back in his Moscow study, Steinbach arranges his papers, preparing to elude a police dragnet that has closed in on his radical friends. He is leaving his factory to the workers and his land to the peasants. Disguised as his own senile uncle—whose haggard countenance and preternatural sensitivity have throughout the novel symbolized the Jew's spiritual power and biological vulnerability—Steinbach slips out of the house and disappears. Mania and Nelidov are dead, the radicals have been arrested or are in flight, and the Jew has vanished. What about the "keys to happiness"? Mania has achieved fame and artistic success in one of the few careers that offered women public visibility and renown. She has enjoyed public admiration, romantic attention, and sexual excitement. In the end, however, this "new woman" cannot tear herself from the bosom of traditional Russian society, from the fatal power of the impoverished nobleman who ultimately triumphs over the crafty, capitalistic Jew and the allure of modern pleasures. Freedom is too much for her. She does not want to be a man.

Yet the boulevard novel carried a more ambivalent message for its female readers than this depressing summary might suggest. For even though the fictional characters failed to achieve fame and femininity at the same time, to be women in public, their authors emphatically did manage to do so. The story of her own life, which Verbitskaia retailed to an eager public, as well as the manifest evidence of her own commercial success, testified to the possibility of enacting that very contradiction. So did the real-life prototypes that inspired Mania's story.

Life as Art: An Aside on Isadora Duncan

It was easy for cultivated contemporaries to mock Verbitskaia's florid rhetoric and her thirst for popular acclaim as products of debased commercial culture. Perhaps their irritation was increased by the extent to which the themes and even the tone of Verbitskaia's work had close affinities with the world of the sophisticated avant-garde, for the boulevard differed from the gutter precisely in its close association with "higher" cultural forms. Gippius and Merezhkovskii were at least as preoccupied with Nietzsche and Greek myth as Verbitskaia's heroine was. Merezhkovskii wrote fictional treatments of classical subjects that were widely read. The struggle between the principles of Dionysus and Apollo continued on a variety of artistic fronts. In 1907, for only one of many possible examples, Viacheslav Ivanov could praise the effects of Dionysian ecstasy for "liberat[ing] the Psyche from the power and tutelage of our conscious principles."[116] Moreover, the revolution in dance and theater that set the stage for Mania's fictional triumphs reflected actual developments in the performing arts. The story of the real Isadora Duncan reveals how little Verbitskaia had to invent.

Duncan was not a remote, foreign idol. Having established her reputation in Europe, she made her Russian debut in 1904, returned for an encore in 1905 in the midst of the revolution, and reappeared in 1907. A review of her first performance in a Russian theater magazine reads like an exerpt from *The Keys to Happiness* in both substance and style: "A half-naked girl makes her appearance in a light, semitransparent Greek tunic, giving full freedom to her movements and not concealing the form of her body. . . . Before us [however] was not a woman creating a sensation. Before us was an artist."[117] And consider Duncan's own ac-

---

[116]Quoted in Read, *Religion and Revolution*, 29.
[117]Quoted in Fredrika Blair, *Isadora: Portrait of the Artist as a Woman* (New York, 1986), 106.

count of her 1905 appearance, written in 1927: "How strange it must have been to those dilettantes of the gorgeous Ballet, with its lavish decorations and scenery, to watch a young girl, clothed in a tunic of cobweb, appear and dance before a simple blue curtain to the music of Chopin; dance her soul as she understood the soul of Chopin! Yet even for the first dance there was a storm of applause."[118]

Duncan was both a popular and a critical sensation; Moscow's cultural elite led the general public in fervid admiration.[119] The overtly erotic element in her art did nothing to dampen her success. Avant-garde critics praised her for "resurrect[ing] the creative side of the dance" and called her a Bacchante who "abandons herself to love, [whose] wild dance intoxicates." Every movement of her body was "an incarnation of a spiritual act," "sinless and pure," "a victory of light over darkness."[120] Duncan herself called dancing a "Dionysian ecstasy" and performed in a work called *Bacchanale*, just as Mania does in the novel.[121] The almost naked body, the physical enactment of erotic ecstasy, the explicit link between artistic and sexual expression, and Duncan's bohemian personal life all reflected the prominence of the sexual thematic in Russian performing art of the period. In 1908, for example, the artist Nikolai Kalmakov designed a stage set that depicted female genitalia for the Temple of the Goddess of Love in the first act of *Salomé*. The design was rejected as offensive and the performance was never staged, but Kalmakov's idea fitted the current penchant for eroticism in the graphic arts. Two exhibitions in St. Petersburg in 1907 featured erotic canvases. In 1908 Mikhail Kuzmin's close friend Konstantin Somov, one of Diaghilev's associates, illustrated a collection of French erotic verse, *Le livre de la marquise*, with titillating images of aristocratic licentiousness.[122]

The provocative elements in Duncan's approach to the dance thus expressed the innovative impulse that shaped the aesthetic of Diaghilev's modernist ballets and that fed on the symbolist mood in Russian art and literature. The conflicting strands in Duncan's personal life story also belong to the cultural moment, reflecting the contradictory elements in the figure of the "new woman," both literary and actual. It is not surprising, therefore, that Mania's invented experience should bear a resemblance to the biography of her prototype. Born in San Francisco

---

[118]Isadora Duncan, *My Life* (New York, 1927), 163.

[119]Ibid., 167.

[120]Opinions from the journal *Vesy*, quoted in Blair, *Isadora*, 112, 118.

[121]Ibid., 86.

[122]See John E. Bowlt, "Through the Glass Darkly: Images of Decadence in Early Twentieth-Century Russian Art," *Journal of Contemporary History* 17 (1982): 101–3; *Le Livre de la marquise: Recueil de poésie et de prose* (St. Petersburg, [1908]).

Illustrations in Konstantin Somov, *Le Livre de la marquise* (St. Petersburg, 1908). Department of Graphics, M. E. Saltykov-Shchedrin State Public Library, St. Petersburg.

in 1878, Duncan built a career on natural talent and the force of her personality. Like Mania, she traveled the globe and shocked respectable society with her unconventional sexual life. Mistress to a number of men, she bore two illegitimate children and preached the virtues of free love.[123] Torn between ambition and the desire for personal happiness, she found herself in the predicament that tormented Verbitskaia's heroine: "My life has known but two motives, Love and Art—and often Love destroyed Art, and often the imperious call of Art put a tragic end to Love. For these two have no accord, but only constant battle."[124] The dancer and her literary shadow also shared an impulse toward cultural democracy. Invited to attend the Russian ballet during her 1905 visit, Duncan described an audience (in terms she might have copied from Verbitskaia) consisting of "the most beautiful women in the world, in

[123]Blair, *Isadora*, 122.

[124]Duncan, *My Life*, 239; see also 208–9 (quoted in part in Blair, *Isadora*, 167). Another prototype for Mania was the Russian memoirist Marie Bashkirtsev, whose example helps the fictional Mania (sometimes known as Marie) establish a literary-historical genealogy for her own self-creation. First published in French, her journal was widely translated; see *Marie Bashkirtseff: The Journal of a Young Artist, 1860–1884*, trans. Mary J. Serrano (New York, 1919).

marvellous décolleté gowns, covered with jewels, escorted by men in distinguished uniforms." But impressing high society was not her mission. "After the performance," she remembered, "I was invited to supper . . . [where] I met Grand Duke Michael, who listened with some astonishment as I discoursed on the plan of a school of dancing for the children of the people."[125]

Just as Verbitskaia denied the conflict between democratic social values and aesthetic sophistication, so Fedor Sologub found Duncan's political pretensions consistent with the message of her art. His comments suggest that the Russian literary tradition of social relevance had survived even the aggressive assault of abstract modernism. "Artistic creations," wrote Sologub, "may be divided between those that respond to external constraints and those that express internal needs. The first kind suits an authoritarian social order. Duncan's creations, by contrast, correspond to the type of society that tries most energetically to liberate the individual, in which public institutions serve the people rather than exercising power over them. Such art I call democratic."[126] After the Bolshevik Revolution, Duncan did return to Russia to found a people's school of the dance under the aegis of Anatolii Lunacharskii, the education commissar. She died in France in 1927, a year before Verbitskaia's death in the Soviet Union, where the novelist had been trying to convince the new cultural authorities that her heart had always been in the right place.[127]

In the years before World War I, high art and the boulevard were not so much in opposition as in dialogue. Democratization took the form of vulgarization, which produced a hierarchy of its own. Hits on the popular stage in 1911 included dramatizations of The Kreutzer Sonata, The Pit, and Sanin (in a version called How to Live). Other titles also suggest the widespread appeal of sexual themes: The Free Love League, Sinful Night, The Bacchante—Vampire of Love, and Living Goods (a reference to white slavery) shared the boards with dozens of farces translated from the French and other European languages.[128] The Keys to Happiness was itself made into a film in 1913.[129] Cinema of course constituted the

[125]Duncan, My Life, 164.

[126]Fedor Sologub, "Ocharovanie vzorov (Aisedora Dunkan)," Teatr i iskusstvo, no. 4 (January 27, 1913): 90.

[127]See Zorkaia, Na rubezhe, 165–67.

[128]See, e.g., lists in Teatr i iskusstvo, no. 1 (January 4, 1909); no. 5 (January 30, 1911); no. 4 (January 23, 1913). Texts available in the St. Petersburg Library of Theatrical Art include Mark Gol'dshtein, Greshnaia noch' (passed by the censor in May 1909); Ia. Gordon, "Kreitserova Sonata" (passed by the censor in October 1910); S. Trefilov, Kak zhit'?! (passed by the censor in May 1910); Petr Olenin-Boldar, Vakkhanka ("Vampukha" liubvi) (St. Petersburg, n.d.).

[129]Unfortunately, no prints of this film have survived. On its popularity, see Denise

Still photograph from the film *The Keys to Happiness* (1913), directed by Iakov Protazanov, in S. Ginzburg, *Kinematografiia dorevoliutsionnoi Rossii* (Moscow, 1963). Firestone Library, Princeton University.

ultimate boulevard—a place of public congregation designed for the mass absorption of images intended to entertain and capable of displaying the private core of intimate relations. Russians in various walks of life also thrilled to American jazz and new dance styles, and Igor Stravinsky incorporated jazz motifs in serious music.[130]

This meager list does not begin to exhaust the repertoire of texts and performances through which the sexual theme reached into the corners of urban society (and perhaps beyond it). Though some scholars have begun to penetrate the world of popular literary pleasures and entertainments, it is still impossible to draw a cultural map of pre–World War I Russian society which includes the variety of institutions and genres

Youngblood, "The Return of the Native: Yakov Protazanov and Soviet Cinema," in *Inside the Film Factory*, ed. Richard Taylor and Ian Christie (London, 1991). I thank Professor Youngblood for allowing me to see her manuscript before its publication. For the relationship between popular literature and film in these years, see Zorkaia, *Na rubezhe*.

[130]S. Frederick Starr, *Red and Hot: The Fate of Jazz in the Soviet Union, 1917–1980* (New York, 1983), 31.

that served a diverse public world.[131] Aside from engaging the professional classes in earnest consideration of political and social values, the subject of sex was a commercial success—a success that itself became the subject of earnest intellectual analysis. No matter how diverting the pretext, politics was never far behind.

[131]The pioneering work is Brooks, *When Russia Learned to Read*. See also Zorkaia, *Na rubezhe*; and Louise McReynolds, *The News under Russia's Old Regime: The Development of a Mass-Circulation Press* (Princeton, N.J., 1991).

Conclusion

In the years leading up to 1917, sexual disarray at the pinnacle of power came to stand for what was wrong with the tsarist regime. Instead of adopting responsible principles of statecraft, the emperor clung to archaic images of rule and let himself be swayed by idiosyncratic figures, among whom the most notorious was Grigorii Rasputin. Presumably the embodiment of popular wisdom and godly strength, Rasputin instead demonstrated the extent to which traditional symbols had lost their force. For this man of the people—a demonic version of folk simplicity—was devious and debauched. A caricature of spiritual authority, he used his powers of seduction to attract female admirers, whom he lured into sensual excesses. An almost blasphemous inversion of sexual propriety, Rasputin's conduct only diminished the tsar's prestige, reminding the court and the public of the moral vacuum at the center of the autocratic state. Rasputin's supposed tyranny over the royal family symbolized the disorder at the heart of the absolute order, impugning the myth of both the monarch and the folk.

But the reaction to Rasputin, culminating in his murder by a small band of ultraconservative men determined to save the monarchy from itself, merely dramatized the connections between sexuality, social class, cultural values, and political principles already well established in public discourse during the last decades of imperial rule. His case both reinforced and contradicted current clichés: since Rasputin was a member of court society, his sexual indulgence could be understood as a consequence of privilege; yet here was a representative of the common folk whose abuses offended the stereotype of popular virtue, and the corruption he spread certainly could not be ascribed to modern ways. Clearly autocracy could no longer guarantee either sexual or political virtue.

Confident in the monarchic principle and distrusting systematic change, Rasputin's murderers believed that removal of the man would alleviate the problem. Liberals, by contrast, had for years been working to reorganize social and political institutions so as to establish public virtue on a new footing. Eager to make a place for themselves in the management of social relations and the exercise of political power, professionals opposed the arbitrary working of administrative control in the name of legal rationality and disciplinary expertise. In promoting individual autonomy and equality before the law, redrawing the relationship between state and society, and redefining personhood itself, they examined the norms and categories of sex and gender.

As historians have shown, the power hierarchies of the Western liberal order depended as much on gender as on class, as much on sexual discipline as on social norms. The tension between the subordinations effected by gender, class, and sex and the official values of the liberal polity was as evident in the thinking of reformers and professionals in Russia as it was elsewhere, though it took somewhat different forms. Like liberals everywhere, these Russians persisted in their custodial attitude toward women while insisting on expanded autonomy for men. Enemies of patriarchy and partisans of modernity, they were yet loath to recognize the individuality and sexual agency of peasants and workers. No less squeamish about sexual perversion than their colleagues abroad, they long considered the problem irrelevant to Russian society. Deeply ambivalent about modernity—with its ambiguous gifts of urban anonymity, commercial culture, materialist values, and selfish individualism—they nevertheless used the framework of political and cultural modernity as the basis of their opposition to absolutist rule.

The vision of political and sexual modernity that Russian professionals counterposed to the administrative rigidity of the old regime entailed two kinds of regulation: first, a guarantee of legal rights, a function of state power based on but also limited by positive law; and second, social discipline exercised by the custodians of scientific and cultural authority. In this imagined liberal system, social autonomy and the conditions of public order would have been underwritten by the force of law. The principles of law and discipline were conceived not as contradictory but as joint alternatives to anarchy and coercion. Had tsarism been succeeded by a liberal regime, the professionals who aspired to such an ideal would undoubtedly have imposed the same kind of normative values, social inequities, and disciplinary constraints as those that operated in the bourgeois West, but they did not get the chance to experiment with that particular combination of freedom and control. At best, those who remained after 1917 got an opportunity to serve as expert advisers to a reinvigorated administrative state that, like its tsarist predecessor, ignored guarantees of legal protection, disdained individual rights, and suppressed political and social autonomy, and thus, to the degree that it was more thoroughgoing than the tsarist regime, even more stringently prohibited sexual as well as political discourse and diversity.

Primary Sources Cited

Archival Sources

TsGIA, f. 759, op. 41, d. 2053, ll. 16–40. Formuliarnyi spisok o sluzhbe direktor S. Peterburgskogo rodovspomogatel'nogo zavedeniia vedomstva uchrezhdenii Imperatritsy Marii, tainogo sovetnika doktora meditsiny i akushera Ippolita Mikhailovicha Tarnovskogo (January 16, 1897).

TsGIA, f. 1297, op. 6, ed. khr. 28 (1850), ll. 1–7. Otchet Moskovskogo vrachebnogo politseiskogo komiteta za 1849 g.

TsGIA, f. 1297, op. 6, ed. khr. 31 (1852), ll. 2–15. Otchet o deistviiakh Moskovskogo vrachebnogo politseiskogo komiteta v 1851–53 gg.

TsGIA, f. 1335, op. 1, d. 24, ll. 20–69. Alfavitnyi spisok chlenov Rossiiskogo obshchestva zashchity zhenshchin s 1901 goda.

TsGIA, f. 1405, op. 543, d. 512. O vvedenii v deistvie glavy XXVII novogo (1903 g.) ugolovnogo ulozheniia, kasaiushcheisia prestupnogo vovlecheniia v razvrat zhenshchin (torg zhenshchinami) (November 18, 1906–January 26, 1917).

TsGVIA, f. 546, op. 2, d. 7945, ll. 57–69. Posluzhnye i attestatsionnye spiski vrachei na bukvy "t", chast' 1 (on V. M. Tarnovskii).

Published Sources

Abrashkevich, M. M. *Preliubodeianie s tochki zreniia ugolovnogo prava: Istoriko-dogmaticheskoe issledovanie.* Odessa, 1904.

Afanas'ev, N. I. *Sovremenniki: Al'bom biografii.* 2 vols. St. Petersburg, 1909, 1910.

Aggeev, K. "Tiazhelyi vopros." *Tserkovno-obshchestvennyi vestnik,* no. 49 (December 12, 1913): 1–3.

Aleksandrov, S. A. "Chetvertyi s"ezd rossiiskikh ginekologov i akusherov v S.-Peterburge, 16–19 dekabria 1911 goda." *Prakticheskii vrach,* no. 1 (1912): 15–16.

Aleksii, Episkop. "Moral' Talmuda." *Pribavleniia k Tserkovnym vedomostiam,* no. 44 (November 2, 1913): 2025–30.

Aletrino, A. "La Situation sociale de l'uraniste." In Congrès international d'anthropologie criminelle, *Compte rendu des travaux de la cinquième session tenue à Amsterdam du 9 au 14 septembre 1901.* Amsterdam, 1901.

Anderson, N. E. "Zabolevaemost' sifilisom sredi fabrichnogo, remeslennogo i torgovo-promyshlennogo muzhskogo naseleniia g. Moskvy po dannym ambulatorii Miasnitskoi bol'nitsy za 1910 g.: Mery bor'by protiv sifilisa." In *Trudy vtorogo vserossiiskogo s"ezda fabrichnykh vrachei i predstavitelei fabrichno-zavodskoi promyshlennosti, izdannye pravleniem Moskovskogo obshchestva fabrichnykh vrachei,* ed. I. D. Astrakhan, vyp. 2. Moscow, 1911.

Andreev, L. N. "V tumane." *Zhurnal dlia vsekh,* no. 12 (1902): 1411–50.

Andreevskii, I. E. *Politseiskoe pravo.* 2d ed., rev. St. Petersburg, 1876.

Andronov, S. V. "Doklad po vnesennomu ministrom iustitsii zakonoproektu o merakh k presecheniiu torga zhenshchinami v tseliakh razvrata." In *Prilozheniia k stenograficheskim otchetam Gosudarstvennoi Dumy,* tretii sozyv, sessiia vtoraia, vol. 2. St. Petersburg, 1909.

Anuchin, E. N. *Issledovaniia o protsente soslannykh v Sibir' v period 1827–1846 godov.* St. Petersburg, 1866.

Arkhangel'skii, G. I. "Zhizn' v Peterburge po statisticheskim dannym." Pts. 1–2. *Arkhiv sudebnoi meditsiny i obshchestvennoi gigieny,* 1869: no. 2, pt. 3, 33–85; no. 3, pt. 3, 84–143.

"Arkhiepiskop Stefan ob aborte." *Russkoe slovo,* no. 64 (March 18/31, 1914): 5.

Artsybashev, M. P. "Sanin." *Sovremennyi mir,* nos. 1–4 (1907). Published in English as Michael Petrovich Artzibashev, *Sanine,* trans. Percy Pinkerton. New York, 1926.

——. "Uzhas: Rasskaz." *Obrazovanie,* no. 3, sec. 1 (1905): 1–20.

Asheshov, N. "Pozornaia glubina (V. Rozanov, *Opavshie list'ia, korob vtoroi i poslednii,* Petrograd, 1915)." *Rech',* no. 224/3247 (August 16/29, 1915): 2.

Ashkinazi, I. "Otto Veininger." *Obrazovanie,* no. 11, sec. 1 (1908): 119–26.

A—vich, N. Review of M. Kuzmin, "Seti." *Obrazovanie,* no. 5, sec. 3 (1908): 107–8.

Babikov, K. I. *Prodazhnye zhenshchiny: Kartiny publichnogo razvrata (prostitutsiia) na vostoke, v antichnom mire, v srednye veka i v nastoiashchee vremia vo Frantsii, Anglii, Rossii i dr. gosudarstvakh Evropy, sostavleno po Paran-Diu-Shatele, Zhanneliu, Sherru, Lakrua i dr.* Moscow, 1870.

Bakhtin, N. "Russo i ego pedagogicheskie vozzreniia." Pts. 1–4. *Russkaia shkola,* 1912: no. 9, 75–89; no. 10, 134–50; no. 11, 119–33; no. 12, 86–121.

Balov, A. "Prostitutsiia v derevne." *Vestnik obshchestvennoi gigieny, sudebnoi i prakticheskoi meditsiny,* no. 12 (1906): 1864–68.

Bashilov, P. P. "O khuliganstve, kak prestupnom iavlenii, ne predusmotrennom zakonom." *Zhurnal Ministerstva Iustitsii,* no. 2 (1913): 222–39.

Bazhenov, N. N. *Psikhologiia i politika.* Moscow, 1906.

Bekhterev, V. M. "Lechenie vnusheniem prevratnykh polovykh vlechenii i onanizma." *Obozrenie psikhiatrii, nevrologii i eksperimental'noi psikhologii,* no. 8 (1898): 587–97.

——. "O lechenii onanizma vnusheniiami v gipnoze." *Obozrenie psikhiatrii, nevrologii i eksperimental'noi psikhologii,* no. 3 (1899): 186–89.

——. *O polovom ozdorovlenii.* St. Petersburg, 1910.

——. "O vneshnykh priznakakh privychnogo onanizma u podrostkov muzhskogo pola." *Obozrenie psikhiatrii, nevrologii i eksperimental'noi psikhologii,* no. 9 (1902): 658–62.

——. "Voprosy nervno-psikhicheskogo zdorov'ia v naselenii Rossii." In *Trudy tret'ego s"ezda otechestvennykh psikhiatrov (s 27-go dekabria 1909 g. po 5-oe ianvaria 1910 g.).* St. Petersburg, 1911.

Belyi, Andrei [Boris Bugaev]. "Na perevale: Veininger o pole i kharaktere." *Vesy*, no. 2 (1909): 77–81.

——. "Publitsistika: V. Rozanov, *Kogda nachal'stvo ushlo.*" *Russkaia mysl'*, no. 11, pt. 2 (1910): 374–76.

——. Review of M. Kuzmin, *Kryl'ia. Pereval*, no. 6 (1907): 50–51.

——. Review of L. Zinov'eva-Annibal, *Tridtsat'-tri uroda: Povest'. Pereval*, no. 5 (1907): 53.

——. "Simvolism i sovremennoe russkoe iskusstvo." *Vesy*, no. 10 (1908): 38–48.

Bentovin, B. I. "O prostitutsii detei." In *Trudy pervogo vserossiiskogo s"ezda po bor'be s torgom zhenshchinami i ego prichinami, proiskhodivshego v S.-Peterburge s 21 po 25 aprelia 1910 goda*, vol. 2. St. Petersburg, 1912.

——. *Torguiushchie telom: Ocherki sovremennoi prostitutsii.* 2d ed., rev. St. Petersburg, 1909.

Berdiaev, N. A. *Novoe religioznoe soznanie i obshchestvennost'.* St. Petersburg, 1907.

Berg, Alexander. *Juden-Bordelle: Enthüllungen aus dunklen Häusern.* 1890. 4th ed. Berlin, n.d.

Bernshtein, A. N. "Psikhicheskie zabolevaniia zimoi 1905–06 gg. v Moskve." *Sovremennaia psikhiatriia*, no. 4 (1907): 49–67.

——. *Voprosy polovoi zhizni v programme semeinogo i shkol'nogo vospitaniia.* Moscow, 1908.

Bertenson, Lev. *Fizicheskie povody k prekrashcheniiu brachnogo soiuza.* Petrograd, 1917.

B. G. Review of E. Nagrodskaia, *Gnev Dionisa. Istoricheskii vestnik*, 125 (1911): 1160–63.

Binshtok, I. I. "Dvenadtsatyi Pirogovskii s"ezd i uchenie Mal'tusa." *Prakticheskii vrach*, no. 9 (1914): 123–27.

——. "Eshche o nakazuemosti aborta." *Prakticheskii vrach*, no. 15 (1914): 213–15.

Bloch, Iwan. *Das Sexualleben unserer Zeit in seinen Beziehungen zur modernen Kultur.* 6th ed. Berlin, 1908.

Blumenau, L. V. "K patologii polovogo vlecheniia." *Obozrenie psikhiatrii, nevrologii i eksperimental'noi psikhologii*, no. 1 (1902): 17–22.

Bobrovskii, P. O. *Proiskhozhdenie artikula voinskogo i izobrazheniia protsessov Petra Velikogo po ustavu voinskomu 1716 g.* 2d ed., rev. St. Petersburg, 1881.

——. *Voennoe pravo v Rossii pri Petre Velikom.* Pt. 2, *Artikul voinskii.* Vol. 1, *Vvedenie: Manifest, prisiaga i pervye chetyre glavy.* St. Petersburg, 1882.

B[ogdanovich], A. "Kriticheskie zametki." *Mir bozhii*, no. 1, sec. 2 (1903): 1–14.

"Bor'ba s prostitutsiei." *Zhenskii vestnik*, no. 12 (1913): 265–66.

Boriakovskii, A. G. "O vrede sredstv, prepiatstvuiushchikh zachatiiu." *Vrach*, no. 32 (1893): 886–87.

Borodina, A. G. "Tsel' i zadachi Obshchestva zashchity zhenshchin." In *Trudy pervogo vserossiiskogo zhenskogo s"ezda pri Russkom zhenskom obshchestve v S.-Peterburge 10–16 dekabria 1908 goda.* St. Petersburg, 1909.

Borovikovskii, A. L. "Brak i razvod po proektu grazhdanskogo ulozheniia." *Zhurnal Ministerstva Iustitsii*, no. 8, sec. 2 (1902): 1–62.

——. "Konstitutsiia sem'i po proektu grazhdanskogo ulozheniia." *Zhurnal Ministerstva Iustitsii*, no. 9, sec. 2 (1902): 1–38.

Borovitnikov, M. M. *Detoubiistvo v ugolovnom prave.* St. Petersburg, 1905.

Borovskii, V. K. "K voprosu ob istochnikakh zarazheniia sifilisom." *Voenno-meditsinskii zhurnal*, no. 8 (1894): 411–23.

Brodskii, V. A. "Iskusstvennyi vykidysh s meditsinskoi i obshchestvenno-ekonomicheskoi tochek zreniia." Pts. 1–4. *Vrachebnaia gazeta*, nos. 18–20, 22 (1913): 656–60, 684–88, 710–14, 781–85.

Bronzov, A. A. *O khristianskoi sem'e i sviazannykh s neiu voprosakh.* St. Petersburg, 1901.

B—skii, N. "Ocherk prostitutsii v Peterburge." *Arkhiv sudebnoi meditsiny i obshchestvennoi gigieny*, no. 4, sec. 3 (1868): 61–99.

Bulkley, L. Duncan. *Syphilis in the Innocent (Syphilis Insontium), Clinically and Historically Considered, with a Plan for the Legal Control of the Disease.* New York, 1894.

Bulletin abolitionniste: Organe central de la Fédération abolitionniste internationale (Geneva), n.s., no. 42 (June 1905).

Burenin, N. E. "Kriticheskie ocherki." *Novoe vremia*, no. 9666 (January 31/ February 13, 1903): 2.

Burlakov, V. M. "Ob anafrodizii zhenshchin." *Meditsinskaia beseda*, no. 13–14 (1902): 390–93.

Butkevich, T. I. "O smysle i znachenii krovavykh zhertvoprinoshenii v dokhristianskom mire i o tak nazyvaemykh ritual'nykh ubiistvakh." Pts. 1–4. *Vera i razum*, nos. 21–24 (1913): 281–99, 413–37, 553–608, 723–68.

Casper, Johann Ludwig. *Klinische Novellen zur gerichtlichen Medizin, nach eigenen Erfahrungen.* Berlin, 1863.

———. *Practisches Handbuch der gerichtlichen Medizin, nach eigenen Erfahrungen.* 2 vols. Berlin, 1857, 1858. 3d ed. published in English as *A Handbook of the Practice of Forensic Medicine, Based upon Personal Experience*, trans. George William Balfour. 3 vols. London, 1861–64.

Chekhov, A. P. "Baby" (1891). In *Polnoe sobranie sochinenii i pisem*, vol. 7. Moscow, 1977.

———. "The Duel." In *The Russian Master and Other Stories.* Oxford, 1984.

———. *Ostrov Sakhalin.* 1895. Rpt. Moscow, 1984.

———. "Pripadok" (1888). In *Polnoe sobranie sochinenii i pisem*, vol. 6. Moscow, 1962.

Chikhachev, K. "O iuridicheskoi sile i prakticheskom znachenii reshenii kassatsionnykh departamentov Pravitel'stvuiushchego Senata." *Zhurnal iuridicheskogo obshchestva pri Imperatorskom S.-Peterburgskom Universitete*, no. 7, pt. 2 (1896): 40–56.

Chistiakov, M. A. *Protokoly sektsii sifilidologii na pervom s"ezde russkikh vrachei 1885 g. v S. Peterburge.* St. Petersburg, 1886.

———. "Sluchai derevenskogo sifilisa v stolitse." *Prakticheskii vrach*, no. 52 (1907): 939.

Chistovich, Ia. A. *Istoriia pervykh meditsinskikh shkol v Rossii.* St. Petersburg, 1883.

Chizh, V. F. "Znachenie politicheskoi zhizni v etiologii dushevnykh boleznei." Pts. 1–2. *Obozrenie psikhiatrii, nevrologii i eksperimental'noi psikhologii*, nos. 1, 3 (1908): 1–12, 149–62.

Chlenov, M. A. "K kazuistike vnepolovogo sifilisa." Pts. 1–2. *Russkii vrach*, nos. 29–30 (1902): 1060–62, 1093–95.

———. *Polovaia perepis' moskovskogo studenchestva i ee obshchestvennoe znachenie.* Moscow, 1909.

Chubinskii, M. P. "Istreblenie ploda i problema ego nakazuemosti." *Iuridicheskii vestnik*, no. 2 (1913): 112–35.

———. *Trudy etnografichiskoi statisticheskoi ekspeditsii v zapadno-russkii krai.* Vol. 6. St. Petersburg, 1872.

——. "Vopros o vykidyshe v sovremennom prave i zhelatel'naia ego posta-novka." *Zhurnal akusherstva i zhenskikh boleznei*, no. 4 (1912): 461–86.

Chukovskii, K. I. "Geometricheskii roman." *Rech'*, no. 123 (May 17/June 9, 1907): 2.

——. *Kniga o sovremennykh pisateliakh*. St. Petersburg, 1914.

Congrès international d'anthropologie criminelle. *Actes du troisième congrès international d'anthropologie criminelle, Bruxelles, 1892*. Brussels, 1893.

——. *Compte-rendu des travaux de la quatrième session tenue à Genève du 24 au 29 août 1896*. Geneva, 1897.

——. *Compte-rendu des travaux de la cinquième session tenue à Amsterdam du 9 au 14 septembre 1901*. Amsterdam, 1901.

——. *Comptes-rendus du sixième congrès international d'anthropologie criminelle (Turin, 28 avril–3 mai 1906)*. Turin, 1908.

Dadonov, V. A. *Verbitskaia i ee romany "Kliuchi schast'ia" i "Dukh vremeni": Kriticheskii ocherk*. Moscow, 1911.

Dal', Vladimir I. *Tolkovyi slovar' velikorusskogo iazyka*. 4th ed., rev. St. Petersburg, 1912.

——. *Zapiska o ritual'nykh ubiistvakh: Rozyskanie ob ubienii evreiami khristianskikh mladentsev i upotreblenii krovi ikh*. 1844. Rpt. St. Petersburg, 1913.

Danilov, N. P. "Pervyi vserossiiskii s"ezd po bor'be s torgom zhenshchinami." *Izvestiia Moskovskoi gorodskoi dumy*, no. 9, otdel obshchii (1910): 94–115.

Daya-Berlin, Werner. "Die sexuelle Bewegung in Russland." *Zeitschrift für Sexualwissenschaft*, no. 8 (1908): 493–502.

Delo Beilisa: Stenograficheskii otchet. 3 vols. Kiev, 1913.

Demkov, M. I. *Istoriia russkoi pedagogii*. Pt. 3, *Novaia russkaia pedagogiia (XIX vek)*. Moscow, 1909.

Deriuzhinskii, V. F. "Mezhdunarodnaia bor'ba s torgovleiu zhenshchinami." *Zhurnal Ministerstva Iustitsii*, no. 8, sec. 2 (1902): 174–209.

——. "Pamiati A. A. Saburova." In *Rossiiskoe obshchestvo zashchity zhenshchin v 1915 godu*. Petrograd, 1916.

——. *Politseiskoe pravo: Posobie dlia studentov*. 3d ed. St. Petersburg, 1911.

"Desiatyi s"ezd Russkoi gruppy mezhdunarodnogo soiuza kriminalistov." *Pravo*, no. 10 (1914): 809–40.

Diday, Paul. *Le Péril vénérien dans les familles*. Paris, 1881.

Ditman, V. "Tainyi porok." Pts. 1–3. *Pedagogicheskii sbornik*, nos. 3–5 (1871): 367–75, 551–56, 654–63.

D. L. "Proekt pravil o razreshenii razdel'nogo zhitel'stva suprugov." *Vestnik prava*, no. 9 (1899): 141–53.

Dnevnik gosudarstvennogo sekretaria A. A. Polovtsova. 2 vols. Moscow, 1966.

Dobrotvorskii, N. I. "Detoubiistvo: Sudebno-psikhiatricheskaia ekspertiza." *Arkhiv psikhiatrii, neirologii i sudebnoi psikhopatologii*, no. 3 (1893): 91–96.

——. "Ubiistvo mater'iu svoego nezakonnorozhdennogo rebenka vo vremia rodov." *Voprosy nervno-psikhicheskoi meditsiny*, no. 1 (1905): 139–43.

"Doklad komissii po bor'be s iskusstvennymi vykidyshami Omskogo meditsinskogo obshchestva dvenadtsatomu Pirogovskomu s"ezdu vrachei." *Obshchestvennyi vrach*, no. 6, sec. 6 (1913): 683–92.

"Doklad psikhiatricheskoi komissii obshchestva neiropatologov i psikhiatrov po voprosu o psikhozakh v sviazi s poslednimi politicheskimi sobytiiami." *Russkii vrach*, no. 23 (1906): 709–11.

Dostoevskii, F. M. *Unizhennye i oskorblennye* (1861). In *Polnoe sobranie sochinenii*, vol. 3. Leningrad, 1972.

Dril', D. A. "Antropologicheskaia shkola i ee kritiki (Zametki po povodu statei g. Obninskogo)." *Iuridicheskii vestnik*, no. 4 (1890): 579–99.

——. "O zabroshennosti detstva, kak mogushchestvennaia prichina detskoi prostitutsii." In *Trudy pervogo vserossiiskogo s"ezda po bor'be s torgom zhenshchinami i ego prichinami, proiskhodivshego v S.-Peterburge s 21 po 25 aprelia 1910 goda*, vol. 1. St. Petersburg, 1911.

Duncan, Isadora. *My Life*. New York, 1927.

Dvenadtsatyi Pirogovskii s"ezd. Peterburg, 29 maia–5 iiunia 1913 g. Vyp. 2. St. Petersburg, 1913.

Efimenko, Aleksandra. *Issledovaniia narodnoi zhizni*. Vol. 1, *Obychnoe pravo*. Moscow, 1884.

Efimov, A. I. *Sifilis v russkoi derevne, ego kharakternye cherty i vliianie na sanitarnoe polozhenie naseleniia*. Kazan, 1902.

——. "Sravnitel'naia otsenka raznykh sposobov izucheniia derevenskogo sifilisa." *Vrach*, no. 51 (1900): 1550–54.

Elistratov, A. I. *O prikreplenii zhenshchiny k prostitutsii: Vrachebno-politseiskii nadzor*. Kazan, 1903.

——. "Prostitutsiia v Rossii do 1917 goda." In *Prostitutsiia v Rossii*, ed. V. M. Bronner and A. I. Elistratov. Moscow, 1927.

Elpat'evskii, S. Ia. "Samoistreblenie chelovechestva: Po povodu s"ezda kriminalistov v Peterburge." *Russkoe bogatstvo*, no. 4 (1914): 262–78.

El'tsina, Z. Ia. "K voprosu o rasshirenii mer bor'by s sifilisom." Pts. 1–2. *Russkii vrach*, nos. 26–27 (1902): 969–71, 999–1000.

——. "Nedostatochnost' nadzora za maloletnimi v artel'nykh masterskikh i neobespechennost' detei sifilitikov bol'nichnymi mestami." *Vrach*, no. 19 (1900): 577–79.

——. "Priobretennyi sifilis detei, ego etiologiia i bor'ba s nim." In *Trudy pervogo vserossiiskogo s"ezda detskikh vrachei v S.-Peterburge, s 27–31 dekabria 1912 goda*. Ed. G. B. Konukhes. St. Petersburg, 1913.

——. "Sifilis i kozhnye bolezni sredi zhenskogo rabochego naseleniia Peterburga." Pts. 1–4. *Vrach*, nos. 42–45 (1896): 1175–79, 1203–7, 1237–41, 1271.

——. "Vybor prislugi." In *Pervyi zhenskii kalendar' na 1903 god*, ed. P. N. Arian. St. Petersburg, 1903.

——. "Zhelatel'nye sposoby vskarmlivaniia grudnykh sifiliticheskikh detei." *Vrach*, no. 4 (1894): 101–3.

Empe. "Neskol'ko slov o sifilise s sanitarnoi tochki zreniia i o polozhenii etogo voprosa v Peterburge." *Russkii vrach*, no. 13 (1903): 498–500.

E. P. "Opyt osvedomleniia v polovom voprose devochki i mal'chika." *Vestnik vospitaniia*, no. 3, sec. 1 (1908): 104–19.

Erikson, E. V. "Nervnye i dushevnye bolezni u evreev." *Nevrologicheskii vestnik*, no. 2 (1913): 217–65.

——. "O polovom razvrate i neestestvennykh polovykh snosheniiakh v korennom naselenii Kavkaza." *Vestnik obshchestvennoi gigieny, sudebnoi i prakticheskoi meditsiny*, no. 12 (1906): 1868–93.

Faingar, I. M. "Detskaia prostitutsiia." *Vestnik psikhologii, kriminal'noi antropologii i pedologii*, no. 3 (1913): 29–47.

Falevich, Ia. Kh. "Itogi tomskoi studencheskoi polovoi perepisi: Doklad, chitannyi 18 fevralia 1910 g. na zasedanii Pirogovskogo studencheskogo meditsinskogo obshchestva pri Tomskom Universitete." Pts. 1–13. *Sibirskaia vrachebnaia gazeta*, nos. 17–29 (1910): 197–98, 209–11, 221–34, 245–47, 257–59, 269–71, 281–82, 293–95, 305–6, 317–19, 329–31, 341–43.

Favorskii, A. V. "O polovom vozderzhanii." *Russkii zhurnal kozhnykh i venericheskikh boleznei*, no. 3 (1905): 257–61.

Fedchenko, N. P. "O zarazhenii sifilisom pri brit'e." *Meditsinskoe obozrenie*, no. 1 (1890): 19–26.

Fedorov, A. I. "Deiatel'nost' S.-Peterburgskogo vrachebno-politseiskogo komiteta za period 1888–95 gg." *Vestnik obshchestvennoi gigieny, sudebnoi i prakticheskoi meditsiny*, no. 11 (1896): 178–94.

———. "Pozornyi promysl'." *Vestnik obshchestvennoi gigieny, sudebnoi i prakticheskoi meditsiny*, no. 8 (1900): 1175–85.

———. "Prostitutsiia v S.-Peterburge i vrachebno-politseiskii nadzor za neiu." *Vestnik obshchestvennoi gigieny, sudebnoi i prakticheskoi meditsiny*, no. 1 (1892): 36–75.

Filipov, M. "Vzgliad na russkie grazhdanskie zakony." Pts. 1–2. *Sovremennik*, 1861: no. 2, 523–62; no. 3, 217–66.

Filippov, Osip A. "Vzgliad na ugolovnoe pravo po predmetu oskorbleniia chesti zhenshchin." *Iuridicheskii vestnik*, no. 2 (1862): 20–33.

Filits, S. V. "Sovremennaia polovaia zhizn' s meditsinskoi tochki zreniia." *Meditsinskaia beseda*, no. 3 (1900): 65–80.

Flirt. St. Petersburg, 1906–8.

Foinitskii, I. Ia. *Kurs ugolovnogo prava: Chast' osobennaia.* 4th ed. St. Petersburg, 1901.

———. "Programma dlia sobiraniia narodnykh iuridicheskikh obychaev: Ugolovnoe pravo." In *Zapiski Imperatorskogo russkogo geograficheskogo obshchestva po otdeleniiu etnografii*, vol. 8. St. Petersburg, 1878.

———. "Zhenshchina-prestupnitsa." Pts. 1–2. *Severnyi vestnik*, 1893: no. 2, 123–44; no. 3, 111–40.

Fournier, Alfred. *Traité de la syphilis.* 3 vols. Paris, 1899–1906.

Frank, F. *Ritual'noe ubiistvo.* Trans. of *Der Ritual-Mord vor dem Gerichtshof der Wahrheit.* Kiev, 1912.

Friche, V. Review of Ruf' Bre, *Pravo na materinstvo,* trans. 1905. *Obrazovanie*, no. 2, sec. 3 (1905): 134–36.

Fridlender, Iu. *Nravstvennye epidemii.* St. Petersburg, 1901.

Gal'perin, V. I. "Prostitutsiia detei." In *Deti-prestupniki*, ed. M. N. Gernet. Moscow, 1912.

Gerasimov, L. "Nasha literatura i pressa posle revoliutsii." *Obrazovanie*, no. 2, sec. 3 (1908): 1–12.

German, I. S. "O psikhicheskom rasstroistve depressivnogo kharaktera, razvivshemsia y bol'nykh na pochve perezhivaemykh politicheskikh sobytii." *Zhurnal nevropatologii i psikhiatrii imeni S. S. Korsakova*, no. 3 (1906): 313–23.

Gernet, M. N. "Detoubiistvo." In *Entsiklopedicheskii slovar' T-va Br. A. i I. Granat i Ko.*, 19: 303–14. 7th ed., rev. Moscow, 1910.

———. "Detoubiistvo: Sotsiologicheskoe i sravnitel'no-iuridicheskoe issledovanie." 1911. In *Uchenye zapiski Imp. Moskovskogo Universiteta*, Otdel iuridicheskii, vyp. 40. Moscow, 1912.

———. *Detoubiistvo: Sotsiologicheskoe i sravnitel'no-iuridicheskoe issledovanie.* Moscow, 1911.

———. "Istreblenie ploda s ugolovno-sotsiologicheskoi tochki zreniia." *Vestnik prava*, no. 3 (1914): 233–38.

———. *Izbrannye proizvedeniia.* Ed. M. M. Babaev. Moscow, 1974.

———. "K voprosu o nakazuemosti plodoizgnaniia: Otvet moim kritikam." *Vestnik prava*, no. 16 (1914): 489–92.

———. "Prestupnost' i zhilishcha bedniakov." *Pravo*, no. 43 (1903): 2393–401.

Gershun, T. M. "K kazuistike vnepolovogo zarazheniia sifilisom: Redkii sluchai pervichnoi skleroznoi iazvy iazyka." *Vrachebnaia gazeta*, no. 24 (1908): 720–21.

Gertsenshtein, G. M. "K statistike sifilisa v Rossii." Pts. 1–2. *Vrach*, nos. 18–19 (1886): 335–37, 358–61.

——. "Peredvizhnye vrachebnye otriady dlia bor'by s sifilisom." *Vestnik obshchestvennoi gigieny, sudebnoi i prakticheskoi meditsiny*, no. 10 (1896): 14–44.

——. [G. M. G.] "Prostitutsiia." In *Entsiklopedicheskii slovar' Brokgauz-Efron*, 25A:479–86. St. Petersburg, 1898.

——. "Sifilis v Novgorodskoi gubernii i voprosy o bor'be s nim na VII i IX s"ezdakh zemskikh vrachei 1888–1895 gg." *Vestnik obshchestvennoi gigieny, sudebnoi i prakticheskoi meditsiny*, no. 4 (1896): 28–61.

Gessen, I. V. *Razdel'noe zhitel'stvo suprugov*. St. Petersburg, 1914.

Ginzburg, A. A. "Izgnanie ploda." *Zhurnal Ministerstva Iustitsii*, no. 7 (1912): 35–70.

Gippius, Z. N. [Anton Krainyi]. "Bratskaia mogila." *Vesy*, no. 7 (1907): 57–64.

——. *Zhivye litsa*. 2 vols. Prague, 1925. Rpt. with Introduction by Temira Pachmuss. Munich, 1971.

Glebovskii, S. "Detoubiistvo v Lifliandskoi gubernii." Pts. 1–3. *Vestnik obshchestvennoi gigieny, sudebnoi i prakticheskoi meditsiny*, nos. 9–11 (1904): 1269–83, 1397–1438, 1702.

Gogel', S. K. "Iuridicheskaia storona voprosa o torgovle belymi zhenshchinami v tseliakh razvrata." *Vestnik prava*, no. 5 (1899): 108–19.

Golubtsov, A. "Po povodu zakona 3 iiunia 1902 goda." *Zhurnal Ministerstva Iustitsii*, no. 2, sec. 2 (1903): 193–96.

Goncharov, K. V. *O venericheskikh bolezniakh v S. Peterburge*. St. Petersburg, 1910.

Gornfel'd, A. G. *Knigi i liudi: Literaturnye besedy*. St. Petersburg, 1908.

Gorovits, L. M. "K voprosu o nakazuemosti aborta." *Sovremennik*, no. 5 (1914): 36–44.

Gosudarstvennaia Duma. *Stenograficheskie otchety*. III sozyv, sessiia II, zasedanie 109 (May 8, 1909).

Gosudarstvennyi Sovet, 1801–1901. St. Petersburg, 1901.

Govorkov, P. A. "Polovaia zhizn' garnizona." Pts. 1–2. *Vrach*, nos. 37–38 (1896): 1015–18, 1049–55.

Gratsianov, P. A. "Bor'ba s sifilisom, kak predmet obshchestvennoi gigieny." In *Vos'moi Pirogovskii s"ezd: Avtoreferaty i polozheniia dokladov po sektsiiam*, vol. 6. Moscow, 1902.

——. "K voprosu o reorganizatsii nadzora za prostitutsiei v Rossii." *Vestnik obshchestvennoi gigieny, sudebnoi i prakticheskoi meditsiny*, no. 11 (1895): 139–69.

——. "Po povodu proekta novogo 'Polozheniia o S.-Peterburgskom vrachebno-politseiskom komitete.'" *Russkii meditsinskii vestnik*, no. 1 (1904): 4–14.

Gratsianskii, P. I. "Nevinnye puti i sposoby zarazheniia i rasprostraneniia sifilisa." *Zhurnal Russkogo obshchestva okhraneniia narodnogo zdraviia*, no. 11 (1892): 769–822.

Gregorovich, N. *Voprosy tak nazyvaemogo ugolovnogo prava*. Kazan, 1897.

Grekhi molodykh liudei: Nastol'naia kniga. 3d ed. Moscow, 1906.

Gremiachenskii, D. *Sovremennyi stroi i prostitutsiia*. Moscow, 1906.

Griaznov, K. *Prostitutsiia, kak obshchestvennyi nedug i mery k ego vrachevaniiu*. Moscow, 1901.

Grigorovskii, Sergei. *Sbornik tserkovnykh i grazhdanskikh zakonov o brake i razvode i sudoproizvodstvo po delam brachnym*. St. Petersburg, 1896.

Gurari, D. L. "K voprosu o neprofessional'noi prostitutsii." *Gigiena i sanitariia*, no. 4 (1910): 284–87.

Gurevich, A. Ia. "O zhenskom fabrichnom trude i prostitutsii." In *Trudy pervogo vserossiiskogo s"ezda po bor'be s torgom zhenshchinami i ego prichinami, proiskhodivshego v S.-Peterburge s 21 po 25 aprelia 1910 goda*, vol. 1. St. Petersburg, 1911.

Gurko, V. I. *Features and Figures of the Past: Government and Opinion in the Reign of Nicholas II*. Stanford, Calif., 1939.

Iablonovskii, Sergei. "Prava nerozhdennykh." *Russkoe slovo*, no. 129 (June 6/19, 1913): 2.

Iakobson, V. L. "Sovremennyi vykidysh s obshchestvennoi i meditsinskoi tochki zreniia." *Zhurnal akusherstva i zhenskikh boleznei*, no. 3 (1912): 305–18.

Iakobzon, L. Ia. "Kakimi merami sleduet borot'sia s rasprostraneniem venericheskikh boleznei sredi uchashchikhsia." *Russkii vrach*, no. 43 (1903): 1509–12.

———. "Polovoe vozderzhanie pered sudom meditsiny." *Russkii vrach*, no. 18 (1905): 588–94.

Iakovenko, V. I. "Zdorovye i boleznennye proiavleniia v psikhike sovremennogo russkogo obshchestva." *Zhurnal obshchestva russkikh vrachei v pamiat' N. I. Pirogova*, no. 4 (1907): 269–87.

Iakshevich, V. S. *Plody razvrata*. St. Petersburg, 1904.

Ianchevskaia, M. M. "Zhenshchina u Veiningera." In *Trudy pervogo vserossiiskogo zhenskogo s"ezda pri Russkom zhenskom obshchestve v S.-Peterburge, 10–16 dekabria 1908 goda*. St. Petersburg, 1909.

Iaroshevskii, S. I. "Materialy k voprosu o massovykh nervnopsikhicheskikh zabolevanii." *Obozrenie psikhiatrii, nevrologii i eksperimental'noi psikhologii*, no. 1 (1906): 1–9.

Igumnov, S. N. "Zemskaia meditsina i narodnichestvo." In *Trudy odinnadstatogo Pirogovskogo s"ezda*, ed. P. N. Bulatov, vol. 1. St. Petersburg, 1911.

Iokhved, Grigorii. "Pederastiia, zhizn' i zakon." *Prakticheskii vrach*, no. 33 (1904): 871–73.

"Iskusstvennyi vykidysh s printsipial'noi tochki zreniia." *Sibirskaia vrachebnaia gazeta*, no. 10 (1911): 115–16.

Issaly, Léon. *Contribution à l'étude de la syphilis dans les campagnes*. Paris, 1895.

Iuridicheskoe obshchestvo pri Imperatorskom S.-Peterburgskom Universitete za dvadtsat' piat' let (1877–1902). St. Petersburg, 1902.

Izgoev, A. S. [A. S. Lande]. "Ob intelligentnoi molodezhi: Zametki ob ee byte i nastroeniiakh." In *Vekhi: Sbornik statei o russkoi intelligentsii*. 2d ed. Moscow, 1909. Rpt. Frankfurt am Main, 1967.

Jakob, Ludwig Heinrich von. *Entwurf eines Criminal-Gesetzbuches für das russische Reich. Mit Anmerkungen über die bestehenden russischen Criminalgesetze. Nebst einem Anhange, welcher enthält: Kritische Bemerkungen über den von der Gesetzgebungs-Commission zu St. Petersburg herausgegebenen Criminal-Codex*. Halle, 1818.

"*Kak smotrit obshchestvo na tserkovnyi ili grazhdanskii brak?*" Otvety na anketu, postavlennuiu knigoizdatel'stvom, so stat'eiu I. Tertychnogo "*Ot chego ne prochnye nashi braki?*" Kiev, 1908.

Kalachov, N. *Ob otnoshenii iuridicheskikh obychaev k zakonodatel'stvu*. St. Petersburg, 1877.

Kalashnikov, S. V. *Alfavitnyi ukazatel' deistvuiushchikh i rukovodstvennykh kanonicheskikh postanovlenii, ukazov, opredelenii i rasporiazhenii Sviateishego Pravitel'stvuiushchego Sinoda (1721–1901 gg. vkliuchitel'no) i grazhdanskikh zakonov, otnosiashchikhsia k dukhovnomu vedomstvu pravoslavnogo ispovedaniia.* 3d ed., rev. St. Petersburg, 1902.

Kalinkovitskii, A. D. "Eshche ob iskusstvennom vykidyshe (V zashchitu amfibiopodobnogo zarodysha)." *Vrachebnaia gazeta,* no. 43 (1913): 1533–36.

Kamenskii, Anatolii. *Zhenshchina: Rasskaz, pamiati Otto Veiningera.* St. Petersburg, 1909.

Kanel', V. Ia. "Polovoi vopros v zhizni detei." *Vestnik vospitaniia,* no. 4, pt. 1 (1909): 138–71.

Kankarovich, I. I. "O prichinakh prostitutsii." In *Trudy pervogo vserossiiskogo s"ezda po bor'be s torgom zhenshchinami i ego prichinami,* vol. 1. St. Petersburg, 1911.

———. *Prostitutsiia i obshchestvennyi razvrat: K istorii nravov nashego vremeni.* St. Petersburg, 1907.

Karlinsky, Simon, ed. *Anton Chekhov's Life and Thought: Selected Letters and Commentary.* Berkeley, Calif., 1973.

Karmina, M. M. "Doklad." In *Trudy pervogo s"ezda deiatelei po voprosam suda dlia maloletnikh: S.-Peterburg, dekabr' 1913 g.* Petrograd, 1915.

Katunskii, E. Ia. "K voprosu o prave roditelei na zhizn' ploda." *Meditsinskaia beseda,* no. 7 (1900): 177–84.

Kauffman, Reginald Wright. *The House of Bondage.* New York, 1910.

Khokhlov, N. I. "Pis'mo v redaktsiiu." *Russkii vrach,* no. 38 (1913): 1341–42.

———. "Po povodu postanovleniia otdela akusherstva 2-go iiunia na dvenadtsatom Pirogovskom s"ezde 'Ob iskusstvennom prestupnom vykidyshe.'" *Russkii vrach,* no. 29 (1913): 1048–49.

Kholevinskaia, M. M. "Otchet ob osmotrakh prostitutok na Samokatskom smotrovom punkte Nizhegorodskoi iarmarki za 1893 god." *Vrach,* no. 17 (1894): 487–91.

"Khronika." *Pravo,* no. 16 (April 21, 1913): 1024.

"Khronika." *Sovremennaia psikhiatriia,* nos. 9–10 (1913): 754–58, 837–38.

"Khronika: Iz deiatel'nosti iuridicheskikh obshchestv. Ugolovnoe otdelenie S.-Peterburgskogo obshchestva, 7 dekabria 1902." *Zhurnal Ministerstva Iustitsii,* no. 1 (1903): 234–35.

"Khuliganstvo." *Novoe vremia,* no. 13318 (April 9/22, 1913): 4.

Kievlianin. "K voprosu ob istinnoi osnove very v ritual'nuiu legendu." *Tserkovno-obshchestvennyi vestnik,* no. 47 (November 28, 1913): 9–13.

Kistiakovskii, A. F. "K voprosu o tsenzure nravov u naroda." In *Zapiski Imperatorskogo russkogo geograficheskogo obshchestva po otdeleniiu etnografii,* vol. 8. St. Petersburg, 1878.

"K kievskomu protsessu." *Tserkovnyi vestnik,* no. 41 (October 10, 1913): 1265–66.

Kokhmanskii, Pavel. *Polovoi vopros: Razbor sovremennykh form polovykh otnoshenii.* Moscow, 1912.

Kollontai, A. M. "Novaia zhenshchina." *Sovremennyi mir,* no. 9 (1913): 151–85.

Kolomoitsev, S. V. "K voprosu o vnepolovom sifilise." Pts. 1–2. *Prakticheskii vrach,* nos. 39–40 (1907): 697–98, 713–14.

Koltonovskaia, E. "Problema pola i ee osveshchenie u neorealistov (Vedekind i Artsybashev)." *Obrazovanie,* no. 1, sec. 2 (1908): 114–32.

Koni, A. F. "Antropologicheskaia shkola v ugolovnom prave." In *Poslednie gody: Sudebnye rechi (1888–1896), iuridicheskie soobshcheniia i zametki, vospominaniia i biograficheskie ocherki, prilozheniia.* 2d ed., rev. St. Petersburg, 1898.

Koptev, D. A., and S. M. Latyshev, eds. *Ugolovnoe ulozhenie (stat'i vvedennye v deistvie).* St. Petersburg, 1912.

Kosorotov, D. P. "O nesposobnosti k brachnomu sozhitiiu." *Zhurnal Ministerstva Iustitsii*, no. 5 (1916): 76–106.

Kovalevskii, P. I. "Menstrual'noe sostoianie i menstrual'nye psikhozy." *Arkhiv psikhiatrii, neirologii i sudebnoi psikhopatologii*, no. 1 (1894): 71–131.

——. [Paul Kovalevsky]. *Psychopathologie légale*. Vol. 1, *La Psychologie criminelle*. Paris, 1903.

——. "Sifilitiki, ikh neschast'e i spasenie." *Arkhiv psikhiatrii, neirologii i sudebnoi psikhopatologii*, no. 1 (1897): 60–88.

Kozhukhov, S. "O praktike Pravitel'stvuiushchego Senata po voprosu o vydache krest'ianskim zhenam otdel'nykh vidov na zhitel'stvo." *Zhurnal Ministerstvo Iustitsii*, no. 3 (1901): 158–68.

Krafft-Ebing, Richard von. *Grundzüge der Criminalpsychologie auf Grundlage des Strafgesetzbuchs des deutschen Reichs für Ärzte und Juristen*. Erlangen, 1872.

——. *Lehrbuch der gerichtlichen Psychopathologie, mit Berücksichtigung der Gesetzgebung von Österreich, Deutschland und Frankreich*. Stuttgart, 1886.

——. *Psychopathia sexualis, mit besonderer Berücksichtigung der conträren Sexualempfindung: Eine klinisch-forensische Studie*. Stuttgart, 1886.

Kranikhfel'd, Vladimir. "Literaturnye otkliki." *Sovremennyi mir*, no. 5, sec. 2 (1907): 126–35.

——. "Literaturnye otkliki." *Sovremennyi mir*, no. 6, sec. 2 (1909): 90–108.

——. "Literaturnye otkliki: O novykh liudiakh A. Verbitskoi." *Sovremennyi mir*, no. 8, sec. 2 (1910): 68–82.

——. "Literaturnye otkliki: Stavka na sil'nykh." *Sovremennyi mir*, no. 5, sec. 2 (1909): 73–84.

Krasnozhen, M. E. *Tserkovnoe pravo*. 2d ed. Iuriev, 1906.

Krestovskii, V. V. "Peterburgskie trushchoby." In *Sobranie sochinenii Vsevoloda Vladimirovicha Krestovskogo*, vol. 2. St. Petersburg, 1899.

Kuprin, A. I. "Iama." In *Zemlia sborniki*, bk. 3 (1909), bk. 15 (1914), bk. 16 (1915). Rpt. in *Sobranie sochinenii v deviati tomakh*, vol. 6. Moscow, 1964. Published in English as Alexander Kuprin, *Yama: The Pit*, trans. Bernard Guilbert Guerney. London, 1930.

Kuzmin, M. A. "Kryl'ia." *Vesy*, no. 11 (1906). Rpt. separately, Moscow, 1907. Published in English as *Wings: Prose and Poetry*, trans. and ed. Neil Granoien and Michael Green. Ann Arbor, Mich., 1972.

——. "Zametki o russkoi belletristike." *Apollon*, no. 9, pt. 2 (1910): 33–35.

Kuznetsov, Mikhail. "Istoriko-statisticheskii ocherk prostitutsii i razvitiia sifilisa v Moskve." *Arkhiv sudebnoi meditsiny i obshchestvennoi gigieny*, no. 4 (1870): 84–201.

——. *Prostitutsiia i sifilis v Rossii: Istoriko-statisticheskie issledovaniia*. St. Petersburg, 1871.

"K voprosu o polovykh snosheniiakh." Pts. 1–2. *Vrach*, nos. 1–2 (1894): 8–12, 38–42.

Lancereaux, Etienne. *Traité historique et pratique de la syphilis*. 2 vols. Paris, 1866. Published in English as *A Treatise on Syphilis: Historical and Practical*. 2 vols. London, 1868, 1969.

Lazovskii, N. "Lichnye otnosheniia suprugov po russkomu obychnomu pravu." *Iuridicheskii vestnik*, no. 6–7 (1883): 358–414.

L. G. "Belye rabyni." *Novoe vremia*, no. 9553 (December 8/21, 1902): 2–3.

Liass, S. A. "Izvrashchenie polovogo vlecheniia." *Obozrenie psikhiatrii, nevrologii i eksperimental'noi psikhologii*, no. 6 (1898): 415–16.

Lichkus, L. G. "Iskusstvennyi prestupnyi vykidysh." *Russkii vrach*, no. 39 (1913): 1358–66.

——. "Vykidysh s sudebno-meditsinskoi tochki zreniia." *Russkii vrach*, no. 4 (1912): 109–18.

——. "Vynuzhdennyi otvet na korrespondentsiiu N. I. Khokhlova: Po povodu postanovleniia otdela akusherstva 2-go iiunia na dvenadtsatom Pirogovskom s"ezde, 'Ob iskusstvennom prestupnom vykidyshe.'" *Russkii vrach*, no. 33 (1913): 1181–82.

"Lichnaia statistika polovoi potrebnosti v primenenii k issledovaniiu polovoi zhizni obshchestva." *Vrach*, no. 6 (1898): 154–58.

Likhachev, A. "Novye raboty v oblasti ugolovnoi statistiki i antropologii." *Zhurnal grazhdanskogo i ugolovnogo prava*, no. 3 (1883): 1–24.

Lindenberg, Viktor. *Materialy k voprosu o detoubiistve i plodoizgnanii v Vitebskoi gubernii (Po dannym vitebskogo okruzhnogo suda za desiat' let, 1897–1906)*. Iuriev, 1910.

Listov, S. V. "Zhenskaia domashniaia prisluga, prostitutsiia i venericheskie bolezni." *Vestnik obshchestvennoi gigieny, sudebnoi i prakticheskoi meditsiny*, no. 4 (1910): 485–93.

Liszt, Franz v. "Die Kindestötung." In *Vergleichende Darstellung des deutschen und ausländischen Strafrechts. Vorarbeiten zur deutschen Strafrechtsreform*, ed. Karl von Birkmeyer et al., 15 vols., Besonderer Teil, vol. 5. Berlin, 1905–8.

Le Livre de la marquise: Recueil de poésie et de prose. St. Petersburg, [1908].

Lokhvitskii, Aleksandr. *Kurs russkogo ugolovnogo prava*. 2d ed. St. Petersburg, 1871.

Lombroso, Cesare. *L'Antisémitisme*. Trans. from 2d Italian ed. Paris, 1899.

——. "Du parallélisme entre l'homosexualité et la criminalité innée." In *Congrès international d'anthropologie criminelle, Comptes-rendus du sixième congrès international d'anthropologie criminelle (Turin, 28 avril–3 mai 1906)*. Turin, 1908.

Lombroso, Cesare, and Guglielmo Ferrero. *La Femme criminelle et la prostituée*. Paris, 1896.

L. R. "Sovremennye zametki: Uchebno-vospitatel'noe zavedenie dlia nesovershennoletnikh prostitutok." *Otechestvennye zapiski*, no. 9 (1869): 125–58.

Lucka, Emil. *Otto Weininger, sein Werk und seine Persönlichkeit*. 1905. Rev. ed. Berlin, 1921.

Ludmer, Iakob. "Bab'i stony." Pts. 1–2. *Iuridicheskii vestnik*, nos. 11–12 (1884): 446–67, 658–79.

Lukian. "Ocheredi." *Birzhevye vedomosti*, no. 15143 (October 12/25, 1915): 3.

——. "Rozanovshchina." *Birzhevye vedomosti*, no. 15543 (May 7/20, 1916): 3.

Lutokhin, D. A. "Vospominaniia o Rozanove." *Vestnik literatury*, no. 4–5 (1921): 5–7.

L'vov-Rogachevskii, V. "M. Artsybashev." *Sovremennyi mir*, no. 11, sec. 2 (1909): 26–48.

Makovskii, F. "Chto takoe russkoe dekadentstvo." *Obrazovanie*, no. 9, sec. 1 (1905): 125–42.

Maksimov, S. V. *Sibir' i katorga*. 3d ed. St. Petersburg, 1900.

Malygin, N. P. "Iz itogov studencheskoi perepisi v Iur'eve (Derpte)." *Zhurnal obshchestva russkikh vrachei v pamiat' N. I. Pirogova*, no. 1 (1907): 20–31.

Martineau, Louis. *Leçons sur les déformations vulvaires et anales produites par la masturbation, le saphisme, la défloration et la sodomie*. 1883. 2d ed., rev. Paris, 1886.

——. *La Prostitution clandestine*. Paris, 1885.

Matiushenskii, A. I. *Polovoi rynok i polovye otnosheniia*. St. Petersburg, 1908.

"Melochi iz zhizni i gazet." *Zemshchina*, no. 1618 (March 22, 1914).

Men'shikov, M. O. "Pis'ma k blizhnim: Evrei o evreiakh." Pt. 1. *Novoe vremia*, no. 11815 (February 1/14, 1909): 4.

——. "Tragikomicheskoe plemia." *Novoe vremia*, no. 13501 (October 12/25, 1913): 4–5.

Merezhkovskii, D. S. *Bylo i budet: Dnevnik, 1910–14.* Petrograd, 1915.

Merzheevskii, I. P. "Ob usloviiakh, blagopriiatstvuiushchikh razvitiiu dushevnykh i nervnykh boleznei v Rossii i o merakh, napravlennykh k ikh umen'-sheniiu." In *Trudy pervogo s"ezda otechestvennykh psikhiatrov, proiskhodivshego v Moskve s 5-go po 11-oe ianvaria 1887 g.* St. Petersburg, 1887.

Merzheevskii, Vladislav. *Sudebnaia ginekologiia: Rukovodstvo dlia vrachei i iuristov.* St. Petersburg, 1878.

Messarosh, P. I. "K voprosu o rasprostranenii sifilisa v Rossii." *Vestnik obshchestvennoi gigieny, sudebnoi i prakticheskoi meditsiny*, no. 7 (1896): 49–70.

M. G. "O detoubiistve." *Arkhiv sudebnoi meditsiny i obshchestvennoi gigieny*, no. 1, sec. 2 (1868): 21–55.

Miagkov, M. I. "Nekotorye zadachi vospitaniia v sviazi s polovoi zhizn'iu chelovecheskogo organizma." In *Trudy pervogo s"ezda ofitserov-vospitatelei kadetskikh korpusov (22–31 dekabria 1908 g.)*, ed. P. V. Petrov. St. Petersburg, 1909.

Mikhailovskii, I. P. "Chetvertyi s"ezd obshchestva rossiiskikh ginekologov i akusherov (Spb. 16–19 dekabria 1911 g.)." *Sibirskaia vrachebnaia gazeta*, no. 4 (1912): 44–47.

Mikhailovskii, N. "Literatura i zhizn': O g. Rozanove, ego velikikh otkrytiiakh, ego makhanal'nosti i filosoficheskoi pornografii." *Russkoe bogatstvo*, no. 8, sec. 2 (1902): 76–99.

Mikhailovskii, N. K. [N. M—ii]. "Bor'ba s polovoi raspushchennost'iu v shkole." Pts. 1–2. *Russkaia shkola*, nos. 7–9 (1907): 23–52.

Mikhel'son, Bruno. *Pol i krasota: Beseda o krasote po povodu stat'i Otto Veiningera "Erotika i estetika."* Moscow, 1909.

Mikhnevich, Vladimir. *Iazvy Peterburga: Opyt istoriko-statisticheskogo issledovaniia nravstvennosti stolichnogo naseleniia.* St. Petersburg, 1886.

"Ministerstvo Iustitsii: O vvedenii v deistvie novogo ugolovnogo ulozheniia." *Gosudarstvennyi Sovet v soedinennykh departamentakh (4 i 27 aprelia 1905 goda)*, no. 23. *Obshchee sobranie Gosudarstvennogo Soveta*, May 30, 1905.

Ministerstvo Iustitsii, Pervyi departament. Chast' iuriskonsul'tskaia, no. 8228 (marta 14 dnia 1898 goda). *Po proektu novogo ugolovnogo ulozheniia.* St. Petersburg, 1898.

Ministerstvo Iustitsii za sto let, 1802–1902: Istoricheskii ocherk. St. Petersburg, 1902.

Mitskevich, S. I. *Na grani dvukh epokh: Ot narodnichestva k marksizmu.* Moscow, 1937.

——. *Zapiski vracha-obshchestvennika (1888–1918).* 2d ed., rev. Moscow, 1969.

Mittermaier, Wolfgang. "Verbrechen und Vergehen wider die Sittlichkeit. Entführung. Gewerbsmässige Unzucht." In *Vergleichende Darstellung des deutschen und ausländischen Strafrechts. Vorarbeiten zur deutschen Strafrechtsreform*, ed. Karl von Birkmeyer et al., 15 vols., Besonderer Teil, vol. 4. Berlin, 1905–8.

Möbius, Paul. *Geschlecht und Unbescheidenheit: Beurteilung des Buches von Otto Weininger "Ueber Geschlecht und Charakter."* Halle, 1904. Trans. from 9th German ed. as *Pol i neuchtivost': Kriticheskii razbor knigi "Pol i kharakter" Otto Veiningera.* Moscow, 1909.

——. *Ueber den physiologischen Schwachsinn des Weibes.* Halle, 1900. Trans. as *Fiziologicheskoe slaboumie zhenshchiny.* Moscow, 1909.

Mordovtsev, D. G. *Zhivoi tovar: Postydnaia mezhdunarodnaia torgovlia molodost'iu i krasotoi i mery protiv beznravstvennykh sovratitelei zhenshchin.* Moscow, 1893.

Mukalov, M. K. *Deti ulitsy: Maloletnie prostitutki.* St. Petersburg, 1906.

Nabokov, V. D. "Chezare Lombrozo." *Pravo,* no. 43 (1909): 2292–97.

——. "Desiatyi s"ezd kriminalistov." *Pravo,* no. 9 (1914): 655–67.

——. *Elementarnyi uchebnik osobennoi chasti russkogo ugolovnogo prava.* Vyp. 1. St. Petersburg, 1903.

——. [Vladimir v. Nabokoff]. "Die Homosexualität im russischen Strafgesetzbuch." *Jahrbuch für sexuelle Zwischenstufen,* no. 2 (1903): 1159–71.

——. "Plotskie prestupleniia, po proektu ugolovnogo ulozheniia." *Vestnik prava,* no. 9–10 (1902). Rpt. in V. D. Nabokov, *Sbornik statei po ugolovnomu pravu.* St. Petersburg, 1904.

Nabokov, Vladimir. *Speak Memory: An Autobiography Revisited.* New York, 1966.

Nagrodskaia, E. A. *Gnev Dionisa.* St. Petersburg, 1910.

Nakashidze, Il'ia. "Bor'ba s nizshimi instinktami cheloveka: K voprosu o polovom samovospitanii." In *Polovoe vospitanie: Sbornik statei, sostavlennykh uchiteliami, roditeliami i vospitateliami.* Moscow, 1913.

Nastol'naia kniga dlia molodykh suprugov s polnym izlozheniem pravil supruzheskoi zhizni. Moscow, 1909.

Naturshchitsa. No. 1. Moscow, 1914.

Nazar'ev, V. N. "Sovremennaia glush': Iz vospominanii mirovogo sud'i." *Vestnik Evropy,* no. 2 (1872): 604–36.

Nekliudov, N. A. *Rukovodstvo osobennoi chasti russkogo ugolovnogo prava.* St. Petersburg, 1887.

——. *Ugolovno-statisticheskie etiudy,* I: *Statisticheskii opyt issledovaniia fiziologicheskogo znacheniia razlichnykh vozrastov chelovecheskogo organizma po otnosheniiu k prestupleniiu.* St. Petersburg, 1865.

——, ed. *Materialy dlia peresmotra nashego ugolovnogo zakonodatel'stva.* 7 vols. St. Petersburg, 1880–83.

Net bolee onanizma, venericheskoi bolezni, poliutsii, muzhskogo bessiliia i zhenskogo besplodiia: Prakticheskie sredstva snova vosstanovliat' i ukrepliat' zdorov'e, rasstroennoe etimi bolezniami. 2d ed. Moscow, 1865.

Nikol'skii, D. "Pamiati vrachei-antropologov: N. V. Gil'chenko i P. N. Tarnovskoi." *Prakticheskii vrach,* no. 13 (1911): 220–23.

Nikol'skii, V. I. "Neskol'ko zamechanii o krest'ianskom sifilise v Tambovskom uezde." *Vrach,* no. 41 (1886): 735–38.

Novopolin, G. S. *Pornograficheskii element v russkoi literature.* St. Petersburg, 1909.

Obolonskii, N. A. *Izvrashchenie polovogo chuvstva.* St. Petersburg, 1898.

Oboznenko, P. E. "Obshchestvennaia initsiativa S.-Peterburga v bor'be s prostitutsiei." Pts. 1–2. *Vestnik obshchestvennoi gigieny, sudebnoi i prakticheskoi meditsiny,* nos. 11–12 (1905): 1671–90, 1864–99.

——. *Podnadzornaia prostitutsiia S.-Peterburga.* St. Petersburg, 1896.

——. "Po povodu novogo proekta nadzora za prostitutsieiu v Peterburge, vyrabotannogo komissieiu Russkogo sifilidologicheskogo obshchestva." *Vrach,* no. 12 (1899): 347–50.

——. "Vopros ob uporiadochenii prostitutsii i o bor'be s neiu na dvukh mezhdunarodnykh soveshchaniiakh 1899 goda." *Vrach,* no. 30 (1900): 909–15.

"Obozrenie ugolovnoi kassatsionnoi praktiki za 1897 g.: K voprosu o rastlenii." *Pravo,* no. 9 (February 27, 1899): 434–40.

"Obsuzhdenie voprosa o detskoi prostitutsii: Preniia po dokladu B. I. Bentovina." In *Trudy pervogo vserossiiskogo s"ezda po bor'be s torgom zhenshchinami i ego prichinami, proiskhodivshego v S.-Peterburge s 21 po 25 aprelia 1910 goda,* vol. 2. St. Petersburg, 1912.

"Odobrennyi Gosudarstvennym Sovetom i Gosudarstvennoi Dumoiu Vyso-chaishe utverzhdennyi zakon: O merakh k presecheniiu torga zhenshchinami v tseliakh razvrata." *Sobranie uzakonenii i rasporiazhenii pravitel'stva*, no. 10, sec. 1 (January 12, 1910): 91–94.

Official Report of the Jewish International Conference on the Suppression of the Traffic in Girls and Women, Held on April 5th, 6th and 7th, 1910, in London, Convened by the Jewish Association for the Protection of Girls and Women. London, 1910.

Okinchits, L. L. "Kak borot'sia s prestupnym vykidyshem." *Zhurnal akusherstva i zhenskikh boleznei*, no. 3 (1912): 319–22.

"Okolo dela Beilisa." *Tserkov': Staroobriadcheskii tserkovno-obshchestvennyi zhurnal*, no. 42 (October 20, 1913): 1001–3.

Okonchatel'noe zakliuchenie komiteta dlia rassmotreniia proekta voinskogo ustava o nakazaniiakh, po voprosu ob otmene telesnykh nakazanii. N.p., [1862].

Okorokov, V. P. *Vozvrashchenie k chestnomu trudu padshikh devushek.* Moscow, 1888.

Oleinik, M. L. "Prestupnyi abort v doktrine i zakonodatel'stve." In *Trudy kruzhka ugolovnogo prava pri Spb. Universitete*, ed. M. M. Isaev. St. Petersburg, 1913.

Olikhov, S. A. "K voprosu o plodovitosti krest'ianok Kineshemskogo uezda Kostromskoi gubernii." *Zemskii vrach*, no. 52 (1890): 823–24.

"O prostitutsii v Rossii: I. Iz otcheta o deistviiakh v S.-Peterburge komissii dlia razbora brodiachikh zhenshchin razvratnogo povedeniia v techenie 5-ti let, s 27-go aprelia 1847 g. po 27 aprelia 1852 goda." *Arkhiv sudebnoi meditsiny i obshchestvennoi gigieny*, no. 1, pt. 3 (1869): 102–8.

O rodstve i svoistve, kak prepiatstviiakh k zakliucheniiu brakov, po deistvuiushchim zakonopolozheniiam. Moscow, 1908.

Orshanskii, I. G. *Evrei v Rossii: Ocherki ekonomicheskogo i obshchestvennogo byta russkikh evreev.* St. Petersburg, 1877.

———. "Uchenie Lombrozo o tipe prestupnika." In *Sudebnaia psikhopatologiia dlia vrachei i iuristov.* Pt. 1. St. Petersburg, 1900.

Osipov, V. P. "O politicheskikh ili revoliutsionnykh psikhozakh." *Nevrologicheskii vestnik*, no. 3 (1910): 437–92.

"Osoboe mnenie chlena s"ezda Z. Ia. El'tsinoi." In *Trudy pervogo vserossiiskogo s"ezda po bor'be s torgom zhenshchinami i ego prichinami, proiskhodivshego v S.-Peterburge s 21 po 25 aprelia 1910 goda*, vol. 2. St. Petersburg, 1912.

"Osoboe mnenie chlenov Gosudarstvennogo Soveta . . . po delu o vvedenii v deistvie ugolovnogo ulozheniia." *Obshchee sobranie Gosudarstvennogo Soveta*, May 30, 1905.

Ostrogorskii, A. N. "Pedagogicheskie ekskursii v oblast' literatury ('Sanin' Artsybasheva: K voprosu po besedakh po polovomu voprosu)." *Russkaia shkola*, no. 3, sec. 1 (1908): 1–22.

Ostrogorskii, S. A. "K voprosu o polovom sozrevanii (ego fiziologiia, patologiia i gigiena)." In *Trudy pervogo vserossiiskogo s"ezda detskikh vrachei v S.-Peterburge, s 27–31 dekabria 1912 goda*, ed. G. B. Konukhes. St. Petersburg, 1913.

Otchet desiatogo obshchego sobraniia Russkoi gruppy mezhdunarodnogo soiuza kriminalistov, 13–16 fevralia 1914 g. v Petrograde. Petrograd, 1916.

Ozerov, Ivan. "Sravnitel'naia prestupnost' polov v zavisimosti ot nekotorykh faktorov." *Zhurnal iuridicheskogo obshchestva pri Imperatorskom S.-Peterburgskom Universitete*, no. 4 (1896): 54–83.

Parent-Duchâtelet, Alexandre. *De la prostitution dans la ville de Paris, considérée sous le rapport de l'hygiène publique, de la morale et de l'administration.* 1836. 3d ed. 2 vols. Paris, 1857.

Pavlov, P. A. "Ob otnoshenii vnepolovogo zarazheniia sifilisom k polovomu mezhdu srednim klassom g. Moskvy." *Meditsinskoe obozrenie*, no. 1 (1890): 12–17.

Pavlovskaia, L. S. "Dva sluchaia dushevnogo zabolevaniia pod vliianiem obshchestvennykh sobytii." *Obozrenie psikhiatrii, nevrologii i eksperimental'noi psikhologii*, no. 6 (1906): 418–22.

——. "Neskol'ko sluchaev dushevnogo zabolevaniia pod vliianiem obshchestvennykh sobytii." *Obozrenie psikhiatrii, nevrologii i eksperimental'noi psikhologii*, no. 9 (1907): 522–58.

Peshekhonov, A. V. "Na ocherednye temy: 'Sanintsy' i 'Sanin.'" Pts. 1–2. *Russkoe bogatstvo*, no. 5, sec. 2, and no. 6, sec. 2 (1908): 104–30, 146–75.

Petersen, O. V. "O sifilise i venericheskikh bolezniakh v gorodakh Rossii." In *Trudy Vysochaishe razreshennogo s"ezda po obsuzhdeniiu mer protiv sifilisa v Rossii*, vol. 1. St. Petersburg, 1897.

——. "O sifilise i venericheskikh bolezniakh v gorodakh Rossii." In *Obshchie doklady po otdelam po izucheniiu rasprostraneniia sifilisa i venericheskikh boleznei v Rossii*. St. Petersburg, [1897].

Petrishchev, A. B. "Iz oblasti shchekotlivykh voprosov." *Russkoe bogatstvo*, no. 9 (1907): 95–126.

Petrovskii, A. G. "Bor'ba s sifilisom v gorodakh." Pts. 1–2. *Izvestiia Moskovskoi gorodskoi dumy*, no. 5, Obshchii otdel (1905): 1–23; no. 7, Obshchii otdel (1905): 1–25.

Piatnitskii, B. I. *Polovye izvrashcheniia i ugolovnoe pravo*. Mogilev, 1910.

Pil'skii, Petr. *Problema pola, polovye avtory i polovoi geroi*. St. Petersburg, 1909.

Pirozhkova, O. P. "K voprosu o vykidyshe." *Zhurnal akusherstva i zhenskikh boleznei*, no. 4 (1912): 519–24.

Pisarevskii, N., Protoierei. "Neskol'ko slov k 'Tiazhelomu voprosu.'" *Tserkovno-obshchestvennyi vestnik*, no. 50 (December 25, 1913): 14–15.

Pismennyi, N. N. "K voprosu o sifilise v fabrichnom naselenii." *Vestnik obshchestvennoi gigieny, sudebnoi i prakticheskoi meditsiny*, no. 11 (1906): 1725–36.

Platonov, I. V. "Ob"ekt prestupleniia izgnaniia ploda." *Vestnik prava*, no. 7 (1899): 155–67.

Pokrovskaia, M. I. "Bor'ba s prostitutsiei." *Zhurnal Russkogo obshchestva okhraneniia narodnogo zdraviia*, no. 4 (1900): 399–428.

——. "Edinaia polovaia nravstvennost'." *Zhenskii vestnik*, no. 4 (1910): 89–92.

——. "Iarmochnaia prostitutsiia." *Novoe vremia*, no. 9130 (August 5/18, 1901): 4.

——. "'Kak muzhchina.'" *Zhenskii vestnik*, no. 12 (1910): 266–68.

——. "Kak zhenshchiny dolzhny borot'sia s prostitutsiei." In *Trudy pervogo vserossiiskogo zhenskogo s"ezda pri Russkom zhenskom obshchestve v S.-Peterburge 10–16 dekabria 1908 goda*. St. Petersburg, 1909.

——. "Kreitserova Sonata." *Zhenskii vestnik*, no. 9 (1908): 193–96.

——. "K voprosu ob aborte." *Zhenskii vestnik*, no. 4 (1914): 102–5.

——. "Mery, preduprezhdaiushchie rasprostraneniia sifilisa." Pts. 1–3. *Russkii vrach*, nos. 10–12 (1903): 372–75, 413–15, 453–55.

——. *O polovom vospitanii i samovospitanii*. St. Petersburg, 1913.

——. "O prostitutsii maloletnikh." *Zhenskii vestnik*, no. 10 (1912): 194–97.

——. *O zhertvakh obshchestvennogo temperamenta*. St. Petersburg, 1902.

——. "Prostitutsiia i bespravie zhenshchin." *Zhenskii vestnik*, no. 10 (1907): 225–31.

——. "Sovremennyi erotizm s fiziologicheskoi tochki zreniia." *Voprosy pola*, no. 4 (March 23, 1908): 30–32.

———. "Spetsial'naia i lichnaia gigiena: Torgovlia zhenshchinami." *Zhenskii vestnik*, no. 1 (1907): 20–24.

———. *Vrachebno-politseiskii nadzor za prostitutsiei sposobstvuet vyrozhdeniiu naroda.* St. Petersburg, 1902.

———. "Zhenskaia bezzashchitnost'." *Zhenskii vestnik*, no. 2 (1905): 45–52.

Pokrovskii, F. "O semeinom polozhenii krest'ianskoi zhenshchiny v odnoi iz mestnostei Kostromskoi gubernii po dannym volostnogo suda." *Zhivaia starina: Periodicheskoe izdanie otdeleniia etnografii Imperatorskogo russkogo geograficheskogo obshchestva*, no. 3–4 (1896): 457–76.

Polnoe sobranie zakonov Rossiiskii Imperii. 1st, 2d, 3d ser. St. Petersburg, 1830–1911.

"Polovaia zhizn' iur'evskogo studenchestva." *Vestnik obshchestvennoi gigieny, sudebnoi i prakticheskoi meditsiny*, no. 7 (1907): 1162–63.

Polovtseva, V. N. "Polovoi vopros v zhizni rebenka." *Vestnik vospitaniia*, no. 9, pt. 1 (1903): 16–29.

Polovtsov, A. A. *Russkii biograficheskii slovar'.* 25 vols. St. Petersburg/Petrograd, 1896–1918.

Ponomarev, S. "Semeinaia obshchina na Urale." *Severnyi vestnik*, no. 1, pt. 2 (1887): 1–38.

Popov, Ardalion. *Sud nakazaniia za prestupleniia protiv very i nravstvennosti po russkomu pravu.* Kazan, 1904.

Popov, T. *Russkaia narodno-bytovaia meditsina: Po materialam etnograficheskogo biuro kn. V. N. Tenisheva.* St. Petersburg, 1903.

Portugalov, O. "Prestuplenie nepotrebstva po novomu ugolovnomu ulozheniiu." *Iuridicheskaia gazeta*, no. 62 (September 14, 1903): 2.

Pospelov, A. I. *O vnepolovom zarazhenii sifilisom sredi liudei chernorabochego klassa g. Moskvy.* St. Petersburg, 1889.

"Po voprosu o neomal'tuzianstve." *Sibirskaia vrachebnaia gazeta*, no. 38 (1908): 411–12.

Poznyshev, S. V. "Kritika i bibliografiia: M. N. Gernet, *Detoubiistvo* (M, 1911)." *Voprosy prava*, no. 1 (1912): 178–92.

Prais, A. A. "Polovaia zhizn' uchashchikhsia." *Meditsinskaia beseda*, no. 23 (1902): 665–71.

Pravdin, V. "Ritual'naia legenda." *Tserkovno-obshchestvennyi vestnik*, no. 43 (October 31, 1913): 1–3.

Priklonskii, I. I. *Prostitutsiia i ee organizatsiia: Istoricheskii ocherk.* Moscow, 1903.

"Priniataia s"ezdom rezoliutsiia o roli sotsial'nykh i politicheskikh faktorov v etiologii nervno-psikhicheskikh zabolevanii." In *Trudy vtorogo s"ezda otechestvennykh psikhiatrov, proiskhodivshego v g. Kieve s 4-go po 11-oe sentiabria 1905 goda.* Kiev, 1907.

Probst, Ferdinand. "Der Fall Otto Weininger: Eine psychiatrische Studie." In *Grenzfragen des Nerven- und Seelenlebens*, vol. 31 (Wiesbaden, 1904).

Probst, Ferdinand, and Greta Meisel-Hess. *Anti-Veininger: Izlozhenie i kriticheskii razbor knigi "Pol i kharakter."* St. Petersburg, 1909.

Prodazha devushek v doma razvrata i mery k ee prekrashcheniiu. Moscow, 1899.

"Proekt osobennoi chasti ugolovnogo ulozheniia v obsuzhdenii Moskovskogo iuridicheskogo obshchestva." *Iuridicheskii vestnik*, no. 10 (1886): 338–86.

Proekt ugolovnogo ulozheniia Rossiiskoi Imperii. St. Petersburg, 1813. In *Arkhiv Gosudarstvennogo Soveta*, vol. 4. St. Petersburg, 1874.

Proekt ulozheniia o nakazaniiakh ugolovnykh i ispravitel'nykh, vnesennyi v 1844 godu v Gosudarstvennyi Sovet, s podrobnym oznacheniem osnovanii kazhdogo iz vnesennykh v sei proekt postanovlenii. St. Petersburg, 1871.

"Protokol zasedaniia chetvertogo s"ezda obshchestva rossiiskikh akusherov i ginekologov." Pts. 1–2. *Zhurnal akusherstva i zhenskikh boleznei*, nos. 3–4 (1912): 386–88, 539–48

"Protokol zasedaniia obshchepedagogicheskogo otdela pedagogicheskogo muzeia voenno-uchebnykh zavedenii 8 fevralia 1908 g." *Pedagogicheskii sbornik*, no. 5 (1908): 431–32.

Purishkevich, V. M. *Materialy po voprosu o razlozhenii sovremennogo russkogo universiteta*. St. Petersburg, 1914.

Quételet, Adolphe. *Sur l'homme et le développement de ses facultés; ou Essai de physique sociale*. Paris, 1835.

Radbruch, Gustav. "Abtreibung." In *Vergleichende Darstellung des deutschen und ausländischen Strafrechts. Vorarbeiten zur deutschen Strafrechtsreform*, ed. Karl von Birkmeyer et al., 15 vols., Besonderer Teil, vol. 5. Berlin, 1905–8.

Radishchev, Aleksandr. *A Journey from St. Petersburg to Moscow*. Trans. Leo Wiener, ed. Roderick Page Thaler. Cambridge, Mass., 1958.

Ragozin, L. F., ed. *Svod uzakonenii pravitel'stva po vrachebnoi i sanitarnoi chasti v Imperii*. 3 vols. St. Petersburg, 1895–96.

Resheniia ugolovnogo kassatsionnogo departamenta Pravitel'stvuiushchego Senata za 1867 god; 1871 god. Ekaterinoslav, 1910.

Resheniia ugolovnogo kassatsionnogo departamenta Pravitel'stvuiushchego Senata za 1869 god; 1870 god; 1872 god. St. Petersburg, n.d.

Resheniia ugolovnogo kassatsionnogo departamenta Pravitel'stvuiushchego Senata za 1875 god; 1876 god; 1881 god; 1888 god; 1895 god; 1897 god; 1904 god. Ekaterinoslav, 1911.

Ribbing, Seved. *Om den sexuela hygienen och några af dess etiska konsequenses*. Lund, 1888. Published in Russian as *Polovaia gigiena i ee nravstvennye posledstviia*, trans. Leinenberg. Odessa, 1891.

Ricord, Philippe. *Lettres sur la syphilis*. 2d ed., rev. Paris, 1856.

Rohleder, Hermann. *Die Masturbation: Eine Monographie für Ärzte und Pädagogen*. Berlin, 1899. Published in Russian as Roleder, *Onanizm: Prichiny, sushchnost', preduprezhdenie, lechenie*, trans. Shekhter. St. Petersburg, 1901.

Rokov, G. "Bol'noi vopros vospitaniia." *Vestnik vospitaniia*, no. 7, pt. 1 (1902): 53–93.

Rossiiskii meditsinskii spisok. 27 vols. St. Petersburg/Petrograd, 1890–1916.

Rossiiskoe obshchestvo zashchity zhenshchin v 1913 godu. Petrograd, 1914.

Rotman, E. A. "K kasuistike vnepolovogo shankra: Vnepolovoi shankr na penis'e." *Russkii zhurnal kozhnykh i venericheskikh boleznei*, no. 10 (1905): 34.

Rozanov, V. V. "A. L. Borovikovskii o brake i razvode." *Novoe vremia*, no. 9604 (November 28/December 11, 1902).

——. "*Angel Iegovy*" u evreev: Istoki Izrailia. St. Petersburg, 1914.

——. *Apokalipsis nashego vremeni*. Sergiev Posad, 1917–18. Rpt. in Vasilii Rozanov, *Izbrannoe*, ed. Evgeniia Zhiglevich. Munich, 1970.

——. *Dekadenty*. St. Petersburg, 1904.

——. *Evropa i evrei*. St. Petersburg, 1914.

——. *Four Faces of Rozanov: Christianity, Sex, Jews, and the Russian Revolution*. Trans. and ed. Spencer E. Roberts. New York, 1978.

——. "Iz pripominanii i myslei ob A. S. Suvorine." In *Pis'ma A. S. Suvorina k V. V. Rozanovu*. St. Petersburg, 1913.

——. *Kogda nachal'stvo ushlo*. St. Petersburg, 1910.

——. *Liudi lunnogo sveta: Metafizika khristianstva*. 1911. 2d ed. St. Petersburg, 1913.

——. "Nasha 'koshernaia pechat'.'" *Zemshchina*, no. 1477 (October 22, 1913): 3.

——. *Oboniatel'noe i osiazatel'noe otnoshenie evreev k krovi*. St. Petersburg, 1914.
——. *Opavshie list'ia*. 2 vols. St. Petersburg, 1913, 1915. Rpt. in Vasilii Rozanov, *Izbrannoe*, ed. Evgeniia Zhiglevich. Munich, 1970.
——. *Semeinyi vopros v Rossii*. 2 vols. St. Petersburg, 1903.
——. *Uedinennoe*. St. Petersburg, 1911. Rpt. in Vasilii Rozanov, *Izbrannoe*, ed. Evgeniia Zhiglevich. Munich, 1970.
——. "Vazhnyi istoricheskii vopros." *Novoe vremia*, no. 13485 (September 26/ October 9, 1913), 4.
——. *V mire neiasnogo i ne reshennogo*. 2d ed. St. Petersburg, 1904.
——. *Voina 1914 goda i russkoe vozrozhdenie*. Petrograd, 1915.
——. *V sosedstve Sodoma: Istoki Izrailia*. St. Petersburg, 1914.
——. "V svoem uglu: Iudaizm," pts. 10–13, 17–20. *Novyi put'*, October 1903, 96–131; December 1903, 101–22.
Rozenbakh, P. Ia. "K kazuistike polovogo izvrashcheniia." *Obozrenie psikhiatrii, nevrologii i eksperimental'noi psikhologii*, no. 9 (1897): 652–56.
Rozenkvist, A. I. "K statistike vnepolovogo zarazheniia sifilisom." *Biblioteka vracha*, no. 8 (1898): 509–37.
——. "Redkii sluchai vnepolovogo zarazheniia sifilisom: Iz ambulatorii Miasnitskoi bol'nitsy v Moskve." *Vrach*, no. 9 (1899): 244–45.
Rozin, N. N. *Ob oskorblenii chesti: Ugolovno-iuridicheskoe issledovanie*. 2d ed., rev. Tomsk, 1910.
Rubakin, N. A. *Sredi knig*. 3 vols. Moscow, 1911–13.
Rubel', A. N. "Zhilishcha bednogo naseleniia Peterburga." *Vestnik obshchestvennoi gigieny, sudebnoi i prakticheskoi meditsiny*, no. 4 (1899): 424–45.
Rubinovskii, A. L. "Povinnost' razvrata." *Vestnik prava*, no. 8 (1905): 155–77.
Rumiantsev, N. E. "K voprosu o polovom vospitanii." In *Trudy pervogo s"ezda ofitserov-vospitatelei kadetskikh korpusov (22–31 dekabria 1908 g.)*, ed. P. V. Petrov. St. Petersburg, 1909.
Runova, O. P. "Bez zaveta: Rasskaz." *Sovremennyi mir*, no. 10 (1913): 29–69.
Russkii erot ne dlia dam. N.p., 1879.
Rutenberg, L. "K voprosu o sovremennom vzgliade na dopustimost' sotsial'nogo vykidysha." Pts. 1–3. *Vrachebnaia gazeta*, nos. 30–32 (1916): 476–78, 489–93, 504–7.
Ryan, Michael. *Philosophy of Marriage*. London, 1839.
——. *Prostitution in London*. London, 1839.
Rybakov, F. E. "Dushevnye rasstroistva v sviazi s sovremennymi politicheskimi sobytiiami." Pts. 1–2. *Russkii vrach*, nos. 3 and 8 (1906): 65–67, 221–22.
——. "Dushevnye rasstroistva v sviazi s tekushchimi politicheskimi sobytiiami." *Russkii vrach*, no. 51 (1905): 1593–95.
——. "O prevratnykh polovykh oshchushcheniiakh." Pts. 1–2. *Vrach*, nos. 22–23 (1898): 640–43, 664–67.
——. "Psikhozy v sviazi s poslednimi politicheskimi sobytiiami v Rossii." *Russkii vrach*, no. 20 (1907): 677–80.
Sabinin, A. Kh. *Prostitutsiia: Sifilis i venericheskie bolezni*. St. Petersburg, 1905.
Sandberg, D. D. "Sifilis v derevne." *Vrach*, no. 26 (1894): 740–44.
Sangailo, K. P. "Polovoi vopros i shkola." *Pedagogicheskii sbornik*, no. 3 (1913): 327–43.
Selivachev, A. "Psikhologiia iudofil'stva: V. V. Rozanov." *Russkaia mysl'*, no. 2, sec. 2 (1917): 49–64.
Sergeevskii, N. D. [N. S.]. "Antropologicheskoe napravlenie v issledovaniiakh o prestuplenii i nakazanii." *Iuridicheskii vestnik*, no. 2 (1882): 209–21.

——. *Russkoe ugolovnoe pravo: Posobie k lekstiiam, chast' osobennaia.* 8th ed. St. Petersburg, 1910.

Sergiev, V. S. "K ucheniiu o fiziologicheskikh proiavleniiakh polovoi zhizni zhenshchiny-krest'ianki Kotel'nicheskogo uezda Viatskoi gubernii." In *Trudy antropologicheskogo obshchestva pri Imperatorskoi Voenno-Meditsinskoi Akademii,* vol. 5. St. Petersburg, 1901.

Shabad, T. O. "Iskusstvennyi vykidysh s printsipial'noi tochki zreniia." In *Trudy odinnadtsatogo Pirogovskogo s"ezda,* ed. P. N. Burlatov, vol. 3. St. Petersburg, 1913.

Sharapov, S. F., ed. *Sushchnost' braka.* Moscow, 1901.

Shashkov, S. S. "Detoubiistvo." Pts. 1–3. *Delo,* 1868: no. 4, 69–118; no. 5, 1–40; no. 6, 25–75.

——. *Istoricheskie sud'by zhenshchiny: Detoubiistvo i prostitutsiia.* St. Petersburg, 1871.

Shchipillo, A. A. "Sostav prestuplenii, predusmotrennykh zakonom 25 dekabria 1909 goda o merakh k presecheniiu torga zhenshchinami v tseliakh razvrata." *Zhurnal Ministerstva Iustitsii,* no. 10 (1911): 56–102.

Shchit. Petrograd, 1916.

Shidlovskii, K. I., ed. *Svodka khodataistv Pirogovskogo obshchestva vrachei pered pravitel'stvennymi uchrezhdeniiami za 20 let (1883–1903 gg.).* Moscow, 1904.

Shiriaev, P. A. "Organizatsiia vrachebnoi pomoshchi pri sifilise i venericheskikh bolezniakh sredi rabochego naseleniia v bol'shikh promyshlennykh i torgovykh tsentrakh." *Meditsinskoe obozrenie,* no. 13–14 (1902): 151–58.

Shmelev, M. "Predokhranitel'nye mery protiv sifilisa." *Sbornik sochinenii po sudebnoi meditsiny,* no. 2 (1872): 217–52.

Sholomovich, A. S. "K voprosu o dushevnykh zabolevaniiakh, voznikaiushchikh na pochve politicheskikh sobytii." *Russkii vrach,* no. 21 (1907): 715–20.

Shperk, Eduard [Edouard Léonard Sperk]. *Oeuvres complètes: Syphilis, prostitution, études médicales diverses.* 2 vols. Paris, 1896.

——. "O merakh k prekrashcheniiu rasprostraneniia sifilisa u prostitutok." *Arkhiv sudebnoi meditsiny i obshchestvennoi gigieny,* no. 3, sec. 3 (1869): 67–84.

——. "Otvet na stat'iu: 'Zhenskii nadzor za prostitutsiei.'" *Arkhiv sudebnoi meditsiny i obshchestvennoi gigieny,* no. 1, sec. 5 (1870): 4–11.

Shtil'man, G. N. "Abort na s"ezde kriminalistov." *Pravo,* no. 10 (1914): 774–81.

Shtiurmer, K. L. *Sifilis v sanitarnom otnoshenii.* St. Petersburg, 1890.

Shumakov, S. A. Review of D. A. Dril', *Uchenie o prestupnosti i merakh bor'by s neiu. Voprosy prava,* no. 3 (1912): 197–203.

Shvarts, A. "K voprosu o priznakakh privychnoi passivnoi pederasti (Iz nabliudenii v aziatskoi chasti g. Tashkenta)." *Vestnik obshchestvennoi gigieny, sudebnoi i prakticheskoi meditsiny,* no. 6 (1906): 816–18.

"Sifilis v Kostromskom uezde za 16 let (1895–1910)." *Vestnik obshchestvennoi gigieny, sudebnoi i prakticheskoi meditsiny,* no. 4 (1912): 630.

Simonov, I. S. "Shkola i polovoi vopros: Iz dnevnika byvshego ofitsera-vospitatelia." Pts. 1–3. *Pedagogicheskii sbornik,* nos. 1, 4–5 (1908): 16–26, 271–90, 401–30.

Siniavskii, A. D. *"Opavshie list'ia" V. V. Rozanova.* Paris, 1982.

Skliar, N. I. "Eshche o vliianii tekushchikh politicheskikh sobytii na dushevnye zabolevaniia." *Russkii vrach,* no. 15 (1906): 448–49.

——. "O vliianii tekushchikh politicheskikh sobytii na dushevnye zabolevaniia." *Russkii vrach,* no. 8 (1906): 222–24.

Sl—, G. "Nravstvennoe i beznravstvennoe v proizvedeniiakh iskusstva." *Iuridicheskii vestnik,* no. 8 (1889): 536–43.

Slovtsova, L. V. "Polovoe vospitanie detei." In *Trudy pervogo vserossiiskogo zhenskogo s"ezda pri Russkom zhenskom obshchestve v S.-Peterburge 10–16 dekabria 1908 goda.* St. Petersburg, 1909.

Slushatel'nitsy S.-Peterburgskikh vysshikh zhenskikh (Bestuzhevskikh) kursov. St. Petersburg, 1912.

Smirnov, A. A. "O polednei knige Rozanova: V. Rozanov, *Oboniatel'noe i osiazatel'noe otnoshenie evreev k krovi* (St. Petersburg, 1914)." *Russkaia mysl'*, no. 4, sec. 3 (1914): 44–47.

Smirnov, Aleksandr. "Ocherki semeinykh otnoshenii po obychnomu pravu russkogo naroda." Pts. 1–6. *Iuridicheskii vestnik*, 1877: no. 1–2, 43–74; no. 3–4, 98–128; no. 5–6, 92–131; no. 7–8, 177–224; no. 9–10, 118–76; no. 11–12, 92–142.

S. M. K. "Raspad sem'i." *Zemshchina*, no. 602 (March 30, 1911): 3.

Sodman, M. M. "K voprosu o sifilise v ispravitel'nykh zavedeniiakh dlia maloletnikh." *Russkii vrach*, no. 40 (1909): 1353–54.

Sologub, Fedor. "Melkii bes." *Voprosy zhizni*, nos. 6–11 (1905). Rpt. St. Petersburg, 1907. Published in English as *The Petty Demon*, trans. S. D. Cioran. Ann Arbor, Mich., 1983.

———. "Ocharovanie vzorov (Aisedora Dunkan)." *Teatr i iskusstvo*, no. 4 (January 27, 1913): 90–93.

———. *Tvorimaia legenda*. Serialized 1907–13. Rpt. St. Petersburg, 1914. Published in English as *The Created Legend*, trans. S. D. Cioran. Pts. 1–3. Ann Arbor, Mich., 1979.

Solov'ev, E. T. *Grazhdanskoe pravo: Ocherk narodnogo iuridicheskogo byta.* Vyp. 1. Kazan, 1888.

"Sostav podpischikov." *Sovremennyi mir*, no. 12, sec. 2 (1910): 156.

"Sovremennaia negrotorgovlia." *Zemshchina*, no. 1620 (March 24, 1914): 1.

[Speranskii, M. M.] *Obozrenie istoricheskikh svedenii o svode zakonov.* St. Petersburg, 1833.

Speranskii, N. S. *K statistike sifilisa v sel'skom naselenii Moskovskoi gubernii.* Moscow, 1901.

Spisok chinam vedomstva Ministerstva Iustitsii. St. Petersburg, 1900.

"Spisok chlenov Russkoi gruppy mezhdunarodnogo soiuza kriminalistov k 4 ianvaria 1909 goda." In *Russkaia gruppa mezhdunarodnogo soiuza kriminalistov: Obshchee sobranie gruppy v Moskve, 4–7 ianvaria 1909 goda.* St. Petersburg, 1909.

"Spisok chlenov Russkoi gruppy mezhdunarodnogo soiuza kriminalistov k 21 aprelia 1910 goda." In *Russkaia gruppa mezhdunarodnogo soiuza kriminalistov: Obshchee sobranie gruppy v Moskve, 21–23 aprelia 1910 goda.* St. Petersburg, 1911.

"Spisok chlenov Russkoi gruppy mezhdunarodnogo soiuza kriminalistov na 1 ianvaria 1914 goda." *Zhurnal ugolovnogo prava i protsessa*, no. 4 (1913): 136–44.

"Spisok deistvitel'nykh chlenov iuridicheskogo obshchestva pri Imperatorskom S.-Peterburgskom Universitete so vremeni uchrezhdeniia obshchestva." In *Iuridicheskoe obshchestvo pri Imperatorskom S.-Peterburgskom Universitete za dvadtsat' piat' let (1877–1902).* St. Petersburg, 1902.

Spisok g.g. chlenam Gosudarstvennogo Soveta (k 22 fevralia 1910 g.). St. Petersburg, 1910.

"Spisok ritual'nykh ubiistv." *Zemshchina*, no. 1369 (June 29, 1913): 5–8.

Spisok vysshim chinam gosudarstvennogo, gubernskogo i eparkhial'nogo upravleniia. St. Petersburg, 1891, 1903.

Spravka po voprosu ob otnoshenii tserkovnogo zakonodatel'stva k gosudarstvennomu. St. Petersburg, 1914.

Starchenko, S. N. "K voprosu o bor'be s venericheskimi bolezniami." *Vrachebnaia gazeta*, no. 33 (1913): 1137–41.

——. "Popularizatsiia svedenii o venericheskikh bolezniakh." *Vrachebnaia gazeta*, no. 18 (1912): 713–15.

Starokotlitskii, N. N. "K voprosu o vozdeistvii polovogo instinkta na religiiu (V sviazi s opisaniem sluchaia religiozno-erotomanicheskogo pomeshatel'-stva)." *Zhurnal nevropatologii i psikhiatrii imeni S. S. Korsakova*, no. 2–3 (1911): 259–302.

Statistique de l'Empire de Russie. Vol. 13. *La Prostitution d'après l'enquête du 1er (13) août 1889*. Publication du Comité central de statistique, Ministère de l'intérieur. St. Petersburg, 1891.

Steklov, Iu. M. *Literaturnyi raspad: Kriticheskii sbornik*. 2 vols. St. Petersburg, 1908, 1909.

Stoianovskaia, E. L. "Otchet: Dvenadtsatyi Pirogovskii s"ezd, otdel akusherstva i zhenskikh boleznei, 2-oe zasedanie 1-go iiunia." *Russkii vrach*, no. 28 (1913): 1010–12.

Struve, Petr. "Na raznye temy: Bol'shoi pisatel' s organicheskim porokom: Neskol'ko slov o V. V. Rozanove." *Russkaia mysl'*, no. 11, sec. 2 (1910): 138–46.

Sukhanov, S. A. "K kazuistike seksual'nykh izvrashchenii." *Nevrologicheskii vestnik*, no. 2 (1900): 164–68.

Suvorov, N. S. *Grazhdanskii brak*. 2d ed. Iuridicheskaia biblioteka, no. 11. St. Petersburg, 1896.

Svod ustavov blagochiniia. Vol. 14 of *Svod zakonov Rossiiskoi Imperii*. St. Petersburg, 1836.

Svod zakonov ugolovnykh. St. Petersburg, 1835.

Tagantsev, N. S. *Russkoe ugolovnoe pravo: Lektsii*. 2d ed., rev. St. Petersburg, 1902.

——, ed. *Ugolovnoe ulozhenie 22 marta 1903 g.* St. Petersburg, 1904.

——, ed. *Ulozhenie o nakazaniiakh ugolovnykh i ispravitel'nykh 1885 goda*. 11th ed., rev. St. Petersburg, 1901.

——, ed. *Ustav o nakazaniiakh, nalagaemykh mirovymi sud'iami. Izdanie 1885 goda. S dopolneniiami po svodnomu prodolzheniiu 1912 goda, s prilozheniem motivov i izvlechenii iz reshenii kassatsionnykh departamentov Senata*. 21st ed., rev. St. Petersburg, 1913.

——, ed. *Ustav o nakazaniiakh, nalagaemykh mirovymi sud'iami. Izdanie 1914 goda. S prilozheniem motivov i izvlechenii iz reshenii kassatsionnykh departamentov Pravitel'stvuiushchego Senata*. 22d ed., rev. Petrograd, 1914.

"'Taina krovi' y evreev: Ekspertiza kuratora-ksendza I. E. Pranaitisa po delu ob ubiistve Andriushi Iushchinskogo." *Missionerskoe obozrenie*, no. 12 (1913): 559–97.

Tardieu, Ambroise. *Etude médico-légale sur les attentats aux moeurs*. 1857. 7th ed. Paris, 1878.

Tarnovskaia, P. N. "Antropometricheskie issledovaniia prostitutok, vorovok i zdorovykh krest'ianok-polevykh rabotnits (zasedanie 21 noiabria 1887 g.)." In *Protokoly zasedanii obshchestva psikhiatrov v S.-Peterburge za 1887 god*. St. Petersburg, 1888. Rpt. in appendix to *Vestnik klinicheskoi i sudebnoi psikhiatrii*, no. 2 (1889).

—— [Pauline Tarnowsky]. "Criminalité de la femme." In Congrès international d'anthropologie criminelle, *Compte-rendu des travaux de la quatrième session tenue à Genève du 24 au 29 août 1896*. Geneva, 1897.

—— [Pauline Tarnowsky]. *Etude anthropométrique sur les prostituées et les voleuses*. Paris, 1889.

———. "Klassy vyrozhdaiushchikhsia v sovremennom obshchestve." In *Protokoly zasedanii obshchestva psikhiatrov v S.-Peterburge za 1886 god.* St. Petersburg, 1887. Rpt. in appendix to *Vestnik klinicheskoi i sudebnoi psikhiatrii*, no. 1 (1887).

———. *Zhenshchiny-ubiitsy: Antropologicheskoe issledovanie.* St. Petersburg, 1902. Published in French as *Les Femmes homicides.* Paris, 1908.

Tarnovskii, E. N. "Dvizhenie chisla nesovershennoletnikh, osuzhdennykh v sviazi s obshchim rostom prestupnosti v Rossii za 1901–1910 gg." *Zhurnal Ministerstva Iustitsii*, no. 10 (1913): 40–90.

———. *Itogi russkoi ugolovnoi statistiki za 20 let (1874–1894 gg.).* St. Petersburg, 1899.

———. "Raspredelenie prestupnosti po professiiam." *Zhurnal Ministerstva Iustitsii*, no. 8 (1907): 54–100.

Tarnovskii, I. M. *Izvrashchenie polovogo chuvstva u zhenshchin.* St. Petersburg, 1895.

"Tarnovskii, I. M." In *Entsiklopedicheskii slovar' Brokgauz-Efron*, 32A:651. St. Petersburg, 1901.

Tarnovskii, V. M. *Izvrashchenie polovogo chuvstva: Sudebno-psikhiatricheskii ocherk dlia vrachei i iuristov.* St. Petersburg, 1885.

———. *Polovaia zrelost', ee techenie, otkloneniia i bolezni.* 1886. 2d ed. St. Petersburg, 1891.

———. *Prostitutsiia i abolitsionizm.* St. Petersburg, 1888.

———. *Sifiliticheskaia sem'ia i ee niskhodiashchee polozhenie: Biologicheskii ocherk.* Kharkov, 1902.

"Tarnovskii, V. M." In *Entsiklopedicheskii slovar' Brokgauz-Efron*, 32A:650–51. St. Petersburg, 1901.

Taxil, Léo [Gabriel Antoine Jogand-Pagès]. *La Corruption fin-de-siècle.* Paris, 1891.

Taylor, Robert W. *A Practical Treatise on Genito-Urinary and Venereal Diseases and Syphilis.* 3d ed., rev. New York, 1904.

———. "Some Unusual Modes of Infection with Syphilis." *Journal of Cutaneous and Genito-Urinary Diseases*, no. 6 (1890): 201–16.

Tenishev, V. V. *Pravosudie v russkom krest'ianskom bytu.* St. Petersburg, 1907.

Timashev, N. S. "Ugolovnoe ulozhenie i volostnoi sud." Pts. 1–2. *Pravo*, nos. 19–20 (1914): 1529–36, 1618–25.

Tissot, S. A. *Onanizm ili rassuzhdenie o bolezniakh, proiskhodiashchikh ot rukobludiia.* Trans. Aleksandr Nikitin. 5th ed. St. Petersburg, 1852.

Tiutriumov, I. M. "O zashchite detstva (zasedanie komiteta 8 ianvaria 1900 g.)." In *Mezhdunarodnyi soiuz kriminalistov, russkaia gruppa, 1899–1902.* St. Petersburg, 1902.

Tolstaia, Sofiia. "Pis'mo v redaktsiiu." *Novoe vremia*, no. 9673 (February 7/20, 1903): 4.

Tolstoi, L. N. [Leo Tolstoy]. *The Kreutzer Sonata.* In *The Death of Ivan Ilych and Other Stories.* New York, 1960.

———. "Posleslovie." In *Polnoe sobranie sochinenii L'va Nikolaevicha Tolstogo*, vol. 10. Moscow, 1913.

——— [L. N. Tolstoy]. *Resurrection.* Trans. Rosemary Edmonds. New York, 1966.

———. "Tak chto zhe nam delat'?" 1886. In *Polnoe sobranie sochinenii L'va Nikolaevicha Tolstogo*, vol. 17. Moscow, 1913.

Trainin, A. N. "Na s"ezde kriminalistov: Fakty i vpechatleniia." *Russkoe bogatstvo*, no. 4 (1914): 248–62.

Trakhtenberg, A. G. "Anomalii polovogo chuvstva v shkol'nom vozraste i sistema fizicheskogo vospitaniia." *Voprosy pola*, no. 4 (1908): 25–30.

——. "Polovoi vopros v sem'e i shkole." *Voprosy pola*, no. 1 (1908): 27–31.

Trotsky, Leon. *My Life*. New York, 1970.

Trubetskoi, E. N. *Entsiklopediia prava*. Kiev, 1901.

——. "Konets revoliutsii v sovremennom romane: Po povodu 'Sanina' Artsybasheva." *Moskovskii ezhenedel'nik*, no. 17 (1908): 1–15.

Trudy komissii sostavleniia zakonov. Vol. 1, *Postanovleniia ob obrazovanii komissii*. 2d ed. St. Petersburg, 1822.

Trudy pervogo vserossiiskogo s"ezda po bor'be s torgom zhenshchinami i ego prichinami, proiskhodivshego v S.-Peterburge s 21 po 25 aprelia 1910 goda. 2 vols. St. Petersburg, 1911, 1912.

Trudy pervogo vserossiiskogo zhenskogo s"ezda pri Russkom zhenskom obshchestve v S.-Peterburge, 10–16 dekabria 1908 goda. St. Petersburg, 1909.

Trudy Vysochaishe razreshennogo s"ezda po obsuzhdeniiu mer protiv sifilisa v Rossii, byvshego pri Meditsinskom Departamente s 15 po 22 ianvaria 1897 goda. 2 vols. St. Petersburg, 1897.

Tsenovskii, A. A. *Abolitsionizm i bor'ba s sifilisom*. Odessa, 1903.

Tulinov, A. I. "Pervichnaia sifiliticheskaia iazva vnepolovogo proiskhozhdeniia na polovykh chastiakh devochki 9 let." *Detskaia meditsina*, no. 4 (1899): 191–202.

Tutyshkin, P. P. "Zadachi tekushchego momenta russkoi obshchestvennoi psikhiatrii." In *Trudy tret'ego s"ezda otechestevennykh psikhiatrov (s 27-go dekabria 1909 g. po 5-oe ianvaria 1910 g.)*. St. Petersburg, 1911.

Tyrkova-Williams, Ariadna [A. V.]. "Krasnyi petukh." *Rech'*, no. 111 (May 13/26, 1907): 2.

Ugolovnoe ulozhenie: Proekt, izmennyi ministrom iustitsii po soglasheniiu s predsedatelem Vysochaishe uchrezhdennoi redaktsionnoi komissii. St. Petersburg, 1898.

Ugolovnoe ulozhenie: Proekt redaktsionnoi komissii i ob"iasneniia k nemu. 8 vols. St. Petersburg, 1895–97.

Ugolovnoe ulozhenie, Vysochaishe utverzhdennoe 22 marta 1903 goda. St. Petersburg, 1903.

Ulozhenie o nakazaniiakh ugolovnykh i ispravitel'nykh. St. Petersburg, 1845.

Uranus. *Ubiistvo Iushchinskogo i kabbala*. 2d ed. St. Petersburg, 1913.

Uspenskii, A. N. "K kazuistike anomalii polovogo chuvstva." *Obozrenie psikhiatrii, nevrologii i eksperimental'noi psikhologii*, no. 12 (1898): 927–28.

"Ustav voennyi: Voinskie artikuly." 1716. In *Polnoe sobranie zakonov Rossiiskoi Imperii*, vol. 5. St. Petersburg, 1830.

Uvarov, M. S. "Sifilis sredi sel'skogo naseleniia." In *Trudy Vysochaishe razreshennogo s"ezda po obsuzhdeniiu mer protiv sifilisa v Rossii*, vol. 1. St. Petersburg, 1897.

Vengerov, S. A. *Istochniki slovaria russkikh pisatelei*. 4 vols. St. Petersburg/Petrograd, 1900–1917.

——. *Kritiko-biograficheskii slovar' russkikh pisatelei i uchenykh: Istoriko-literaturnyi sbornik*. 6 vols. St. Petersburg, 1889–1904.

Verbitskaia, A. A. *Dukh vremeni*. Moscow, 1907.

——. *Igo liubvi*. Moscow, 1914.

——. *Kliuchi schast'ia*. 6 vols. Moscow, 1910–13.

——. *Moemu chitateliu: Avtobiograficheskie ocherki s dvumia portretami*. Moscow, 1908.

Vereshchagin. "K voprosu o registratsii brakov raskol'nikov." *Iuridicheskii vestnik*, no. 2 (1885): 288–303.

——. "O bab'ikh stonakh." *Iuridicheskii vestnik*, no. 4 (1885): 750–61.

Ves' Peterburg na 1913 god: Adresnaia i spravochnaia kniga g. S.-Peterburga. St. Petersburg, 1913.

Vigdorchik, N. A. "Detskaia smertnost' sredi peterburgskikh rabochikh." *Obshchestvennyi vrach*, no. 2 (1914): 212–53.

——. "Vrachebnye otkliki: Ereticheskie mysli o prestupnykh vykidyshakh i o preduprezhdenii beremennosti." *Prakticheskii vrach*, no. 15 (1912): 242–46.

Viktorov, P. P. *Gigiena i etika braka v sviazi s voprosom o polovoi zhizni iunoshestva.* Moscow, 1904.

Virenius, A. S. "Beseda po voprosu o bor'be s polovymi anomaliiami (onanizmom) uchashchikhsia, dlia roditelei i vospitatelei." *Meditsinskaia beseda*, no. 13–14 (1902): 373–89.

——. "Period polovogo razvitiia v antropologicheskom, pedagogicheskom i sotsiologicheskom otnoshenii." Pts. 1–2. *Russkaia shkola*, 1902: no. 10–11, 113–26; no. 12, 120–30.

——. "Polovaia raspushchennost' v shkol'nom vozraste." *Vrach*, no. 41 (1901): 1261–65.

——. "Polovoi vopros: Po povodu sochinenii prof. Avg. Forel', 'Polovoi vopros.' St. Petersburg, 1906." *Zhurnal Ministerstva Narodnogo Prosveshcheniia*, no. 5 (1907): 73–91.

——. "Zhiznennye soblazny i bor'ba s nimi s tochki zreniia gigieny i pedagogii." Pts. 1–2. *Pedagogicheskii sbornik*, nos. 11–12 (1904): 409–24, 469–85.

Vishnevskii. "Bogoslovskaia ekspertiza." *Tserkovno-obshchestvennyi vestnik*, no. 45 (November 14, 1913): 3–5.

Vladimirov, L. E. *Ugolovnyi zakonodatel', kak vospitatel' naroda.* Moscow, 1903.

Vladimirskii-Budanov, M. F. *Obzor istorii russkogo prava.* 7th ed. Petrograd, 1915.

"V obshchestve psikhiatrov." *Rech'*, no. 294 (October 27/November 9, 1913): 6.

Volkov, A. F., and I. Iu. Filipov. *Slovar' iuridicheskikh i gosudarstvennykh nauk.* St. Petersburg, 1901.

Volynskii, A. "'Fetishizm melochei.'" Pt. 2. *Birzhevye vedomosti*, no. 15348 (January 27/February 9, 1916): 2.

Vseslav. "Prizrenie maloletnikh prostitutok v Rossii." *Novoe vremia*, no. 9137 (August 12/25, 1901): 4.

Vsesviatskii, P. V. "Prestupnost' i zhilishchnyi vopros v Moskve." *Pravo*, no. 20 (1909): 1264–78.

Vspomogatel'naia kniga dlia muzhchin, strazhdushchikh [sic] rasslableniem polovykh organov, proiskhodiashchem ot slishkom rannego ili slishkom chastogo udovletvoreniia fizicheskikh pobuzhdenii liubvi, ili ot onanizma, ili ot preklonnoi starosti, ili vsledstvie boleznei: Sochinenie prakticheskogo vracha. Moscow, 1857.

Vvedenskii, A. A. "Prostitutsiia sredi sel'skogo (vne-gorodskogo) naseleniia." In *Obshchie doklady po otdelam po izucheniiu rasprostraneniia sifilisa i venericheskikh boleznei v Rossii.* St. Petersburg, [1897].

——. "Prostitutsiia sredi sel'skogo (vne-gorodskogo) naseleniia." In *Trudy Vysochaishe razreshennogo s"ezda po obsuzhdeniiu mer protiv sifilisa v Rossii*, vol. 1. St. Petersburg, 1897.

Vygodskii, Ia. E. "Iskusstvennyi vykidysh s obshchestvennoi i vrachebnoi tochki zreniia." In *Dvenadtsatyi Pirogovskii s"ezd. Peterburg, 29 maia-5 iiunia 1913 g.*, vyp. 2. St. Petersburg, 1913.

"V zashchitu aborta i preduprezhdeniia beremennosti." *Sibirskaia vrachebnaia gazeta*, no. 17 (1912): 209–10.

Weininger, Otto. *Geschlecht und Charakter: Eine prinzipielle Untersuchung.* 1903. Rpt. Munich, 1980.

——. *Taschenbuch und Briefe an einen Freund.* Ed. Arthur Gerber. Leipzig, 1919.

Zagorovskii, A. I. "O proekte semeistvennogo prava." *Zhurnal Ministerstva Iustitsii*, no. 2, sec. 2 (1903): 55–109.

Zagoskin, N. P., ed. *Za sto let: Biograficheskii slovar' professorov i prepodavatelei Imperatorskogo Kazanskogo Universiteta (1804–1904) v dvukh chastiakh.* 2 vols. Kazan, 1904.

Zak, G. Ia. "Umershchvlenie ploda i ugolovnoe pravo." Pts. 1–2. *Pravo*, nos. 46–47 (1910): 2751–57, 2834–42.

Zakharov, N. A. "Prichiny rasprostraneniia prostitutsii nakhodiatsia ne stol' v ekonomicheskikh, skol' v moral'nykh usloviiakh." In *Trudy pervogo vserossiiskogo s″ezda po bor'be s torgom zhenshchinami i ego prichinami, proiskhodivshego v S.-Peterburge s 21 po 25 aprelia 1910 goda*, vol. 1. St. Petersburg, 1911.

Zalesskii, S. *Polovoi vopros s tochki zreniia nauchnoi meditsiny: Gigienicheskii etiud.* Krasnoiarsk, 1909.

Zamengof, M. F. "Brak, sem'ia, i prestupnost'." *Zhurnal Ministerstva Iustitsii*, no. 2 (1916): 143–74.

Zaozerskii, N. "Tsennost' verdikta prisiazhnykh po delu Beilisa." *Tserkovno-obshchestvennyi vestnik*, no. 45 (November 14, 1913): 6–8.

Zarechnaia, Sof'ia. "Neomal'tuzianstvo i zhenskii vopros." *Zhenskoe delo*, no. 27–28 (August 10, 1910): 10–12.

Zarudnyi, M. I. *Zakony i zhizn': Itogi issledovaniia krest'ianskikh sudov.* St. Petersburg, 1874.

Zeland, Nikolai. *Zhenskaia prestupnost'.* St. Petersburg, 1899.

"Zemskaia zhizn': V iudeiskoi kabale." *Zemshchina*, no. 1347 (June 6, 1913): 4.

Zhbankov, D. N. *Bab'ia storona: Statistiko-etnograficheskii ocherk.* Kostroma, 1891.

——. "Izuchenie voprosa o polovoi zhizni uchashchikhsia." Pts. 1–3. *Prakticheskii vrach*, nos. 27–29 (1908): 470–74, 486–90, 503–7.

——. "K voprosu o plodovitosti zamuzhnikh zhenshchin." *Vrach*, no. 13 (1889): 309–11.

——. "K voprosu o vykidyshakh." Pts. 1–5. *Prakticheskii vrach*, nos. 31–38 (1914): 423–25, 432–34, 442–44, 452–54, 463–67.

——. *Materialy o rasprostranenii sifilisa i venericheskikh zabolevanii v Smolenskoi gubernii.* Smolensk, 1896.

——. *O deiatel'nosti sanitarnykh biuro i obshchestvenno-sanitarnykh uchrezhdenii v zemskoi Rossii.* Moscow, 1910.

——. "O detoubiistve." *Prakticheskii vrach*, no. 17 (1909): 316–18.

——. "O dopushchenii zhenshchin v universitet." *Russkii vrach*, no. 6 (1902): 209–13.

——. "O s″ezde pri Meditsinskom Departamente po obsuzhdeniiu meropriiatii protiv sifilisa v Rossii." Pts. 1–2. *Vrach*, nos. 29–30 (1897): 799–803, 829–34.

——. "Polovaia prestupnost'." *Sovremennyi mir*, no. 7, sec. 2 (1909): 54–91.

——. "Polovaia vakkhanaliia i polovye nasiliia: Pir vo vremia chumy." Pts. 1–3. *Prakticheskii vrach*, nos. 17–19 (1908): 308–10, 321–23, 340–42.

Zhbankov, D. N., and V. I. Iakovenko. *Telesnye nakazaniia v Rossii v nastoiashchee vremia.* Moscow, 1899.

Zhdanov, V. A. "Iz pisem k Tolstomu (Po materialam Tolstovskogo arkhiva)." *Literaturnoe nasledstvo*, no. 37–38 (1939): 369–96.

Zhukovskii, A. A. "Detoubiistvo v Poltavskoi gubernii i predotvrashchenie ego." *Arkhiv sudebnoi meditsiny i obshchestvennoi gigieny*, no. 3, sec. 2 (1870): 1–13.

Zhukovskii, M. N. *O vliianii obshchestvennykh sobytii na razvitie dushevnykh zabolevanii.* St. Petersburg, 1907.

Zhukovskii, V. P. "K voprosu o rannei detskoi smertnosti v S.-Peterburge i o merakh bor'by s neiu." *Vrachebnaia gazeta*, no. 24 (1910): 734–37.

Zhurnal obshchego sobraniia Gosudarstvennogo Soveta po proektu ugolovnogo ulozheniia. St. Petersburg, 1903.

Zhurnal osobogo prisutstviia Gosudarstvennogo Soveta Vysochaishe uchrezhdennogo dlia obsuzhdeniia proekta ugolovnogo ulozheniia. St. Petersburg, 1902.

Zinov'eva, E. "V zashchitu prav rozhdennykh: Pis'mo v redaktsiiu." *Sovremennyi mir*, no. 8 (1913): 248–56.

Zinov'eva-Annibal, L. *Tridtsat'-tri uroda: Povest'*. St. Petersburg, 1907. Published in English as Lydia Zinovieva-Annibal, "Thirty-three Abominations," trans. Samuel Cioran. *Russian Literature Triquarterly*, no. 9 (Spring 1974): 94–116.

Zlotnikov, L. "Iudei v iskusstve." *Zemshchina*, no. 767 (September 21, 1911): 3.

Zolotarev, L. A. *Chto govorit nauka o polovoi potrebnosti: Populiarno-nauchnyi ocherk dlia roditelei, vospitatelei i uchashcheisia molodezhi*. Moscow, 1900.

———. *Gigiena supruzheskoi zhizni*. Moscow, 1901.

Zolotarev, S. "Deti revoliutsii." *Russkaia shkola*, no. 3 (1907): 1–23.

Zuk. "O protivozakonnom udovletvorenii polovogo pobuzhdeniia i o sudebnomeditsinskoi zadache pri prestupleniiakh etoi kategorii." *Arkhiv sudebnoi meditsiny i obshchestvennoi gigieny*, no. 2, sec. 5 (1870): 8–13.

Index

Page numbers in *italics* refer to illustrations.

Abortion: as criminal offense, 4, 266; 337–40, 342, 343, 349–58; and debate over fetus, 352, 354; incidence of, 106, 225, 341–42; and Jews, 305, 351; legal view of, 349–58; and male continence, 221; medical view of, 340–49; and modernization, 334–35; physicians' view of, 339–40; therapeutic, 343; as urban vs. rural phenomenon, 113–14, 266; and women's rights, 337, 341–44
Abrashkevich, M. M., 93n
Abstinence, 220–22, 231, 250
"Abyss, The" (Andreev), 372–74
Academy of Military Medicine, faculty and staff of, 50, 133, 134, 174, 343n
—opinions of: on homosexuality, 229; on prostitution, 169, 194; on venereal disease, 169, 194, 271
Adultery: and criminal code, 37–38, 51–53; and divorce, 52; jurisdiction over, 35, 51–53; in penal colonies, 30; prevalence of, 100, 101; and prostitution, 88; as violation of intent of marriage, 41, 51, 55, 93
Advertising, sexual, 360–67
Age of consent, 36, 76–77, 80–81, 89–90
Alexander I (tsar of Russia), 21, 39
Alexander II (tsar of Russia), 17, 18, 22, 33, 144, 176
Alexander III (tsar of Russia), 23, 58n
Alexis (tsar of Russia), 21
All-Russian Congress of Educational Psychology, 246
Alphonse pogroms, 309
Anal sex. *See* Bestiality; Sodomy
Andreev, Leonid, stories by: criticism of, 368, 371–72n; students' opinions of, 373, 374; themes in, 236, 237, 372, 375
Anna Karenina (Tolstoy), 1, 32
Annulment, 32, 54

Anti-Semitism, 279, 301, 308–9, 332. *See also* Beilis, Mendel; Dal', Vladimir I.; Jews; Men'shikov, Mikhail O.; Rozanov, V. V.
Anuchin, Evgenii N., 100, 103, 109–10, 306–7n
Apocalypse of Our Time (Rozanov), 329
Article 44, charter of justices of peace, 88, 89
Artsybashev, Mikhail, *384*; compared to Kuzmin, 388, 390, 397
—works by: criticism of, 368, 371–72n, 383, 385–86; popularity of, 1, 376, 387; students' opinion of, 373, 374; themes in, 1, 372, 383, 385–86, 397. *See also Sanin*
Autonomy, personal, 7, 8, 151–52
Avant-garde fiction, 368, 369, 372, 374, 387, 388, 391, 397, 414

Bacchanale (dance), 415
Bacchante—Vampire of Love, The (play), 418
Bakst, Lev, 391–92n
Balov, A., 272–73
Barthélemy, Toussaint, 176
Bashkirtsev, Marie, 417n
Bazhenov, Nikolai N., 7, 256n
Bebel, August, 251
Beilis, Mendel, 300–301, 321, 322n, 324–26, 334
Bekhterev, Vladimir M., *244, 246*; on Beilis trial, 326n; on capitalism and pathology, 263; on homosexuality, 229; and Ministry of Internal Affairs medical council, 50n; on morality, 140; on penis size, 235n; on regulation of marriage, 243; and Society for Protection of Women, 281; on women's equality, 248
Belyi, Andrei (Boris Bugaev), 256, 311, 314n, 320n, 391, 393

452

Library of Congress Cataloging-in-Publication Data

Engelstein, Laura.
　The keys to happiness : sex and the search for modernity in fin-de-siècle Russia / Laura
Engelstein.
　　p.　cm.
　Includes bibliographical references and index.
　ISBN 0-8014-2664-2 (alk. paper)
　　1. Sex customs—Soviet Union—History—19th century. 2. Soviet Union—Social con-
ditions—1801–1917. 3. Soviet Union—Moral conditions. I. Title.
HQ18.S65E54　1992
306.7'0947—dc20　　　　　　　　　　　　　　　　　　　　　　92-52751